Functional Materials

Functional Materials

Advances and Applications in Energy Storage and Conversion

edited by

Toshio Naito

PAN STANFORD PUBLISHING

Published by

Pan Stanford Publishing Pte. Ltd.
Penthouse Level, Suntec Tower 3
8 Temasek Boulevard
Singapore 038988

Email: editorial@panstanford.com
Web: www.panstanford.com

British Library Cataloguing-in-Publication Data
A catalogue record for this book is available from the British Library.

Functional Materials: Advances and Applications in Energy Storage and Conversion

ISBN 978-981-4800-09-9 (Hardcover)
ISBN 978-0-429-46813-1 (eBook)

Contents

Preface

This book is intended for students and researchers who are interested in, but not necessarily familiar with, solid-state properties. As a comprehensive introduction to the selected research field of interest, each chapter contains historical perspectives, practical examples, and detailed descriptions of particular experiments with plenty of illustrations. The main subject of this book concerns cooperative phenomena such as electrical conduction and magnetic, thermal, and optical properties of organic, inorganic, and metal-complex compounds. From a viewpoint of basic and/or applied research, some chapters concentrate on one of these properties on a series of related materials, while other chapters review a wide range of compounds for a particular physical property. All the authors are affiliated to Ehime University, Japan, but have different backgrounds. Some are interested in synthetic chemistry, while others are interested in solid-state physics. All of the authors have been long devoted to materials science with enthusiasm. Without doubt, all of them will be overjoyed if this book can help in the understanding of the interesting and profound world of materials science, which is always ready to welcome a new young generation.

This book would never have come into being without the kind and continuous help of Dr. Stanford Chong, who gave me an opportunity to write and edit this book. Here, on behalf of all the authors, I would like to express our sincere thanks to him. During the initial correspondence, he was willing to accept my proposal that the subject of the book should be diverse but not divergent, for there were already abundant books on this particular subject. The wider the variety of subjects become, the wider the spectrum of the readers becomes. This will hopefully encourage more and more people to join materials science and add their own new ideas and different points of view. Last but not least, I would like to thank all the staff of Pan Stanford Publishing for their long-term and friendly collaboration.

Toshio Naito

2018

Chapter 1

Control of Magnetism and Conduction in Organic Materials by Light

Toshio Naito
Department of Chemistry and Biology, Graduate School of Science and Engineering, Ehime University, 2-5 Bunkyo-cho, Matsuyama, Ehime 790-8577, Japan
tnaito@ehime-u.ac.jp

Chapter 1 deals with photoconductors of a new type found in 2012. Elemental metals such as aluminum and copper are highly opaque and reflective as well as highly conductive, and some are magnetic. Thus they do not exhibit an evident response to light in terms of electrical and magnetic properties. However, if one can realize a nonmagnetic insulator that becomes a metal and a magnet simultaneously and reversibly under light, such a material might open a new horizon for electro-, magneto-, and optical devices. This kind of material is the very subject in the opening chapter of this book. The chapter begins with a general introduction of molecular materials, which will help the readers understand the succeeding chapters. Now that the molecular materials are diverse and cover too wide a variety of functions to review in a single chapter, we

Functional Materials: Advances and Applications in Energy Storage and Conversion
Edited by Toshio Naito

ISBN 978-981-4800-09-9 (Hardcover), 978-0-429-46813-1 (eBook)
www.panstanford.com

should mainly limit ourselves to the electrical, magnetic, optical, and structural aspects of organic and metal-organic crystalline materials.

1.1 Introduction

1.1.1 Historical Background

Until ~50 years ago, it had been taken for granted that molecular materials were insulating and nonmagnetic and it was considered to be impossible to utilize them as (semi)conducting and/or magnetic materials. Now they are used as parts of semiconducting devices in our daily lives, such as liquid crystal displays, organic electroluminescence displays, and touch screens. What has made them evolve in such a way? As is often the case, necessity is the mother of invention. So are accidental events.

(a)

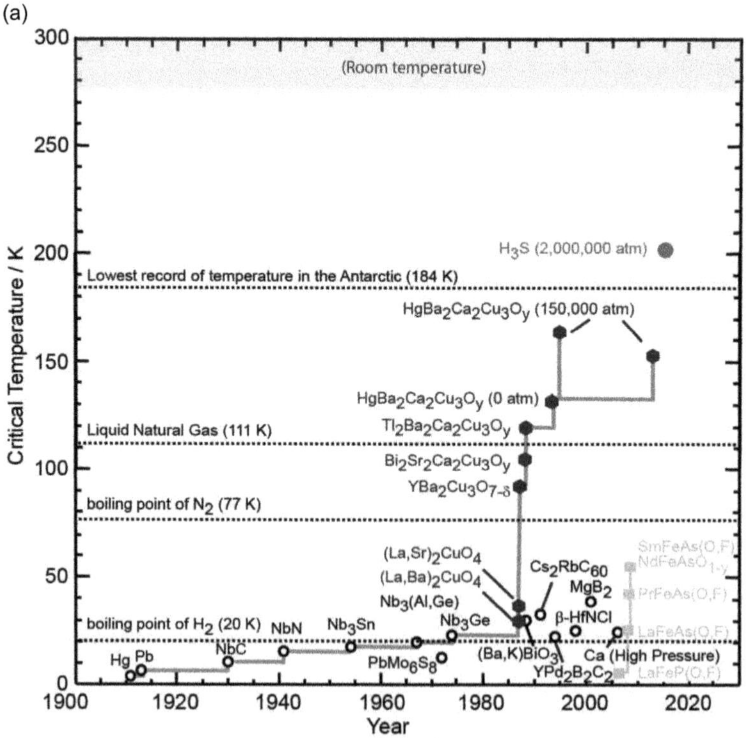

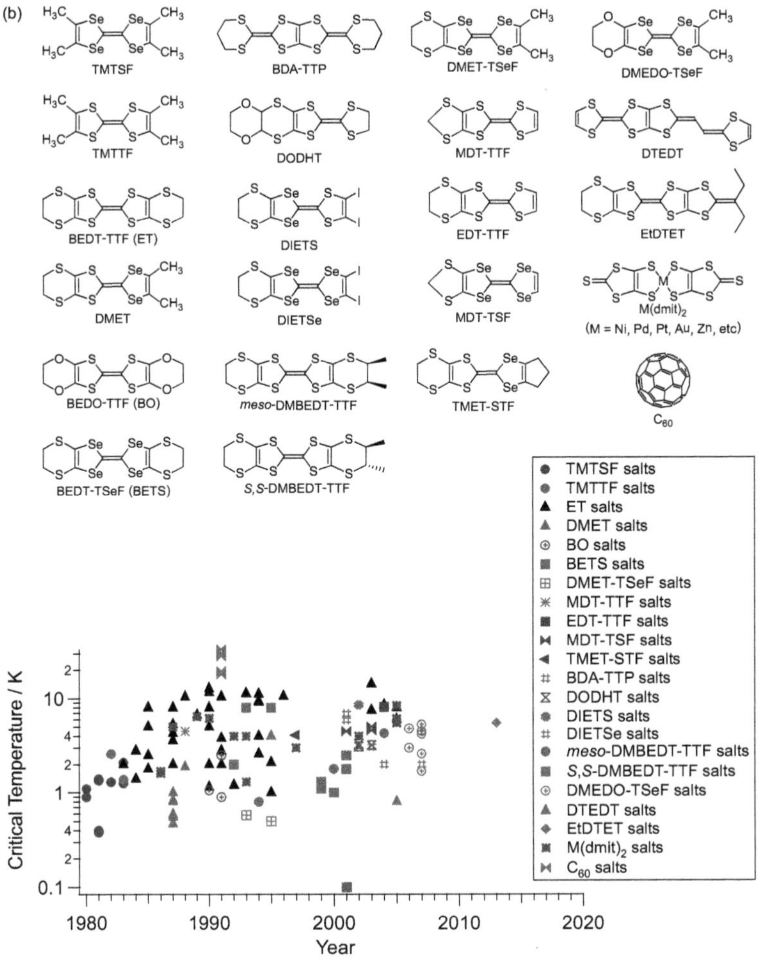

Figure 1.1 History of critical temperatures of selected superconductors (not an exhaustive list). (a) Inorganic and (b) organic or molecular superconductors. For more details on organic superconductors, see Ref. [162].

As illustrated with such names as the Stone Age, the Bronze Age, and the Iron Age, human history is characterized by the evolution of the materials mankind has obtained. Still, future historians might be astonished and bewildered to learn of an outburst of new materials around the beginning of the 21st century. The finding of a new material and/or a new property always happens suddenly. Yet, once it happens, nothing succeeds like success. The history of superconductors (SCs) well illustrates this situation (Fig. 1.1). Heike Kamerlingh Onnes, a physicist in Netherlands, found in 1911 that mercury conducts electricity with a zero-resistance state below 4 K [1], which is now called superconductivity. Then, some pure metals of simple substances were successively found to exhibit superconductivity. All of them become superconductive below certain temperatures peculiar to the substances, called critical temperatures (T_Cs). T_Cs remained very low for well more than half a century. In September 1986, Bednorz and Müller broke the long stagnation of raising T_Cs by reporting "Possible high T_C superconductivity in the Ba-La-Cu-O system" [2]. Shortly thereafter different groups broke the record one after another by finding even higher T_C superconductivity in the related systems [3–7], which set fire to a worldwide competitive search for even-higher-T_C SCs. Thereafter a series of cuprates were found to be SCs one after another, and the highest record of T_C rapidly increased as if it would reach room temperature (RT) within a few years [8–12]. However, some of them are not reproducible. The highest T_C now (in June 2018) is 203 K, observed in H_3S under 200 GPa [13–16]. Subjects related to organic SCs are discussed in Chapters 2 and 5.

In the same year that superconductivity was found, organic compounds also became a focus of contemporary material science research, mainly due to their established synthetic methodologies [17]. However, few people had dreamed of the appearance of a metallic one, which was realized in a certain kind of organic charge transfer (CT) complex in 1973 [18–20]. In 1979 an organic selenium compound attracted the world's attention because its radical salts became the first organic compounds to exhibit superconductivity [21]. In the succeeding three years, researchers were kept excited by successive discoveries of superconductivity in a series of related organic CT salts [22–27].

In the meantime, in the late 1980s some organic polymers were claimed to be ferromagnets even at RT [28–36]. In 1991 a polymeric vanadium complex was found to be attracted to a permanent magnet up to its thermal decomposition temperature (>350 K) [34, 35]. Yet few researchers would have had the slightest idea that the first simple and well-defined organic radical ferromagnet would coincide with the polymeric vanadium complex [36–38].

Then a soccer-ball-shaped all-carbon macromolecule, or rather a carbon cluster, C_{60}, extracted from carbon soot, was highlighted for a potentially wide range of applications in addition to its peculiar shape and preparation method. For example, it has been shown to exhibit ferromagnetism [39, 40] and superconductivity [41–52], depending upon how heavily and with what it is doped. Here "doped" means that the original material was treated or mixed with a trace amount of foreign substance. This brings about a redox reaction in a portion of the material to produce unpaired electrons and/or holes. Thus produced electrons/holes serve as carriers, making the original material (semi)conductive. The conduction mechanism will be discussed in more detail later in this chapter.

1.1.2 Molecular Materials

Now the word "molecular materials" includes a wide range of compounds. They are comprising monomers, oligomers, and polymers of organic molecules and/or metal complexes with organic ligands and sometimes include inorganic ions as well. Organic-inorganic hybrid compounds also may belong to molecular materials. Their functions also vary. Even if we limit ourselves to consider the function concerning energy transformation and communication, the compounds might be used for electro-optical-magnetic devices (Chapter 1), thermoelectric power generators or cooling devices (Chapter 3), rechargeable batteries (Chapter 4), fuel cells (Chapter 7), solar cells (Chapter 8), and so on. Some organic compounds become superconducting, as stated above. In this chapter, we will focus on the electronic functions of molecular materials, crystalline materials in particular, such as conduction, magnetism, and optical properties. They are generally called cooperative phenomena, meaning that intermolecular interactions dominate the bulk properties. They are dependent on details of the crystal structures. Owing to the highly developed techniques of synthetic chemistry, one could design and synthesize almost any elaborate molecule now. However, no one could predict and precisely control molecular arrangements in solid states. The design of molecules is one thing, and the design of molecular solids is another.

1.1.3 Different Solid States for Different Purposes

Research on the cooperative phenomena in molecular materials requires various kinds of analyses. This is because there are many kinds of interactions involved in assembling and arranging molecules to make them crystalline phases. Thus it is usually effective and almost required to examine the crystal structure of the material in question before studying its physical properties. If one needs to clarify the details of the structure and arrangement of the molecules in the solid-state, single-crystal X-ray structural analysis is the most straightforward way. Once the X-ray structural analysis gives us the atomic parameters that describe the precise positions of all the atoms included, we can calculate molecular orbitals, band structures, physical properties, and so on. Generally speaking, low-molecular-weight (LMW) compounds more frequently produce single crystals, which allow a higher-quality X-ray structural analysis than do macromolecules or polymers. Single crystals are generally homogenous and the purest form of the solid state, with a well-defined chemical formula and three-dimensionally ordered atomic arrangements. Thus LMW crystalline compounds enable a detailed discussion about the relationship between structures and properties. The drawbacks of molecular crystals generally include their small sizes (typically ~1 mm or less), mechanical fragility, and tedious workup to obtain them. On the other hand, it is generally much easier to fabricate thin films, for example, on appropriate substrates having desired dimensions and shapes. Such films are much easier to handle. Yet, a thin film is an inhomogeneous assembly of molecules, being similar to powder samples. It consists of randomly aggregated domains, particles, or microcrystals of ~10^0–10^2 μm, which may consist of even smaller domains, particles, or microcrystals. The sizes, shapes, and orientations of thin films vary without any regularity. Thus thin films usually contain a number of voids and cracks on the nanometer-micrometer scale, depending on how they are made. The thickness and density differ from part to part on a microscopic scale (≤micrometers). Accordingly it would be impossible to prepare two sheets of thin films sharing exactly the same structure. Electrical and magnetic properties can be affected by such structural differences, which often lowers the reproducibility of physical property measurements. At the same time, this situation sometimes makes it difficult to distinguish intrinsic properties from extrinsic properties. Because of these advantages and disadvantages, crystalline phases of LMW compounds are suitable for scientific study with rigorous discussion, while polymer thin films, for example, are suitable for application in the industry. This trend is largely true also for other types of compounds. This book covers both crystalline and noncrystalline solid-state properties of organic and inorganic compounds.

1.1.4 Properties and Structures

As is discussed below, to develop conducting and/or magnetic materials, one almost always requires π-conjugated molecules. Paying particular attention to LMW compounds with π-conjugated systems, this chapter briefly surveys the subtle relation between crystal structures and electrical properties. There are two requirements to materials for high electrical conduction: one is having carriers, and the other is having conduction pathways. Carriers are usually unpaired electrons and/or holes and expected to carry the charge in the form of electric current when voltages are applied. The overlaps between two neighboring molecular orbitals make conduction pathways for carriers to travel in the solid. The detailed mechanism of the conduction will be discussed in the next section.

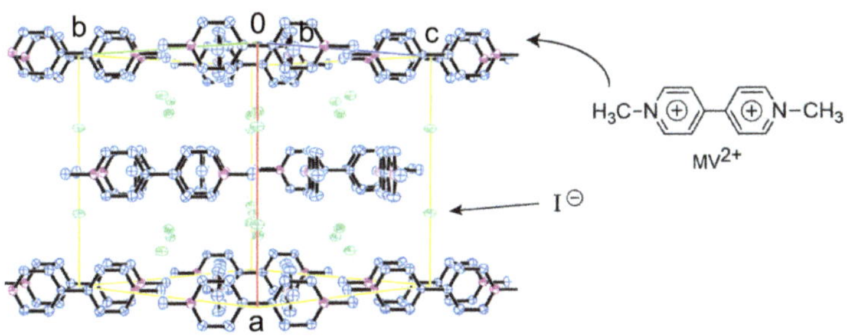

Figure 1.2 A crystal comprising organic cations and inorganic anions, $(MV)I_2$. Hydrogen atoms are omitted for simplicity. Blue, pink, and green spheres designate carbon, nitrogen, and iodine atoms, respectively. MV, methyl viologen.

In Figs. 1.2–1.5, a typical series of crystal structures of molecular materials is shown. Note that they share a simple feature in the molecular arrangements; planar (parts of the) molecules are apt to stack themselves and aggregate themselves. The reason is discussed in the next section in connection with electrical conduction mechanism.

1.1.4.1 Crystal structure without unpaired electrons or conduction pathways

The first example is $(MV)I_2$ (MV = methyl viologen), which is comprising aromatic molecular dications MV^{2+} and halide anions I^- [53–56]. MV^{2+} is well known for its strong photoinduced redox activity. In Fig. 1.2 all the anions I^- and cations MV^{2+} have closed-shell structures, that is, there are no carriers in $(MV)I_2$. Thus $(MV)I_2$ is an insulator. Substances without unpaired electrons, like $(MV)I_2$, exhibit diamagnetism. The name originates from the magnetic behavior; diamagnetic substances move away ("dia" is a prefix meaning separation and movement in the opposite direction) from the magnetic field. These properties are independent of temperature because the binding energy between two unpaired electrons is far higher than the thermal energy gained from the surroundings in the temperature range where this salt can exist without decomposition.

1.1.4.2 Crystal structure with localized unpaired electrons

The conductivity and magnetic susceptibility of $[Ru(bpy)_3][Ni(dmit)_2]_2$ shown in Fig. 1.3 depend on temperature. Note that every two $[Ni(dmit)_2]^-$ anions are particularly close to each other and face each other in Fig. 1.3. There is an unpaired electron on each of the $[Ni(dmit)_2]^-$ anions. Yet, due to strong antiferromagnetic interaction between them, unpaired electrons on the neighboring $[Ni(dmit)_2]^-$ anions make a pair as if they have formed a weak covalent bond between them. This kind of intermolecular interaction leads to a molecular pair often called a dimer in the research fields of molecular conductors and magnets. Note that the term "dimer" in general chemistry formally means a single molecule comprising originally two independent molecules covalently bonded with each other. In other words, the unpaired electrons cannot move as carriers in this solid state. In their ground state, that is, at ~0 K, all the $[Ni(dmit)_2]^-$ dimers practically possess no unpaired electrons. Neither do the $[Ru(bpy)_3]^{2+}$ cations at any temperature. At a finite temperature, the unpaired electrons receive energy as heat from their surroundings and some are excited to behave like unpaired electrons. With increasing temperature, an increasing number of pairs break into unpaired electrons. The resultant unpaired electrons still cannot move so freely but can serve as carriers to produce semiconducting properties. They have also magnetic moments called spins, which exhibit magnetism. In this way, $[Ru(bpy)_3][Ni(dmit)_2]_2$ exhibits temperature-dependent conducting and magnetic properties and behaves as a diamagnetic insulator in the ground state.

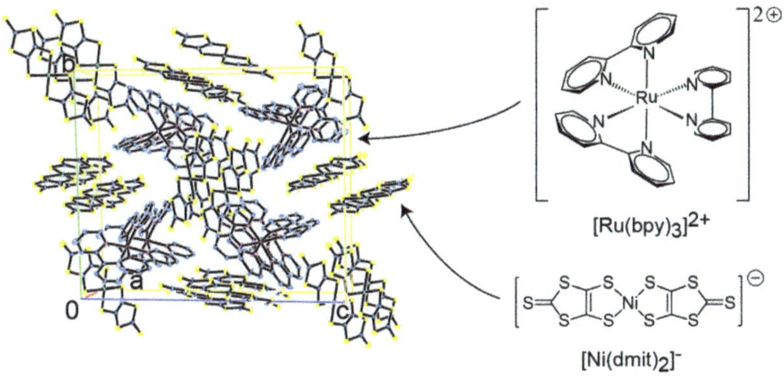

Figure 1.3 A crystal comprising Ru(II)-complex cations and Ni(II)-complex radical anions, $[Ru(bpy)_3][Ni(dmit)_2]_2$, where bpy = bipyridyl and $dmit^{2-}$ = 1,3-dithiole-2-thione-4,5-dithiolate. Hydrogen atoms are omitted for simplicity. Blue, pink, yellow, and dark-pink spheres designate carbon, nitrogen, sulfur, and ruthenium atoms, respectively.

1.1.4.3 Crystal structure of two-dimensional conductors

The material shown in Fig. 1.4, κ-$(ET)_2Cu[N(CN)_2]Br$, is a SC with a critical temperature T_C of ~12 K, which is still one of the highest T_Cs in organic compounds [57]. On cooling the material down from RT (~300 K), the electrical resistivity suddenly drops to zero at T_C. This is called superconducting transition. The resistivity remains zero below T_C, which is characteristic of SCs. Above T_C this compound exhibits a peculiar metallic behavior [58]. SCs usually exhibit a monotonous decrease in resistivity with a decreasing temperature, while κ-$(ET)_2Cu[N(CN)_2]Br$ does not. Recently, in some organic CT complexes, the increase in resistivity immediately above T_C has been associated with fluctuation, which is considered to have an important connection with superconductivity. The interrelation of fluctuation and various degrees of freedom characteristic of molecular conductors are discussed in Chapter 2. In accordance with the peculiar metallic behavior of κ-$(ET)_2Cu[N(CN)_2]Br$, its magnetic behavior is also dependent on temperature in a complicated manner. The polymeric anions $\{Cu[N(CN)_2]Br\}^-$ do not have any role in conduction and magnetism, simply because they do not have any unpaired electrons. Thus the above-mentioned conducting and magnetic properties are governed by the organic molecules called ET [59]. κ-$(ET)_2Cu[N(CN)_2]Br$ contains the organic radical cations nominally described as $ET^{0.5+}$. As an electron cannot be divided in two, $ET^{0.5+}$ is actually a dimer cation $\{(ET)_2\}^+$, where a single unpaired electron is shared or *delocalized* between two ET molecules. The cations have molecular structures similar to those of the $[Ni(dmit)_2]^-$ anions in Fig. 1.3 in that they have π-conjugated planar systems with many sulfur atoms. The cations strongly interact with each other to form dimers. This trend of $ET^{0.5+}$ to form dimers is also similar to that of the $[Ni(dmit)_2]^-$ anions in $[Ru(bpy)_3][Ni(dmit)_2]_2$. Yet there is a striking difference between the two—the number of unpaired electrons each dimer possesses. Because there is only one unpaired electron, or more exactly a single hole, on every $ET^{0.5+}$ dimer, the dimerization could not form a strong chemical bond between the two ET molecules. This contrasts with the case of the $[Ni(dmit)_2]^-$ dimers in $[Ru(bpy)_3][Ni(dmit)_2]_2$; the $[Ni(dmit)_2]$ dimer possesses two unpaired electrons, a situation suitable for single-bond formation. Furthermore, the characteristic molecular arrangement (Fig. 1.5) enables rather isotropic and close *interdimer* interactions in a two-dimensional (2D) way, which do not exist in $(MV)I_2$ and $[Ru(bpy)_3][Ni(dmit)_2]_2$. This characteristic pattern of molecular arrangement is named "κ-type structure" [60, 61]. Because molecular crystals often exhibit polymorphs, where different kinds of materials share the chemical formula with different crystal structures and different physical properties, the Greek letter classifying the crystal structure is designated with the chemical formula. The resultant intermolecular interactions in this salt make the material metallic, in contrast with insulating $[Ru(bpy)_3][Ni(dmit)_2]_2$. $[Ru(bpy)_3][Ni(dmit)_2]_2$ can be regarded as a zero-dimensional (0D) electronic system, having *localized* valence electrons within the $[Ni(dmit)_2]^-$ dimers.

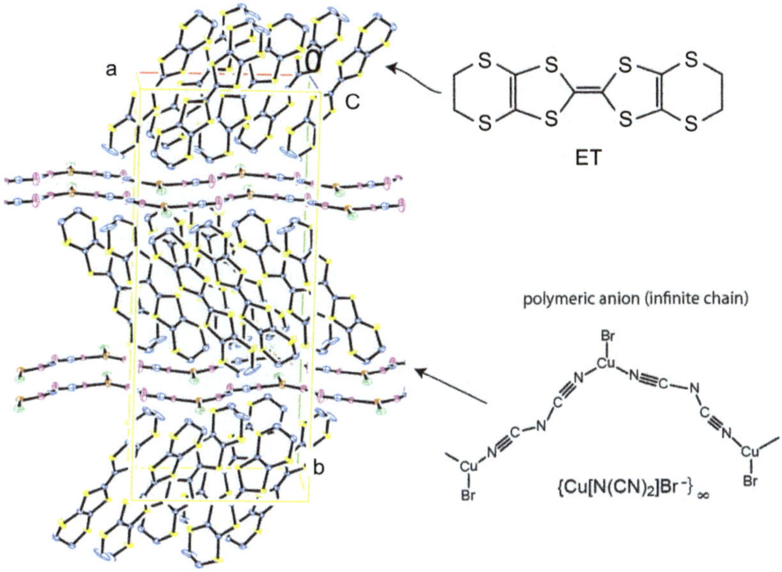

Figure 1.4 A crystal comprising organic radical cations and inorganic polymeric anions, κ-$(ET)_2[Cu[N(CN)_2]Br$. ET = bis(ethylenedithio)tetrathiofulvalene. Hydrogen atoms are omitted for simplicity. Blue, purple, orange, green, and yellow spheres designate carbon, nitrogen, copper, bromine, and sulfur atoms, respectively.

1.1.4.4 Crystal structures of one-dimensional and three-dimensional conductors

In Fig. 1.6 all the organic anions coordinate with the Ag^+ ions in a tetrahedral way, leading to all of the cations and anions being interconnected with each other in a three-dimensional (3D) way by covalent bonds, as in a covalent crystal. However, $Ag(DMe\text{-}DCNQI)_2$ is metallic [62–107], that is, exhibiting totally different properties from the insulating properties of covalent crystals such as diamond and quartz. The organic molecules DMe-DCNQI are solely responsible for conduction and magnetism. The inorganic cations Ag^+ do not have unpaired electrons, playing no role in the conduction and magnetism of this compound. In other words, electrical conduction occurs through the apparently nonbonded one-dimensional (1D) array, that is, the *stacking columns* of DMe-DCNQI instead of 3D network of covalent bonds. Consistently, $Ag(DMe\text{-}DCNQI)_2$ exhibits

Pauli paramagnetism, which is typical magnetic behavior common to metallic electronic systems. What is more, this compound brings about a metal-insulator transition at a low temperature (~100 K), a behavior not observed for elemental metals such as iron, aluminum, and silver. Metal instability is the characteristic behavior of 1D and some of the 2D metallic systems. Interestingly, the slightest chemical modifications, such as replacement of the substituent (methyl group, CH_3) by the methoxy group (CH_3O) or halogen atoms (Cl, Br, I), Ag by Cu, H by D, and/or ^{12}C by ^{13}C, lead to a qualitative change in conducting and magnetic properties, though the crystal and molecular structures remain practically unchanged [62–107]. For example, simply by replacing the silver ions with copper ions, 3D metallic electronic structures are realized. This is because the copper 3d orbitals are close to the π-bands formed by the DCNQI radical anions in terms of energy, which is not the case for the silver ions. Such close proximity makes the band and the orbitals hybridize with each other, making the copper ions a part of conduction pathways. Thus the entire 3D crystal structure serves as conduction pathways in $Cu(DMe\text{-}DCNQI)_2$ but not in $Ag(DMe\text{-}DCNQI)_2$. The slightest change in pressure and/or temperature often results in a qualitative change in conducting and magnetic properties as well as the dimensionality in the band structures. A similar effect is observed by alloying of the two different kinds of the DCNQI salts. For example, the single crystals of $Cu[(DMe)_{1-x}(MeBr)_x\text{-}DCNQI]_2$ and $Cu_{1-x}Li_x(MeBr\text{-}DCNQI)_2$ (MeBr-DCNQI is one of the DCNQI derivatives obtained by the replacement of one of the methyl groups CH_3 by bromine Br in DMe-DCNQI) exhibit qualitatively different conduction behaviors from the behavior of either of the component compounds of the alloy [108, 109]. Such qualitative change in the conduction property has also been observed under the pressure caused by the weight of a few pieces of a Japanese coin (~10 g/piece) [110]. The copper 3d orbitals are narrow in energy dispersion and susceptible to Jahn–Teller distortion. As the π-d-hybridized band structures are based on the balance of relative energy levels between the copper 3d orbitals and the DCNQI-π bands, the hybridization is extremely sensitive to various conditions and perturbations affecting the levels of the 3d orbitals.

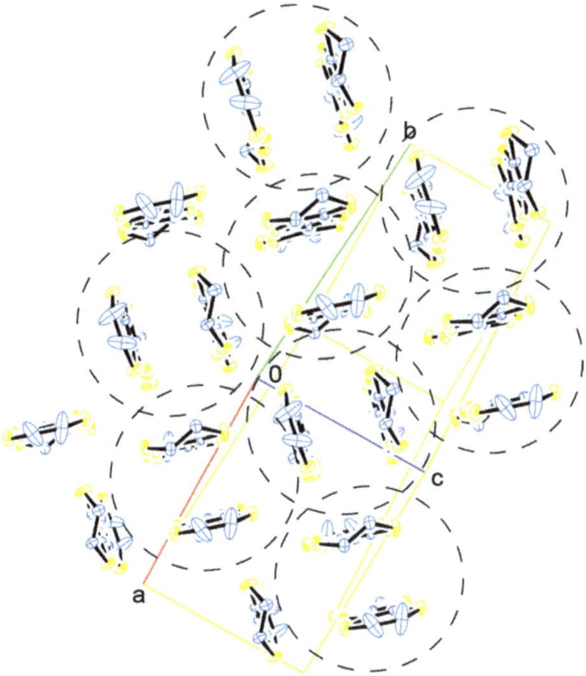

Figure 1.5 A conduction sheet comprising organic radical cations in κ-$(ET)_2[Cu[N(CN)_2]Br$. ET = bis(ethylenedithio)tetrathiofulvalene. Hydrogen atoms are omitted for simplicity. Blue and yellow spheres designate carbon and sulfur atoms, respectively. Note that pairs of planar ET radical cations facing each other, that is, dimers of ET radical cations (circled), arrange themselves in an orthogonal way to form a 2D sheet. This type of molecular arrangement is generally called κ-type and is often found in organic superconductors with the highest T_Cs (~10 K) in organic compounds.

The organic compounds or molecular crystals are supposed to be nonmagnetic insulators, and so are most of them in fact. However, as illustrated by these examples above, the electrical and magnetic properties of molecular materials are diverse and surprisingly sensitive to the detail of the crystal structures. Many of simple substances originate from metallic elements. They, in fact, exhibit metallic properties under sufficiently high pressure, while most of the compounds are insulators under practically available thermodynamic conditions. Yet, whether a given material is metallic is independent of whether it is an organic or inorganic compound and whether it contains metallic elements. To understand the essential origin of metallic properties, we will discuss how conduction and magnetism occur next.

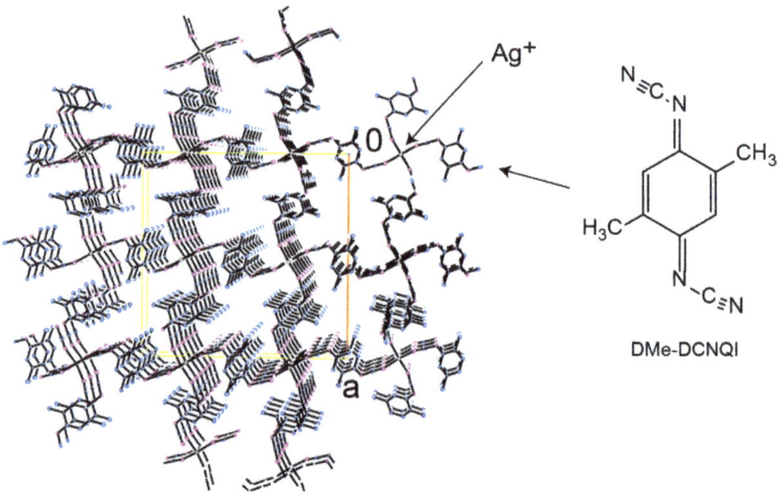

Figure 1.6 A crystal comprising inorganic cations and organic radical anions, Ag(DMe-DCNQI)$_2$. DMe-DCNQI = 2,5-dimethyl-*N,N'*-dicyanoquinonediimine. Hydrogen atoms are omitted for simplicity. Blue, pink, and blank spheres designate carbon, nitrogen, and silver atoms, respectively. Although the DCNQI salts are designated as (DCNQI)$_2$M (M = metal cations) in most of the papers in the reference list, they are designated as M(DCNQI)$_2$ in this book, since cations come first in chemical formulae in this book.

1.2 Cooperative Phenomena in Molecular Crystals

In molecular materials cooperative phenomena such as conduction and magnetism require intermolecular interactions without exception. They sometimes produce unpaired electrons/holes from neutral or closed-shell molecules, while they more often cancel out the existing unpaired electrons by making them pairs by strong intermolecular interactions such as dimerization, trimerization, and tetramerization. Thus intermolecular interactions are more dominant factors in the cooperative phenomena than the individual properties of component molecules. Intermolecular interactions are governed by the overlap of the molecular orbitals involved and thus are sensitively dependent on the relative positions of the two neighboring or facing molecules and their energy levels. To specify this important and basic rule, we should make a close comparison of two substances again: the insulator in Fig. 1.3 and the conductor in Fig. 1.4.

1.2.1 Prerequisites for Conductors

Electric charge can be carried in solids by different mechanisms. Elemental metals conduct electricity via what is called the band conduction mechanism, which requires two factors: carriers and their pathways. The carriers are generally electrons and/or holes, while the conduction pathways comprise orbital overlaps between adjacent chemical species. If the overlaps of orbitals extend all through the solids, the resultant merged orbitals are generally called bands. The more extended the merged orbitals become, the wider the bands become. When unpaired electrons/holes are accommodated in wide bands, they should be delocalized in the solids, being capable of carrying electric charge under an applied voltage. In short, the solids should be conducting when they possess unpaired electrons/holes in a wide band. If the overlaps are insufficient for the band conduction, and/or if there are formally no unpaired electrons/holes, carriers could still occur and travel among adjacent chemical species with the aid of thermal excitation energy. This mechanism is universally observed in semiconducting materials, called thermally-activated-type conduction or simply "hopping." In some inorganic ionic solids, ions may travel through the solids under an applied voltage at elevated temperatures. This also results in thermally-activated-type conduction called ionic conduction. Band (metallic) and semiconducting conductions are often referred to in the molecular conductors (appearing in the following chapters in this book), while ionic conduction is important in solid oxide fuel cells, discussed in Chapter 7. In compaction pellet samples, semiconductor devices with junction structures, and thin films, the resistivity due to the interfaces/boundaries between powder particles and domains (grain boundaries) is sometimes dominant rather than their intrinsic resistivity. The intrinsic resistivity, including anisotropy, is always most reliably measured using single-crystal samples.

1.2.2 Example of an Insulator

In the salt shown in Fig. 1.3, $[Ru(bpy)_3][Ni(dmit)_2]_2$, there are no infinite molecular arrays of the same species, either of cations or anions. Here the infinite molecular arrays indicate extended orbital overlaps between cations or between anions all through the crystal. Such a network occurs, for example, when the molecular species are so closely arranged that their π-conjugated parts are parallel and that the distance between the nearest atoms in the π-conjugation on the neighboring molecular species is within the sum of van der Waals radii. Thus even if $[Ru(bpy)_3][Ni(dmit)_2]_2$ had unpaired electrons/holes as potential carriers, it still would be an insulator as there are no conduction pathways.

1.2.3 Example of Molecular Metals and Superconductors

In the salt shown in Fig. 1.4, κ-$(ET)_2[Cu[N(CN)_2]Br$, it is the balance of intermolecular interactions that realizes metallic properties. Isotropic and homogeneous interactions are crucial for metallic properties, whether the substance is made of atoms (elemental metals) or molecules (those in this chapter). In κ-$(ET)_2[Cu[N(CN)_2]$ Br, ET molecules make a pair called a dimer, which forms an effective unit of the crystal structure. Thus formed dimers are arranged in an orthogonal way to produce a 2D sheet, allowing an overlap of molecular orbitals between the neighboring dimers, as schematically shown by the broken circles in Fig. 1.5. This 2D network leads to a thermodynamically stable metallic 2D band. In fact, the nearly square lattice ($a \approx c \approx 10$ Å, $\beta = 90°$) is the common feature shared by the κ-type ET salts. In such a network, the many peripheral sulfur atoms in the π-conjugation play an indispensable role, which extend the π orbital in the molecular plane in addition to the direction normal to the molecular plane. Usually, the π-conjugated orbitals extend only in the direction normal to the molecular planes. Accordingly, many other π-conjugated molecules without peripheral chalcogen atoms (O, S, Se, Te), such as aromatic hydrocarbons, do not have π orbitals extended in the molecular plane. However, a 2D network of interaction requires this type of isotropic extension in π orbitals. The 2D network provides another advantage for conduction. The 2D band structure is basically free of the metal-to-insulator transitions frequently observed at low temperatures, which is characteristic of 1D and some of the 2D band structure metals. Such transitions originate from the thermodynamic instability of 1D and low-dimensional metals, theoretically predicted by Peierls in 1955 [111], and have been actually observed in a number of compounds having 1D metallic bands [112]. The stable metallic band structures often lead to superconducting transition at low temperatures, instead of metal-to-insulator transitions. In short, the strategy for developing molecular (super)conductors can be summarized as follows. Firstly, one should realize 2D or 3D π-bands comprising molecular orbitals and their overlaps extended all through the crystals. This would be realized by close packing of π-conjugated radical species with many peripheral chalcogen atoms or by hybridizing of the molecular orbitals with transition metal d orbitals. Next, one should make a sufficient number of carriers delocalize in the bands. Then how should we secure the carriers?

1.2.4 Doping of Molecular Materials

As regards carriers, it is sometimes more complicated to decide whether a given system should possess carriers or not than to decide whether the given system should possess conduction pathways or not. This is because intermolecular interactions often occur as they make the unpaired electrons pair as much as possible and intermolecular interactions do not always appear so obviously in the molecular arrangements as illustrated in Figs. 1.2 and 1.3. The most straightforward way to obtain information on carriers is based on direct measurements of resistivity using single crystals, or in some cases, it can be also estimated by the use of band structure calculations. Both examples will appear repeatedly in this book. As for the κ-type ET salts, some are insulators and others are metallic at RT, in spite of almost identical crystal structures (Figs. 1.4 and 1.5). The mechanism for insulating properties is complicated and still under study in the research field of molecular (super)conductors. Independently of mechanisms, production of carriers in a given material, generally called *doping*, is a difficult but indispensable issue to design conducting materials. For molecular conductors in particular, it would be too demanding to design new conductors having desired band structures and carrier densities beginning with a molecule, unless one can precisely control the details of molecular arrangements as desired. At present, it is impossible to precisely control the molecular arrangements in crystals. Recently more and more organic and metal-organic compounds are attracting worldwide attention as new materials for semiconducting devices such as rechargeable batteries (Chapter 4), solar cells/photovoltaics (Chapter 8), and field-effect transistors (Chapter 8). Some of them do not need doping in advance in device application, while others do. In spite of the substantial body of research results accumulated during nearly a century, few of the molecular CT salts have been put to practical use. The difficulty of doping in molecular materials has always prevented us from trying it. This is partly because molecular materials are generally fragile and vulnerable, unlike inorganic materials such as silicon and gallium arsenide. These inorganic substances are insoluble, are never volatile, and have remarkably high melting points, which undoubtedly make it difficult to purify them into large single crystals and to dope them. In spite of the difficulty, suitable doping methods have been established to become a part of modern fabrication technology of semiconductor devices. On the other hand, a versatile method for doping molecular crystals has been elusive for the entire history of the related research.

1.3 Photoconduction and Related Phenomena in Molecular Materials: A Tutorial

1.3.1 Ground vs. Photoexcited States: A Structural Aspect

Thus far we have mainly discussed *ground states* of molecular materials. In the ground states of molecular conductors, carriers have been doped by chemical reactions during the formation of the CT complexes. The reactions typically include spontaneous redox reactions between electron-donor and electron-acceptor molecules in solution (Fig. 1.7) and electrochemical redox reactions of donor and/or acceptor molecules with appropriate supporting electrolytes in solution (Fig. 1.8). In any case, these initial redox reactions produce all the carriers in the ground states, and there is no additional opportunity for doping the CT complexes. During the same reactions, crystal and band structures are fixed depending on the reaction conditions. In other words, every factor affecting conduction properties should be controlled, if possible, at the same time in the single step of the formation reaction. In addition *polymorphs* are frequently observed in the molecular crystals, where different crystal structures and physical properties occur with identical chemical formulae to each other. This is generally ascribed to low symmetries of molecules and weak van der Waals interactions forming the molecular crystals. Such a situation may often produce different molecular arrangements having close energies to each other. The polymorphs make it more difficult to selectively synthesize a particular CT complex with the desired conduction property. In short, crystal structures cannot be controlled precisely in a desired way, even though molecules can be designed in detail and synthesized in an exactly desired way. This directly means that it is impossible to control or predict precisely the conducting and magnetic properties of unknown crystals in the ground states on the basis of the given constituent molecules.

Figure 1.7 A typical diffusion method for single-crystal growth using H-shaped glassware. Donor and acceptor compounds are respectively added to the bottom of each compartment, and an appropriate solvent is added slowly so as not to dissolve them as far as possible. The two compartments are separated by glass frits with fine porosity. The whole system is tightly sealed, keeping the atmosphere inert, by purging nitrogen gas, and kept still at a constant temperature usually in a dark and dry place. In a month or so, the two compounds slowly dissolve and diffuse into the solution until they meet and react with each other. This type of spontaneous redox reaction would sometimes, but not always, produce single crystals of resultant/unknown compounds.

Figure 1.8 Typical glassware called electrochemical cell for single-crystal growth. Donor and/or acceptor compounds are respectively added to each compartment, and an appropriate solvent and supporting electrolyte are added to dissolve all as much as possible. The two compartments are separated by glass frits with fine porosity. Typical electrodes are wires or plates made of platinum. The whole system is tightly sealed, keeping the atmosphere inert, by purging nitrogen gas, and kept still at a constant temperature usually in a dark and dry place. In a month or so, electrochemical redox reaction slowly proceeds under a constant current or voltage until single crystals appear.

However, once the crystal structure is given, one can know the band structure in detail by calculation. This is also ascribed to the weak van der Waals interactions, which do not seriously perturb the electronic structures of the constituent molecules. In addition, if the conduction, magnetic, and/or spectroscopic properties are actually known by measurement, one can check whether and how the calculated band structure is valid and accurate. Such a situation may enable us to control and predict precisely the conducting and magnetic properties in the excited states under photoirradiation, since the crystal structure would remain the same under irradiation as long as the Franck–Condon principle is valid. On the basis of the observed solid-state spectra, one can exactly know the wavelengths of light the material absorbs, in addition to the oscillator strength of each optical transition. This means that one may control the electron densities at the Fermi level E_F and/or the excited states of a given material by irradiating with light of appropriate intensity and wavelengths. Here the "Fermi level" indicates the highest energy the electrons in the material can take in the ground state (Fig. 1.9) [113]. If attention is paid to the excited/unoccupied states instead of the ground/occupied states, one does not need to control the complex structures of molecular crystals in order to control the physical properties.

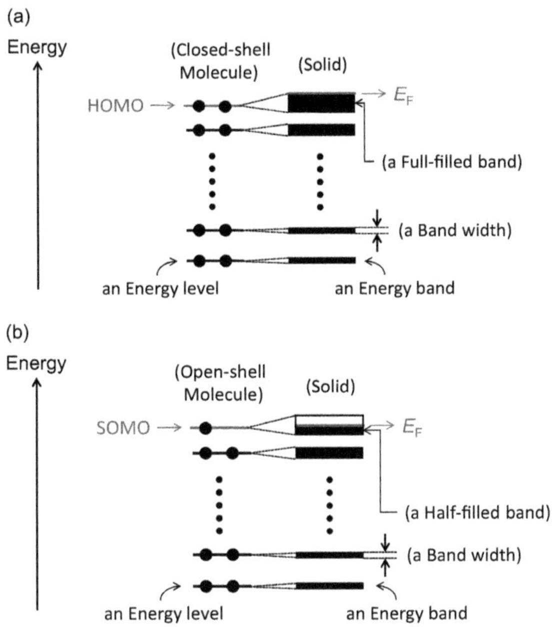

Figure 1.9 A schematic description of the relation between molecular orbitals and energy bands; (a) a solid comprising closed-shell molecules and (b) a solid comprising open-shell molecules. Black circles, horizontal lines, and rectangles designate electrons, energy levels, and energy bands, respectively. The black-filled area in each energy band designates the occupied states by electrons, while the remaining open area designates unoccupied states. Different from an isolated state, all the chemical species form an energy state comprising what are called "bands," characteristic of solid states. Each band consists of a great number of practically degenerated molecular orbitals and has a certain degree of freedom in energy called "band width." The band width is designated by the height of each rectangle above, and it depends on how closely the original molecular orbitals interact with each other; the closer the interaction, the wider becomes the band. In a solid state containing more than one chemical species, mixing of the bands more or less occurs due to the interaction between different species. As a result, a given band does not always correspond to a single molecular orbital, like the formation of a molecular orbital from different kinds of atomic orbitals. The occupancy of a given band, called band-filling, depends on the occupancy of the original ("parent") molecular orbitals due to the conservation of the number of electrons of the system.

1.3.2 Ground vs. Photoexcited States: An Electronic Aspect

There is another contrasting situation generally lying between ground and photoexcited states of solids. In isolated molecules, such as those in a gas phase and diluted solution, the unpaired electrons should be stabilized within each molecule in a thermodynamic way to be stable radical species. Thus if the molecule has no way for stabilization, such as delocalization of the unpaired electrons, it would not remain for a long time, even though the molecules are unlikely to be attacked by other chemical species. In solids, which generally have the highest densities among all states of matter, molecules are closely packed to interact closely with each other. This means that it is difficult for unpaired electrons to remain intact, even if they are sufficiently stabilized in a thermodynamic way. In short, because of the short distances and close interactions among molecules in solid states, the molecules with unpaired electrons are often made to form a dimer, a trimer, a tetramer, etc., to make an electron pair, that is, a closed-shell supramolecular species, as discussed in the introduction of Section 1.2. This is based on a mechanism similar to that of chemical bond formation, and thus the stabilization energy by the pairing generally corresponds to that of a weak covalent bond formation, being typically in the energy scale of 10^1–10^3 K. Considering that RT is ~300 K, some of the molecular solids containing radical species are diamagnetic insulators independently of temperature (≤300 K), while others exhibit temperature-dependent conducting and magnetic properties, including thermodynamic phase transitions (Fig. 1.10). This situation makes it difficult to retain unpaired electrons in solids in a wide temperature range ($0 < T \leq 300$ K).

In contrast, the energies given by photoexcitation are by far larger than those by thermal excitation, as stated above. Therefore, photoexcitation always produces unpaired electrons by breaking the electron pairs, as long as the solid absorbs the light. In other words, one can assume the existence of unpaired electrons in photoexcited states, being ready to serve as carriers for conduction or localized spins for magnetism. The number and energy of such photoexcited unpaired electrons can be controlled by irradiation conditions, which can be finely and more easily controlled than the crystal structures can be. Thus if the photoexcited unpaired electrons should have a sufficiently long life to contribute to conduction and magnetism, photoconductors/photomagnets would be realized based on molecular crystals.

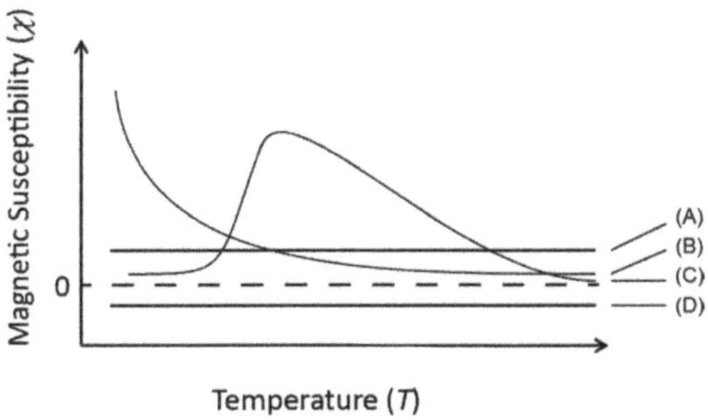

Figure 1.10 Schematic description of temperature dependence of magnetic susceptibility $\chi(T)$ often observed for molecular CT complexes. (A) Temperature-independent small positive χ characteristic of metals (Pauli paramagnetism), (B) χ inversely proportional to T for (nearly) isolated spin systems (Curie or Curie–Weiss paramagnetism), (C) χ with the maximum characteristic of low-dimensional antiferromagnetic spin systems, and (D) temperature-independent small negative χ for insulators (diamagnetism). The sudden decrease in χ observed for (C) could be due to a phase transition, depending on the substance. As regards (D), the diamagnetism originates from the core electrons in the closed-shell structures and thus is exhibited by every substance independent of its physical properties and thermodynamic states (gases, liquids, and solids). Therefore, in a narrow sense, diamagnetism is not included in magnetism and diamagnetic substances are sometimes called *nonmagnetic* substances.

As per the discussion thus far it appears that every material can become a conductor and/or a magnet under appropriate irradiation. However, this is not the case. If a given material absorbs a particular wavelength in a resonant way, this generally means that the relaxation time τ, that is, the averaged lifetime of the photoexcited (unpaired) electrons should approximately coincide with the frequency (more exactly, the period) of the incident light, for example, 10^{-14}–10^{-16} s for near-infrared-ultraviolet-visible (NIR-UV-Vis) light, which is too short to exhibit conduction or magnetism. Even if there are (photoexcited) unpaired electrons, they require sufficiently close interaction among themselves to exhibit conduction and/or magnetism. However, the value of τ generally decreases when interaction increases. Thus the next problem is how to make the relaxation time of photoexcited states much longer while retaining the close intermolecular interactions. It is not a problem where you require to make τ a few times longer; rather you require to make it more than 10^{12} times longer, since even quickest magnetic or electrical measurements generally take 10^{-2}–10^{-3} s [114]. Could it be possible? The answer is yes, as shown below.

1.3.3 CT Interaction between Different Components: Net Carrier and Spin Injection

To stabilize radical (ionic) states, which leads to the retention of unpaired electrons, CT interaction between different kinds of molecules plays an important role (Fig. 1.11). CT interaction generally stabilizes the orbitals accommodating an unpaired electron, whether

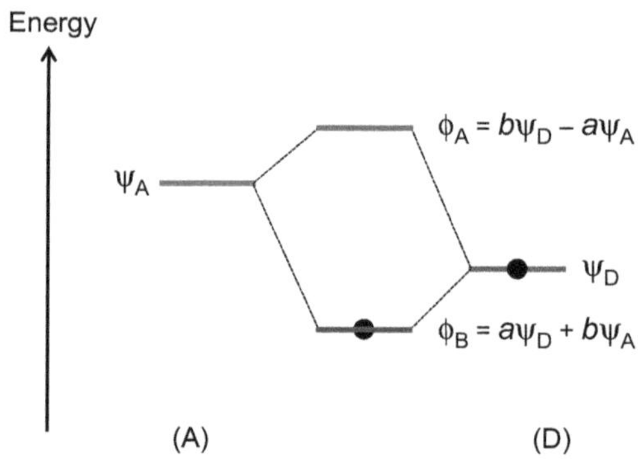

Figure 1.11 A schematic description of interaction between two orbitals of electron-donor (D) and –acceptor (A) species. Red, blue, purple, and violet horizontal lines designate the two orbitals before interaction (red ψ_A and blue ψ_D) and those after interaction (purple ϕ_A and violet ϕ_B), respectively. A black-filled circle designates an unpaired or a photoexcited electron. By the interaction, both orbitals are mixed with each other to form two new orbitals. The mixing ratios in the new orbitals depend on the proximity of the original levels; the closer orbital in energy should have the larger contribution in the new orbitals, that is, $|a| > |b|$. Thus the purple orbital has a larger coefficient of the red orbital, and the violet orbital has a larger coefficient of the blue orbital. Before interaction, the chemical species D possesses a single electron in the blue orbital ψ_D, while A has none in ψ_A. After interaction, the one (unpaired) electron in ϕ_B is shared by both A and D, which formally means that a part of the electron should transfer from D to A by the interaction. This is why the interaction is called charge transfer (CT) interaction. The discussion here is valid independently of relative energy levels between ψ_A and ψ_D or electron counts, except for the case where both of the orbitals (ψ_A and ψ_D) are fully filled or completely unoccupied.

the system is under dark or irradiation. Thus CT interactions in the photoexcited states may prolong τ. In addition, CT interaction corresponds to a partial redox reaction, leaving unpaired electrons on both chemical species involved in the interaction. Therefore, solids having CT interactions between different components contain a certain number of carriers and/or localized spins already in the ground state, and they will increase in number under irradiation by CT excitation. As a summary, CT interactions between different components produce net unpaired electrons, potentially leading to conduction and/or magnetism, whether the system is under dark or irradiation. Whether the unpaired electrons actually contribute to conduction and/or magnetism depends on the crystal structures, that is, the molecular arrangements in the solid, which will be discussed in the next subsection.

1.3.4 CT Interaction between the Same Components: Forming Conduction Pathways

The strategy for forming conduction pathways in a solid is independent of whether the conduction is under dark or photoconduction. The basic principle of conduction in crystalline materials has been discussed in Section 1.2. In the case of molecular crystals, for example, electrical conduction requires an extended array of the same molecular radical species. Here, the radical species include not only radical anions and cations but also formally neutral molecules in the case of single-component molecular conductors (SCMs) [115–125]. SCMs correspond to molecular versions of elemental semimetals or metals. They consist of apparently closed-shell molecules, yet experiments and calculation unambiguously have revealed that they are practically open-shell, that is, neutral radical molecules, in fact, due to the close intermolecular interactions among them. It should be also noted here that "the same molecular radical species" are in the most exact sense of the words. Even if the chemical formula and the formal charge are the same, they are not adequate. The same molecular radical species designate the crystallographically same molecules, which are exactly identical in every aspect, such as net electric charges and spins, chemical formula, bond lengths and angles, and relative positions and orientations, to the surrounding molecules. In addition, even if there is an extended molecular array of the same species all through the crystal, an irregular arrangement often hinders high conduction as well as magnetism (Fig. 1.12). If one of these conditions is not satisfied, a molecular array in a solid might not serve as a conduction pathway. For example, molecular self-assembly, such as dimerization and tetramerization, could transform the unpaired electrons into paired and localized states, making the solid lose conduction and paramagnetism. Such a solid will often exhibit diamagnetism (Fig. 1.10). So the conduction (as well as magnetism) is sensitively dependent on molecular arrangement.

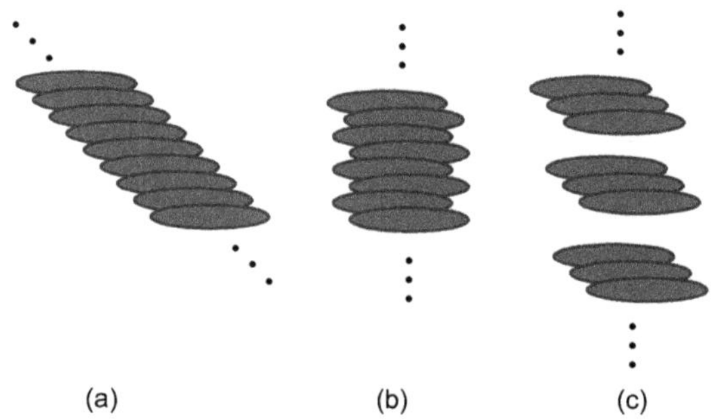

Figure 1.12 A schematic description of different patterns in molecular arrangement. Each oval designates a molecule. (a) A regular array, (b) an irregular array (twofold period, that is, an example of dimerization), and (c) an irregular array (threefold period, that is, an example of trimerization).

The only additional focus in considering conduction pathways in molecular crystals under irradiation is a possible structural change in the photoexcited state. Although the Franck–Condon principle has been considered to be valid in many cases of molecular crystals under irradiation, they will sometimes exhibit different crystal and molecular structures in the photoexcited states. This is because of the strong electron-lattice interaction having the possibility of deforming the lattice, in addition to the weak van der Waals interaction forming the lattice, both of which are characteristic of molecular crystals.

1.3.5 Magnetism vs. Conduction: A Tutorial

Here, we should review a fundamental problem. If both magnetism and conduction originate from unpaired electrons, what creates the differences in their roles in the resultant physical property? The answer is whether they are localized or delocalized. In other words, the role of unpaired electrons depends on the energy levels/bands they occupy. The unpaired electrons in a wide band are delocalized in the solid states and generally exhibit conductivity when a voltage is applied. The delocalized unpaired electrons are often called carriers, as they carry electric charge as electric current under the applied voltage. In contrast, the unpaired electrons in a narrow band or a rather localized level have localized spins and generally exhibit magnetism. Delocalized unpaired electrons also exhibit magnetism. The carriers in a metallic substance generally exhibit paramagnetism of small and temperature-independent magnetic susceptibility. The thermally-activated-type semiconductors often exhibit a different type of paramagnetism, called Curie or Curie–Weiss behavior, depending on the details of temperature dependence. In other words, if there are unpaired electrons in a solid sample, it would likely exhibit slight conduction as a (semi)conductor and some type of magnetism.

Conduction properties are classified into two kinds on the basis of temperature dependence. If resistivity decreases with a decreasing temperature, the substance is metallic independently of the chemical components. If the resistivity increases with a decreasing temperature, the substance is nonmetallic. The nonmetallic substances are usually called semiconducting or insulating materials depending on how high the electrical resistivity is at RT. Typical nonmetals exhibit resistivity ρ, which increases with a decreasing temperature T in an exponential way.

$$\rho \propto e^{\frac{E_\mathrm{a}}{RT}},$$

where E_a and R are the activation energy for thermal carrier excitation and the gas constant, respectively. There is no definite difference between semiconductors and insulators. In fact, there are many substances exhibiting intermediate behavior between metals and nonmetals, that is, exhibiting resistivity with almost no or small temperature dependence. Both semiconductors and insulators have energy gaps at their Fermi levels, which require activation energies (E_a) to produce carriers for electrical conduction. The metallic substances do not have energy gaps at their Fermi levels and thus do not require activation energies for conduction. Naturally occurring metals, that is, elemental metals, such as aluminum and iron, can be chemically unstable toward rusting and corroding yet thermodynamically stable owing to their small anisotropy in crystal and electronic structures. However, synthetic metals, that is, other conductors than elemental metals can be thermodynamically unstable because of their highly anisotropic crystal and electronic structures. They often make transitions between metals and nonmetals on changing temperatures and/or pressures. Thus *dimensionality* of electronic structures, in addition to thermodynamic conditions, is important in electrical and magnetic properties. Even among the elemental metals, tin exhibits such a metal-insulator transition at ~13.2°C (a very broad and slow transition), below which, gray-colored, brittle, and insulating α-tin is stabler than white, shiny, malleable metallic β-tin.

Similarly to conduction, magnetism exhibits rich variety, including diamagnetism, paramagnetism, and ferromagnetism. Both paramagnetism and ferromagnetism can be further classified into many different types. There are also unique patterns of magnetic behavior characteristic of low-dimensional spin systems, as shown in Fig. 1.10. The variety originates from the variety in interactions between spins. The differences in magnetism are identified typically on the basis of the temperature dependence of magnetic susceptibility.

1.3.6 Control of Magnetism and/or Conduction: Thermodynamic vs. Optical Methods

As is discussed above, magnetism and conduction are typically controlled by temperature and pressure in addition to the chemical modification of samples. This is because the interactions between unpaired electrons are sensitive to thermodynamic conditions as well as how they are arranged, that is, structures. Variation in temperature cause variation in the occupancy of electrons in energy levels/bands in solids, generally called thermal excitation. Here the occupancy designates how many electrons occupy each energy level. Instead of the actual number of electrons, the occupancy of an energy band is described by the filling ratio, such as quarter-filled and half-filled. The occupancy and band filling designate the oxidation state or effective charge of the component chemical species. In the first approximation, the arrangement of the energy levels/bands remains the same irrespective of occupancy. In contrast, variation in pressure (at a constant temperature) affects the arrangement of the energy levels/bands, while the occupancy of electrons is retained. Variation in temperature causes variation in the occupancy as thermal excitation, while it may or may not affect the arrangement of energy levels/bands, depending on whether there would be structural changes. If there is a drastic change caused by a chemical reaction or a phase transition in the course of temperature/pressure variation, the crystal and electronic structures, including occupancy, may also change in a drastic way. Sometimes, crystals crack or collapse or even transform into an amorphous state because of the drastic change in structures. Chemical modification utilizes synthetic chemistry to finely tune the molecular structures, which results in variation in crystal and electronic structures in a fine or a rather unexpectedly drastic way. In either case, all these control methods are available under ambient or near-ambient conditions, that is, below the highest temperature the substance can endure and below the highest pressure available. Thus thermodynamic control usually can provide samples with perturbation having an energy scale corresponding to a few hundred kelvins and can cause volume changes within a few percentages. When one is required to control physical properties beyond this limitation, photoirradiation may be the best choice. Ultraviolet (UV), for example, can provide samples with perturbation having an energy scale corresponding to several tens of thousands of kelvins without instantly decomposing them under ambient conditions. In contrast to temperature and pressure control, photoirradiation can affect samples without contact with them. In addition, photoirradiation is more easily controlled than other methods, which enables us to select the phenomena to occur at the selected part of the sample. In other words, optical processes are highly selective, which is a characteristic other methods do not have.

As regards control of conduction and/or magnetism by photoirradiation, there are pioneering pieces of work, called light-induced excited spin-state trapping (LIESST), where the spin states of transition metal complexes are switched between high and low spins by photoexcitation at low temperatures [126–134]. A closely related series of phenomena called photoinduced phase transitions (PIPTs) are also reported for the control of magnetism [135] and other properties, including conduction [136–151]. However, there are two important features in the control method discussed in the latter half of this chapter (now called optical doping) that are different from LIESST or PIPTs. Firstly, the optical doping is effective at any temperature and/or pressure, while both PIPTs and LIESST require a particular thermodynamic condition depending on the samples. Secondly, the optical doping is independent of phase transitions, while both PIPTs and LIESST are based on phase transitions and/or spin crossovers the samples originally exhibit. Thus PIPTs and LIESST are combined control methods of thermodynamic and optical aspects, while optical doping is a purely optical control method.

1.4 Design of Photoconductors of a New Type

1.4.1 Choice of Building Blocks for Forming Conduction Pathways

On the basis of the two necessary conditions for efficient photoconductors, discussed in Sections 1.3.3 and 1.3.4, we should consider now how to realize them simultaneously in a molecular crystal. In principle, it is independent of the resultant conduction and magnetism whether a building block for molecular conductors/magnets is anionic or cationic. However, frontier orbitals, that is, the highest occupied molecular orbital (HOMO) and the lowest unoccupied molecular orbital (LUMO), are generally expanded in anionic species, unlike those of the corresponding neutral or cationic species, because of the increase in electron count and electron-electron coulombic repulsion. On the one hand, electron-rich and electron-donating heteroatoms, like S, Se, and Te atoms (chalcogen atoms), produce expanded molecular orbitals and thus are advantageous for intermolecular interaction when they are incorporated in the peripheral positions of π-conjugated systems. On the other hand, π-conjugated molecules containing many chalcogen atoms are easily oxidized rather than reduced, behaving as electron-donor (D) molecules to become cations. In oxidized states, the frontier orbitals generally shrink compared with those before oxidation. This is disadvantageous for intermolecular interaction between neighboring molecules. In contrast, acceptor (A) molecules, such as O, N, and halogen atoms, which usually become anions after redox reactions, have high electron affinity and thus often contain heteroatoms with high electronegativity. Their strong electron-withdrawing abilities inevitably shrink the frontier orbitals, which is disadvantageous for intermolecular interaction between neighboring molecules, even if the frontier orbitals may somewhat expand after reduction. In short, both A and D molecules have their own advantages and disadvantages as regards intermolecular interactions.

However, there is an exceptional series of metal-complex molecules possessing both advantages above. They are generally described as $[M(dmit)_2]^{n-}$ (M = Ni, Pd, Pt, Au, etc.; dmit = 1,3-dithiole-2-thione-4,5-dithiolate; $0 \leq n \leq 2$; Fig. 1.13) [152–166]. The molecular structure has a close similarity to that of the donor molecule ET, since both of the π-conjugated systems include many S atoms in the periphery. In fact, many of $[M(dmit)_2]^{n-}$ are easily oxidized to yield CT complexes, like ET does. However, even in the oxidized states, $[M(dmit)_2]^{n-}$ usually occur as stable anionic species. Thus $[M(dmit)_2]^{n-}$ are chalcogen acceptors, which have expanded frontier orbitals due to the many peripheral chalcogen atoms in addition to the anionic character. $[M(dmit)_2]^{n-}$ are building blocks of choice for molecular conductors, and they in fact produce metals and SCs when M = Ni and Pd. Because $[Ni(dmit)_2]^{n-}$ usually or most often gave crystalline CT complexes with better chemical stability, better crystal quality, and higher conductivity than other $[M(dmit)_2]^{n-}$, we selected $[Ni(dmit)_2]^{n-}$ as the building block for photoconductors.

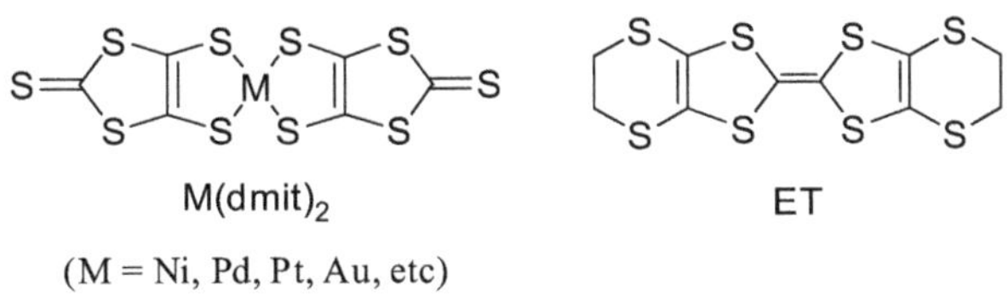

Figure 1.13 Chemical structures of $M(dmit)_2$ and ET. Despite the structural similarity, many $M(dmit)_2$ become stable radical anions, while ET molecules usually become stable radical cations when they form salts.

1.4.2 Choice of Counterionic Species

Next, the countercations for $[Ni(dmit)_2]^{n-}$ should be selected after considering the following conditions:

First, the stoichiometry is important, that is, the ratio between anions and cations. For example, in the monoanion salt of $Ni(dmit)_2$, $(Cation^{m+})[Ni(dmit)_2]_m$, m increases on increasing the charge of the cations. The increase in m leads to denser packing of $[Ni(dmit)_2]^-$ in the unit cell, simply because the volume fraction occupied by $[Ni(dmit)_2]^-$ in the unit cell increases with m. This is favorable for $[Ni(dmit)_2]^-$ to form conduction pathways. On the other hand, the increase in the charge of the cations m is disadvantageous for conduction as the larger charge produces a larger amplitude of periodic potential, which hinders the movement of carriers. Thus it is generally desirable to combine cations and anions with smaller charges from the viewpoint of periodic potential. To reconcile with these two requirements, a small value of m, such as m = 0.5–1, is often selected.

Secondly, one should pay attention to relative molecular sizes, shapes, and symmetries of the anions and the cations. These factors seriously affect the molecular arrangement, the overlapping mode, and the interaction between them.

The most straightforward and simple structure favorable for conduction is a cation-anion segregated structure where cations and anions separately form columns or sheets in crystals. This type of molecular arrangement is often observed in molecular conducting materials (Figs. 1.4–1.6 and 1.14). However, it is unfavorable for cation-anion interactions, which are required for net carrier doping as per the photoconduction mechanism proposed above. Therefore, some other molecular arrangement is required where both cation-$[Ni(dmit)_2]$ and $[Ni(dmit)_2]$-$[Ni(dmit)_2]$ interactions are close enough. One such candidate is the anion-cation-mixed conduction sheet structure shown in Fig. 1.15. Not only the interaction among the molecules governing conduction but also the coulomb attraction between cations and anions stabilizes the anion-cation-mixed conduction sheet structures. This structural motif is basically the same as the honeycomb network often observed in chalcogen donor CT complexes, as shown in Fig. 1.14, and counterionic molecules are inserted while retaining most of the original interaction network. As a result, the structure has features of both ionic crystals and molecular crystals. Such structural flexibility is an advantage of molecular conductors. Although it is difficult to predict or control crystal structures and molecular arrangements, these examples suggest that the prospective candidates for countercations for the $Ni(dmit)_2$ salts should include aromatic ammonium cations slightly smaller than $Ni(dmit)_2$ anions.

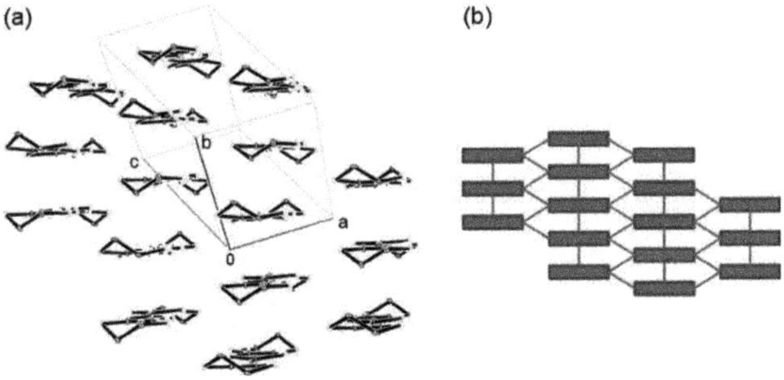

Figure 1.14 An example of chalcogen-donor (ET) arrangement in a conduction sheet (β-$(ET)_2I_3$ at 120 K) [167]. (a) Actual crystal structure and (b) schematic representation of intermolecular interactions among ET molecules. In (a) disordered atoms in ethylene groups and all hydrogen atoms are omitted for clarity. The I_3^- anions also assemble into sheets and they alternate with the conduction sheets along the *c* axis. (b) Each rectangle and thin line, respectively, designates an ET molecule and the interaction between ET molecules.

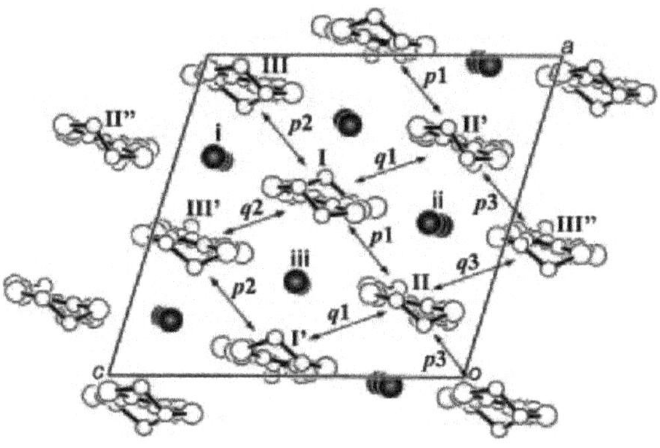

Figure 1.15 Anion-cation mixed conduction sheet found in $(ET)_3(Br_3)_5$. Brown spheres designate bromine atoms. I, II, and III designate crystallographically independent ET molecules, while those such as I', II', II", III', and III" designate symmetry-related ET molecules to I, II, and III, respectively. Arrows with *p*1-*p*3 and *q*1-*q*3 designate possible patterns of intermolecular interactions. A similar molecular arrangement was also observed in $(DED)_2X$ (X = BF_4, PF_6, SbF_6) with clear metallic properties. Reprinted from Ref. [170] with permission from the Chemical Society of Japan.

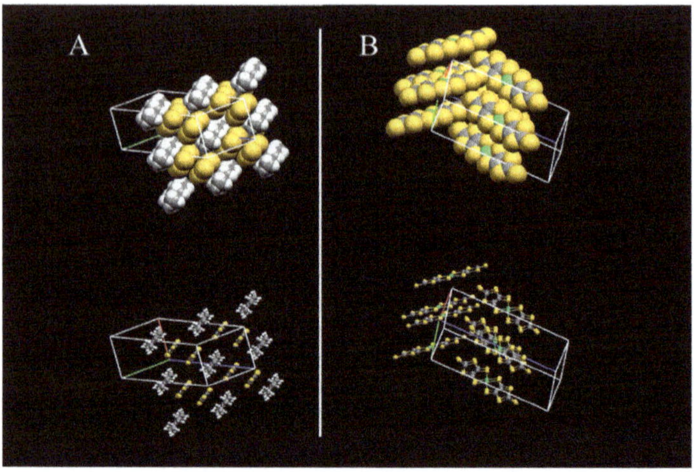

Figure 1.16 Crystal structure of MV[Ni(dmit)$_2$]$_2$. (A) Views of the molecular long axis of the [Ni(dmit)$_2$] anion and (B) views of [110]-direction. The atoms are designated by the following colors: Ni, green; S, yellow; C, gray, H, white; and N, blue. The unit cells are denoted by white lines, where red, green, and blue lines are for the *a*, *b*, and *c* axes, respectively. In (B) the MV cations are omitted for clarity. (A) shows how the MV cation is surrounded by four pairs of [Ni(dmit)$_2$] anions, resulting in a cation-anion-mixed molecular assembly sheet, while (B) shows the adjacent two sheets are connected to each other like "gears." Intermolecular short contacts are found among the [Ni(dmit)$_2$] anions in the same and neighboring sheets. These intermolecular interactions are quantitatively estimated by the calculated intermolecular overlap integrals between the frontier orbitals, that is, the molecular orbitals that have the unpaired electrons of the [Ni(dmit)$_2$] anions. The structural features of this compound directly imply a 3D band structure.

1.5 Examples of New Types of Photoconductors

1.5.1 Photomagnetic Conductors

A number of aromatic ammonium cations, in particular photochemically redox-active species such as methyl viologen and bipyridyl derivatives were examined. The following compounds were found to exhibit novel conducting and magnetic properties. The newly found properties are a combination of photoconductors and "photomagnets," where UV irradiation induces metallic and paramagnetic properties in an instant and reversible manner. Such a property has never been found in any kind of substance. In addition, interaction has been found between magnetism (unpaired electrons on the cations) and conduction (those on the anions) in some of these compounds. Because of the characteristics, these compounds have been named "photomagnetic conductors," a combination of "photoconductors" and "magnetic conductors."

It is generally difficult for localized (immobile) and delocalized (mobile) unpaired electrons to coexist in the same compound as it requires two different energy states with contrasting characters, that is, localized and delocalized states both at E_F. In addition, it is more difficult to retain both types of unpaired electrons interacting with each other as interaction between unpaired electrons often makes them paired electrons, making them lose conduction and magnetism. Therefore, aside from naturally occurring magnetic metals, synthetic magnetic conductors are hardly reported, whether they are responsive to light or not. Naturally occurring magnetic metals do not respond to light to change their conducting and/or magnetic properties.

Generally speaking, magnetic conductors are sometimes limited to designating metallic conductors exhibiting a magnetic order, such as antiferromagnetism and ferromagnetism. However, photomagnetic conductors designate any material having both carriers and localized spins under irradiation, which include paramagnetic semiconductors in this chapter. Such inconsistency in the technical terms or definitions originates from the fact that photoexcited states are not in thermal equilibrium and are always mixed states of thermally and optically excited states. The mixing ratio between thermally and optically excited states depends on the irradiation conditions. The physical properties of a given substance under dark conditions depend only on thermodynamic conditions, while those under irradiated conditions may predominantly depend on irradiation conditions. As for conduction under irradiation, we have found a way of analysis to distinguish the contribution from optically excited carriers and that from thermally excited carriers. This will be discussed in the last section of this chapter.

Figure 1.16 shows the crystal structure of the first example of "photomagnetic conductors" found in 2012, $MV[Ni(dmit)_2]_2$ (salt 1) [168], where MV = methyl viologen (as shown in Fig. 1.2). The basic motif of the molecular arrangement is closely related to that of $(ET)_3(Br_3)_5$ [169, 170] in Fig. 1.15, characterized by a weak but isotropic intermolecular interaction between $Ni(dmit)_2$. This interaction network is reminiscent of metallic crystals. In fact, $(ET)_3(Br_3)_5$ and $(DED)_2X$ (X = BF_4, PF_6, and SbF_6) [171] exhibit metallic conduction. Yet salt 1 does not have a sufficient number of carriers unlike metallic crystals.

Salt 1 possesses features of both ionic and molecular crystals. Such combined structural features are also found in their physical properties. For example, salt 1 is almost transparent in the infrared (IR) region, like alkali halides, except for a limited number of molecular vibration peaks in the IR spectra, as molecular crystals generally exhibit (Fig. 1.17). The IR spectra indicate that the constituent molecular species are practically isolated from each other without particular interactions except for coulombic interaction. As expected from these structural and spectroscopic properties, salt 1 is a diamagnetic insulator under dark conditions, which is a typical property commonly observed in ionic and molecular crystals.

However, there is an important difference in this salt from other compounds discussed thus far; the similarity of anions ($Ni(dmit)_2$) and cations (MV) in their molecular orbitals produces many groups of nearly degenerated energy levels both at occupied and unoccupied states, in addition to CT interaction between $Ni(dmit)_2$ and MV based on the direct overlap between their π orbitals. As a result, a series of CT transitions exists between MV and $Ni(dmit)_2$, corresponding to the photon energies of the UV region. Typical energies of the UV region correspond to $\sim 10^4$ K, which is inaccessible to thermal excitation. Thus there is practically no CT between them without UV irradiation, which leaves both kinds of molecular ions as effectively closed-shell species under dark conditions. If the series of UV excitations occurs, a number of unpaired electrons and holes would be produced on both $Ni(dmit)_2$ and MV. The unpaired electrons and holes on $Ni(dmit)_2$ would travel through the $Ni(dmit)_2$ network, serving as carriers, while those on MV would be localized and stay there, serving as paramagnetic spins. The network is formed by interaction among the excited states/bands. Even if there is no interaction between occupied levels/bands, there could be interaction between unoccupied levels/bands as expected from the band calculation and demonstrated by the experiments. In fact, under UV irradiation, salt 1 exhibits conducting (Fig. 1.18) and magnetic (Fig. 1.19) properties, like transition metals such as cobalt and copper, except for magnetic ordering. This is because UV irradiation excites CT transition between MV and $Ni(dmit)_2$, producing a sufficient number of carriers and localized spins. Thus, salt 1 possesses three aspects in the classification of crystalline states: molecular, ionic, and metallic crystals. Since photoconduction was observed in Se in the 1870s, a great number of photoconductors have been identified to date. Yet none of them become metallic under irradiation except for salt 1.

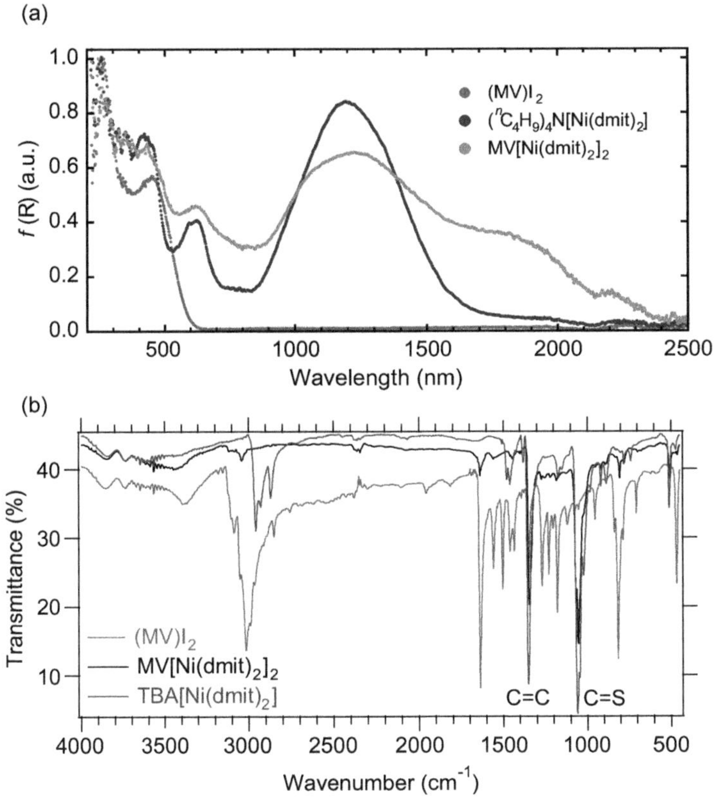

Figure 1.17 (a) Diffuse reflectance spectra of the powder sample and (b) IR spectra in a KBr disk of $\{(^nC_4H_9)_4N^+\}\{[Ni(dmit)_2]^-\}$, MVI_2, and $MV[Ni(dmit)_2]_2$. The agreement of many peaks indicates that MV and $Ni(dmit)_2$ should take the same charges and electronic states as in $(MV^{2+})(I^-)_2$ and $\{(^nC_4H_9)_4N^+\}\{[Ni(dmit)_2]^-\}$, respectively. (a) Adapted from Ref. [168] with permission from John Wiley & Sons. (b) Adapted from Ref. [176] with permission from the Chemical Society of Japan.

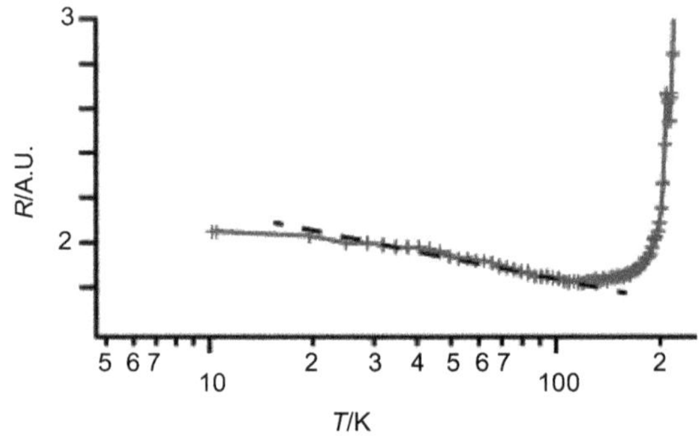

Figure 1.18 Temperature (*T*) dependence of resistance (*R*) of MV[Ni(dmit)$_2$]$_2$ under UV irradiation showing the linear variation of *R* versus log *T* at low *T*, which is characteristic of carriers interacting with localized spins (the Kondo effect). Adapted from Ref. [168] with permission from John Wiley & Sons. Adapted from Ref. [176] with permission from the Chemical Society of Japan.

A related salt BPY[Ni(dmit)$_2$]$_2$ (salt 2, BPY = *N,N'*-ethylene-2,2'-bipyrydyl; Fig. 1.20) also exhibits metallic photoconductivity, while such high photoconduction is not observed in other Ni(dmit)$_2$ salts, where there is no CT interaction between cations and Ni(dmit)$_2$ anions (Fig. 1.21) [172]. The comparison demonstrates that the CT interaction between cations and anions is important for metallic photoconduction, instead of particular components such as Ni(dmit)$_2$ and MV. If this is true, the strategy for developing photomagnetic conductors discussed thus far can be general, not necessarily restricted to the Ni(dmit)$_2$ salts.

1.5.2 Giant Photoconductivity

In the course of the study of photomagnetic conductors, another novel type of photoconductors, even different from those discussed in the previous subsection, has been recently found. This is, again, a simple salt NMQ[Ni(dmit)$_2$] (salt 3, NMQ = *N*-methyl quinolinium) [174, 175]. Tight-binding band calculation indicated that salt 3 belongs to insulating compounds. It actually exhibited high resistivity and diamagnetism, both of which are typical properties of insulators. The crystal structure is shown in Fig. 1.22. There is a honeycomb network comprising weak interaction among [Ni(dmit)$_2$]$^-$. The cation NMQ$^+$ is

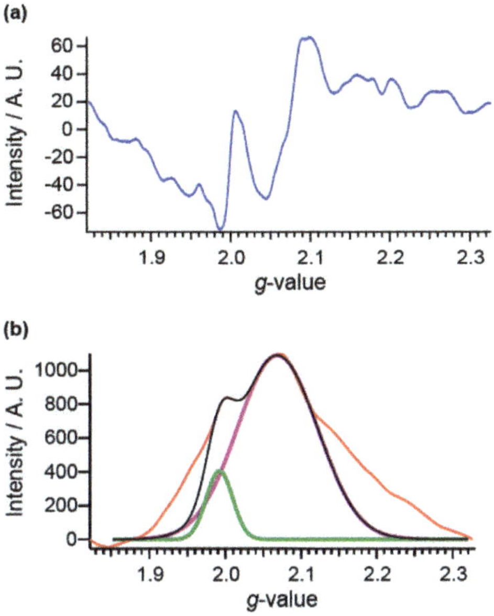

Figure 1.19 ESR spectra during UV irradiation measured on aligned single crystals of MV[Ni(dmit)$_2$]$_2$ at RT; (a) the first derivative of the spectrum after subtracting the spectrum measured under dark conditions on the same sample prior to irradiation; (b) (red curve) absorption spectrum obtained by integrating the spectrum in (a), (green and violet curves) deconvolution of the main two peaks, and (black curve) the total contribution of the two deconvoluted peaks. The analysis in (b) shows that there are at least two kinds of spins (green and violet curves) in the observed spectrum in (a). On the basis of the *g*-values, intensities, and line widths, the green and violet curves can be assigned to the spins localized at the MV cations and those delocalized among the Ni(dmit)$_2$ anions, respectively. (a) Adapted from Ref. [168] with permission from John Wiley & Sons. (b) Adapted from Ref. [176] with permission from the Chemical Society of Japan.

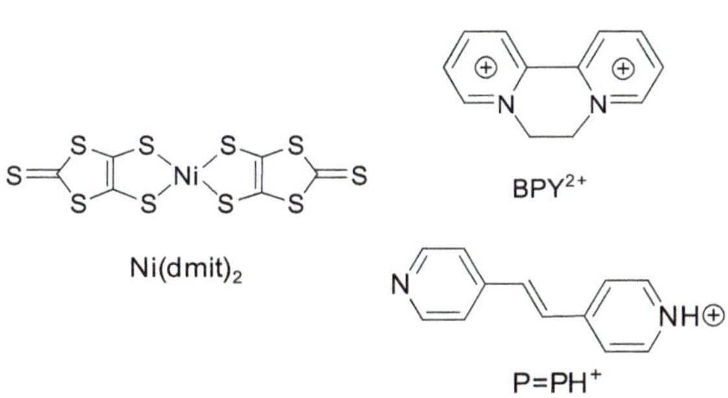

Figure 1.20 Chemical structures of BPY, P=PH, and Ni(dmit)$_2$. For the P=PH salt of Ni(dmit)$_2$, see Ref. [173].

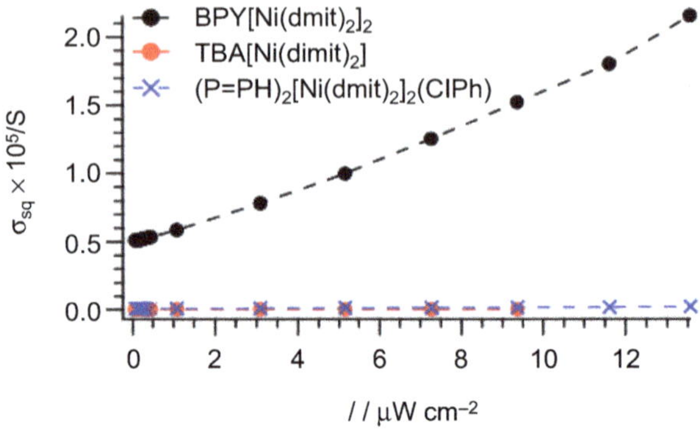

Figure 1.21 Comparison of surface conductivity under UV irradiation σ_{sq} between TBA[Ni(dmit)$_2$], (P=PH)$_2$[Ni(dmit)$_2$]$_2$(ClPh) [173], and BPY[Ni(dmit)$_2$]$_2$ (ClPh = C_6H_5Cl and TBA = *n*-$(C_4H_9)_4N$). For P=PH, see Fig. 1.20). Adapted with permission from Ref. [172]. Copyright (2012) American Chemical Society. Adapted from Ref. [176] with permission from the Chemical Society of Japan.

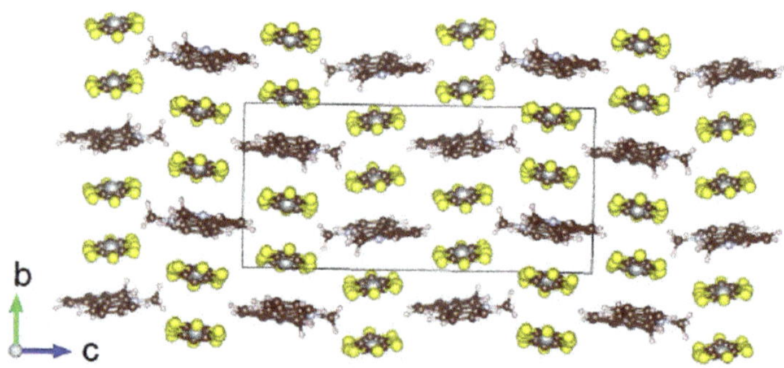

Figure 1.22 Crystal structure of NMQ[Ni(dmit)$_2$]: molecular arrangement in the *bc* plane. Yellow, brown, and gray atoms are sulfur, carbon (filled)/hydrogen (open), and nitrogen (smaller)/nickel (larger), respectively. Adapted from Ref. [175] with permission from John Wiley & Sons. Adapted from Ref. [176] with permission from the Chemical Society of Japan.

accommodated in each cell of the honeycomb, similarly to salts 1 and 2. This molecular arrangement would provide conduction pathways by the honeycomb network and "isolated spin-holders" by the NMQ cations at the same time, when the CT transition between NMQ$^+$ and [Ni(dmit)$_2$]$^-$ takes place. However, contrary to the anticipation, there were no CT transitions between NMQ$^+$ and [Ni(dmit)$_2$]$^-$ observed in the solid-state spectra. The calculation of overlap integrals and X-ray photoelectron spectroscopy (XPS) under photoirradiation supported that there were no CT interactions between NMQ^+ and $[Ni(dmit)_2]^-$. The comparison between the electron spin resonance (ESR) spectra under dark and irradiated conditions did not show any sign of such CT transitions. Thus, it was considered that salt 3 would not exhibit high photoconduction, as there appeared to be no reason for a large amount of carrier production under irradiation. Yet this speculation turned out to be untrue.

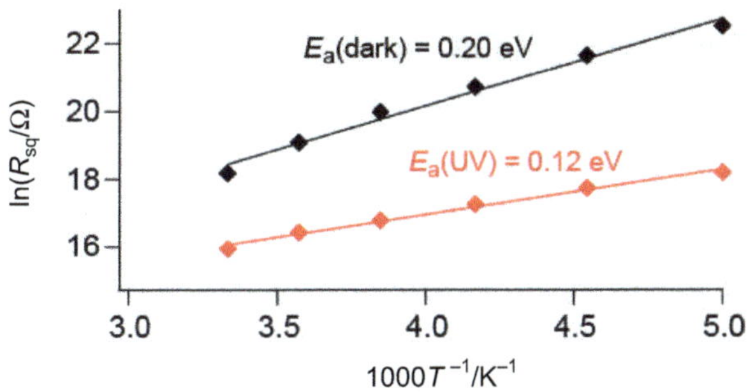

Figure 1.23 Electrical behavior of $NMQ[Ni(dmit)_2]$: comparison of activation energies under dark (black; E_a(dark) = 0.20 eV) and UV-irradiated (red; E_a(UV) = 0.12 eV) conditions. Adapted from Ref. [175] with permission from John Wiley & Sons. Adapted from Ref. [176] with permission from the Chemical Society of Japan.

The unique feature of salt 3 in photoconductivity lies in two properties: firstly, it responds only to UV of 375 nm, that is, it has unusually sharp wavelength selectivity, and secondly, it exhibits an unusually large conductivity ratio ($\sigma_{UV}/\sigma_{dark}$) between dark and UV-irradiated conditions, though the absolute value of photoconductivity (σ_{UV}) is not so high. The ratio $\sigma_{UV}/\sigma_{dark}$ is as high as ~40 and ~ 80 at 300 K and 200 K, respectively, while it is 1–2 for common substances independently of temperature. In common with the salts 1–3, where the ratio $\sigma_{UV}/\sigma_{dark}$ is high and increases with decreasing temperature, the activation energy for conduction varies between dark and irradiated conditions and varies with measurement conditions such as light intensity and the distance between electrical leads on the sample. The former dependence can be explained by the fact that the irradiated sample generally contains both photoexcited and thermally activated carriers. The latter dependence can be also explained by the same fact. Photoexcited carriers have finite lifetimes and thus can transport charges within limited distances. These differences enable us to distinguish the photoexcited carriers from thermally activated carriers on the basis of experiments on activation energy under various irradiation conditions. Details are discussed at the end of this chapter. In the case of the single crystals of salt 3, the activation energy decreases from 0.20 eV to 0.12 eV under UV irradiation (375 nm, 11.6 $mWcm^{-2}$) (Fig. 1.23) [175].

1.5.3 New Types of Photoconduction

In the course of studying the conduction mechanism, we have found that there are some common features shared by all these new types of photoconductors with unusually large ratios $\sigma_{UV}/\sigma_{dark}$. For example, the activation energies clearly decrease under irradiation. The conduction is dependent on the light intensity in a nonlinear way. None of them have been observed in existing photoconductors and cannot be explained within a framework of the present understanding of the photoconduction mechanism. The former feature will be discussed in the next section. We shall discuss the latter feature here, which manifests new types of photoconduction and accounts for the large ratios $\sigma_{UV}/\sigma_{dark}$. The control method of conduction and magnetism by photoirradiation is called optical doping, whether the effect is reversible or not [176].

Figure 1.24 shows the light-intensity (I) dependence of surface conductivity (σ_{sq}) at different temperatures observed for salt 3 [175]. Here σ_{sq} is defined by Eq. 1.1:

$$\sigma_{sq} = \frac{L_1}{RL_2}, \tag{1.1}$$

where R (ohm), L_1, and L_2 designate resistance, the distance between two electrical leads, and the sample dimension perpendicular to the distance L_1, respectively (Fig. 1.25). Note that σ_{sq} has the same unit and dimension as conductance R^{-1} (ohm^{-1}) yet a different physical meaning from R^{-1} (an extensive variable) in that σ_{sq} belongs to intensive variables. Note also that σ_{sq} may contain contribution from thermal carriers. σ_{sq} is independent of the thickness of the sample and the penetration depth of irradiation, while conductivity is dependent on both of them. Since it is difficult to accurately estimate the penetration depth, the discussion based on σ_{sq} will be clearer than that based on conductivity. The observed behavior in Fig. 1.24 is evidently nonlinear for all temperatures of measurement and can be well reproduced by Eq. 1.2.

$$\sigma_{sq} = a + bI + cI^2 + dI^3, \tag{1.2}$$

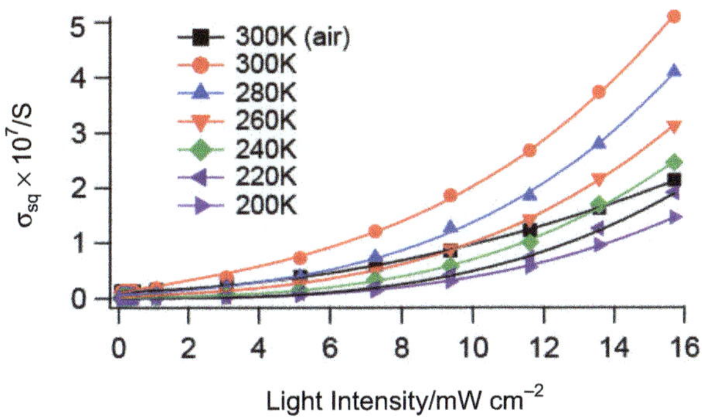

Figure 1.24 Light-intensity (*I*; 375 nm) dependence of surface conductivity (σ_{sq}) at different temperatures observed for NMQ[Ni(dmit)$_2$]: 300 K in air (black squares), 300 K (red circles), 280 K (blue triangles), 260 K (red triangles), 240 K (green diamonds), 220 K (black triangles), and 200 K (purple triangles). Equation 1.2 was used in the curve-fitting analysis of the results, except for the data obtained at 300 K in air. The solid curves are the best fits. Measurements were carried out at 300 K under vacuum (red circles) and under ambient atmosphere (i.e., in air) (black squares), while all other measurements were performed under vacuum. Adapted from Ref. [175] with permission from John Wiley & Sons. Adapted from Ref. [176] with permission from the Chemical Society of Japan.

where *a*, *b*, *c*, and *d* are coefficients as fitting parameters for the observed σ_{sq} at a given temperature. This analysis divides the contribution to conduction in accordance with the power series of *I*. The first term should describe the dark conductivity because it is independent of *I*. As is stated below, this is consistent with the fact that the activation energy derived from this analysis, that is from the temperature dependence of the term *a*, agrees well with that directly derived from the temperature-dependent (dark) conductivity σ_{dark} measurements. Thus the analysis based on Eq. 1.2 is considered to be valid. The second term, *bI*, designates the standard photoconductivity, which is generally proportional to *I*. The third and fourth terms, that is, the nonlinear terms are newly found in these materials and cannot be explained by the known mechanism of photoconduction. In summary, the qualitatively new features of the photomagnetic conductors and giant photoconductors lie in the sharp wavelength selectivity and the nonlinear dependence on the light intensity in their photoresponse.

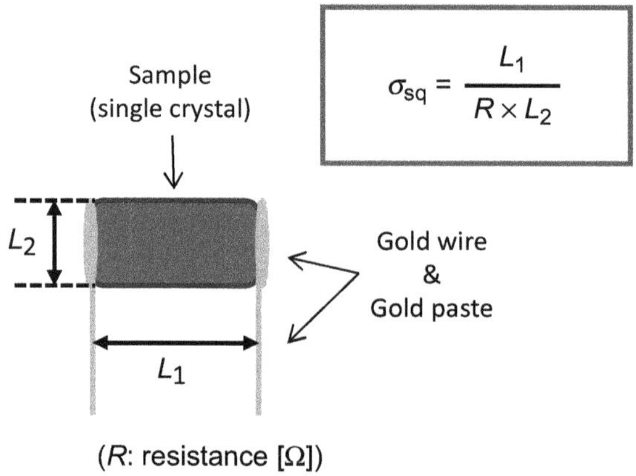

Figure 1.25 Typical sample preparation for the measurement of surface conductivity. Adapted from Ref. [176] with permission from the Chemical Society of Japan.

1.6 How to Distinguish Purely Optical Processes from Thermal Effects

If all of the carriers are produced by optical excitation, photoconduction does not require thermal activation of carriers and thus there should be no activation energy in photoconduction. However, there are some thermally activated carriers in general samples around RT independently of optical excitation, and these carriers also contribute to conduction both under dark and irradiated conditions. For example, under dark conditions, although the activation energy of salt 1 is 0.28 eV ≈ 3250 K, being much higher than RT ≈ 300 K, it conducts electricity as stated above. When the contributions of both types of carriers to conduction are non-negligible compared to each other, the photoconduction mechanism can be complicated.

Photoexcited carriers have their own lifetimes and can vanish during transportation. In the meantime, the number of thermally activated carriers is constant as long as the sample temperature is constant. However, during measurement of photoconduction, joule heat evolves due to the resistivity of the samples as well as heating effects of irradiation. Both inevitably elevate the temperature of the sample. This heating effect often fluctuates and varies from part to part on/in the sample until it reaches thermal equilibrium with the surroundings. If we could distinguish the thermal and optical contributions to conduction using a clear and general method, it would contribute to the understanding of the mechanism of photomagnetic conductors and giant photoconductors and the development of their new related compounds. In this section, we will discuss such a method based on experiments.

1.6.1 Thermal Effects in Irreversible Optical Doping

If irradiation brings about photochemical reactions in the sample that results in change in conduction properties, the heating effects can be directly estimated by standard resistivity measurements.

Figure 1.26 shows time dependence of the resistivity ρ of the single crystal of an organic CT complex, Ag(DMe-DCNQI)$_2$, under continuous UV irradiation at RT [176, 177]. We have discussed the metallic property and the crystal structure of Ag(DMe-DCNQI)$_2$ (see Fig. 1.6). When this compound is irradiated with UV rays, a series of complicated photochemical reactions occurs in the solid state, followed by structural change [176–188]. As a result, the UV irradiation gradually and irreversibly transforms the compound from a metallic to a semiconducting material with a steady increase in resistivity. This process has been examined in detail. During irradiation, ρ increased rapidly at first, then ($t \geq 5 \times 10^3$ s) gradually and constantly. At an arbitrary time ($t = 24 \times 10^3$ s in this case) the irradiation was discontinued long enough (75 min. in this case) to let the sample cool down spontaneously. Then UV irradiation was resumed to confirm that the resistivity immediately recovered the original level before interruption of the irradiation. This experiment enabled us to distinguish thermal (reversible; $\Delta\rho_h(24)$) and optical (irreversible; $\Delta\rho_{UV}(24)$) effects on the observed change in ρ.

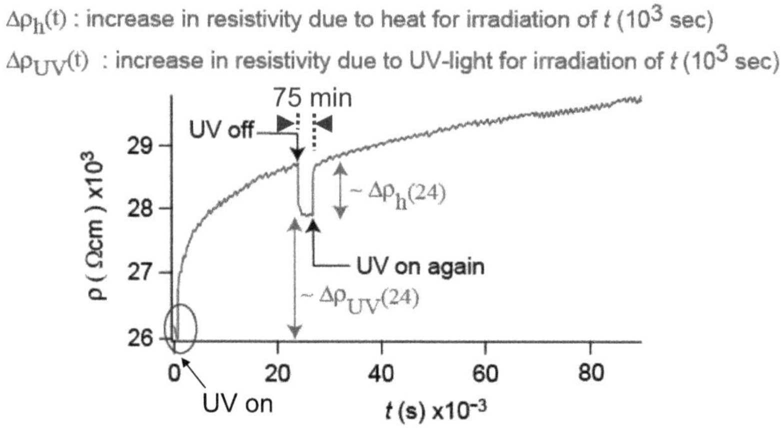

Figure 1.26 Change in the resistivity of $Ag(DMe\text{-}DCNQI)_2$ during UV irradiation. Adapted from Ref. [176] with permission from the Chemical Society of Japan. Cited and modified with permission. From Ref. [177]. © IOP Publishing. Reproduced with permission. All rights reserved.

1.6.2 Thermal Effects in Reversible Optical Doping

1.6.2.1 Problems and difficulties

The thermal effects of irradiation in reversible photoconduction are as transient and complicated as those in time-resolved laser spectroscopy and related physical measurements [126–151, 189–192]. Furthermore, unlike spectroscopy, a significant amount of joule heat inevitably evolves in the measurement of photoconduction since photoconductors generally exhibit high resistivity. As conduction/resistivity is a function of temperature, and the sample temperature is affected by the resistivity through joule heat, it has been extremely demanding to carry out the precise correction for thermal effects in the transient or reversible optical responses. Thus far, many patterns of attempts for the correction or estimation of heating effects have been proposed, for example, measurements of the light-intensity dependence (often called “fluence dependence”) of the photoresponse-enabled extrapolation of the photoresponse at zero intensity. Yet in many cases the photoresponse is not always linear to the intensity of the incident light, and the intensity range is limited to be actually examined without doing damage to samples and with a tolerable signal-to-noise ratio. This situation often makes the extrapolation to zero intensity difficult and ambiguous. Some researchers measure the samples of interest with known samples set in thermal contact with each other under identical measurement conditions. The known samples are assumed to play the role of thermometers, showing the difference in the sample temperature and the sample room's temperature by characteristic behavior such as phase transitions. Yet the observed difference in temperatures cannot be assumed to be constant at all temperatures of measurement. More fundamentally, it is not certain that the known sample is always in perfect thermal equilibrium with the sample of interest, particularly the irradiated part/surface of the sample. Some researchers tried to estimate the heating effects by a calculation that assumes an "effective sample temperature" and an averaged heat capacity. However, a sample under irradiation is never in a homogeneous or thermal equilibrium state, where the temperature is well defined. Under irradiation, the sample temperature should be different between surface and bulk, should often fluctuate, and should depend on the distances from the area of irradiation. Both heat capacity and heat conductivity are functions of temperature and generally exhibit anisotropy. This situation often makes the estimation far from being convincing. Furthermore, all these corrections/estimations suffer from the same weak point: they cannot be checked for validity by any other method. The only available way for cross-checking is to compare thus obtained results with other physical measurement data. If the results appear to be consistent or compatible with other data, the correction/estimation has been considered to be valid. As discussed below, in the course of related study but for a different purpose, a new method has been found to distinguish thermal and optical carriers in a clear way, where the actual heat the carrier system receives is estimated directly from their conduction behavior and a general assumption instead of any of the previously utilized assumptions above.

1.6.2.2 Dependence of photocurrent and activation energy on light intensity

Under dark conditions, the activation energy (E_a) of a given sample is generally constant and does not depend on experimental conditions such as the distance (L_1 in Fig. 1.25) between electrical leads. Similarly under dark conditions, the current under a constant applied voltage is inversely proportional to L_1, obeying Ohm's law, as long as the sample is homogenous along the current. However, under photoirradiation, the photocurrent does not obey Ohm's law (Fig. 1.27) [193]. This is because the number and relaxation time (τ) of photocarriers depend on light intensity, wavelength, and so on. For instance, the longer L_1 becomes, the fewer photocarriers arrive at the opposite lead before they optically relax and disappear. Interestingly, in the course of the examination of the mechanism of the giant photoconductivity, E_a is found to be necessary even in photoconduction and is also found to depend on irradiation conditions (Fig. 1.28) [193]. The former finding appeared to be puzzling because the majority of the carriers should be produced by photoexcitation instead of thermal excitation, implying that the carriers do not require activation energy. The latter finding appeared even more puzzling and could not be explained at first.

1.6.2.3 Model for activation energy in photoconduction

Firstly, we assume here that carriers serve equally to conduction, whether they are thermally or optically excited. On this basis, the effective E_a as a function of light intensity I, $E_a(I)$, can be described as follows:

$$E_a(I) = \frac{N_{th}(I)}{N_{th}(I) + N_{ph}(I)} E_{a,0} \tag{1.3}$$

Here $N_{th}(I)$, $N_{ph}(I)$, and $E_{a,0}$ designate the number of thermally activated carriers, the number of optically excited carriers, and the activation energy under dark conditions (0.24 ± 0.02 eV for NMQ[Ni(dmit)$_2$]), respectively. Secondly, both under dark and irradiated conditions, we assume that conductivity (σ_{dark} and σ_{ph}) is proportional to the current (I_{dark} and I_{ph}) under a constant voltage (Ohm's law) and that the current, in its turn, is proportional to the number of carriers N. As N should be a function of light intensity I under irradiation, these assumptions are described as follows:

$$\begin{cases} \sigma_{dark} \propto I_{dark} \\ \sigma_{ph} \propto I_{ph} \end{cases} \tag{1.4}$$

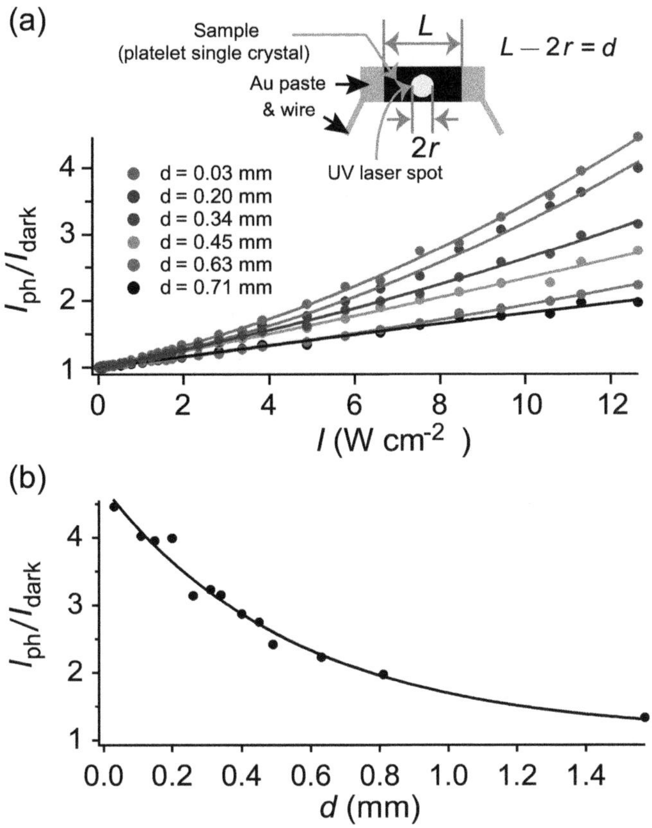

Figure 1.27 (a) I_{ph}/I_{dark} versus I curve at 298 K with different distances between the electrical leads (gold wires) for the same single crystal of NMQ[Ni(dmit)$_2$]. Here I_{ph}, I_{dark}, and I designate photocurrent, dark current, and light intensity, respectively. Closed circles: observed; curves: best-fit curves using the following polynomials:

$$I_{ph}/I_{dark} = K_0' + K_1'I + K_2'I^2 + K_3'I^3,$$

where K_i' (i = 0–3) are fitting parameters. To estimate the net effect of differences in travel path lengths of the photocarriers, d does not include the spot diameter ($2r$ = 0.1 mm) of the UV laser (375 ± 5 nm; see inset). (b) I_{ph}/I_{dark} versus d curve at 298 K (375 ± 5 nm and I = 12.6 Wcm^{-2}). Closed circles: observed; curve: best-fit curve using the following equation:

$$I_{ph}/I_{dark} = y_0 + A\exp\left(-\frac{d}{\tau}\right),$$

where y_0, A, and τ = 1.08, 3.65, and 0.56 mm, respectively [193].

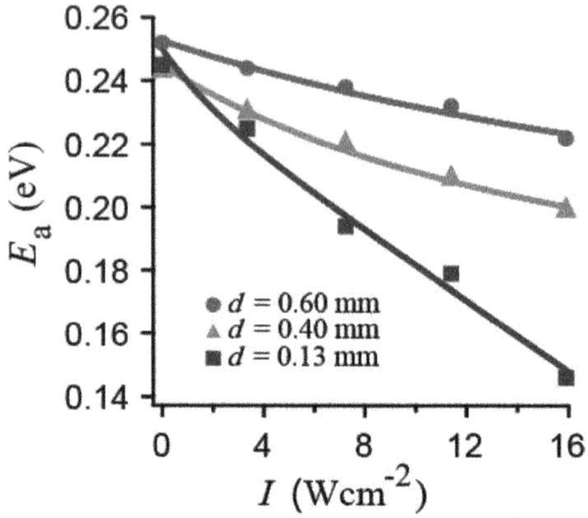

Figure 1.28 *I* dependence of activation energy $E_a(I)$ for NMQ[Ni(dmit)$_2$]. For the meaning of distance *d*, see Fig. 1.27a. Data observed in air. Best-fit curves generated using Eq. 1.10 [193].

and

$$\begin{cases} I_{\text{dark}} \propto N_{\text{th}}(I) \\ I_{\text{ph}} \propto N_{\text{th}}(I) + N_{\text{ph}}(I) \end{cases} \tag{1.5}$$

Here σ_{dark}, σ_{ph}, I_{dark}, and I_{ph} designate dark conductivity, photoconductivity, dark current, and photocurrent, respectively. Under dark conditions (I = 0), the sample is in thermal equilibrium with the surroundings, where the number of thermally excited carriers $N_{\text{th}}(0)$ obeys the Boltzmann distribution. If there are different bands for which thermal and optical excitation should be considered, $N_{\text{th}}(I)$ and $N_{\text{ph}}(I)$ should be replaced with due summation, but the remaining part of the discussion here is still valid. During photoconduction, the heat involved with irradiation as well as the joule heat elevates the temperature of the sample. Then, N_{th} becomes a linear function of I,

$$N_{\text{th}}(I) = a + bI \tag{1.6}$$

where the coefficients a and b are constants depending on experimental conditions and samples. The first constant (a) corresponds to the number of carriers under the dark condition $N_{\text{th}}(0)$, while the latter (b) is primarily dependent upon the heat capacity of the sample, including the electrical contacts, under a given set of irradiation conditions. As a good approximation, $N_{\text{th}}(I)$ can be described by Eq. 1.6 because the heat involved with irradiation is proportional to I under a given set of experimental conditions and additionally because the joule heat is approximately proportional to I under the assumption of Eqs. 1.4 and 1.5. On the other hand, the observed I_{ph} is well reproduced by Eq. 1.7 (Fig. 1.27a).

$$I_{\mathrm{ph}} = \sum_{i=0} K_i I^i \tag{1.7}$$

The coefficients K_i (i = 0, 1, 2, . . .) are fitting parameters dependent on the temperature T and are determined by the curve-fitting analysis of the observed data. Note that K_0 should equal I_{dark}. By Eqs. 1.5 and 1.7,

$$N_{\mathrm{th}}(I) + N_{\mathrm{ph}}(I) = A \sum_{i=0} K_i I^i . \tag{1.8}$$

Here A is a proportionality constant and can be described by

$$A = \frac{L^2}{V \mu e}, \tag{1.9}$$

where L, V, μ, and e indicate the distance between the two electrical leads, applied voltage, mobility of the carriers, and elementary charge, respectively, as shown in Ref. [193]. Substitution of Eqs. 1.6 and 1.8 in Eq. 1.3 gives

$$E_{\mathrm{a}}(I) = \frac{a + bI}{A \sum K_i I^i} E_{\mathrm{a},0} . \tag{1.10}$$

Finally, Eq. 1.10 is transformed into a linear relationship in order to unambiguously fit the observed behavior.

$$E_{\mathrm{a}}(I) \sum_{i=0} K_i I^i = \frac{1}{A}(a + bI) E_{\mathrm{a},0} = a' + b'I$$

$$\begin{cases} a' = \dfrac{E_{\mathrm{a},0}}{A} a \\ b' = \dfrac{E_{\mathrm{a},0}}{A} b \end{cases} \tag{1.11}$$

A comparison of the observed behavior with Eq. 1.11 proves the validity of the discussion thus far (Fig. 1.29) [193]. All the data from three independent measurements show a fair agreement with the behavior expected from Eq. 1.11.

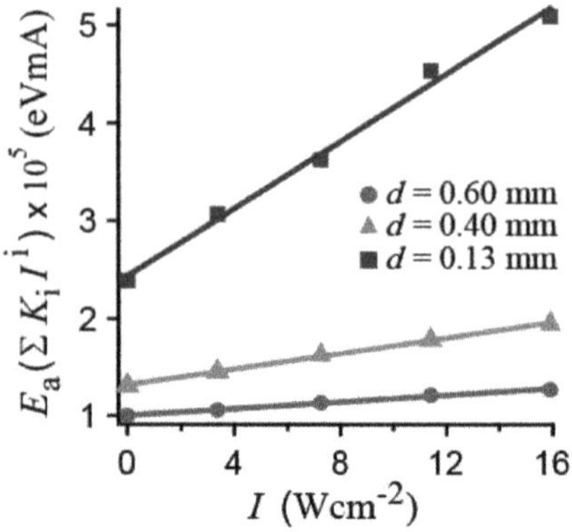

Figure 1.29 I dependence of $E_a(I)(\Sigma K_i I^i)$ for the same sample with that in Fig. 1.28. Data observed in air. Best-fit lines generated using Eq. 1.11. As regards the corresponding data for other samples including different materials (see Ref. [193]), all show a fair agreement with the model and the observation.

1.6.2.4 Separation of thermal effects from optical effects

As stated above, it has been always difficult to distinguish unavoidable thermal effects from purely optical effects in these kinds of experiments. However, if only μ is known, this analysis enables $N_{th}(I)$ and $N_{ph}(I)$ to be found unambiguously, as shown below. By use of Eqs. 1.6 and 1.8,

$$N_{ph}(I) = \{N_{th}(I) + N_{ph}(I)\} - N_{th}(I)$$
$$= A(K_0 + K_1 I + K_2 I^2 + K_3 I^3 + \cdots) - (a + bI)$$
$$= (AK_1 - b)I + AK_2 I^2 + AK_3 I^3 + \cdots \ (\because AK_0 - a = 0). \quad (1.12)$$

All of the parameters K_i in Eq. 1.11 can be determined directly from the observation of I_{ph} versus I, that is, the relationship in Eq. 1.7, as shown in Fig. 1.27a. Similarly, the I dependence of E_a can be directly observed, as shown in Fig. 1.28. The curve-fitting analysis of the data in Fig. 1.28 using Eq. 1.10 gives the value of parameter b. The value of A can be obtained from Eq. 1.9. Thus substituting these values for A, K_i, and b in Eq. 1.12, the number of photoexcited carriers $N_{ph}(I)$, that is, the purely optical effect, can be obtained. Furthermore, as thus obtained N_{ph} is a function of I, one can even know N_{ph} at an arbitrary light intensity I without actual experiments as long as Eq. 1.7 is valid.

In general, a different material and/or a different measurement condition can give a different I dependence of I_{ph} instead of Eq. 1.7. Even is such a case, if Eq. 1.7 is replaced with the appropriate equation, the discussion thus far remains essentially unaltered, except for the details of Eqs. 1.8, 1.10, and 1.11. In fact, the analysis here can be applied to different cases to give similarly consistent results [193].

1.7 Control of Spin Distribution by Light

Since both magnetism and conduction originate from unpaired electrons, it is also possible to control magnetism by the same method above, that is, by controlling the unpaired electrons through optical excitation. Because optical excitation realizes extremely high energy states with unpaired electrons, which would never realize otherwise, novel conduction and magnetic properties are often found. Such an example will be discussed below.

1.7.1 $[Cu(dmit)_2]^{2-}$ Salts: Initial Prospect and Present Status as a Building Block for Molecular Conductors and Magnets

The $[Cu(dmit)_2]^{2-}$ anion (Fig. 1.30) has been known for ~40 years as a member of the $[M(dmit)_2]^{n-}$ series [194, 195]. The feature of this complex is that there is an unpaired electron formally on the metal center because of the d^9 configuration of the copper atom. If the unpaired electron delocalizes over the entire molecule, the solid would behave as a conductor. If the unpaired electron is localized at the metal center, the solid would behave as a magnet. In fact, the coordination geometry of the anion is known to alter flexibly and sensitively to countercations and/or crystal structures from square-planar to (distorted) tetrahedral CuS_4 cores [196–202]. This suggests that the distribution of the unpaired electron should vary depending on the chemical environments. Thus it was expected to be a promising building block for conducting and magnetic materials.

However, generally speaking, various chemical reactions, including oxidation, are required in order to control the spin distribution and make them conductors and/or magnets. Unfortunately, this complex is unstable against chemical reactions, particularly against oxidation. Thus there are a limited number of reports on this complex despite its long history [194–206]. Yet, even if the chemical approach is difficult, optical excitation may be effective. Optical transition between these two states is considered to be possible since there is interaction between the metal and the ligands, which is generally called π–d interaction as per the orbitals involved. This idea has turned out to be true, as is described next.

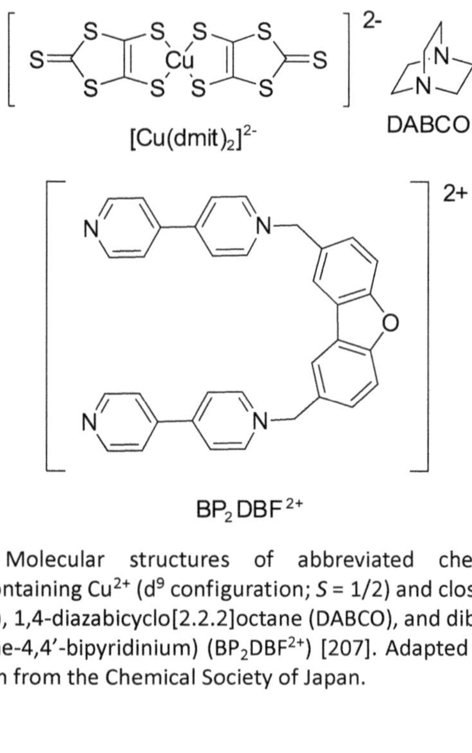

Figure 1.30 Molecular structures of abbreviated chemical species; $[Cu(dmit)_2]^{2-}$ containing Cu^{2+} (d^9 configuration; $S = 1/2$) and closed-shell ligands ($dmit^{2-} = C_3S_5^{2-}$), 1,4-diazabicyclo[2.2.2]octane (DABCO), and dibenzofuran-2,2′-bis(*N*-methylene-4,4′-bipyridinium) (BP_2DBF^{2+}) [207]. Adapted from Ref. [176] with permission from the Chemical Society of Japan.

1.7.2 Spin Distribution vs. Molecular Structures

To clarify whether the unpaired electron actually alters its distribution (spin distribution) over the complex anion, theoretical calculation was carried out [176, 207, 208]. The molecular structures assumed

(a)

(b)

(c)

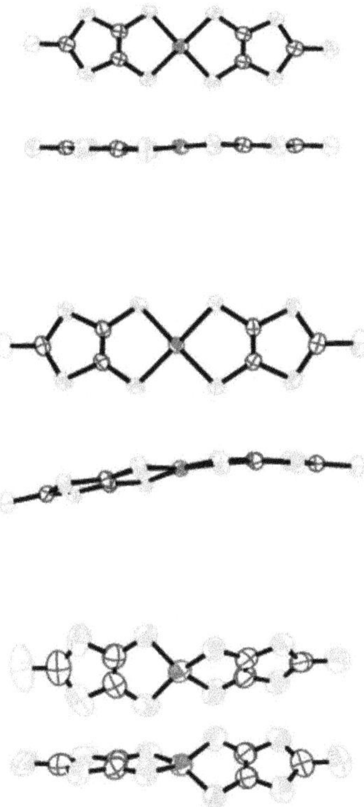

Figure 1.31 Observed molecular structures in the three different $[Cu(dmit)_2]^{2-}$ salts, that is, (a) $(^nC_4H_9)_4N^+$, (b) $(DABCO)H^+$, and (c) BP_2DBF^{2+} salts [207]. Adapted from Refs. [176, 209] with permission from the Chemical Society of Japan.

in the calculation were those obtained from the single-crystal X-ray structural analysis of the $(^nC_4H_9)_4N^+$, $(DABCO)H^+$ and BP_2DBF^{2+} salts. They contained complex anions having square-planar (the $(^nC_4H_9)_4N^+$ salt), slightly (the $(DABCO)H^+$ salt), and obviously (the BP_2DBF^{2+} salt) distorted tetrahedral coordination geometries around the copper ions (Fig. 1.31). Thus the three salts cover all the representative coordination geometries of the anion. Calculation of the molecular orbitals gave us interesting and unexpected results. Firstly, the energies/stabilities of different molecular structures are quite close to each other as isolated molecules. Additionally, structural optimization indicated that both planar and nonplanar coordination geometries are stable and that the stable structure depends on the initial structure assumed [176, 207]. These results clearly indicate that the molecular structure of each salt is not affected by the chemical environment, including the crystal structure, but is dominated by the stability of itself. Thirdly, although the molecular structures of these salts are different from each other, the charge and spin distributions in molecular orbitals are calculated to be very similar to each other (Fig. 1.32). They are all delocalized over the entire molecules, independently of coordination geometry. This denotes that π–d mixing is independent of coordination geometry for $[Cu(dmit)_2]^{2-}$. Similarly, although the calculated spin densities are apparently higher around the metal center than the rest of the molecules, they are still delocalized over the entire molecules (Fig. 1.32). In fact, all three salts exhibited qualitatively identical magnetic susceptibility, including temperature dependence (diamagnetism); all their susceptibilities are independent of temperature [176, 207]. In addition, they are all unstable to chemical reactions. Therefore, it has turned out to be rather difficult to control the spins and resultant properties of the $[Cu(dmit)_2]$ salts by chemical or thermodynamic methods.

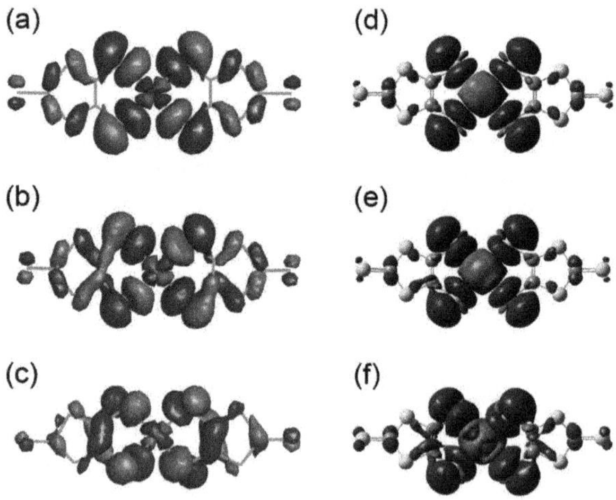

Figure 1.32 Calculated singly occupied molecular orbitals (SOMO) [(a) $(^nC_4H_9)_4N^+$, (b) $(DABCO)H^+$, and (c) BP_2DBF^{2+} salts] via extended Hückel calculation and spin densities [(d) $(^nC_4H_9)_4N^+$, (e) $(DABCO)H^+$, and (f) BP_2DBF^{2+} salts] for $[Cu(dmit)_2]^{2-}$ anions via Gaussian 09 [B3LYP/6-31+G(3d)] [207]. Adapted from Ref. [176] with permission from the Chemical Society of Japan.

1.7.3 Response of Spins to UV: Results

To find which atoms unpaired electrons are located at, ESR spectra are the most powerful tool. We measured and compared ESR spectra of the above-mentioned three salts under dark and photoirradiated conditions. Under dark conditions, the spectra contained a broad signal without hyperfine structures for all three salts. For example, the ESR spectra for the $(^{n}C_4H_9)_4N^+$ salt are shown in Fig. 1.33 [176, 207]. The remaining salts exhibited a similar trend in spectra in a qualitative sense. The copper complexes generally exhibit hyperfine structures in ESR due to the interaction between the magnetic moments of electron and nuclear spins ($I(^{63}Cu) = I(^{65}Cu) = 3/2$; isotope abundance ^{63}Cu = 69.09%, ^{65}Cu = 30.91%). In addition, no correlation is found between the ESR features and the conduction properties; the $(DABCO)H^+$ salt is a semiconductor with high resistivity, and the remaining salts are insulators. Therefore, the features of these ESR spectra are considered to be dominated by the delocalized spins on the nearly isolated Cu-complex anions. This interpretation is supported by the diamagnetic and practically insulating properties in addition to their crystal structures and is also consistent with the molecular orbital calculation (Fig. 1.32) [176, 207].

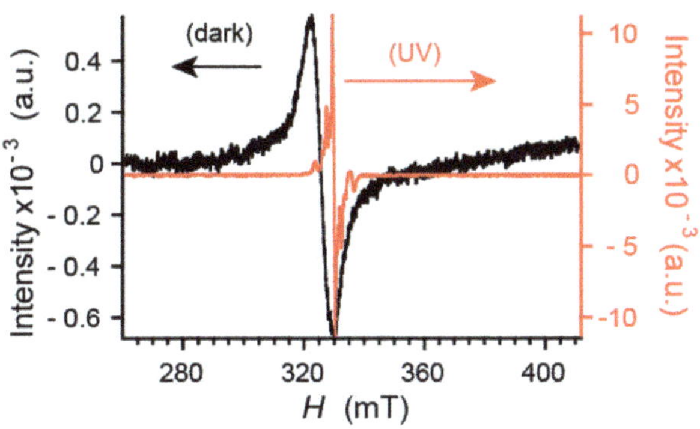

Figure 1.33 Comparison of the ESR spectra under dark (black) and UV-irradiated (red) conditions at 123 K for the $(^{n}C_4H_9)_4N^+$ salt (single crystal) [207]. Adapted from Ref. [176] with permission from the Chemical Society of Japan.

Under UV irradiation, a very intense signal appeared for all these salts as well as hyperfine structures due to the copper ions (Fig. 1.33). Spectral simulation clearly indicates that the feature of each spectrum is dominated by the ligands under dark conditions, while it is dominated by the metal center under UV irradiation [176, 207]. Another feature of the ESR spectra is the g-value under UV irradiation. It becomes unusually small and isotropic for copper complexes, comparable to that of free electrons (g = 2.002319). These features under UV irradiation are shared by all three salts. The interpretation of the observed response will be discussed in the next section.

1.7.4 Response of Spins to UV: Discussion

To explain the observation in the previous section, the molecular orbitals of excited states are examined. Under the dark condition, unpaired electrons are accommodated in the delocalized π orbital of the ligand. Among excited states, there are rather localized orbitals around the copper atoms (Fig. 1.34) [176, 207]. These orbitals have a major contribution from the copper atomic orbitals and are within the excitation energy of incident light. Thus they account for the hyperfine structures due to the copper atoms observed under UV irradiation. Similarly, because these orbitals are highly isotropic, the orbital angular momenta will be practically quenched. Accordingly, the g-values under UV irradiation should be highly isotropic and close to the value of free electrons (g = 2.002319).

In fact, there is a peak in the solution spectra [176, 207] at the corresponding energy (Fig. 1.35). This denotes that the above-mentioned optical excitation actually occurs under UV irradiation in an isolated molecule. Next, the band calculation and solid-state spectra have been examined in order to check whether the corresponding transition should also occur in the solid state. The calculated band structure [176, 207] certainly includes the bands with a major contribution from the copper atomic orbitals at the expected energies (Fig. 1.36). Considering the band mixing, that is, respective orbital contributions from the ligands and the copper atoms to the bands, optical transition is allowed between these two bands under UV irradiation. Regarding the solid-state spectra, the

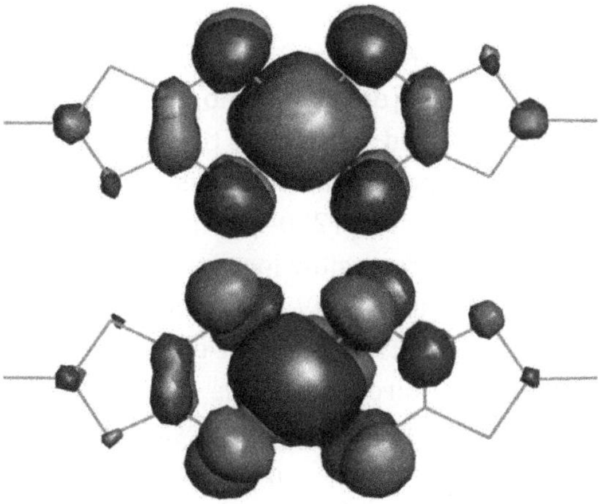

Figure 1.34 Selected calculated molecular orbitals (MOs) of $[Cu(dmit)_2]^{2-}$ in excited states. These are the 52nd MOs from the lowest-energy MOs of the $(^nC_4H_9)_4N^+$ (upper) and $(DABCO)H^+$ (lower) salts, respectively [207] and located within the excitation energies of incident light (250–450 nm = 2.75–5.00 eV) from each E_F. The energies are measured from each E_F (–9.17 and –9.19 eV, respectively) on the basis of the extended Hückel calculation. The calculation was done by assuming the ground state structures obtained from the single-crystal X-ray structural analysis under the dark condition. As the molecular structure of $[Cu(dmit)_2]^{2-}$ does not alter the MO in a qualitative way (Fig. 1.32), this assumption is valid for qualitative discussion, even if the molecular structures in excited states are different from those in the ground states. Adapted from Ref. [176] with permission from the Chemical Society of Japan.

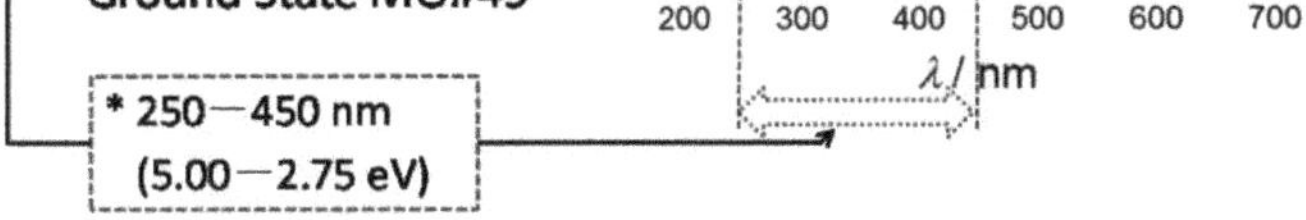

Figure 1.35 UV-Vis spectrum of $[Cu(dmit)_2]^{2-}$ (the $(^nC_4H_9)_4N^+$ salt) in CH_3CN solution at RT. The unpaired electron occupies the 49th MO in the ground state. 4.2 eV = 295 nm.

peak positions hardly differ from each other between the solution and the solid-state spectra (Fig. 1.37) [176, 207]. This is consistent with the fact that the $[Cu(dmit)_2]^{2-}$ anions are practically isolated from each other in the solid states of all three salts. Therefore, it has been shown that the optical transition discussed thus far actually occurs in each salt under UV (250–450 nm) irradiation. Thus the spin distribution can be reversibly localized and delocalized within each Cu-complex molecule by UV irradiation. The remaining salts exhibit similar behavior. The observed optical response does not depend on the relative direction of the magnetic field applied in the ESR measurement, molecular and crystal structures, electrical properties, or countercations in a qualitative sense. A comparison of the spectra under dark and UV-irradiated conditions (Fig. 1.38) clearly shows that under UV irradiation, there are both delocalized and localized spins on the copper complex anions in these salts. As a result, the complex anions possess both carriers and localized spins under UV irradiation. The ratio between carriers and localized spins can be controlled by irradiation conditions. By focusing the UV ray on the sample, one can even select which part of the sample should be controlled by this method, with the remaining part unchanged. Such a spin control has never been realized by any chemical method or under any thermodynamic condition.

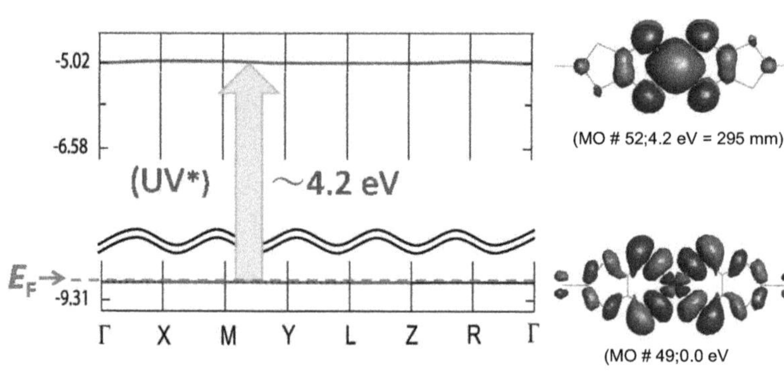

Figure 1.36 Selected excited state from the calculated band structure of $[(^nC_4H_9)_4N]_2[Cu(dmit)_2]$. The unpaired electron occupies the 305th and 306th bands (degenerated) at $E_F \approx -9.17$ eV in the ground state. 4.2 eV = 295 nm. *Incident light = 250–450 nm. The molecular orbitals correspond to each band beside them.

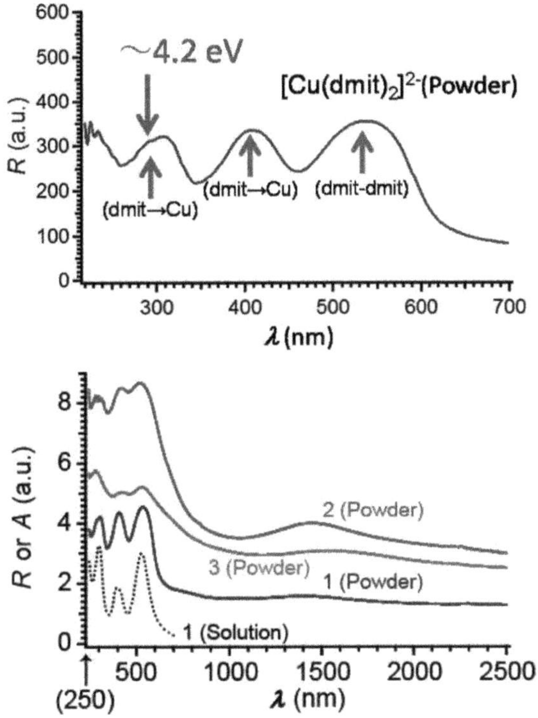

Figure 1.37 Solid-state (diffuse reflectance) spectra of $[(^nC_4H_9)_4N]_2[Cu(dmit)_2]$ (upper) and a comparison between solution (absorption, *A*) and powder (diffuse reflectance, *R*) spectra of $[(^nC_4H_9)_4N]_2[Cu(dmit)_2]$ (1), $[(DABCO)H]_2[Cu(dmit)_2]$ CH_3CN (2), and $BP_2DBF[Cu(dmit)_2]$ (lower) [207]. Adapted from Ref. [176] with permission from the Chemical Society of Japan.

1.8 Summary and Prospects

In this chapter, we have provided an overview of the physical and structural properties of molecular conducting and magnetic materials. Beginning with the close relationship between physical and structural properties, the discussion is focused on how their unique conducting and magnetic properties occur. The latter half of the chapter is devoted to the control method of such conduction and magnetism using photoirradiation, which has been recently

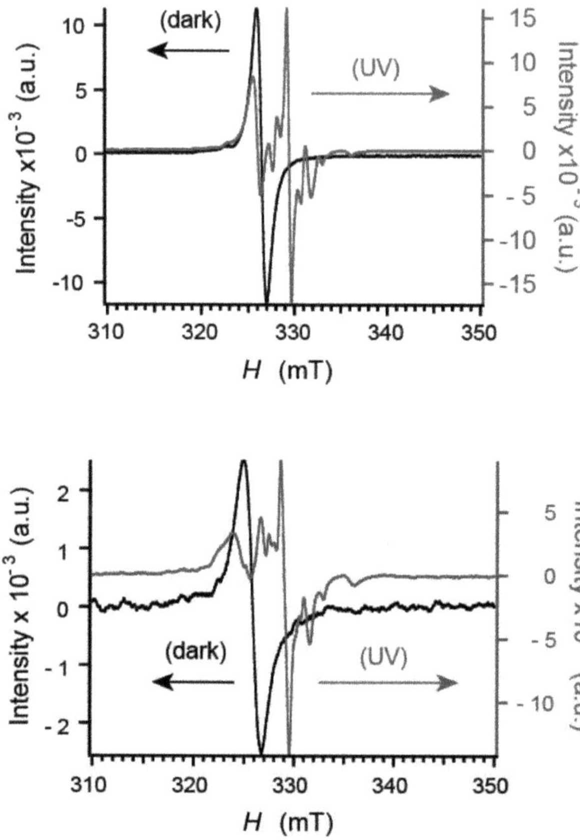

Figure 1.38 Comparison of ESR spectra of $[(DABCO)H]_2[Cu(dmit)_2]CH_3CN$ (upper) and $BP_2DBF[Cu(dmit)_2]$ (lower) between dark and UV-irradiated conditions [207]. Adapted from Ref. [176] with permission from the Chemical Society of Japan.

developed by the present author. It is a nondestructive method for molecular solids such as single crystals of CT complexes. The method is named "optical doping," where appropriate irradiation is utilized under ambient conditions. It can be applied to a wide range of substances while measuring the properties during the control (in situ). In addition, the method adds unique conduction and magnetic properties to common insulators, including unnamed intermediate or mixed properties. Unlike other doping methods, optical doping only affects the properties and/or structures of the irradiated part of a sample, leaving the rest of the sample unchanged. Although there is not much reference to irreversible optical doping in this chapter, it enables us to make a junction structure, where qualitatively different properties coexist in a single crystal by a facile procedure of a single step. The finding of optical doping is an undoubtedly important and novel step toward a wide variety of application and new related research fields such as molecular electronics. Still there is a substantial amount of work required for the practical use of optical doping in the fabrication of molecular semiconductor devices, for example. In the meantime, the future will be brighter if the readers of this book will join us in exploring the fascinating world of materials science.

References

1. Onnes, H. K. (1912). Disappearance of the electrical resistance of mercury at helium temperatures, *Proc. Akad. Wetenshappen (Amsterdam)*, **14**, pp. 113–115.
2. Bednorz, J. G. and Müller, K. A. (1986). Possible high T_C superconductivity in the Ba-La-Cu-O system, *Z. Phys.*, **B64**, pp. 189–193.
3. Cava, R. J., van Dover, R. B., Batlogg, B. and Rietman, E. A. (1987). Bulk superconductivity at 36 K in $La_{1.8}Sr_{0.2}CuO_4$, *Phys. Rev. Lett.*, **58**, pp. 408–410.
4. Tarascon, J. M., Greene, L. H., McKinnon, W. R., Hull, G. W. and Geballe, T. H. (1987). Superconductivity at 40 K in the oxygen-defect perovskites $La_{2-x}Sr_xCuO_{4-y}$, *Science*, **235**, pp. 1373–1376.
5. Chu, C. W., Hor, P. H., Meng, R. L., Gao, L., Huang, Z. J. and Wang, Y. Q. (1987). Evidence for superconductivity above 40 K in the La-Ba-Cu-O compound system, *Phys. Rev. Lett.*, **58**, pp. 405–407.
6. Chu, C. W., Hor, P. H., Meng, R. L., Gao, L. and Huang, Z. J. (1987). Superconductivity at 52.5 K in the lanthanum-barium-copper-oxide system, *Science*, **235**, pp. 567–569.
7. Wu, M. K., Ashburn, J. R., Torng, C. J., Hor, P. H., Meng, R. L., Gao, L., Huang, Z. J. Wang, Y. Q. and Chu, C. W. (1987). Superconductivity at 93 K in a new mixed-phase Y-Ba-Cu-O compound system at ambient pressure, *Phys. Rev. Lett.*, **58**, pp. 908–910.
8. Maeda, H., Tanaka, Y., Fukutomi, M. and Asano, T. (1988). A new high-T_C oxide superconductor without a rare earth element, *Jpn. J. Appl. Phys.*, **27**, pp. L209–L210.

9. Sheng, Z. Z. and Hermann, A. M. (1988). Superconductivity in the rare-earth-free Tl-Ba-Cu-O system above liquid-nitrogen temperature, *Nature*, **332**, pp. 55–58.
10. Sheng, Z. Z. and Hermann, A. M. (1988). Bulk superconductivity at 120 K in the Tl-Ca/Ba-Cu-O system, *Nature*, **332**, pp. 138–139.
11. Kaneko, T., Wada, T., Yamauchi, H. and Tanaka, S. (1990). (Tl, Pb, Bi) $Sr_2Ca_2Cu_3O_z$ superconductors with zero resistance at 120 K, *Appl. Phys. Lett.*, **56**, pp. 1281–1283.
12. Laguës, M., Xie, X. M., Tebbji, H., Xu, X. Z., Mairet, V., Hatterer, C., Beuran, C. F. and Cavellin, C. D. (1993). Evidence superconductivity at 250 K in a sequentially deposited cuprate film, *Science*, **262**, pp. 1850–1852.
13. Drozdov, A. P. Eremets, M. I. and Troyan, I. A. (2014). Conventional superconductivity at 190 K at high pressures, *Preprint arXiv:1412.0460, (1 Dec 2014)*.
14. Drozdov, A. P., Eremets, M. I., Troyan, I. A., Ksenofontov, V. and Shylin, S. I. (2015). Conventional superconductivity at 203 K at high pressures, *Nature*, **525**, pp. 73–76.
15. Einaga, M., Sakata, M., Ishikawa, T., Shimizu, K., Eremets, M. I., Drozdov, A. P., Troyan, I. A., Hirao, N. and Ohishi, Y. (2015). Crystal structure of 200 K-superconducting phase of sulfur hydride system, *Preprint arXiv:1509.03156*.
16. Bianconi, A. and Jarlborg, T. (2015). Lifshitz transitions and zero point lattice fluctuations in sulfur hydride showing near room temperature superconductivity, *Nov. Supercond. Mater.*, **1**, pp. 37–49.
17. McCoy, H. N. and Moore, W. C. (1911). Organic amalgams: substances with metallic properties composed in part of non-metallic elements, *J. Am. Chem. Soc.*, **33**, pp. 273–292.
18. Ferraris, J., Cown, D. O., Walatka, V. and Perlstein, J. H. (1973). Electron transfer in a new highly conducting donor-acceptor complex, *J. Am. Chem. Soc.*, **95**, pp. 948–949.
19. Kistenmacher, T. J., Phillips, T. E. and Cowan, D. O. (1974). The crystal structure of the 1:1 radical cation-radical anion salt of 2,2'-bis-1,3-dihtiole (TTF) and 7,7,8,8-tetracyanoquinodimethan (TCNQ), *Acta Crystallogr.*, **B30**, pp. 763–768.
20. Coleman, L. B., Cohen, M. J., Sandman, D. J., Yamagishi, F. G., Garito, A. F. and Ferraris, A. J. (1973). Superconducting fluctuations and the peierls instability in an organic solid, *Solid State Commun.*, **12**, pp. 1125–1132.
21. Jérome, D., Mazaud, A., Ribault, M. and Bechgaard, K. (1980). Superconductivity in a synthetic organic conductor (TMTSF)2PF6, *J.*

Phys. Lett. (Paris), **41**, pp. 95–98. (Note: The correct chemical formula should be "$(TMTSF)_2PF_6$"; however, the reference is cited by exactly the same title as published. The same/similar notice is applied to Refs. 22 and 23.)

22. Ribault, M., Benedek, G., Jérome, D. and Bechgaard, K. (1980). Diamagnetic AC susceptibility in the quasi-one dimensional organic conductor: (TMTSF)2PF6, *J. Phys. Lett. (Paris)*, **41**, pp. 397–399.

23. Ribault, M., Pouget, J. P., Jérome, D. and Bechgaard, K. (1980). Superconductivity and absence of a Kohn anomaly in the quasi-one-dimensional organic conductor: (TMTSF)2AsF6, *J. Phys. Lett. (Paris)*, **41**, pp. 607–610.

24. Parkin, S. S. P., Ribault, M., Jérome, D. and Bechgaard, K. (1981). Three new superconducting members of the family of tetramethyltetraselenafulvalene (TMTSF) salts: $TMTSF_2ClO_4$, $TMTSF_2SbF_6$, $TMTSF_2TaF_6$, *J. Phys. C*, **14**, pp. L445–L450.

25. Parkin, S. S. P., Jérome, D. and Bechgaard, K. (1982). Pressure dependence of the metal-insulator and superconducting phase transitions in $(TMTSF)_2ReO_4$, *Mol. Cryst. Liq. Cryst.*, **79**, pp. 213–224.

26. Lacoe, R. C., Wolf, S. A., Chaikin, P. M., Wudl, F. and Shalom, E. A–. (1983). Metal-insulator transitions and superconductivity in ditetramethyltetraselenafulvaleneium fluorosulfonate $[(TMTSF)_2FSO_3]$, *Phys. Rev. B*, **27**, pp. 1947–1950.

27. Thorup, N., Rindorf, G., Soling, H. and Bechgaard, K. (1981). The structure of di(2,3,6,7-tetramethyl-1,4,5,8-tetraselenafulvalenium) hexafluorophosphate, $(TMTSF)_2PF_6$, the first superconducting organic solid, *Acta Crystallogr.*, **B37**, pp. 1236–1240.

28. Korshak, Y. V., Ovchinnikov, A. A., Shapiro, A. M., Medvedeva, T. V. and Spector, V. N. (1986). Organic polymer ferromagnetic materials, *Pris'ma Zh. Eksp. Teor. Fiz.*, **43**, pp. 309–311.

29. Torrance, J. B., Oostra, S. and Nazzal, A. (1987). A new, simple model for organic ferromagnetism and the first organic ferromagnet, *Synth. Met.*, **19**, pp. 709–714.

30. Ovchinnikov, A. A. and Spector, V. N. (1988). Organic ferromagnets. New results, *Synth. Met.*, **27**, pp. B615–B624.

31. Miller, J. S., Epstein, A. J. and Reiff, W. M. (1988). Ferromagnetic molecular charge-transfer complexes, *Chem. Rev.*, **88**, pp. 201–220.

32. Ohta, M., Otani, S. and Igarashi, M. (1989). Evolution of ferromagnetic properties by dehydrogenation of triarylmethane resin, *Chem. Lett.*, **18**, pp. 1183–1186.

33. Tanaka, H., Tokuyama, K., Sato, T. and Ota, T. (1990). Preparation and properties of organic polymer ferromagnet with indigo unit, *Chem. Lett.*, **19**, pp. 1813–1816.

34. Manriquez, J. M., Yee, G. T., Mclean, R. S., Epstein, A. J. and Miller, J. S. (1991). A room-temperature molecular/organic-based magnet, *Science*, **252**, pp. 1415–1417.

35. Miller, J. S. and Epstein, A. J. (1994). Organic and organometallic molecular magnetic materials: designer magnets, *Angew. Chem. Int. Ed. Engl.*, **33**, pp. 385–415 (see also *Angew. Chem.*, **106**, pp. 399–432).

36. Kinoshita, M., Turek, P., Tamura, M., Nozawa, K., Shiomi, D., Nakazawa, Y., Ishikawa, M., Takahashi, M., Awaga, K., Inabe, T. and Maruyama, Y. (1991). An organic radical ferromagnet, *Chem. Lett.*, **20**, pp. 1225–1228.

37. Gatteschi, D., Kahn, O., Miller, J. S. and Palacio, F. (eds.) (1991). *Magnetic Molecular Materials*, NATO ASI Series E, Applied Sciences, Vol. 198, Kluwer Academic Publishers, Dordrecht, Netherlands.

38. Iwamura, H. and Miller, J. S. (eds.) (1993). Chemistry and Physics of Molecular Based Magnetic Materials, in *Mol. Cryst. Liq. Cryst. Sci. Technol., Sec. A*, 232.

39. Allemand, P.-M., Khemani, K. C., Koch, A., Wudl, F., Holczer, K., Donovan, S., Grüner, G. and Thompson, J. D. (1991). Organic molecular soft ferromagnetism in a fullerene C_{60}, *Science*, **252**, pp. 301–303.

40. Stephens, P. W., Cox, D., Lauher, J. W., Mihaly, L., Allemand, P. M., Hirsch, A., Holczer, K., Li, Q., Thompson, J. D. and Wudl, F. (1992). Lattice structure of the fullerene ferromagnet TDAE-C_{60}, *Nature*, **355**, pp. 331–332.

41. Hebard, A. F., Rosseinsky, M. J., Haddon, R. C., Murphy, D. W., Glarum, S. H., Palstra, T. T. M., Ramirez, A. P. and Kortan, A. R. (1991). Superconductivity at 18 K in potassium-doped C_{60}, *Nature*, **350**, pp. 600–601.

42. Stephens, P. W., Mihaly, L., Lee, P. L., Whetten, R. L., Huang, S.-M., Kaner, R., Deiderich, F. and Holczer, K. (1991). Structure of single-phase superconducting K_3C_{60}, *Nature*, **351**, pp. 632–634.

43. Holczer, K., Klein, O., Huang, S.-M., Kaner, R. B., Fu, K.-J., Whetten, R. L. and Deiderich, F. (1991). Alkali-fullerene superconductors: synthesis, composition, and diamagnetic shielding, *Science*, **252**, pp. 1154–1157.

44. Tanigaki, K., Ebbesen, T. W., Saito, S., Mizuki, J., Tsai, J. S., Kubo, Y. and Kuroshima, S. (1991). Superconductivity at 33 K in $Cs_xRb_yC_{60}$, *Nature*, **352**, pp. 222–223.

45. Kelty, S. P., Chen, C.-C. and Lieber, C. M. (1991). Superconductivity at 30 K in caesium-doped C_{60}, *Nature*, **352**, pp. 223–225.

46. Rosseinsky, M. J., Ramirez, A. P., Glarum, S. H., Murphy, D. W., Haddon, R. C., Hebard, A. F., Palstra, T. T. M., Kortan, A. R., Zahurak, S. M. and Makhija, A. V. (1991). Superconductivity at 28 K in Rb_xC_{60}, *Phys. Rev. Lett.*, **66**, pp. 2830–2832.

47. McCauley, Jr., J. P., Zhu, Q., Coustel, N., Zhou., O., Vaughan, G., Idziak, S. H. J., Fischer, J. E., Tozer, S. W., Groski, D. M., Bykovetz, N., Lin, L. C., McGhie, A. R., Allen, B. H., Romanow, W. J., Denenstein, A. M. and Smith III, A. B. (1991). Synthesis, structure, and superconducting properties of single-phase Rb_3C_{60}. A new, convenient method for the preparation of M_3C_{60} superconductors, *J. Am. Chem. Soc.*, **113**, pp. 8537–8538.

48. Zhou, O., Fischer, J. E., Coustel, N., Kycia, S., Zhu, Q., McGhie, A. R., Allen, B. H., Romanow, W. J., McCauley, Jr., J. P., Smith III, A. B. and Cox, D. E. (1991). Structure and bonding in alkali-metal-doped C_{60}, *Nature*, **351**, pp. 462–464.

49. Wang, H. H., Kini, A. M., Savall, B. M., Carlson, K. D., Williams, J. M., Lykke, K. R., Wurz, P., Parker, D. H., Pellin, M. J., Gruen, D. M., Welp, U., Kwok, W.-K., Fleshler, S. and Crabtree, G. W. (1991). First easily reproduced solution-phase synthesis and confirmation of superconductivity in the fullerene K_xC_{60} (T_C = 18.0 ±0.1 K), *Inorg. Chem.*, **30**, pp. 2838–2839.

50. Wang, H. H., Kini, A. M., Savall, B. M., Carlson, K. D., Williams, J. M., Lathrop, M. W., Lykke, K. R., Wurz, P., Parker, D. H., Pellin, M. J., Gruen, D. M., Welp, U., Kwok, W.-K., Fleshler, S., Crabtree, G. W., Schirber, J. E. and Overmyer, D. L. (1991). Superconductivity at 28.6 K in a rubidium-C_{60} fullerene compound, Rb_xC_{60}, synthesized by a solution-phase technique, *Inorg. Chem.*, **30**, pp. 2962–2963.

51. Fischer, J. E., Heiney, P. A. and Smith III, A. B. (1992). Solid state chemistry of fullerene-based materials, *Acc. Chem. Res.*, **25**, pp. 112–118.

52. Haddon, R. C. (1992). Electronic structure, conductivity, and superconductivity of alkali metal doped C_{60}, *Acc. Chem. Res.*, **25**, pp. 127–133.

53. Bird, C. L. and Kuhn, A. T. (1981). Electrochemistry of viologens, *Chem. Soc. Rev.*, **10**, pp. 49–82.

54. Russel, J. H. and Wallwork, S. C. (1972). The crystal structures of the dichloride and isomorphous dibromide and diiodide of the N, N'-dimethyl-4,4'-bipyridylium ion, *Acta Cryst.*, **B28**, pp. 1527–1533.

55. Mohammad, M. (1987). Methyl viologen neutral MV:. 1. Preparation and some properties, *J. Org. Chem.*, **52**, pp. 2779–2782.

56. Bockman, T. M. and Kochi, J. K. (1990). Isolation and oxidation-reduction of methylviologen cation radicals. Novel disproportionation in charge-transfer salts by X-ray crystallography, *J. Org. Chem.*, **55**, pp. 4127–4135.

57. Kini, A. M., Geiser, U., Wang, H. H., Carlson, K. D., Williams, J. M., Kwok, W. K., Vandervoort, K. G., Thompson, J. E., Stupka, D. L., Jung, D. and Whangbo, M.-H. (1990). A new ambient-pressure organic superconductor, κ-$(ET)_2Cu[N(CN)_2]Br$, with the highest transition temperature yet observed (inductive onset T_C = 11.6 K, resistive onset = 12.5 K), *Inorg. Chem.*, **29**, pp. 2555–2557.

58. Williams, J. M., Ferraro, J. R., Thorn, R. J., Carlson, K. D., Geiser, U., Wang, H. H., Kini, A. M. and Whangbo, M.-H. (1992). *Organic Superconductors (Including Fullerenes): Synthesis Structure, Properties, and Theory*, Prentice-Hall, Englewood Cliffs, NJ, USA.

59. Mizuno, M., Garito, A. F. and Cava, M. P. (1978). 'Organic metals': alkylthio substitution effects in tetrathiafulvalene–tetracyanoquinodimethane charge-transfer complexes, *J. Chem. Soc., Chem. Commun.*, pp. 18–19.

60. Kobayashi, H., Kato, R., Kobayashi, A., Moriyama, S., Nishio, Y., Kajita, K. and Sasaki, W. (1988). Layered molecular superconductors θ- and κ-$(BEDT\text{-}TTF)_2I_3$, *Synth. Met.*, **27**, pp. A283–A288.

61. Kobayashi, H., Kato, R., Kobayashi, A., Mori, T., Inokuchi, H., Nishio, Y., Kajita, K. and Sasaki, W. (1988). Molecular design and solid state properties of new superconductors and molecular metals with ordered spin structures, *Synth. Met.*, **27**, pp. A289–A297.

62. Aumüller, A. and Hünig, S. (1986). Multistep reversible redox systems, XLVI *N,N'*-dicyanoquinonediimines – a new class of compounds, I: synthesis and general properties, *Liebigs Ann. Chem.*, **1**, pp. 142–164.

63. Aumüller, A., Erk, P., Klebe, G., Hünig, S., Schütz, J. U. and Werner, H.-P. (1986). A radical anion salt of 2,5-dimethyl-*N,N'*-dicyanoquinonediimine with extremely high conductivity, *Angew. Chem. Int. Ed. Engl.*, **25**, pp. 740–741.

64. Kobayashi, A., Kato, R., Kobayashi, H., Mori, T. and Inokuchi, H. (1987). The organic π-electron metal system with interaction through mixed-valence metal cation: electronic and structural properties of radical salts of dicyano-quinodiimine, $(DMe\text{-}DCNQI)_2Cu$ and $(MeCl\text{-}DCNQI)_2Cu$, *Solid State Commun.*, **64**, pp. 45–51.

65. Kato, R., Kobayashi, H., Kobayashi, A., Mori, T. and Inokuchi, H. (1987). Crystal structures of $M(DCNQIs)_2$ (DCNQIs=*N,N'*-dicyanoquinonediimines; M=Li, Na, K, NH_4, Cu, Ag), *Chem. Lett.*, pp. 1579–1582.

66. Mori, T., Imaeda, K., Kato, R., Kobayashi, A., Kobayashi, H. and Inokuchi, H. (1987). Pressure-induced one-dimensional instablity in $(DMDCNQI)_2Cu$, *J. Phys. Soc. Jpn.*, **56**, pp. 3429–3431.

67. Tomić, S., Jérome, D., Aumüller, A., Erk, P., Hünig, S. and von Schütz, J. U. (1988). Pressure-induced metal-to-insulator phase transitions in the organic conductor (2,5 DM-DCNQI)$_2$Cu, *Europhys. Lett.*, **5**, pp. 553–558.

68. Tomić, S., Jérome, D., Aumüller, A., Erk, P., Hünig, S. and von Schütz, J. U. (1988). The pressure-temperature phase diagram of the organic conductor (2,5 DM-DCNQI)$_2$Cu, *J. Phys. C: Solid State Phys.*, **21**, pp. L203–L207.

69. Kobayashi, H., Kato, R., Kobayashi, A., Mori, T. and Inokuchi, H. (1988). The first molecular metals with ordered spin structures, (R_1,R_2-DCNQI)$_2$Cu (R_1, R_2 = CH_3, CH_3O, Cl, Br) – Jahn-Teller distortion, CDW instability and antiferromagnetic spin ordering, *Solid State Commun.*, **65**, pp. 1351–1354.

70. Mori, T., Inokuchi, H., Kobayashi, A., Kato, R. and Kobayashi, H. (1988). Electrical conductivity, thermoelectric power, and ESR of a new family of molecular conductors, dicyanoquinonediimine-metal [$(DCNQI)_2M$] compounds, *Phys. Rev. B*, **38**, pp. 5913–5923.

71. Kobayashi, H., Miyamoto, A., Kato, R., Kobayashi, A., Nishio, Y., Kajita, K. and Sasaki, W. (1989). Reentrant behavior in the pressure-temperature dependence of the resistivity of (DMeO-DCNQI)$_2$Cu, *Solid State Commun.*, **72**, pp. 1–5.

72. Kato, R., Kobayashi, H. and Kobayashi, A. (1989). Crystal and electronic structures of conductive anion-radical salts, (2,5-R_1,R_2-DCNQI)$_2$Cu (DCNQI = *N,N'*-dicyanoquinonediimine; R_1, R_2 = CH_3, CH_3O, Cl, Br), *J. Am. Chem. Soc.*, **111**, pp. 5224–5232.

73. Tajima, H., Ojima, G., Ida, T., Kuroda, H., Kobayashi, A., Kato, R., Kobayashi, H., Ugawa, A., Yakushi, K. (1990). Reflectance spectra of DCNQI salts, in *The Physics and Chemistry of Organic Superconductors*, Saito, G. and Kagoshima, S. (eds.), Springer Proceedings in Physics, Vol. 51, Springer-Verlag Berlin, Heidelberg.

74. Yakushi, K., Ugawa, A., Ojima, G., Ida, T., Tajima, H., Kuroda, H., Kobayashi, A., Kato, R. and Kobayashi, H. (1990). Polarized reflectance spectra of DCNQI salts, *Mol. Cryst. Liq. Cryst.*, **181**, pp. 217–231.

75. Hünig, S., Sinzger, K., Jopp, M., Bauer, D., Bietsch, W., von Schütz, J. U. and Wolf, H. C. (1992). Deuterium-induced phase transition of an organic metal: an unusual isotope effect, *Angew. Chem. Int. Ed. Engl.*, **31**, pp. 859–862.

76. Nishio, Y., Kajita, K., Sasaki, W., Kato, R., Kobayashi, A. and Kobayashi, H. (1992). Thermal and magnetic properties in organic metals (DMe-DCNQI)$_2$Cu, (DMeO-DCNQI)$_2$Cu and (DMe$_{1-x}$MeBr$_x$-DCNQI)$_2$Cu: enhancement of density of states, *Solid State Commun.*, **81**, pp. 473–476.

77. Kobayashi, H., Miyamoto, A., Kato, R., Sakai, F., Kobayashi, A., Yamakita, Y., Furukawa, Y., Tasumi, M. and Watanabe, T. (1993). Mixed valency of Cu, electron-mass enhancement, and the three-dimensional arrangement of magnetic sites in the organic conductors (R_1,R_2-*N,N'*-dicyanoquinonediimine)$_2$Cu (where R_1,R_2 = CH_3,CH_3O,Cl,Br), *Phys. Rev. B*, **47**, pp. 3500–3510.

78. Tamura, M., Sawa, H., Aonuma, S., Kato, R., Kinoshita, M. and Kobayashi, H. (1993). Weak ferromagnetism and magnetic anisotropy in Cu salt of fully deuterated DMe-DCNQI, (DMe-DCNQI-d_8)$_2$Cu, *J. Phys. Soc. Jpn.*, **62**, pp. 1470–1473.

79. Aonuma, S., Sawa, H., Kato, R. and Kobayashi, H. (1993). Giant metal-insulator-metal transition induced by selective deuteration of molecular conductor, (DMe-DCNQI)$_2$Cu (DMe-DCNQI = 2,5-dimethyl-*N,N'*-dicyanoquinonediimine), *Chem. Lett.*, pp. 513–516.

80. Uji, S., Terashima, T., Aoki, H., Brooks, J. S., Kato, R., Sawa, H., Aonuma, S., Tamura, M. and Kinoshita, M. (1994). Coexistence of one- and three-dimensional Fermi surfaces and heavy cyclotron mass in the molecular conductor (DMe-DCNQI)$_2$Cu, *Phys. Rev. B*, **50**, pp. 15597–15601.

81. Sawa, H., Tamura, M., Aonuma, S., Kinoshita, M. and Kato, R. (1994). Charge-transfer-controlled phase transition in a molecular conductor, (DMe-DCNQI)$_2$Cu-doping effect, *J. Phys. Soc. Jpn.*, **63**, pp. 4302–4305.

82. Tanaka, A., Chainani, A., Yokoya, T., Takahashi, T., Miyazaki, T., Hasegawa, S. and Mori, T. (1995). Electronic structure of the quasi-one-dimensional organic conductors DCNQI (*N,N'*-dicyanoquinonediimine)-Cu salts, *Phys. Rev. B*, **52**, pp. 7951–7959.

83. Sekiyama, A., Fujimori, A., Aonuma, S., Sawa, H. and Kato, R. (1995). Fermi-liquid versus Luttinger-liquid behavior and metal-insulator transition in *N,N'*-dicyanoquinonediimine-Cu salt studied by photoemission, *Phys. Rev. B*, **51**, pp. 13899–13902.

84. Uji, S., Terashima, T., Aoki, H., Kato, R., Sawa, H., Aonuma, S., Tamura, M. and Kinoshita, M. (1995). Fermi surface and absence of additional mass enhancement near the insulating phase in (DMe-DCNQI)$_2$Cu, *Solid State Commun.*, **93**, pp. 203–207.

85. Tamura, M., Kashimura, Y., Sawa, H., Aonuma, S., Kato, R. and Kinoshita, M. (1995). Enhanced magnetic susceptibility of $(DI\text{-}DCNQI)_2Cu$, *Solid State Commun.*, **93**, pp. 585–588.

86. Hünig, S. (1995). *N,N′*-dicyanoquinone diimines (DCNQIs): unique acceptors for conducting materials, *J. Mater. Chem.*, **5**, pp. 1469–1479.

87. Aonuma, S., Sawa, H. and Kato, R. (1995). Chemical pressure effect by selective deuteration in the molecular-based conductor, 2,5-dimethyl-*N,N′*-dicyano-*p*-benzoquinone diimine copper salt, $(DMe\text{-}DCNQI)_2Cu$, *J. Chem. Soc., Perkin Trans. 2*, pp. 1541–1549.

88. Miyazaki, T., Terakura, K., Morikawa, Y. and Yamasaki, T. (1995). First-principles theoretical study of metallic states of DCNQI-(Cu,Ag) systems: simplicity and variety in complex systems, *Phys. Rev. Lett.*, **74**, pp. 5104–5107.

89. Miyazaki, T. and Terakura, K. (1996). First-principles theoretical study of metallic states of DCNQI-(Cu,Ag,Li) systems, *Phys. Rev. B*, **54**, pp. 10452–10464.

90. Hiraki, K. and Kanoda, K. (1996). Electronic state of the organic salt $(DI\text{-}DCNQI)_2Ag$, where DI-DCNQI is 2,5-diiodo-*N,N′*-dicyanoquinonediimine, *Phys. Rev. B*, **54**, pp. R17276–R17279.

91. Sekiyama, A., Fujimori, A., Aonuma, S. and Kato, R. (1997). Observation of anisotropy in the electronic states of *N,N′*-dicyanoquinonediimine-Cu salt by x-ray-absorption spectroscopy, *Phys. Rev. B*, **56**, pp. 9937–9941.

92. Hiraki, K. and Kanoda, K. (1998). Wigner crystal type of charge ordering in an organic conductor with a quarter-filled band: $(DI\text{-}DCNQI)_2Ag$, *Phys. Rev. Lett.*, **80**, pp. 4737–4740.

93. Akaki, O., Chainani, A., Takahashi, T., Kashimura, Y. and Kato, R. (1998). Temperature-dependent core-level x-ray photoemission spectroscopy of the organic conductors $(MeCl\text{-}DCNQI)_2Cu$, $(MeBr\text{-}DCNQI)_2Cu$, and $(DI\text{-}DCNQI)_2Cu$, *Phys. Rev. B*, **57**, pp. 11846–11849.

94. Hünig, S., Kemmer, M., Meixner, H., Sinzger, K., Werner, H., Bauer, T., Tillmanns, E., Lux, F. R., Hollstein, M., Groß, H.-G., Langohr, U., Werner, H.-P., von Schütz, J. U. and Wolf, H.-C. (1999). 2,5-disubstituted *N,N′*-dicyanobenzoquinonediimines (DCQNIs): charge-transfer complexes and radical-anion salts and copper salts with ligand alloys: syntheses, structures and conductivities, *Eur. J. Inorg. Chem.*, pp. 899–916.

95. Miyagawa, K., Kawamoto, A. and Kanoda, K. (1999). π-d orbital hybridization in the metallic state of organic-inorganic complexes seen by ^{13}C and ^{15}N NMR at selective sites, *Phys. Rev. B*, **60**, pp. 14847–14851.

96. Kato, R. (2000). Conductive copper salts of 2,5-disubstituted *N,N'*-dicyanobenzoquinonediimines (DCNQIs): structural and physical properties, *Bull. Chem. Soc. Jpn.*, **73**, pp. 515–534.

97. Nishio, Y., Tamura, M., Kajita, K., Aonuma, S., Sawa, H., Kato, R. and Kobayashi, H. (2000). Thermodynamic study of $(DMe\text{-}DCNQI)_2Cu$ system - mechanism of reentrant metal-insulator transition, *J. Phys. Soc. Jpn.*, **69**, pp. 1414–1422.

98. Nakazawa, Y., Seki, M., Saito, K., Hiraki, K., Takahashi, T., Kanoda, K. and Sorai, M. (2002). Thermodynamic investigation of the charge-ordered insulating state of the quasi-one-dimensional organic system $(DI\text{-}DCNQI)_2Ag$, *Phys. Rev. Lett.*, **88**, pp. 076402 (4 pages).

99. Nakazawa, Y., Sato, A., Seki, M., Saito, K., Hiraki, K., Takahashi, T., Kanoda, K. and Sorai, M. (2003). Spin-Peierls transition of the quasi-one-dimensional electronic system $(DMe\text{-}DCNQI)_2M$ (M = Li, Ag) probed by heat capacity, *Phys. Rev. B*, **68**, pp. 085112 (8 pages).

100. Shinohara, Y., Kazama, S., Mizoguchi, K., Hiraoka, M. and Sakamoto, H. (2007). Spin density distribution and electronic states in $(DMe\text{-}DCNQI)_2M$ (M = Li, Ag, Cu) from high-resolution solid-state NMR, *Phys. Rev. B*, **76**, pp. 035128 (9 pages).

101. Kakiuchi, T., Wakabayashi, Y., Sawa, H., Itou, T. and Kanoda, K. (2007). Wigner crystallization in $(DI\text{-}DCNQI)_2Ag$ detected by synchrotron radiation x-ray diffraction, *Phys. Rev. Lett.*, **98**, pp. 066402 (4 pages).

102. Ito, T., Kanoda, K., Hiraki, K., Takahashi, T., Murata, K. and Matsumoto, T. (2005). Collapse of the charge order in $(DI\text{-}DCNQI)_2Ag$ by dimensional crossover, *Phys. Rev. B*, **72**, pp. 113109 (4 pages).

103. Ito, T., Kanoda, K., Murata, K., Matsumoto, T., Hiraki, K. and Takahashi, T. (2004). Collapse of charge order in a quasi-one-dimensional organic conductor with a quarter-filled band, *Phys. Rev. Lett.*, **93**, pp. 216408 (4 pages).

104. Sakurai, T., Nakagawa, N., Okubo, S., Ohta, H., Kanoda, K. and Hiraki, K. (2001). ESR measurements of $(DI\text{-}DCNQI)_2Ag$ single crystals, *J. Phys. Soc. Jpn.*, **70**, pp. 1794–1800.

105. Haruyama, Y., Nath, K. G., Kimura, S., Ufuktepe, Y., Kinoshita, T., Hiraki, K. and Kanoda, K. (1999). Electronic structures of organic salts $(DI\text{-}DCNQI)_2M$ (M=Cu and Ag) using photoelectron spectromicroscopy, *Solid State Commun.*, **110**, pp. 17–22.

106. Imamura, Y., Ten-no, S., Yonemitsu, K. and Tanimura, Y. (1998). Theoretical study on electron correlation of 1-D $(DCNQI)_2M$ (M=Li, Ag) salts, *Chem. Phys. Lett.*, **298**, pp. 15–20.

107. Yonemitsu, K. (1997). Renormalization-group approach to the metal-insulator transitions in $(DCNQI)_2M$ (DCNQI is *N,N'*-dicyanoquinonediimine and M = Ag, Cu), *Phys. Rev. B*, **56**, pp. 7262–7276.

108. Kobayashi, A., Kato, R. and Kobayashi, H. (1989). Reentrant behavior of the temperature dependence of resistivity of DCNQI-Cu alloy system, $[(DMe)_{1-x}(MeBr)_x\text{-}DCNQI]_2Cu$, *Chem. Lett.*, **18**, pp. 1843–1846.

109. Kobayashi, H., Miyamoto, A., Moriyama, H., Kato, R. and Kobayashi, A. (1991). Suppression of the metal instability and continuous change of the carrier density in $(MeBr\text{-}DCNQI)_2Cu_{1-x}Li_x$, *Chem. Lett.*, **20**, pp. 863–866.

110. Miyamoto, A., Kobayashi, H., Kato, R., Kobayashi, A., Nishio, Y., Kajita, K. and Sasaki, W. (1992). Metal instability of $(DMe\text{-}DCNQI)_2Cu$ induced by uniaxial stress and enhancement of electron mass, *Chem. Lett.*, **21**, pp. 115–118.

111. Peierls, R. E. (1955). *Quantum Theory of Solids*, Clarendon Press, Oxford, p. 108, Chapter 5.

112. Saito, G. and Kagoshima, S. (eds.) (1990). *The Physics and Chemistry of Organic Superconductors*, Springer Proceedings in Physics, Vol. 51, Springer-Verlag Berlin, Heidelberg.

113. Because the electrons at E_F, sometimes called valence electrons, almost exclusively respond to perturbation to dominate chemical and physical reactions of the systems, they are important in electrical, magnetic and optical properties. The Fermi level originates from the energy level of the highest occupied molecular orbital (HOMO) in an isolated molecule with a closed-shell structure, while it originates from that of the singly occupied molecular orbital (SOMO) in a radical species. In a solid, the Fermi level coincides with the chemical potential at $T = 0$ [K], i.e. in the ground state. Exactly speaking, "the ground state" only indicates the state at $T = 0$ [K], and does not include any state at a finite temperature. However, in the discussion on photoexcited states in this chapter, we shall include any state under a dark condition at a finite temperature in "ground states", unless it would be misleading. As the photon energies in a UV-vis range, for example, correspond to an order of 10^4 K, a thermal excitation at ~300 K, for example, can be negligible in many photoexcited states.

114. Saiki, T., Mori, S., Ohara, K. and Naito, T. (2014). Capacitor-like behavior of molecular crystal β-DiCC$[Ni(dmit)_2]$, *Chem. Lett.*, **43**, pp. 1119–1121.

115. Tanaka, H., Okano, Y., Kobayashi, H., Suzuki, W. and Kobayashi, A. (2001). A three-dimensional synthetic metallic crystal composed of single-component molecules, *Science*, **291**, pp. 285–287.

116. Kobayashi, A., Tanaka, H. and Kobayashi, H. (2001). Molecular design and development of single-component molecular metals, *J. Mater. Chem.*, **11**, pp. 2078–2088.

117. Kobayashi, A., Fujiwara, E. and Kobayashi, H. (2004). Single-component molecular metals with extended-TTF dithiolate ligands, *Chem. Rev.*, **104**, pp. 5243–5264.

118. Yamamoto, K., Fujiwara, E., Kobayashi, A., Fujishiro, Y., Nishibori, E., Sakata, M., Tanaka, M., Tanaka, H., Okano, Y. and Kobayashi, H. (2005). Single-component molecular conductor [Zn(tmdt)$_2$] and related Zn complexes, *Chem. Lett.*, **34**, pp. 1090–1091.

119. Kobayashi, A., Okano, Y. and Kobayashi, H. (2006). Molecular design and physical properties of single-component molecular metals, *J. Phys. Soc. Jpn.*, **75**, pp. 051002 (12 pages).

120. Zhou, B., Shimamura, M., Fujiwara, E., Kobayashi, A., Higashi, T., Nishibori, E., Sakata, M., Cui, H. B., Takahashi, K. and Kobayashi, H. (2006). Magnetic transitions of single-component molecular metal [Au(tmdt)$_2$] and its alloy systems, *J. Am. Chem. Soc.*, **128**, pp. 3872–3873.

121. Zhou, B., Kobayashi, A., Okano, Y., Nakashima, T., Aoyagi, S., Nishibori, E., Sakata, M., Tokumoto, M. and Kobayashi, H. (2009). Single-component molecular conductor [Pt(tmdt)$_2$] (tmdt = trimethylenetetrathiafulvalenedithiolate) – an advanced molecular metal exhibiting high metallicity, *Adv. Mater.*, **21**, pp. 3596–3600.

122. Zhou, B., Yajima, H., Kobayashi, A., Okano, Y., Tanaka, H., Kumashiro, T., Nishibori, E., Sawa, H. and Kobayashi, H. (2010). Single-component molecular conductor [Cu(tmdt)$_2$] containing an antiferromagnetic heisenberg chain, *Inorg. Chem.*, **49**, pp. 6740–6747.

123. Zhou, B., Idobata, Y., Kobayashi, A., Cui, H.-B., Kato, R., Takagi, R., Miyagawa, K., Kanoda, K. and Kobayashi, H. (2012). Single-component molecular conductor [Cu(dmdt)$_2$] with three-dimensionally arranged magnetic moments exhibiting a coupled electric and magnetic transition, *J. Am. Chem. Soc.*, **134**, pp. 12724–12731.

124. Idobata, Y., Zhou, B., Kobayashi, A. and Kobayashi, H. (2012). Molecular alloy with diluted magnetic moments - molecular kondo system, *J. Am. Chem. Soc.*, **134**, pp. 871–874.

125. Cui, H. B., Kobayashi, H., Ishibashi, S., Sasa, M., Iwase, F., Kato, R. and Kobayashi, A. (2014). A single-component molecular superconductor, *J. Am. Chem. Soc.*, **136**, pp. 7619–7622.

126. Tissot, A. (2014). Photoswitchable spin crossover nanoparticles, *New J. Chem.*, **38**, pp. 1840–1845.

127. Hatcher, L. E., Lauren, E. and Raithby, P. R. (2014). Dynamic single-crystal diffraction studies using synchrotron radiation, *Coord. Chem. Rev.*, **277–278**, pp. 69–79.

128. Gütlich, P., Gasper, A. B. and Garcia, Y. (2013). Spin states switching in iron coordination compounds, *Beil. J. Org. Chem.*, **9**, pp. 342–391.

129. Letard, J.-F. (2006). Photomagnetism of iron(II) spin crossover complexes – the *T*(LIESST) approach, *J. Mater. Chem.*, **16**, pp. 2550–2559.

130. Varret, F., Boukheddaden, K., Codjovi, E., Enachescu, C. and Linares, J. (2004). On the competition between relaxation and photoexcitations in spin crossover solids under continuous irradiation, in *Spin Crossover in Transition Metal Compounds II*, Topics in Current Chemistry, Vol. 234, Springer-Verlag Berlin, Germany, pp. 199–229.

131. Gütlich, P., Garcia, Y. and Spering, H. (2003). Spin transition phenomena, in *Magnetism: Molecules to Materials IV*, Miller, J. S. and Drillon, M. (eds.), Wiley-VCH Verlag GmbH & Co., Weinheim, FRG, pp. 271–344.

132. Spiering, H., Kohlhaas, T., Romstedt, H., Hauser, A., Bruns-Yilmaz, C., Kusz, J. and Gütlich, P. (1999). Correlations of the distribution of spin states in spin crossover compounds, *Coord. Chem. Rev.*, **190–192**, pp. 629–647.

133. Hauser, A., Jeftic, J., Romstedt, H., Hinek, R. and Spiering, H. (1999). Cooperative phenomena and light-induced bistability in iron(II) spin-crossover compounds, *Coord. Chem. Rev.*, **190–192**, pp. 471–491.

134. Gütlich, P., Ensling, J. and Tuczek, F. (1994). Metastable electronic states induced by nuclear decay and light, *Hyperfine Interact.*, **84**, pp. 447–469.

135. Tokoro, H. and Ohkoshi, S. (2010). Photo-induced phase transition in RbMnFe Prussian blue analog, in *Molecular Electronic and Related Materials-Control and Probe with Light*, Naito, T. (ed.), Research Signpost, Trivandrum, India, pp. 203–222.

136. Kumagai, S., Takaishi, S., Iguchi, H. and Yamashita, M. (2015). Charge-bistable Pd(III)/Pd(II,IV) coordination polymers: phase transitions and their applications to optical properties, *Dalton Trans.*, **44**, pp. 8590–8599.

137. Komitov, L. (2015). Molecular shape related effects in liquid crystals, *Mol. Cryst. Liq. Cryst.*, **608**, pp. 246–257.

138. Cho, H. J., Chung, M. and Shim, M. S. (2015). Engineered photo-responsive materials for near-infrared-triggered drug delivery, *J. Ind. Eng. Chem.*, **31**, pp. 15–25.

139. Grossmann, P., Rajkovic, I., More, R., Norpoth, J., Techert, S., Jooss, C. and Mann, K. (2012). Time-resolved near-edge x-ray absorption fine structure spectroscopy on photo-induced phase transitions using a tabletop soft-x-ray spectrometer, *Rev. Sci. Instrum.*, **83**, pp. 053110.

140. Wu, J. M. and Liou, L. B. (2011). Room temperature photo-induced phase transitions of VO_2 nanodevices, *J. Mater. Chem.*, **21**, pp. 5499–5504.

141. Chollet, M., Guerin, L., Uchida, N., Fukaya, S., Shimoda, H., Ishikawa, T., Matsuda, K., Hasegawa, T., Ota, A., Yamochi, H., Saito, G., Tazaki, R., Adachi, S. and Koshihara, S. (2005). Gigantic photoresponse in 1/4-filled-band organic salt (EDO-TTF)2PF6, *Science*, **307**, pp. 86–89.

142. Tajima, N., Fujisawa, J., Naka, N., Ishihara, T., Kato, R., Nishio, Y. and Kajita, K. (2005). Photo-induced insulator-metal transition in an organic conductor α-(BEDT-TTF)$_2$I$_3$, *J. Phys. Soc. Jpn.*, **74**, pp. 511–514.

143. Iimori, T., Naito, T. and Ohta, N. (2007). A memory effect controlled by a pulsed voltage in photoinduced conductivity switching in an organic charge transfer salt, *J. Am. Chem. Soc.*, **129**, pp. 3486–3487.

144. Iimori, T., Ohta, N. and Naito, T. (2007). Molecular-based light-activated thyristor, *Appl. Phys. Lett.*, **90**, pp. 262103 (3 pages).

145. Yamamoto, K., Iwai, S., Boyko, S., Kashiwazaki, A., Hiramatsu, F., Okabe, C., Nishi, N. and Yakushi, K. (2008). Strong optical nonlinearity and its ultrafast response associated with electron ferroelectricity in an organic conductor, *J. Phys. Soc. Jpn.*, **77**, pp. 074709 (6 pages).

146. Iwai, S., Nakaya, H. and Kawakami, Y. (2010). Ultrafast photo-induced insulator to metal transition in layered BEDT-TTF based salts, in *Molecular Electronic and Related Materials-Control and Probe with Light*, Naito, T. (ed.), Research Signpost, Trivandrum, India, pp. 37–58.

147. Okamoto, H. (2010). Ultrafast photoinduced phase transitions in one-dimensional organic correlated electron systems, in *Molecular Electronic and Related Materials-Control and Probe with Light*, Naito, T. (ed.), Research Signpost, Trivandrum, India, pp. 59–98.

148. Tajima, N., Fujisawa, J. and Kato, R. (2010). Photo-switching between charge-ordered insulator and metal phases in an organic conductor α-(BEDT-TTF)$_2$I$_3$, in *Molecular Electronic and Related Materials-Control*

and Probe with Light, Naito, T. (ed.), Research Signpost, Trivandrum, India, pp. 155–165.

149. Iimori, T., Naito, T. and Ohta, N. (2010). Syneregy effects of photoirradiation and applied voltage on electrical conductivity of α-$(BEDT\text{-}TTF)_2I_3$, in *Molecular Electronic and Related Materials-Control and Probe with Light*, Naito, T. (ed.), Research Signpost, Trivandrum, India, pp. 167–184.

150. Yamamoto, K. and Yakushi, K. (2010). Second-harmonic generation study of ferroelectric organic conductors α-$(BEDT\text{-}TTF)_2X$ (X = I_3, I_2Br), in *Molecular Electronic and Related Materials-Control and Probe with Light*, Naito, T. (ed.), Research Signpost, Trivandrum, India, pp. 185–201.

151. Yonemitsu, K. (2010). Theory of photoinduced phase transitions in quasi-one-dimensional organic conductors, in *Molecular Electronic and Related Materials-Control and Probe with Light*, Naito, T. (ed.), Research Signpost, Trivandrum, India, pp. 305–320.

152. Alvarez, S., Vicente, R. and Hoffmann, R. (1985). Dimerization and stacking in transition-metal bisdithiolenes and tetrathiolates, *J. Am. Chem. Soc.*, **107**, pp. 6253–6277.

153. Cassoux, P., Valade, L., Kobayashi, H., Kobayashi, A., Clark, R. A. and Underhill, A. E. (1991). Molecular metals and superconductors derived from metal complexes of 1,3-dithiole-2-thione-4,5-dithiolate (dmit), *Coord. Chem. Rev.*, **110**, pp. 115–160.

154. Williams, J. M., Schultz, A. J., Geiser, U., Carlson, K. D., Kini, A. M., Wang, H. H., Kwok, W.-K., Whangbo, M.-H. and Schirber, J. E. (1991). Organic superconductors – new benchmarks, *Science*, **252**, pp. 1501–1508.

155. Olk, R.-M., Olk, B., Dietzsch, W., Kirmse, R. and Hoyer, E. (1992). The chemistry of 1,3-dithiole-2-thione-4,5-dithiolate (dmit), *Coord. Chem. Rev.*, **117**, pp. 99–131.

156. Svenstrup, N. and Becher, J. (1995). The organic chemistry of 1,3-dithiole-2-thione-4,5-dithiolate (dmit), *Synthesis*, pp. 215–235.

157. Canadell, E. (1997). Electronic structure of two-band molecular conductors, *New. J. Chem.*, **21**, pp. 1147–1159.

158. Kato, R., Liu, Y.-L., Hosokoshi, Y., Aonuma, S. and Sawa, H. (1997). Se-substitution and cation effects on the high-pressure molecular superconductor, β-$Me_4N[Pd(dmit)_2]_2$ – a unique two-band system, *Mol. Cryst. Liq. Cryst.*, **296**, pp. 217–244.

159. Rosa, A., Ricciardi, G. and Baerends, E. J. (1998). Structural properties of $M(dmit)_2$-based (M = Ni, Pd, Pt; $dmit^{2-}$ = 2-thioxo-1,3-dithiole-

4,5-dithiolato) molecular metals. Insights from density functional calculations, *Inorg. Chem.*, **37**, pp. 1368–1379.

160. Pullen, A. E. and Olk, R.-M. (1999). The coordination chemistry of 1,3-dithiole-2-thione-4,5-dithiolate (dmit) and isologs, *Coord. Chem. Rev.*, **188**, pp. 211–262.

161. Kato, R. (2004). Conducting metal dithiolene complexes: structural and electronic properties, *Chem. Rev.*, **104**, pp. 5319–5346.

162. Mori, H. (2006). Materials viewpoint of organic superconductors, *J. Phys. Soc. Jpn.*, **75**, pp. 051003 (15 pages).

163. Valade, L. and Tanaka, H. (2010). Molecular inorganic conductors and superconductors, *Mol. Mater.*, pp. 211–280.

164. Mercuri, M. L., Deplano, P., Pilia, L., Serpe, A. and Artizzu, F. (2010). Interaction modes and physical properties in transition metal chalcogenolene-based molecular materials, *Coord. Chem. Rev.*, **254**, pp. 1419–1433.

165. de Bonneval, B. G., Ching, K. I. M.-C., Alary, F., Bui, T.-T. and Valade, L. (2010). Neutral d^8 metal bis-dithiolene complexes: synthesis, electronic properties and applications, *Coord. Chem. Rev.*, **254**, pp. 1457–1467.

166. Papavassiliou, G. C., Anyfantis, G. C. and Mousdis, G. A. (2012). Neutral metal 1,2-dithiolenes: preparations, properties and possible applications of unsymmetrical in comparison to the symmetrical, *Crystals*, **2**, pp. 762–811.

167. Leung, P. C. W., Emge, T. J., Beno, M. A., Wang, H. H., Williams, J. M., Petricek, V. and Coppens, P. (1985). Novel structural modulation in the ambient-pressure sulfur-based organic superconductor β-(BEDT-TTF)$_2$I$_3$: origin and effects on its electrical conductivity, *J. Am. Chem. Soc.*, **107**, pp. 6184–6191.

168. Naito, T., Karasudani, T., Ohara, K., Takano, T., Takahashi, Y., Inabe, T., Furukawa, K. and Nakamura, T. (2012). Simultaneous control of carriers and localized spins with light in organic materials, *Adv. Mater.*, **24**, pp. 6153–6157.

169. Minemawari, H., Naito, T. and Inabe, T. (2007). (ET)$_3$(Br$_3$)$_5$: a metallic conductor with an unusually high oxidation state of ET (ET = bis(ethylenedithio)tetrathiafulvalene), *Chem. Lett.*, **36**, pp. 74–75.

170. Minemawari, H., Jose, J. F. F., Takahashi, Y., Naito, T. and Inabe, T. (2012). Structural characteristics in a stable metallic ET salt with unusually high oxidation state (ET: bis(ethylenedithio)tetrathiafulvalene), *Bull. Chem. Soc. Jpn.*, **85**, pp. 335–340.

171. Naito, T., Kobayashi, H., Kobayashi, A. and Underhill, A. E. (1996). New synthetic metals based on a thiadiazole network, *Chem. Commun.*, pp. 521–522.

172. Naito, T., Karasudani, T., Mori, S., Ohara, K., Konishi, K., Takano, T., Takahashi, Y., Inabe, T., Nishihara, S. and Inoue, K. (2012). Molecular photoconductor with simultaneously photocontrollable localized spins, *J. Am. Chem. Soc.*, **134**, pp. 18656–18666.

173. Hirose, T., Imai, H., Naito, T. and Inabe, T. (2002). Charge carrier doping in the $Ni(dmit)_2$ simple salts by hydrogen-bonding pyridinium cations (dmit=1,3-dithiol-2-thione-4,5-dithiolate), *J. Solid State Chem.*, **168**, pp. 535–546.

174. Cornelissen, J. P., Creyghton, E. J., de Graaff, R. A. G., Haasnoot, J. G. and Reedijk, J. (1991). Crystal structure and electrical conductivity of *N*-methylquinolinium bis(1,3-dithiole-2-thione-4,5-dithiolato) nickelate(III), *Inorg. Chim. Acta*, **185**, pp. 97–102.

175. Naito, T., Karasudani, T., Nagayama, N., Ohara, K., Konishi, K., Mori, S., Takano, T., Takahashi, Y., Inabe, T., Kinose, S., Nishihara, S. and Inoue, K. (2014). Giant photoconductivity in $NMQ[Ni(dmit)_2]$, *Eur. J. Inorg. Chem.*, **2014**, pp. 4000–4009.

176. Naito, T. (2017). Development of control method of conduction and magnetism in molecular crystals, *Bull. Chem. Soc. Jpn.*, **90**, pp. 89–136.

177. Naito, T., Sugawara, H. and Inabe. T. (2007). Mechanism of spatially resolved photochemical control of resistivity of a molecular crystalline solid, *Nanotechnology*, **18**, pp. 424008 (8 pages).

178. Naito, T., Inabe, T., Niimi, H. and Asakura, K. (2004). Light-induced transformation of molecular materials into devices, *Adv. Mater.*, **16**, pp. 1786–1790.

179. Naito, T., Sugawara, H., Inabe, T., Miyamoto, T., Niimi, H. and Asakura, K. (2004). Photochemical fabrication of molecular devices, *J. Non-Cryst. Solids*, **352**, pp. 2628–2630.

180. Naito, T., Sugawara, H., Inabe, T., Kitajima, Y., Miyamoto, T., Niimi, H. and Asakura, K. (2006). Photochemical control of dark conductivity: a new approach to devices based on molecular crystals, *J. Low Temp. Phys.*, **142**, pp. 383–386.

181. Naito, T., Sugawara, H., Inabe, T., Miyamoto, T., Niimi, H. and Asakura, K. (2006). Photochemical method of device fabrication starting from molecular crystals, *Mol. Cryst. Liq. Cryst.*, **455**, pp. 311–316.

182. Naito, T., Sugawara, H., Inabe, T., Kitajima, Y., Miyamoto, T., Niimi, H. and Asakura, K. (2007). Uv-vis induced vitrification of a molecular crystal, *Adv. Funct. Mater.*, **17**, pp. 1663–1670.

183. Miyamoto, T., Niimi, H., Chun, W.-J., Kitajima, Y., Sugawara, H., Inabe, T., Naito, T. and Asakura, K. (2007). Chemical states of Ag in Ag(DMe-DCNQI)$_2$ photoproducts and a proposal for its photoinduced conductivity change mechanism, *Chem. Lett.*, **36**, pp. 1008–1009.

184. Tsutsumi, T., Miyamoto, T., Niimi, H., Kitajima, Y., Sakai, Y., Kato, M., Naito, T. and Asakura, K. (2007). Energy-filtered X-ray photoemission electron microscopy and its applications to surface and organic materials, *Solid-State Electron.*, **51**, pp. 1360–1366.

185. Naito, T., Kakizaki, A., Wakeshima, M., Hinatsu, Y. and Inabe, T. (2009). Photochemical modification of magnetic properties in organic low-dimensional conductors, *J. Solid State Chem.*, **182**, pp. 2733–2742.

186. Miyamoto, T., Kitajima, Y., Sugawara, H., Naito, T., Inabe, T. and Asakura, K. (2009). Origin of photochemical modification of resistivity of Ag(DMe-DCNQ)$_2$ studied by X-ray absorption fine structure, *J. Phys. Chem. C*, **113**, pp. 20476–20480.

187. Miyamoto, T., Niimi, H., Kitajima, Y., Naito, T. and Asakura, K. (2010). Ag L$_3$-edge X-ray absorption near-edge structure of 4d^{10} (Ag$^+$) compounds: origin of the edge peak and its chemical relevance, *J. Phys. Chem. A*, **114**, pp. 4093–4098.

188. Naito, T., Kakizaki, A., Inabe, T., Sakai, R., Nishibori, E. and Sawa, H. (2011). Growth of nanocrystals in a single crystal of different materials: a way of giving function to molecular crystals, *Cryst. Growth Des.*, **11**, pp. 501–506.

189. Iimori, T., Naito, T. and Ohta, N. (2007). Photoinduced phase transition in the organic conductor α-(BEDT-TTF)$_2$I$_3$ at temperatures near the metal-insulator phase transition, *Chem. Lett.*, **36**, pp. 536–537.

190. Iimori, T., Naito, T. and Ohta, N. (2009). Unprecedented optoelectronic function in organic conductor: memory effect of photoswitching controlled by voltage pulse width, *J. Phys. Chem. C*, **113**, pp. 4654–4661.

191. Iimori, T., Naito, T. and Ohta, N. (2010). Time-resolved measurement of the photoinduced change in the electrical conductivity of the organic superconductor α-(BEDT-TTF)$_2$Cu[N(CN)$_2$]Br, *J. Phys. Chem. C*, **114**, pp. 9070–9075.

192. Iimori, T., Sabeth, F., Naito, T. and Ohta, N. (2011). Time-resolved photoresponse measurements of the electrical conductivity of the quasi-two-dimensional organic superconductor β-(BEDT-TTF)$_2$I$_3$

using a nanosecond laser pulse, *J. Phys. Chem. C*, **115**, pp. 23998–24003.

193. Nagayama, N., Yamamoto, T. and Naito, T. (2015). Activation energy for photoconduction in molecular crystals, *CheM*, **2015**, pp. 74–80.

194. Steimecke, G., Sieler, H.-J., Kirmse, R. and Hoyer, E. (1979). 1,3-dithiol-2-thion-4,5-dithiolat aus schwefelkohlenstoff und alkalimetall, *Phosphorus Sulfur*, **7**, pp. 49–55.

195. Steimecke, G., Kirmse, R. and Hoyer, E. (1975). Dimercapt-isotrithion – ein neuer, ungesättigter 1,2-ditholatligand, *Z. Chem.*, **15**, pp. 28–29.

196. Stach, J., Kirmse, R., Dietzsch, W., Olk, R. M. and Hoyer, E. (1984). Single-crystal EPR spectra of tetra-n-butylammonium bis(isotrithione-3,4-dithiolato)cuprate(II) (copper-63), *Inorg. Chem.*, **23**, pp. 4779–4780.

197. Matsubayashi, G., Takahashi, K. and Tanaka, T. (1988). X-ray crystal structure of bis(*N*-ethylpyridinium) bis[4,5-dimercapto-1,3-dithiole-2-thionate(2-)]copper(II) and electrical properties of its oxidized salts, *J. Chem. Soc. Dalton Trans.*, pp. 967–972.

198. Guo, W. F., Sun, X. B., Sun, J., Yu, W. T., Wang, X. Q., Zhang, G. H. and Xu, D. (2007). Preparation, single crystal growth and characterization of bis(tetrabutylammonium)bis(4,5-dithiolato-1,3-dithiole-2-thione) copper, *Cryst. Res. Technol.*, **42**, pp. 349–335.

199. Guo, W. F., Sun, X. B., Sun, J., Yu, W. T., Wang, X. Q., Zhang, G. H. and Xu, D. (2007). Synthesis, crystal structure and third order nonlinear optical properties of bis(tetra-n-propylammonium) bis(2-thioxo-1,3-dithiole-4,5-dithiolato)cuprate(II), *Cryst. Res. Technol.*, **42**, pp. 522–528.

200. Wang, X.-Q., Yu, W.-T., Xu, D., Wang, Y.-L., Li, T.-B., Zhang, G.-H., Sun, X.-B. and Ren, Q. (2005). Bis(tetraethylammonium) bis(2-thioxo-1,3-dithiole-4,5-dithiolato)cuprate(II), *Acta Cryst. E*, **61**, pp. m717–m719.

201. Li, T., Hu, Y., Ma, C., He, G., Zhao, R. and Li, J. (2011). Crystal structure and third-order nonlinear optical property study of a copper complex constructed by DMIT ligand, *Mater. Chem. Phys.*, **130**, pp. 835–838.

202. Teschmit, G., Strauch, P., Barthel, A., Reinhold, J. and Kirmse, R. (1999). EPR-Einkristall-Untersuchungen an $(n\text{-}Bu_4N)_2[Cu(dmit)_2]$ im antiferromagnetisch gekoppelten Wirtsgitter $(n\text{-}Bu_4N)_2$[(dmit)Cu(tto)Cu(dmit)]: ein Beitrag zur Aufklärung der sogenannten „paramagnetischen Verunreinigung“ im Wirtsgitter / a single crystal EPR investigation on $(n\text{-}Bu_4N)_2[Cu(dmit)_2]$ in the antiferromagnetically coupled host lattice $(n\text{-}Bu_4N)_2$[(dmit)Cu(tto)Cu(dmit)]: a contribution to the nature of the so-called "paramagnetic impurities", *Z. Naturforsch. B*, **54**, pp. 832–838.

203. Wang, Y.-L., Xu, D., Yu, W.-T., Sun, X.-B., Ren, Q., Zhang, G.-H., Ding, Y.-Q., Wang, X.-Q., Guo, W.-F., Chen, H.-Y. and Feng, L. (2008). Synthesis, crystal structure and saturable absorption in the near-IR regions of a new copper complex of dmit: hexadecyltrimethylammonium *bis* (2-thioxo-1,3-dithiole-4,5-dithiolato)-copper, *J. Coord. Chem.*, **61**, pp. 768–775.

204. Xiao, Y., Yao, Z. and Jin, D. (1993). Mixed conducting Langmuir-Blodgett films of a novel charge-transfer complex TBT-TTF-Cu(dmit)$_2$, *J. Phys. Chem.*, **97**, pp. 8519–8523.

205. Hoffmann, S. K., Goslar, J., Lijewski, S. and Zalewska, A. (2013). EPR and ESE of CuS_4 complex in Cu(dmit)$_2$: *g*-factor and hyperfine splitting correlation in tetrahedral Cu-sulfur complexes, *J. Mag. Res.*, **236**, pp. 7–14.

206. Hoffmann, S. K., Goslar, J., Lijewski, S., Tadyszak, K., Zalewska, A., Jankowska, A., Florczak, P. and Kowalak, S. (2014). EPR and UV-vis study on solutions of Cu(II) *dmit* complexes and the complexes entrapped in zeolite A and ZIF-Cu(IM)$_2$, *Microporous Mesoporous Mater.*, **186**, pp. 57–64.

207. Noma, H., Ohara, K. and Naito, T. (2016). Direct control of spin distribution and anisotropy in Cu-dithiolene complex anions by light, *Inorganics*, **4**, pp. 7 (21 pages).

208. Gaussian 09, Revision C.01, Frisch, M. J., Trucks, G. W., Schlegel, H. B., Scuseria, G. E., Robb, M. A., Cheeseman, J. R., Scalmani, G., Barone, V., Mennucci, B., Petersson, G. A., Nakatsuji, H., Caricato, M., Li, X., Hratchian, H. P., Izmaylov, A. F., Bloino, J., Zheng, G., Sonnenberg, J. L., Hada, M., Ehara, M., Toyota, K., Fukuda, R., Hasegawa, J., Ishida, M., Nakajima, T., Honda, Y., Kitao, O., Nakai, H., Vreven, T., Montgomery, J. A., Jr., Peralta, J. E., Ogliaro, F., Bearpark, M., Heyd, J. J., Brothers, E., Kudin, K. N., Staroverov, V. N., Kobayashi, R., Normand, J., Raghavachari, K., Rendell, A., Burant, J. C., Iyengar, S. S., Tomasi, J., Cossi, M., Rega, N., Millam, J. M., Klene, M., Knox, J. E., Cross, J. B., Bakken, V., Adamo, C., Jaramillo, J., Gomperts, R., Stratmann, R. E., Yazyev, O., Austin, A. J., Cammi, R., Pomelli, C., Ochterski, J. W., Martin, R. L., Morokuma, K., Zakrzewski, V. G., Voth, P. Salvador, G. A., Dannenberg, J. J., Dapprich, S., Daniels, A. D., Farkas, O., Foresman, J. B., Ortiz, J. V., Cioslowski, J. and Fox, D. J. Gaussian, Inc., Wallingford CT, 2009.

209. Noma, H., Ohara, K. and Naito, T. (2014). $[Cu(dmit)_2]^{2-}$ building block for molecular conductors and magnets with photocontrollable spin distribution, *Chem. Lett.*, **43**, pp. 1230–1232.

Chapter 2

Diversity in the Electronic Phase due to Interchange of MO Levels in $[M(dmit)_2]$ Anion Salts (M = Pd and Pt)

Takashi Yamamoto
Department of Chemistry and Biology, Graduate School of Science and Engineering, Ehime University, 2-5 Bunkyo-cho, Matsuyama, Ehime 790-8577, Japan
yamataka@ehime-u.ac.jp

Intermolecular interaction in the two-dimensional and triangular lattice of the molecular conductor $X[M(dmit)_2]_2$ (dmit = 1,3-dithiole-2-thione-4,5-dithiolate; X = monovalent cation; and M = Pd and Pt) is significantly different from that of the molecular conductors consisting of TTF derivatives (TTF = tetrathiafulvalene) because of the interchange in the energy levels of molecular orbitals near Fermi energy. Electrons in the triangular lattice exhibit attractive interaction due to the cooperation between the valence bond formation and intersite coulomb repulsion. This attractive interaction produces a rich variety of electron configurations, resulting in diversity in the ground states, including spin frustration and charge ordered state. Various electronic configurations and intermolecular interactions

Functional Materials: Advances and Applications in Energy Storage and Conversion
Edited by Toshio Naito

ISBN 978-981-4800-09-9 (Hardcover), 978-0-429-46813-1 (eBook)
www.panstanford.com

are not always detected by diffraction methods. Instead they can be analyzed by the vibrational spectroscopy focused on the C=C stretching modes.

2.1 Introduction

Molecular crystals consisting of metal-dithiolene complexes have been attracting attention from the viewpoint of the physical property characteristic of the hybrid orbitals composed of π orbitals at ligands and d orbitals at a central metal [1–6]. $X[M(dmit)_2]_2$ salts [dmit = 1,3-dithiole-2-thione-4,5-dithiolate; X = monovalent cation; and M = Pd and Pt], shown in Fig. 2.1, belong to the metal-dithiolene complexes. When the energy levels of the ligand orbitals and the d orbitals near the Fermi level are close to each other, molecular orbitals (MOs) are modified by a subtle change in the intermolecular distance and an external field [7–9]. The single-component metals belong to the metal-dithiolene complexes, whose ligands are extended [10, 11]. The metallic behavior of the single-component metals is ascribed to the partially occupied bands due to the proximity of the energy bands derived from the highest occupied molecular orbital (HOMO) and the lowest unoccupied molecular orbital (LUMO) [10–13]. A similar conducting mechanism has been proposed for the TTP derivatives (TTP = tetrathiapentalene) [14–17]. In the molecular crystals consisting of metal-dithiolene complexes, the molecular symmetry and/or the intermolecular distance can be changed by the substitution of the central atoms or the ligands, which leads to modification in the MOs and their energy levels. Therefore, metal-dithiolene complexes are the candidates of the multifunctional material whose physical properties are affected by the interchange of the energy levels or the degeneracy of the MOs. In this sense, it is of considerable importance to understand the roles of the modification of the MOs and the energy levels in the physical properties of the metal-dithiolene complexes. It is widely accepted that the MO levels near the Fermi level in the $X[M(dmit)_2]_2$ salts are interchanged owing to the tight dimerization shown in Fig. 2.1 [7–9, 18]. HOMOs and LUMOs of monomers constitute the LUMO and HOMO of a tight dimer (TDM), respectively. The interchange in the MO levels had been reported more than 25 years ago [7, 9, 18]. Nevertheless, the effect of the interchange on the physical properties has not been examined in detail so far, which is introduced in this review based on our recent experimental results [19–23].

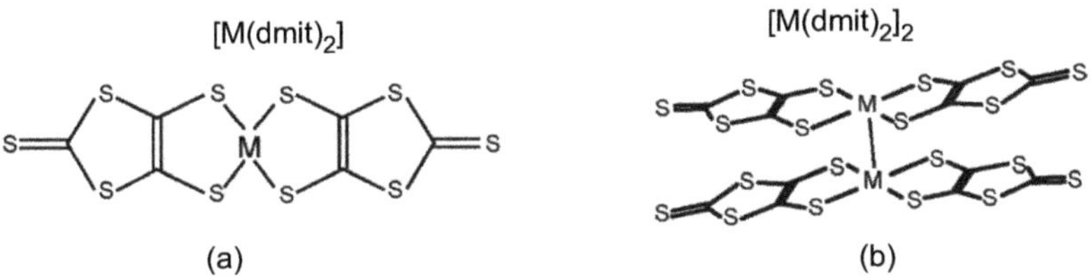

Figure 2.1 (a) Monomer $[M(dmit)_2]$ and (b) tight dimer $[M(dmit)_2]_2$ (M = Pt and Pd).

Table 2.1 shows the abbreviations of the $X[M(dmit)_2]_2$ salts introduced in this review, the three-dimensional (3D) crystal structure, the structure of the two-dimensional (2D) layer and the ground state, and/or the temperature of the phase transition. The details of the 3D and 2D structures are described in the next section. For all $X[M(dmit)_2]_2$ salts shown in Table 2.1, the 2D layer is the triangular lattice consisting of a TDM of $[M(dmit)_2]_2$. From the chemical composition, one electron is formally accommodated in the TDM. In this sense, $X[M(dmit)_2]_2$ salts are formally regarded as half-filled Mott insulators. Indeed, the ground states of several $X[M(dmit)_2]_2$ salts under ambient pressure belong to antiferromagnetic (AF) states and superconducting (SC) transition is observed under hydrostatic or uniaxial pressures [24, 25]. This result might support that the physical properties of $X[M(dmit)_2]_2$ salts can be described from the viewpoint of Mott insulators. More strictly, however, it is not always evident that the valence electron is resonated between two neighboring molecules in a TDM. Under the assumption that intradimer resonance is restricted by intermolecular interaction, the $X[M(dmit)_2]_2$ salts cannot always be regarded as Mott insulators but regarded as a quarter-filled system where two molecules in a dimer exhibit natures different from each other [26, 27]. Indeed, the observation of several kinds of charge ordered (CO) state and valence bond formation indicates that a picture of the quarter-filled system is non-negligible

[19–23, 28–35]. In the same manner as the CO state of the β''-type bis(ethylenedithio)tetrathiafulvalene (BEDT-TTF) salts, the neutral-like molecule and the ionic molecule are arranged in the 2D layer of the CO state [36, 37]. However, the mechanism of the CO state of the X[M(dmit)$_2$]$_2$ salts is significantly different from that of the BEDT-TTF salts: the intermolecular coulomb repulsion cooperates with the valence bond formation in the same direction of X[M(dmit)$_2$]$_2$ salts, whereas this cooperative mechanism is not applied to the CO state of the BEDT-TTF salts. This cooperative mechanism, which is described in the next sections, is characteristic of the interchange in MO levels. Some X[M(dmit)$_2$]$_2$ salts exhibit the SC state and metallic states under hydrostatic pressure although the ground states under ambient pressure are nonmagnetic (NM) insulators [38–40]. In addition, some X[M(dmit)$_2$]$_2$ salts exhibit behavior suggesting quantum spin liquid (SL) even below the liquid helium temperature [41–44]. It should be noted that diversity of the ground states is realized by subtle differences in the triangular lattice [3, 45, 46]. Therefore, X[M(dmit)$_2$]$_2$ salts are among the best model compounds to examine the relationship between the ground states and the intermolecular interactions.

Table 2.1 Summary of the crystal structure and ground state of X[M(dmit)$_2$]$_2$

Material	Abbreviation	3D structure	2D layer	Ground state
β'-Me$_4$P [Pd(dmit)$_2$]$_2$	b-04P	Solid crossing [47]	Isosceles triangular	T_{AF} = 42 [42]
β'-Et$_2$Me$_2$P [Pd(dmit)$_2$]$_2$	b-22P	Solid crossing [48]	Isosceles triangular	T_{AF} = 17 [42]
β'-Me$_4$Sb [Pd(dmit)$_2$]$_2$	b-04Sb	Solid crossing [46]	Isosceles triangular	T_{AF} = 16 [42]
β'-EtMe$_3$Sb [Pd(dmit)$_2$]$_2$	b-13Sb	Solid crossing [41]	Equilateral triangular	SL [41–44]
monoclinic-EtMe$_3$P [Pd(dmit)$_2$]$_2$	m-13P	Nonsolid crossing [39]	Equilateral triangular	T_{CO} = 20 [20, 38]

Material	Abbreviation	3D structure	2D layer	Ground state
triclinic-$EtMe_3P$ $[Pd(dmit)_2]_2$	t-13P	Nonsolid crossing [21]	Isosceles triangular	T_{CO} = 50 [21]
β′-Cs $[Pd(dmit)_2]_2$	b-Cs	Solid crossing [9]	Scalene triangular	T_{CO} = 56 [28]
β′-Et_2Me_2Sb $[Pd(dmit)_2]_2$	b-22Sb	Solid crossing [28]	Equilateral triangular	T_{CO} = 70 [28, 35]
β′-Me_4P $[Pt(dmit)_2]_2$	—	Solid crossing [49]	Isosceles triangular	T_{CO} = 218 [49]

Note: AF, SL, and CO denote the antiferromagnetic, spin liquid, and charge ordered states. The 3D structure denotes the relationship of the 2D layers separated by a countercation layer, and the 2D layer is designated as the arrangement of the tight dimer estimated by the calculated interdimer transfer integrals [23]. Details of the 3D structure and the 2D layer are introduced in Sections 2.2 and 2.3, respectively. The fifth column denotes the ground state or the temperature of the ground-state phase transition (kelvin).

This review is organized as follows: The crystal structures are introduced in Section 2.2. The details of the 2D structure from the viewpoint of the triangular lattice are described in Section 2.3. The diversity in the CO state due to the interchange of the MO levels is introduced in Section 2.4. The electronic transition observed in the reflectance spectra is discussed in Section 2.5. The methodology for analyzing the electronic state by use of the vibrational spectroscopy is described in Section 2.6. The vibrational spectra and the electronic state of several $X[M(dmit)_2]_2$ salts are introduced in Section 2.7. The experimentally obtained parameters classifying the ground states are proposed in Section 2.8.

2.2 Crystal Structure

Similarly to the BEDT-TTF salts and the Bechgaard salts, $X[M(dmit)_2]_2$ salts introduced in this review exhibit a segregated stacking structure [3]. The 2D layer consisting of $[M(dmit)_2]$ and the 2D layer composed of the counterion X are alternately arranged. Figure 2.2 shows the 2D layer of β'-$Et_2Me_2Sb[Pd(dmit)_2]_2$ (abbreviated to b-22Sb) at room temperature [28]. The intermolecular overlap due to the metal-ligand orbital in $[Pd(dmit)_2]_2$ contributes to the conducting pathways. In the same way as the κ-type BEDT-TTF salts, a closed Fermi surface is obtained due to sulfur-sulfur contacts between neighboring dimers under the assumption that the intermolecular coulomb repulsions are negligible [48, 50, 51]. A pair of $[M(dmit)_2]$ molecules (M = Pd and Pt) forms a TDM whose degree of dimerization is enhanced as compared with that of the κ-type BEDT-TTF salts [3, 52]. The tight dimerization is ensured from the intradimer transfer integral estimated from the optical conductivity spectra obtained from the IR reflectance spectra [35]. The estimated transfer integral is ca. 700 meV for M = Pd, which is 10 times as large as those of the interdimer transfer integrals [35]. In addition, this intradimer transfer integral is significantly larger than that of the κ-type BEDT-TTF salts, 200–300 meV [52]. The interdimer transfer integrals in the stacking (S), transverse, and diagonal directions, t_B, t_S, and t_r, are almost comparable (Fig. 2.2) [46]. As a result, a triangular lattice is formed in the 2D layers. According to the results of the structural analyses at room temperature, the 2D layers of b-22Sb, along with those of *monoclinic*-$EtMe_3P[Pd(dmit)_2]_2$ (abbreviated to m-13P), are the closest to the equilateral triangular lattice [28, 39, 46]. The 3D structures of the $X[M(dmit)_2]_2$ salts are classified into two groups, solid-crossing structure and non-solid-crossing structure [4], shown in Figs. 2.2 and 2.3, respectively. In the solid-crossing structure, the stacking directions (S directions) between the 2D layers separated by the countercation layer are related with the glide plane. The diagonal directions are also related with the glide plane. On the other hand, the transverse directions are identical to each other. In the non-solid-crossing structure, the S, diagonal, and transverse directions in each 2D layer are identical to each other. As shown in Table 2.1, m-13P and t-13P belong to the non-solid-crossing structure [21, 39]. The ground states are not AF states but NM insulators [21, 38].

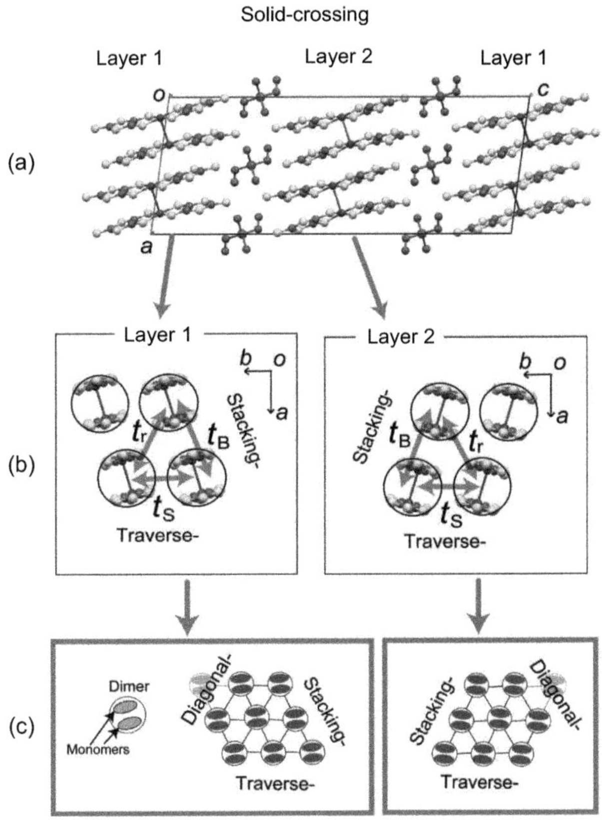

Figure 2.2 (a) Crystal structure of β'-$Et_2Me_2Sb[Pd(dmit)_2]_2$ (abbreviated to b-22Sb) at room temperature viewed along the *b* axis. (b) End-on projection of the 2D layers separated by the layer of the countercation. (c) Schematic views of molecular arrangements in Layers 1 and 2.

In general, the substance names of the solid-crossing structures begin with "β'-" and those of the non-solid-crossing structures begin with the crystal system. Hereafter, we use the following abbreviations expressing the substance names: β'-$Cs[Pd(dmit)_2]_2$ is denoted as b-Cs and β'-$Et_nMe_mY[Pd(dmit)_2]_2$ is denoted as b-*nm*Y, where *n* and *m* are the number of ethyl- and methyl- groups bounded to Y (Y = N, P, As, and Sb). In this definition, β'-$Et_2Me_2Sb[Pd(dmit)_2]_2$ is abbreviated to b-22Sb. In the same manner, *monoclinic*-$EtMe_3P[Pd(dmit)_2]_2$ and *triclinic*-$EtMe_3P[Pd(dmit)_2]_2$ are denoted as m-13P and t-13P, respectively.

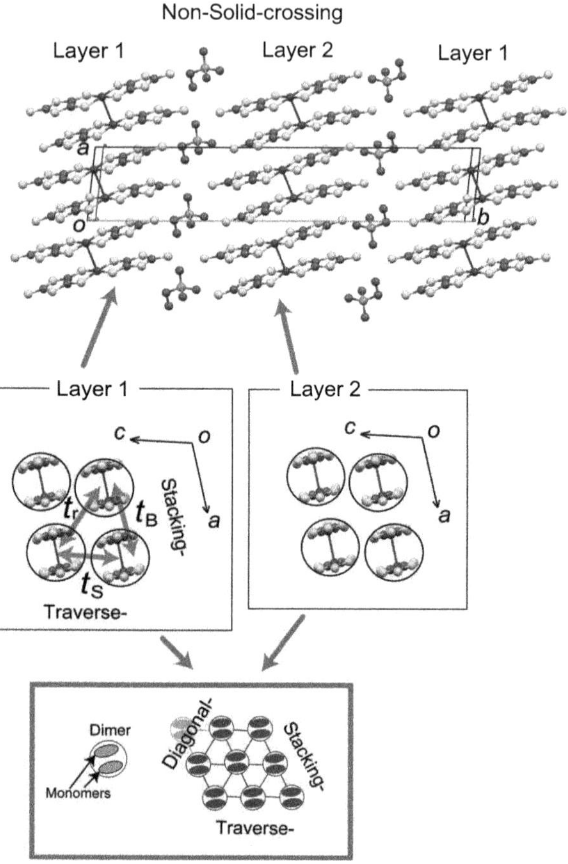

Figure 2.3 (a) Crystal structure of *triclinic*–$EtMe_3P[Pd(dmit)_2]_2$ (abbreviated to t-13P) viewed along the *c* axis. (b) End-on projection of the 2D layers separated by the layer of the countercation. (c) Schematic views of the molecular arrangement in a 2D layer.

In both 3D structures, the overlapping modes in the S, diagonal, and transverse directions in a triangle are not related with any symmetric operation. In this sense, t_B, t_r, and t_S shown in Figs. 2.2 and 2.3 are independent of one another. In general, the length of the side in the triangular lattice expresses the magnitude of the triangular lattice: the length decreases on increasing the transfer integral. The 2D layer of the $X[Pd(dmit)_2]_2$ salts is schematically comprised of a scalene triangle, which is different from the 2D structure of the κ-type BEDT-TTF salts [4, 52, 53]. For most of the κ-type BEDT-TTF salts, the 2D layer is comprised of an isosceles triangle whose legs are shorter than that of the base, $t_{leg} \geq t_{base}$ [52, 54]. By neglecting t_{base}, the 2D layer becomes a square lattice. $t_{leg} \leq t_{base}$ can be realized by substituting a counterion or by applying uniaxial pressure. Indeed, κ-type BEDT-TTF salts exhibiting $t_{leg} \leq t_{base}$ were reported [54]. Nevertheless, it is difficult to induce an asymmetric dimer due to valence bond formation as long as the 2D layer remains an isosceles triangular lattice. On the other hand, the 2D layer of the $X[Pd(dmit)_2]_2$ salts can become equilateral, isosceles, and scalene triangular lattices depending on the countercations [21, 28, 39, 45, 46]. This property enables us to examine the physical properties not only in square and triangular lattices but also in an anisotropic lattice exhibiting a lattice deformation. Therefore, the 2D layer of κ-type BEDT-TTF salts is specific. The absence of the restriction in the 2D structure of $X[M(dmit)_2]_2$ salts suggests diversity in the ground states.

The CO state is observed in several compounds belonging not only to the solid-crossing structure but also to the non-solid-crossing structure [20, 21, 28]. The pressure-induced SC state is also observed in compounds belonging to each structure [3, 45]. The crystal structures of β'-$Me_4P[Pd(dmit)_2]_2$ (abbreviated to b-04P), β'-$Et_2Me_2P[Pd(dmit)_2]_2$ (abbreviated to b-22P), and β'-$Me_4Sb[Pd(dmit)_2]_2$ (abbreviated to b-04Sb), those that exhibit the AF ground state, belong to the solid-crossing structure. The transfer integrals exhibiting the AF state show the relationship of $t_B \approx t_S > t_r$, which indicates that the scalene triangle is close to the isosceles triangle [46]. This result indicates that transfer integrals in the S and transverse directions are more or less averaged in the solid-crossing structure. Interestingly, the relationship of $t_B \approx t_S \approx t_r$ does not necessary hold for β'-$EtMe_3Sb[Pd(dmit)_2]_2$ (abbreviated to b-13Sb) although the ground state is SL [46, 55]. The deviation from the equilateral triangular lattice indicates that the SL state of b-13Sb might be ascribed to a mechanism other than geometrical frustration.

The AF transition temperatures are roughly correlated with the magnitude of interdimer transfer integrals [46]. With increasing t_B and t_S as well as decreasing t_r, the transition temperature increases [46]. Furthermore, the magnitude of the transfer integral of the β'-type $[Pd(dmit)_2]_2$ salts is more or less correlated with cation size.

Among all β'-X[Pd(dmit)$_2$]$_2$ salts, the cation sizes of b-22Sb and b-13Sb are, respectively, the largest and the second largest, which means that t_B and t_S are small. These ground states are not AF states but CO and SL states. The cation sizes of the β'-X[Pd(dmit)$_2$]$_2$ salts exhibiting AF transition are smaller than those of b-22Sb and b-13Sb. Therefore, the cation sizes have a non-negligible relationship with the ground states, which suggests that the intermolecular interaction is correlated with the cation size: with decreasing cation size, the 2D layer changes from an equilateral triangular lattice to a square lattice composed of isosceles triangles.

The SC transition is also relevant to the interdimer interactions. According to the temperature dependence of the electrical resistivity, β'-Me$_4$As[Pd(dmit)$_2$]$_2$ (b-04As), b-22P, and b-04Sb exhibit SC transitions under uniaxial pressure applied in the transverse direction, that is, the direction of t_S, whereas nonmetallic behavior is robust under the *a* axis pressure [3, 4]. As shown in Fig. 2.2, the *a* axis corresponds to the averaged direction between S and diagonal directions. Interestingly, with increasing cation sizes, critical pressures decrease and transition temperatures increase [3, 4]. Because 2D layers at ambient pressure are regarded as pseudosquare lattices, the uniaxial-pressure and cation-size dependencies indicate that the SC transitions are enhanced when the 2D layer becomes a scalene triangular lattice rather than an isosceles or equilateral triangular lattice.

On the contrary, no remarkable relationship is obtained between the CO transition temperatures and t_B, t_S, and t_r. Particularly, the 2D layers of both b-22Sb and m-13P are closest to the equilateral triangular lattice whereas the transition temperature of b-22Sb, 70 K, is significantly higher than that of m-13P, 20K [28, 39, 46]. This result means that the ground state is not always governed by the degree of deviation from the equilateral triangular lattice.

2.3 Degree of Deviation from the Equilateral Triangular Lattice

The 2D layers of the β'-X[Pd(dmit)$_2$]$_2$ salts have attracted attention because the AF transition temperatures increase when they approach the square lattice and because SL behavior is observed when they approach the equilateral triangular lattice [46]. However, this trend is not exactly the case with all of the X[Pd(dmit)$_2$]$_2$ salts shown in Table 2.1. In this section, the degree of deviation from the equilateral triangular lattice is introduced. The author also introduces the puzzling phenomena, that the X[Pd(dmit)$_2$]$_2$ salts close to the equilateral triangular lattice do not exhibit the SL state and the 2D layer of the SL state non-negligibly deviates from the equilateral triangular lattice.

The degree of deviation from the equilateral triangular lattice has been evaluated on the basis of different assumptions and definitions proposed by different groups. One of the simplest definitions is $\Delta = 2t_r/(t_S + t_B)$ [3, 46]. Because there is no experimental method to obtain the values of t_B, t_r and t_S, the interdimer transfer integrals are estimated by the extended Hückel calculation on the basis of the result of the X-ray structural analysis [3, 46]. In this method, the degree of deviation depends on the quality of the X-ray structural analysis. By comparing the previous and recent results, the molecular deformations along the short axis of the molecule are taken into consideration in the recent result [3, 46]. The difference in the structural analysis means that the t_r and t_S are underestimated in the previous data. The values of Δ without parenthesis in Table 2.2 are obtained from the transfer integrals, where molecular deformations are taken into consideration. The values with parentheses are obtained from the transfer integrals without the effect of molecular deformation described above [21, 28, 39, 46]. For the equilateral triangular lattice, Δ becomes 1. On deviating from the equilateral triangular lattice, Δ is larger or smaller than 1. However, using this definition, the scalene triangular lattice cannot be distinguished from the square lattice. Furthermore, evaluation by this definition is based on the conjecture that the natures of the interdimer interactions in two of the three directions are identical. Nevertheless, Δ reveals some non-negligible and systematic trends, which are described in the following paragraphs.

As far as the β'-X[Pd(dmit)$_2$]$_2$ salts exhibiting AF ground states are concerned, a systematic trend is observed: with decreasing AF transition temperatures, the value of Δ increases and approaches unity [46]. Furthermore, Δ of b-13Sb is closer to unity than of those exhibiting AF ground states. One may speculate that the SL state is induced by purely geometrical frustration because this trend indicates that the AF transition temperature is reduced to zero when approaching the equilateral triangular lattice. On the other hand, Δ of b-22Sb is the closest to unity whereas the ground state is a CO state [28, 46]. A similar puzzling phenomenon is also observed in the X[Pd(dmit)$_2$]$_2$ salts belonging to the non-solid-crossing structure [39]. Because the 2D layers of t-13P and m-13P are also classified into the triangular lattice, the values of $\Delta = 2t_r / (t_S + t_B)$ were also obtained from the calculated transfer integrals. Interestingly, Δ of m-13P is very close to that of the equilateral triangular lattice; however, the ground state is a CO state [39]. In addition, Δ of b-Cs predicts that the ground state belongs to the AF state but the observed ground state is a CO state [28]. The inconsistency between Δ and the observed ground state suggests the following scenarios: (i) the definition is insufficient for evaluation of the 2D layer, (ii) the calculated transfer integral is insufficient, and (iii) any additional mechanism other than the geometrical frustration is applied to the triangular lattice of the X[Pd(dmit)$_2$]$_2$ salts. The remainder of this section discusses scenarios (i) and (ii).

Table 2.2 Degree of deviation from the equilateral triangular lattice

Abbreviation of the material	$\Delta = 2t_r/(t_S + t_B)$ obtained from the extended Hückel calculation
b-04P	0.62
b-22P	0.84
b-04Sb	0.86
b-13Sb	0.91
m-13P	(1.05)
t-13P	(0.29)
b-Cs	(0.74)
b-22Sb	1.01

Let us examine scenario (i) under the assumption that the calculated transfer integrals are reflected on the genuine crystal structure. The previous definition assumed that the difference between t_S and t_B is not taken into consideration. The difference among t_r, t_S, and t_B can be evaluated by use of the ternary figure shown in Fig. 2.4 [23]. The equilateral triangular lattice is located in the center of the ternary figure. Three kinds of one-dimensional (1D) lattices, denoted as S, R, and B, are distinguished from each other, and three kinds of square lattices, denoted as SQ1, SQ2, and SQ3, are also distinguished from each other. The pseudosquare lattice can be distinguished from the anisotropic lattice, whereas those are not necessarily distinguished from each other by the use of Δ. Figure 2.4b shows the enlarged view around the center of the ternary figure. Concerning the β′-X[Pd(dmit)$_2$]$_2$ salts exhibiting AF ground states, the plotted points are located near the line bounded between the equilateral triangular lattice and SQ1. This result indicates that the AF transition temperature is more or less correlated with the degree of deviation and Δ is a good parameter for the β′-X[Pd(dmit)$_2$]$_2$ salts exhibiting AF ground states. On the other hand, Δ values of b-04P and b-22P were intermediate, although whereas the ground state is the CO state. From the viewpoint of Fig. 2.4b, b-Cs can be regarded as a slightly anisotropic lattice rather than a pseudosquare lattice. This fact is in agreement with the robustness of the lattice instability above the CO transition temperature. However, the 2D layers of b-22Sb and m-13P are closer to the equilateral triangular lattice. Furthermore, t-13P is the closest to the square lattice although the CO state is a NM insulator. These inconsistencies support scenarios (ii) and/or (iii) rather than scenario (i). t_r, t_S, and t_B of b-22Sb are obtained by the old and new methods. Nevertheless, there is no remarkable difference between the positions of b-22Sb in Fig. 2.4b, which indicates that scenario (iii) is nonnegligible.

Let us examine scenario (ii) from the experimental viewpoint. The transfer integral can be experimentally estimated from the plasma frequency obtained from the reflectance spectra [55]. For the metallic materials, finite reflectivity is observed from zero frequency to the midinfrared, near infrared, or visible region. For most of the conducting molecular crystals, the reflectivity is decreased around the midinfrared or near-infrared (NIR) region. Concerning the X[Pd(dmit)$_2$]$_2$ salts discussed in this article, electronic transitions due to interdimer and intradimer charge transfers (CTs) are observed in the midinfrared and NIR region, respectively. The reflectivity around the boundary between two regions is decreased to a few percentage points, which means that the interdimer and intradimer CT transitions do not overlap. The absence of overlapping is ascribed to the tight dimerization. Therefore, the intradimer and interdimer interactions can be independently analyzed from the reflectance spectra. The frequency where reflectivity is decreased down to a few percent near the boundary between the midinfrared and NIR regions is defined as the plasma edge. Because the principal axis of the solid-crossing structure corresponds to the *a* and *b* axes in Fig. 2.2, the transfer integrals are analyzed from the *a*- and *b*-polarized reflectance spectra. The polarized reflectance spectra are transformed into the conductivity spectra by use of the Kramers–Kronig analysis. The plasma frequencies of the *a* and *b* axes are obtained from the integration of the polarized conductivity spectra. The region of the integration begins with zero and finishes at the plasma edge. The plasma frequencies of non-solid-crossing structure are obtained from the reflectance spectra polarized in the S direction and the perpendicular direction.

The principal axes in the 2D layer lie in the S and perpendicular directions. Concerning the solid-crossing structure, however, these directions do not correspond to the *a* and *b* directions. To evaluate the degree of deviation from the equilateral triangular lattice in the solid-crossing structure, the plasma frequencies require to be transformed into those of the S and perpendicular directions, ω_S and ω_P, by use of the angle between the *a* axis and the S directions [55]. Concerning the non-solid-crossing structure, on the other hand, there is no need to transform the direction of the plasma frequencies [55]. For m-13P, ω_S and ω_P are directly obtained from the integration of the polarized conductivity spectra. In both solid-crossing and non-solid-crossing structures, t_B contributes to ω_S solely but t_S and t_r contribute to both ω_S and ω_P because of the triangular lattice. However, to estimate the degree of deviation, it is required to obtain the interdimer distances and interior angles at a low temperature. Therefore, quantitative analysis using plasma frequency is a task for the future. Nevertheless, the polarization dependencies in the ground states give us helpful hints for interpreting the interdimer interactions. The conductivity spectra of t-13P exhibit an anisotropic polarization dependence compared with m-13P [19, 20]. The polarization dependence in the CO state of b-Cs is significantly different from that above the transition temperature [9].

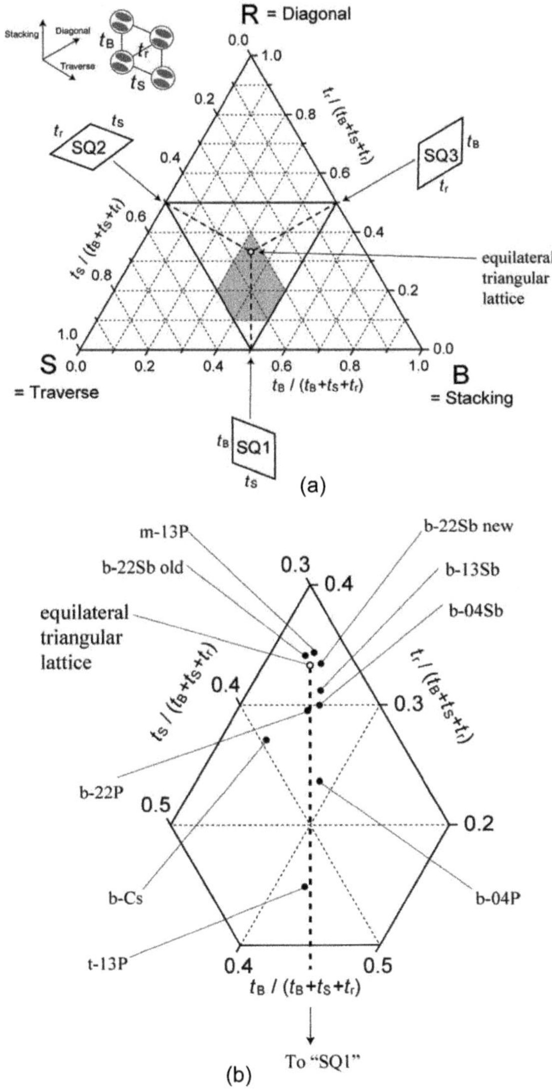

Figure 2.4 (a) Ternary figure examining the degree of deviation from the equilateral triangular lattice. (b) Enlarged view of the gray area of (a). Transfer integrals of b-22Sb in Table 2.2 correspond to "b-22Sb new." The transfer integrals in the old literature correspond to "b-22Sb old." Open circles in both (a) and (b) denote the equilateral triangular lattice. The bold broken lines in both (a) and (b) correspond to the isosceles triangular lattice.

2.4 Charge Separation and Self-Organization due to the Interchange of MO Levels

In this section, we introduce the interchange of the MO levels because this phenomenon plays a crucial role in the intermolecular interaction. The monomer of [M(dmit)$_2$] consists of the same ligands and central metal. Owing to the coordinate bonds between two ligands and central atoms, the energy levels between the HOMO and the LUMO are close to each other. Proximity in the energy bands leads to the multiband system near the Fermi level, where the bands due to HOMO and LUMO overlap in the reciprocal space [3]. It is expected that the multiband system is advantageous to high conductivity. Therefore, a series of dithiolene-complexes, including single-component metals, has attracted attention from the viewpoint of metallic conduction.

On the other hand, the proximity of the energy level does not necessarily lead to a metallic behavior when the MO is reconstructed, due to self-organization such as dimerization and tetramerization. When MOs near the Fermi level are reconstructed, the electronic state is also reconstructed. Figure 2.5 shows the energy levels of a monomer, a loosely bound dimer, and a tightly bound dimer. Hereafter, loosely bound dimers and tightly bound dimers are denoted as loose dimers (LDMs) and tight dimers (TDMs), respectively. Concerning a LDM, HOMOs of two monomers form the bonding and antibonding orbitals, denoted as LDM1 and LDM2, respectively, where the number appended to "LDM" indicates the energy level. LUMOs also constitute LDM3 and LDM4. In this case, there is no interchange in the energy levels between a monomer and a LDM. With decreasing intermolecular distance within a dimer, the intradimer interaction is enhanced, which leads to a decrease in the levels of LDM1 and LDM3 and an increase in the levels of LDM2 and LDM4. When the degree of dimerization exceeds the critical point, LDM2 and LDM3 become TDM3 and TDM2, respectively. The HOMO of a neutral and tight dimer, that is, [tight dimer]^0, is composed of LUMOs of monomers and the LUMO of [tight dimer]^0 is composed of HOMOs of monomers. The interchanges of HOMO and LUMO between LDM and TDM are named as HOMO-LUMO (HL) inversion [7, 8]. HL inversion is supported from the optical conductivity spectra in the NIR region [9, 56].

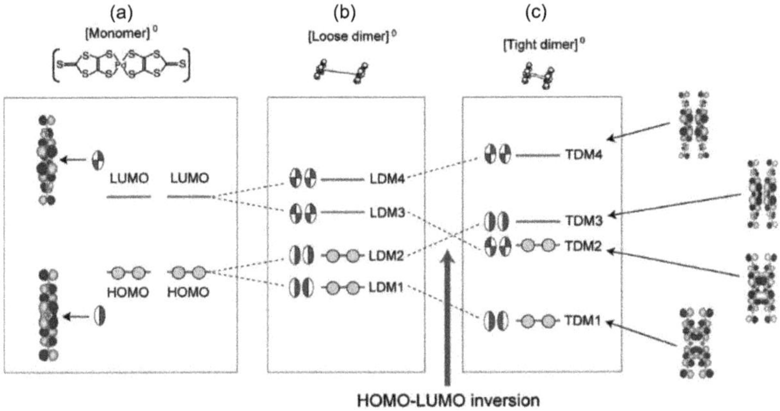

Figure 2.5 Energy diagrams of (a) a monomer, (b) a loose dimer, and (c) a tight dimer near the Fermi level. LDM1–LDM4 belong to the loose dimer. TDM1–TDM4 belong to the tight dimer. LDM2 and LDM3 become TDM3 and TDM2, respectively. Filled circles denote the electrons in the fully occupied orbital.

It is noteworthy that TDM2 is the bonding orbital whereas LDM2 is the antibonding orbital (Fig. 2.5). This property indicates that the neutral and tight dimer becomes more stable than the neutral and loose dimer when the degree of dimerization is enhanced. On the other hand, the radical anion of the TDM, that is, [tight dimer]$^-$, is not always stable because TDM3 is the antibonding orbital whereas LDM3 is the bonding orbital. Therefore, the energy level of [tight dimer]$^-$ can be modified by the intermolecular interactions, the CTs, the deformation of the dimer, etc. Figure 2.6 shows the modified MO levels when one electron is formally accommodated in a dimer. Figure 2.6c denotes the energy diagram without any modification, where dimers are uniformly arranged in the 2D layer. When any modification is applied, the energy diagrams become those shown in Fig. 2.6a, 2.6b, and 2.6d [20, 21, 23, 29, 30, 33–35].

In the loose and tight dimers mechanism shown in Fig. 2.6a, one half of the TDMs in the 2D layer become LDMs whereas the other half remain TDMs [23]. Because LDMr3 is bonding but TDMp3 is antibonding, two valence electrons are accommodated into LDMr3 whereas TDMp3 is unoccupied, where "r" and "p" denote "charge-rich" and "charge-poor," respectively. The intradimer distance of a LDM is larger than that of a TDM, which contributes to reducing the intermolecular coulomb repulsion within a dimer in LDMr2. In a

TDM, TDMp2 is an occupied orbital whereas TDMp3 is unoccupied. The bonding orbital of TDMp2 becomes stable because the short intradimer distance is enhanced. Among the occupied orbital, LDMr3 is the highest occupied orbital and LDMr2 and TDMp3 are the next- and third-highest occupied orbitals, respectively. As a result, the 2D layer is comprised of [loose dimer]$^{2-}$ and [tight dimer]^{0}. Therefore, redistribution of valence electrons is induced. Valence bond ordering (VBO) is formed by the highest occupied orbitals of both TDMs and LDMs, that is, TDMp2 and LDMr3, respectively. The intermolecular coulomb repulsion in a dimer also operates in the 2D layer, but the coulomb repulsion is reduced because the intradimer distance of [loose dimer]$^{2-}$ is larger than that of [tight dimer]^{0}. In this redistribution, the intermolecular coulomb repulsion (V) within a LDM cooperates with VBO.

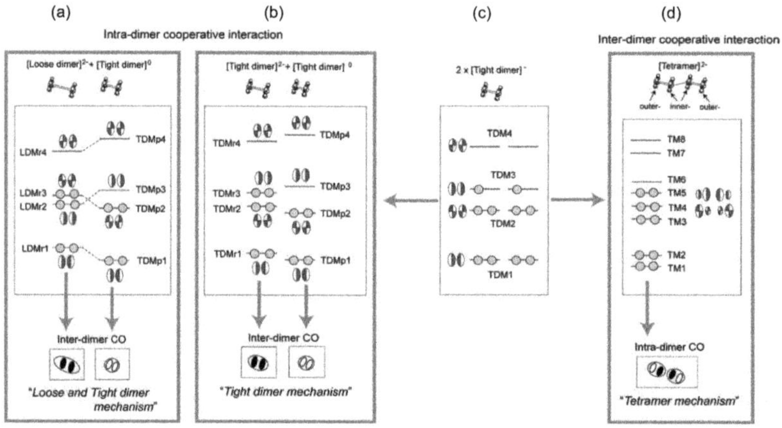

Figure 2.6 Energy diagrams of (a) loose and tight dimers mechanism, (b) tight dimer mechanism, (c) $[M(dmit)_2]_2^-$, and (d) tetramer (TM) mechanism near Fermi energy. For each case, the averaged and formal charges at a dimer are −1. Filled circles denote the electrons in the fully or singly occupied orbital. LDM, TDM, and TM denote "loose dimer," "tight dimer," and "tetramer." "r" and "p" appended to "LDM" and "TDM" denote "charge-rich" and "charge-poor." The numbering appended to LDMr, TDMp, TDMr, and TM corresponds to the energy level. Black and white sections in the bottom panels in (a) and (b) denote the charge-rich and charge-poor molecules, respectively, while those in (d) denote the electron densities of HOMO of $[TM]^{2-}$, that is, TM5.

The charge distribution due to the loose and tight dimers mechanism (Fig. 2.6a) belongs to the CO state. Because the highest occupied orbitals in both [loose dimer]$^{2-}$ and [tight dimer]^0 are the bonding orbitals, this CO state is one of the candidates of the ground states without any additional interdimer interaction except for the Madelung interactions. Under the assumption that the Madelung interaction between dimers and countercations is negligible, the dimers are alternately arranged in two of three directions: S, diagonal, and transverse. A stripe pattern is formed in the rest of the directions. The phase transition from [tight dimer]$^-$ to the CO state due to the loose and tight dimers mechanism is accompanied by the expansion of the charge-rich dimer. This CO transition might be allowed when the contraction of the charge-poor dimer exceeds the expansion of the charge-rich dimer. It should be noted that VBO and V in the different orbitals are cooperatively operating in this mechanism. On the other hand, in the CO state of BEDT-TTF salts VBO and V in the same HOMOs are cooperatively operating in the 2D layer [26, 27].

The tight dimer mechanism shown in Fig. 2.6b is firstly introduced, accounting for the CO transition of b-Cs and b-13Sb [29, 30, 34, 35]. As shown in Fig. 2.6b, the intradimer distances in one half of TDMs decrease whereas those of the other half increase. HL inversion is applied to all dimers. Both TDMr2 and TDMp2 are the bonding orbitals, and both TDMr3 and TDMp3 are the antibonding orbitals. Owing to the difference in the intradimer distances, the energy level of TDMp2 is lower than that of TDMr2 and the level of TDMp3 is higher than that of TDMr3. The two valence electrons are accommodated into TDMr3, whereas TDMp3 is unoccupied. As a result, the 2D layer is comprised of [tight dimer]$^{2-}$ and [tight dimer]^0, whose degrees of dimerization are different from each other. The intradimer interaction of the charge-poor dimer is strengthened because the highest occupied orbital in a charge-poor dimer, that is, TDMp2 of [tight dimer]^0, is the bonding orbital with the short intradimer distance. The intermolecular coulomb repulsion within a charge-rich dimer, that is [tight dimer]$^{2-}$, is reduced because of the long intradimer distance. Therefore, the VBO and the intersite coulomb repulsion (V) are cooperatively operating in the tight dimer mechanism shown in Fig. 2.6b. Similar to Fig. 2.6a, this cooperative interaction is also characteristic of the HL inversion system.

Although the highest occupied orbital at a charge-poor dimer, that is, TDMp2 of [tight dimer]^0, shows a bonding interaction, the

HOMO at the charge-rich dimer, that is, TDMr3 of [tight dimer]$^{2-}$, exhibits an antibonding interaction. This condition is significantly different from that of Fig. 2.6a, where both HOMOs in [loose dimer]$^{2-}$ and [tight dimer]^{0} show bonding interactions. Because of the instability due to the antibonding interaction in TDMr3, the CO state in the tight dimer mechanism is not always realized, which indicates that any additional interaction is required for realizing the CO state due to the tight dimer mechanism. One of the candidates of additional interactions is the Madelung interaction, including the anion-cation interactions. It is noteworthy that the CO states of b-Cs and b-13Sb do not show the checkerboard pattern where the charge-rich and charge-poor dimers are alternately arranged in two of three directions: S, diagonal, and transverse. The CO states of b-Cs and b-13Sb show the similar distributions; two charge-rich dimers and two charge-poor dimers are alternately arranged in the S direction [28]. This distribution is reproduced from the calculation of the Madelung energy [34]. On the other hand, vibrational spectroscopy reveals that both charge-rich and charge-poor dimers exhibit intradimer charge separations [19, 23]. This result suggests that not only the Madelung interaction but also some additional interaction is required for reproducing the charge distribution. We can surmise that any additional interaction might be relevant to the reduction of the instability in TDMr3 of [tight dimer]$^{2-}$ and to compensating the expansion of [tight dimer]$^{2-}$. The additional interaction is discussed in later paragraphs.

Prior to discussing the additional interaction described above, we discuss the CO state due to the interdimer interaction. The interdimer interaction plays a crucial role in the tetramer (TM) mechanism [20, 21]. Figure 2.6d shows the energy diagram of the TM composed of two TDMs. TDM3 in Fig. 2.6c constitutes the bonding and antibonding orbitals, that is, TM5 and TM6 in Fig. 2.6d, respectively. When one electron is formally accommodated in one dimer, two valence electrons occupy TM5 whereas TM6 is unoccupied. TM5 is the highest occupied orbital of [TM]$^{2-}$. Because TM5 is a bonding orbital between dimers, the electron densities at the inner molecules are higher than those of the outer molecules. TDM2 of two TDMs constitute TM4 and TM3. TM4 is the next highest occupied orbital of [TM]$^{2-}$. Because of the antibonding orbital at TM4, the electron densities at the outer molecules are higher than those at the inner molecules. A charge separation is induced by this mechanism. The intermolecular coulomb repulsion (V) between inner molecules, that is, V_{inter}, is operating at TM4. On the other hand, VBO is formed at TM5. An intermolecular coulomb repulsion between two molecules within a dimer, that is, V_{dimer}, is also operating because of no doubly occupied dimer. Therefore, VBO and V are cooperatively operating in the different orbitals due to the TM mechanism. This cooperative interaction is also characteristic of the HL inversion system. Because two dimers form a TM, the 2D layer shows the contraction below the transition temperature. In addition, the highest occupied orbital at $[TM]^{2-}$ is bonding. These properties indicate that no additional interaction is required for the CO transition in the TM mechanism, which is significantly different from the dimer mechanism.

In most of the molecular conductors exhibiting no HL inversion, the charge distribution due to the VBO competes with that due to the intermolecular coulomb repulsion (V). This competition is crucial for the quasi-1D molecular conductors because the direction of the VBO is exactly the same as that of V [57, 58]. This competition might be more or less reduced in the 2D layer when the direction of the largest V is normal to that of VBO [37, 59, 60]. The competitive and cooperative interactions in several quasi-1D and quasi-2D systems were theoretically reproduced on the basis of the extended Hubbard model, including the electron-phonon interaction [26, 27, 61]. From the experimental viewpoint, we also showed that V in the direction of VBO is non-negligible because the intermolecular distance in the direction of VBO is comparable to that along the largest V. Indeed, some β''-type ET salts exhibit the frustration of the different charge distributions required for the Vs along the S and perpendicular directions [36, 37]. Another β''-type ET salt exhibits the frustration of the different charge distributions required for the Vs along the same direction [37]. The frustration of charge distribution is also observed for the superconductors belonging to the β''-type ET salts [37, 62]. These phenomena are ascribed to the fact that both Vs and VBO originate from the same HOMOs. Concerning the HL inversion system, on the contrary, VBO and the charge separation due to V are not competing but cooperative because these interactions are enhanced in the different orbitals. It is suggested that the competition in the quasi-1D system, the square lattice, and the triangular lattice can be resolved by the cooperative interaction characteristic of the HL inversion. The CO transition due to the TM mechanism is applied not only to the anisotropic lattice but also to the equilateral triangular lattice, those that correspond to t-13P and m-13P, respectively [20, 21].

As described in the previous paragraphs, some additional interaction is required for the CO transition of b-Cs and b-22Sb because not only interdimer charge separation but also intradimer charge disproportionation is observed in the CO state. Let us introduce the combination between a TM mechanism and a tight dimer mechanism [23]. Figure 2.7 shows the correlation diagrams of (a) a TM mechanism, (c) a tight dimer mechanism, and (b) a combination of these two mechanisms.

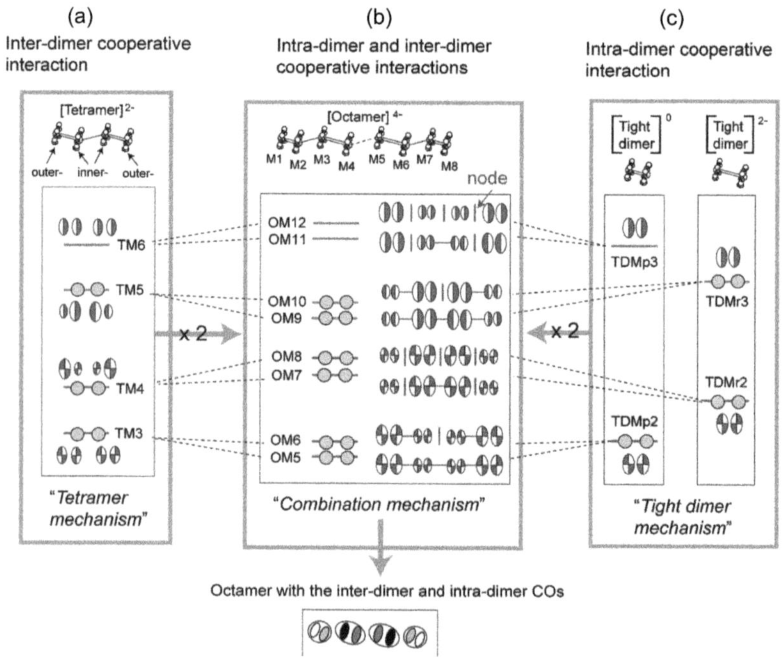

Figure 2.7 Comparison of the energy levels between (a) TM, (b) combination, and (c) tight dimer mechanisms. Filled circles denote electrons in the fully occupied orbital. TM3–TM8 are the same as those in Fig. 2.6d. TDMp2, TDMr2, TDMr3, and TDMp3 are the same as those in Fig. 2.6b. "OM" means octamer (OM), and the numbering to OM denotes the energy level. Black, white, and gray sections in the bottom panel denote the electron densities of OM10.

At first, we examine an octamer (OM) from the viewpoint of TMs. The OM in Fig. 2.7b is composed of two TMs in Fig. 2.7a. Pairs of TM3s, TM4s, TM5s, and TM6s in Fig. 2.7a constitute pairs of OM5 and OM6, OM7 and OM8, OM9 and OM10, and OM11 and OM12, respectively. In Fig. 2.7a, TM1, TM2, TM7, and TM8 are omitted and OM1–OM4 and OM13–OM16 are omitted in Fig. 2.7b. The accumulation of electron densities from OM1 to OM8 does not contribute to charge disproportionation. On the other hand, OM9 and OM10 contribute to charge disproportionation. The number of nodes is zero for OM9 whereas one for OM10. M4 and M5 are the most charge-rich sites in OM9, whereas M3 and M6 become the most charge-rich sites in OM10. A pair of M2 and M7 or a pair of M4 and M5 becomes the secondary anionic site in OM10. The electron densities at M2, M4, M5, and M7 depend on the magnitude of the intradimer and interdimer interactions. Because $[OM]^{4-}$ is composed of two $[TM]^{2-}$, the VBO and the charge separation due to *V*s are cooperatively enhanced in the different orbitals.

Next, we examine an OM in terms of dimers. The OM in Fig. 2.7b is composed of four TDMs, shown in Fig. 2.7c. TDMp1, TDMr1, TDMp4, and TDMr4 are omitted in Fig. 2.7c. Pairs of TDMp2s and TDMr2s constitute OM5 and OM6, respectively. Because the energy level of TDMp2 is different from that of TDMr2, TDMp2s make larger contributions to OM5 and OM6 whereas TDMr2s make fewer contributions to OM5 and OM6. Pairs of TDMr2s, TDMr3s, and TDMp3s make larger contributions to pairs of OM7 and OM8, OM9 and OM10, and OM11 and OM12, respectively. The number of nodes is zero, one, two, and three for OM5, OM6, OM7, and OM8. The same is applicable to OM9–OM12. OM5 is the most stable among OM5–OM8 because all interdimer interactions within an OM are bonding interactions. In the same manner, OM9 is the most stable among OM9–OM12. The energy level of OM8 is the highest among OM5–OM8 because all interdimer interactions within an OM are antibonding. OM12 is the most unstable among OM9–OM12. OM6 and OM10 are more stable than TDMp2 and TDMr3 in Fig. 2.7c, respectively. The levels of OM7 and OM11 are higher than those of TDMr2 and TDMp3, respectively. When four electrons are accommodated in an OM, OM9 and OM10 are occupied whereas OM11 and OM12 are unoccupied. OM10 becomes the highest occupied orbital of $[OM]^{4-}$. Because OM10 exhibits the nature of the bonding orbital, the OM due to the combined mechanism is more stable than the TDMr3 in Fig. 2.7c. In addition, the interdimer distance is decreased due to the interdimer bond alternation. Concerning the tight dimer mechanism (Fig. 2.7c), the contraction of the charge-poor dimer is canceled by the expansion of the charge-rich dimer, which indicates that a CO transition is not necessarily enhanced. On the contrary, contraction in the 2D layer is ensured in Fig. 2.7b because the interdimer distance deceases due to bond formation. Hereafter, the mechanism shown in Fig. 2.7b is denoted as a combination mechanism [23]. The cooperative interaction between VBO and the charge separation due to *V*s, characteristic of the tight dimer mechanism, are valid for the different orbitals in $[OM]^{4-}$. Furthermore, as described in the previous paragraph, the cooperative interaction characteristic of the TM mechanism is also valid for the combination mechanism. Therefore, a CO state due to a combination mechanism is more advantageous than that due to a tight dimer mechanism.

As shown in Figs. 2.7b and 2.7c, a pair of TDMr3s makes a larger contribution to OM10. The inner dimers in OM10 are charge rich, whereas the outer dimers are charge poor. Owing to the interdimer interaction, CT from the inner dimers to the outer dimers is induced. This CT is remarkable in OM10 because a pair of M3 and M2 and a pair of M6 and M7 form a bonding orbital whereas the neighboring charge-rich dimers, that is, M4 and M5, form an antibonding orbital. As a result, the electron densities of two molecules in the charge-rich dimer become inhomogeneous. The electron densities of M3 and M6 are higher than those of M4 and M5. In the same manner, those in the charge-poor dimer become inhomogeneous. The electron densities of M2 and M7 are higher than those of M1 and M8. Therefore, both interdimer charge separation and intradimer charge separation are induced by the combination mechanism (Fig. 2.7b). So the OMs in the CO state of b-22Sb and b-Cs are ascribed to a combination mechanism rather than a tight dimer mechanism.

Figure 2.8 shows the distributions of the molecular charges induced by (a) a loose and tight dimers mechanism (Fig. 2.6a) or a tight dimer mechanism (Fig. 2.6b and Fig. 2.7c), (b) a TM mechanism (Fig. 2.6d and Fig. 2.7a), and (c) a combination mechanism (Fig. 2.7b). The open ellipses in Fig. 2.8a denote the electron densities in [tight dimer]^0. The closed ellipses denote those in [loose dimer]$^{2-}$ or [tight dimer]$^{2-}$. The black, gray, and white ellipses in Figs. 2.8b and 2.8c denote the electron density in the highest occupied orbital [TM]$^{2-}$ and [OM]$^{4-}$, respectively. In Fig. 2.8a, the cooperative interaction operates within dimers whereas no cooperative interaction operates between dimers. When no additional interaction, except for the Madelung interactions between charge-rich dimers, operates in the 2D layer, the simplest distribution due to the dimer mechanism and the loose and tight dimers mechanism is the checkerboard patterns. Three patterns in Fig. 2.8a are equivalent. The charge-rich and charge-poor dimers are alternately arranged along two of three directions: S, diagonal, and transverse. The stripe pattern is formed along the other direction. The CO state of $Me_4P[Pt(dmit)_2]_2$ is consistent with distribution due to a dimer mechanism [49].

According to the X-ray structural analysis in the CO state of b-Cs and b-22Sb, the charge distributions are different from that of $Me_4P[Pt(dmit)_2]_2$. Two charge-rich and two charge-poor dimers are alternately arranged in the S direction. Hereafter, this distribution is denoted as the "double checkerboard pattern." Several charge distributions, including the double checkerboard pattern, were examined from the viewpoint of the Madelung interaction [30]. The calculation of Madelung energies, including a cation-anion interaction, reveals that the energy of the double checkerboard pattern is almost comparable to that of the checkerboard pattern [30], which is in agreement with the experimental result. However, a detailed analysis in the CO state by using vibrational spectroscopy revealed that the intradimer charge separation is superimposed on the interdimer charge separation. This result indicates that the combination mechanism plays an important role in the charge distributions of b-Cs and b-22Sb [19, 23].

Figure 2.8b shows the 2D layers comprised of TMs due to the cooperative interaction between dimers [20, 21]. In the 1D TM, the cooperative interaction operates in the S, diagonal, or transverse direction. In this distribution, no additional interaction, except for the Madelung interactions in the diagonal and transverse directions, operates in the 2D layer. The 2D layer can be regarded as an anisotropic lattice rather than equilateral or isosceles triangular lattices. The CO state of t-13P belongs to 1D TM. Figure 2.8b also shows the other distribution, denoted as 2D TM. In this distribution, the cooperative interaction operates in at least two of three directions: S, diagonal, and transverse. Owing to the cooperative interactions characteristic of the HL inversion, tetramerization is allowed in not only the quasi-1D systems but also the equilateral or isosceles triangular lattices. The phase transition of m-13P is ascribed to this mechanism [20, 21].

The distributions due to the combination mechanism are shown in Fig. 2.8c [23]. In the left panel, the charge distributions in the orange rectangle correspond to those in Fig. 2.7b whereas the distributions in the cyan rectangle do not correspond to those in Fig. 2.7b. The cooperative interaction operates in the S direction, whereas it does not operate in the diagonal direction. In this sense, the OMs in the left panel can develop in the S direction. The center panel shows one of the remaining distributions due to the combination mechanism, where the cooperative interaction operates in the diagonal direction but does not operate in the S direction. The left and center panels are equivalent from the viewpoint of intermolecular interactions. The OMs in the left and center panels have a 1D nature. The left panel corresponds to the CO state of b-22Sb. The polarization dependence of the conductivity spectra of b-Cs suggests the charge distribution of the center panel. The right panel of Fig. 2.8c shows the 2D structure where the cooperative interactions are enhanced in both diagonal and transverse directions. To our best knowledge, however, the distribution of the right panel has not been observed.

In summary, the ground state of the HL inversion system under ambient pressure is not only the AF state but also the CO state originating from the cooperation between VBO and the charge separation due to *V*s. The interdimer cooperative interaction contributes to the intradimer charge inhomogeneity in a TM. The intradimer cooperative interaction induces the interdimer charge inhomogeneity. The combination of the intradimer and interdimer cooperative interactions contributes to both interdimer and intradimer charge inhomogeneities in an OM, respectively. Detailed charge distribution of several $X[M(dmit)_2]_2$ salts are introduced in Section 2.7.

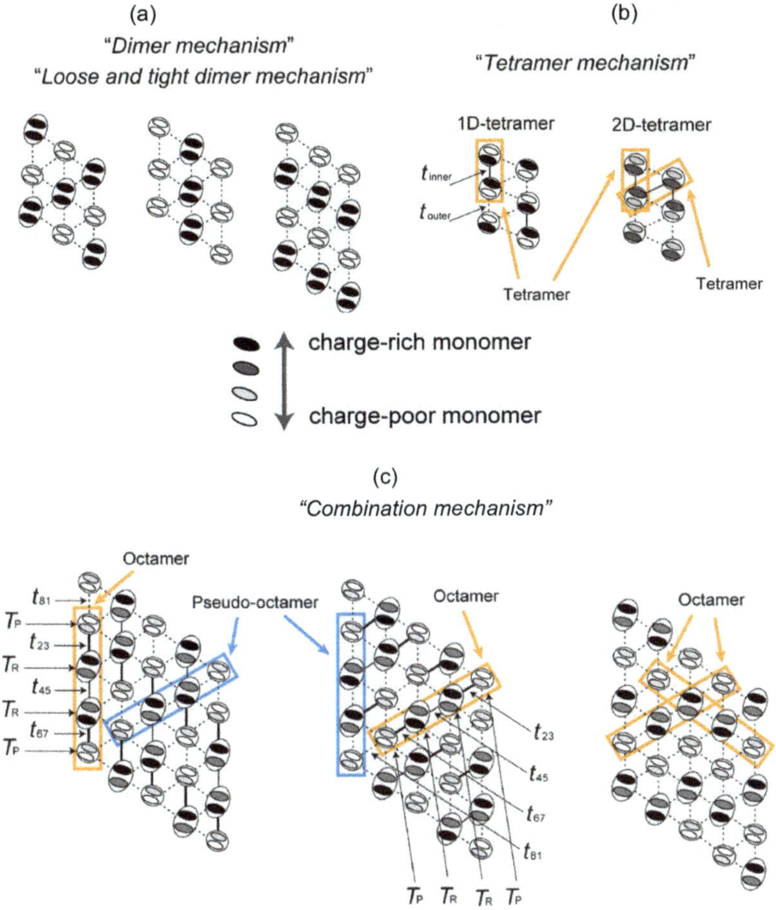

Figure 2.8 Distribution of the molecular charges due to (a) a dimer mechanism and the loose and tight dimers mechanism, (b) a TM mechanism, and (c) a combination mechanism. The orange rectangles in (b) and (c) denote $[TM]^{2-}$ and $[OM]^{4-}$; and the black, white, and gray sections in a monomer denote the electron densities in the HOMOs of $[TM]^{2-}$ and $[OM]^{4-}$. t_{inner} and t_{outer} of (b) denote interdimer transfer integrals within a TM and between TMs. The subscripts of "inner" and "outer" correspond to those in Fig. 2.7a. T_P and T_R in (c) denote intradimer transfer integrals at charge-poor and charge-rich dimers. t_{81}, t_{23}, t_{45}, and t_{67} denote four kinds of interdimer transfer integrals in an OM. 81, 23, 45, and 67 correspond to M8–M1, M2–M3, M4–M5, and M6–M7 in Fig. 2.7b.

2.5 Electronic Spectra in the Charge Ordered State

In this section, we introduce the electronic transition of $X[Pd(dmit)_2]_2$ salts and $X[Pt(dmit)_2]_2$ salts. $X[M(dmit)_2]_2$ salts shown in Table 2.1 are the crystalline materials exhibiting metallic luster. In general, an electronic transition is measured using the reflection method. The transmission method is applied to a thin crystal [49]. The polarized reflectance spectra are transformed into the conductivity spectra by the use of the Kramers–Kronig relationship. Figure 2.9 shows the conductivity spectra in the CO states of t-13P and b-22Sb [19, 21, 35]. The conductivity spectra of b-22Sb have been reported in [35], but the spectra in Fig. 2.9b were newly measured in order to study the vibrational spectra [19]. The CO states of t-13P and b-22Sb are ascribed to a TM mechanism (Figs. 2.6d, 2.7a, and 2.8b) and a combination mechanism (Figs. 2.7b and 2.8c), respectively. In most molecular conductors, electronic transition is often observed in the midinfrared region. This midinfrared transition is ascribed to the intermolecular CT. Also in the present compounds, midinfrared transition, denoted as CT, is observed around 2000 cm^{-1} in both Figs. 2.9a and 2.9b. The multiple and sharp peaks in the low-frequency region of the CT transition belong to the molecular vibration. The properties of the molecular vibration are introduced in the next section. Under the assumption that the molecules in the 1D chain or the 2D layer are uniformly arranged, the CT transition exhibits no peak separation. The frequency of the CT band depends on the intermolecular transfer integral, t, and the intermolecular coulomb repulsion, V. When the degree of dimerization is enhanced, the CT band becomes nondegenerated. The CT band is separated into the interdimer CT band and the intradimer CT band. The frequency of the former depends on the interdimer transfer integral (i.e., the intermolecular transfer integral between neighboring molecules of the neighboring dimers), t_{inter}, and the interdimer coulomb repulsion (i.e., the intersite coulomb repulsion between neighboring molecules of the neighboring dimers), V_{inter}. The frequency of the latter depends on the intradimer transfer integral (the intermolecular transfer integral between neighboring molecules within a dimer), T_{intra}, and the intradimer coulomb repulsion (i.e., the intersite coulomb repulsion between neighboring molecules within a dimer), V_{intra}. On increasing the degree of dimerization, the frequency of the intradimer CT increases, which is distinguished from the interdimer CT [8, 56]. Concerning $X[M(dmit)_2]_2$ shown in Table 2.1, intradimer CT is observed in the NIR region whereas interdimer CT in the midinfrared region.

The energy diagrams of a monomer, a TDM, and a TM are inserted in Fig. 2.9a. Δ denotes the HL transition of a monomer. This transition energy is almost identical to that of Δ in a TDM. HL inversion is supported by the fact that the transition energy of Δ is lower than that of the intradimer CT, which is denoted as Γ in the energy diagram, as shown in the conductivity spectra of t-13P. The intensity of Δ is weak in the conductivity spectra because a HL transition is inherently forbidden owing to the difference in the parities of the ground state and the excited state [35]. The observation of Δ is ascribed to the deformation of the molecules due to a tight dimerization. The Γ band corresponds to two electronic transitions, TDM1→TDM3 and TDM2→TDM4, shown in Fig. 2.9a. In the former transition, both ground and excited states involve the HOMOs of monomers. In the latter transition, both states involve the LUMOs of monomers. Because the parities of the ground and excited states are identical, Γ is the allowed transition [35]. Therefore, the remarkable difference in the intensities of Δ and Γ is ascribed to the difference in the parities mentioned above.

When the 2D layer exhibits a CO transition due to a TM mechanism, the transition energies of Δ, Γ, and HL in the TM are almost identical to those of the 2D layer in which dimers are uniformly arranged. Both energy diagrams are shown as $[TM]^{2-}$ and $[dimer]^{-}$ in the inset of Fig. 2.9a. Because the energy levels of TDM1–TDM4 become separated in the TM, the line widths of Δ, Γ, and HL should increase below the CO transition temperature. However, the analyses of the line widths in the HL and Δ are not straightforward owing to the weak intensities. Concerning the Γ transition, the line width is already broad above the transition temperature because the transition energy of TDM1→TDM3 is not always identical to that of TDM2→TDM4 [9, 21, 35, 56]. Therefore, it is not easy to analyze the tetramerization in terms of the electronic transition. The tetramerization should be analyzed through the use of vibrational spectroscopy, which will be introduced in the next section.

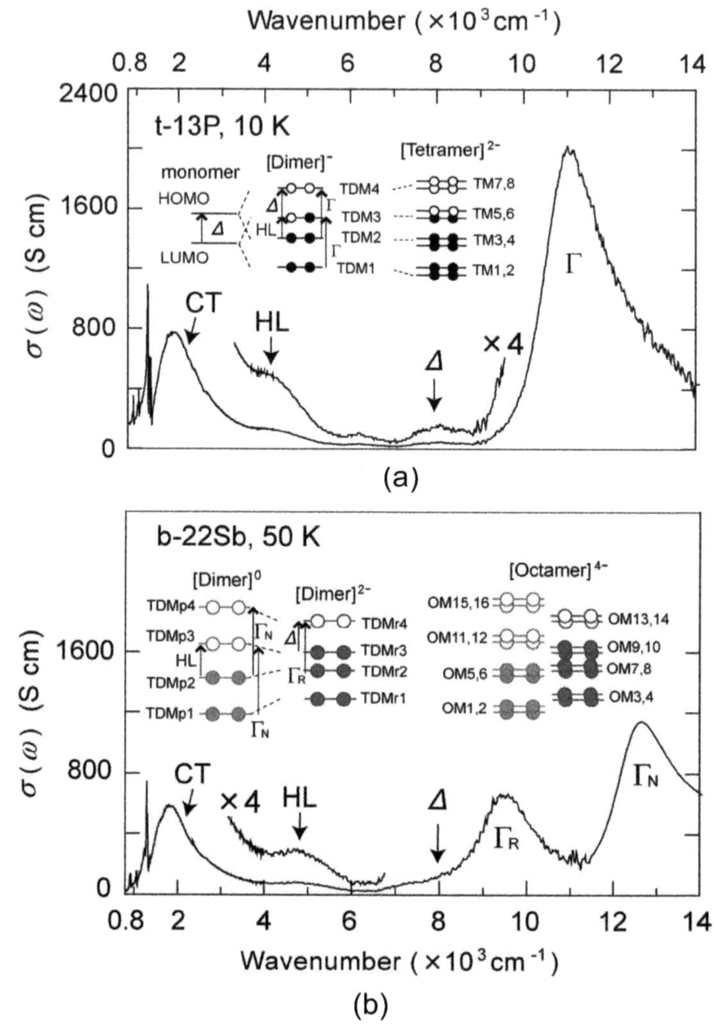

Figure 2.9 Conductivity spectra in the CO states of (a) t-13P and (b) b-22Sb. The directions of the polarization correspond to the *a* directions in Figs. 2.3 and 2.2, and the single crystals are cooled at 10 K and 50 K, respectively. Schematic diagrams of energy levels are shown in both panels. Filled circles denote the electrons in the fully or singly occupied orbital. Open circles denote the unoccupied and virtual electrons.

On the other hand, the conductivity spectra in the CO state of b-22Sb exhibit a characteristic behavior [35]. The energy diagrams of the tight dimer mechanism and the combination mechanism are inserted in Fig. 2.9b. As shown in the energy diagrams, the transition energies due to a combination mechanism are almost identical to those due to a tight dimer mechanism. In the same way as the TM mechanism, however, the analysis of the interdimer interaction is not straightforward. In this section, the electronic transitions of b-22Sb are analyzed from the viewpoint that the 2D layer is composed of charge-rich and charge-poor dimers. The Γ transition of uniformly arranged dimers divides itself in the Γ_R and Γ_N transitions in the CO state. Γ_R and Γ_N belong to the charge-rich and charge-poor dimers, respectively. Because the degree of dimerization in the charge-poor dimer is more enhanced than that in the charge-rich dimer, the transition energy of Γ_N is higher than that of Γ_R. The ratio of the intensities, $I(\Gamma_N)/I(\Gamma_R)$, becomes 2:1 since four electrons per dimer participate in the Γ_N transition whereas two electrons participate in the Γ_R transition [35]. Indeed, the differences in the frequencies and the intensities estimated from the energy diagram are consistent with the conductivity spectra [35]. In addition, the energy diagram predicts that the intensity of the Γ above the CO transition temperature is three times (1.5 times) as large as that of the Γ_R (Γ_N) transition [35]. This prediction is also in agreement with the experimental result [35]. Furthermore, the energy diagrams predict the behavior of the HL transition. The HL transition of the charge-rich dimer is forbidden because TDMr3 is already occupied. On the other hand, the HL transition of the charge-poor dimer is allowed because TDMp2 is occupied and TDMp3 is unoccupied. Comparing the energy diagrams in Figs. 2.9b and 2.9a, the frequency of the charge-poor dimer should be higher than that of the TDM and higher than that of the TM mechanism. Indeed, the frequency of the HL transition of b-22Sb is higher than that of t-13P. The conductivity spectra in the CO state of b-Cs are similar to those of b-22Sb in the NIR and visible regions [9, 19, 23, 35, 56]. Furthermore, the frequency of the HL transition of b-Cs, ca. 5000 cm^{-1}, is almost identical to that of b-22Sb [9, 19, 23, 35, 56]. The consistency in the electronic transitions confirms that the 2D layer in the CO state of b-Cs is comprised of charge-rich and charge-poor dimers.

The electronic transition in the CO state of β'-$Me_4P[Pt(dmit)_2]_2$ was observed from the midinfrared region to the visible region [49]. Owing to the thin single crystals, the electronic spectra were observed by a transmission method rather than a reflectance method. The optical density spectra are composed of the vibrational part in the infrared region, the midinfrared transition, the HL transition around 5000 cm^{-1}, electronic transitions from 8000 to 10,000 cm^{-1}, and a strong electronic transition around 13,000 cm^{-1}, having the shoulder around 15,000 cm^{-1} [49]. The peak separation in the NIR region indicates that the 2D layer is comprised of charge-rich and charge-poor dimers. However, the charge distribution is different from those of b-Cs and b-22Sb. The conductivity spectra of β'-$Me_4N[Pt(dmit)_2]_2$ are also ascribed to the CO state mentioned above, although there is no report on the CO state [8, 56]. Let us tentatively examine the electronic transition from the viewpoint of the energy diagram in Fig. 2.9b. Because the frequency of the HL transition is comparable to that in the CO state of b-22Sb, it is suggested that the HL transition of β'-$Me_4P[Pt(dmit)_2]_2$ belongs to the charge-poor dimer. The frequency as well as the strong intensity of the electronic transition around 13,000 cm^{-1} are consistent with those of the Γ_N transition of b-22Sb. The observation of the shoulder is also consistent with the electronic spectra of b-22Sb. This shoulder is interpreted from the viewpoint that the energy gap between TDMp1 and TDMp3 is not exactly identical to that between TDMp2 and TDMp4. This interpretation is supported by the fact that the Γ transition above the CO transition temperature also has a shoulder in the high-frequency region. Also in this case, the energy gap between TDM1 and TDM3 is not exactly identical to that between TDM2 and TDM4. The Γ transition in the CO state of t-13P also has a shoulder, which is also ascribed to a similar origin. Furthermore, the Γ_R transition of b-22Sb does not have any shoulder. The absence of the shoulder is ascribed to the fact that the electronic transition from TDMr1 to TDMr3 is forbidden because TDMr3 is occupied. These results support the fact that electronic transitions around both 13,000 and 15,000 cm^{-1} in the CO state of β'-$Me_4P[Pt(dmit)_2]_2$ belong to the charge-poor dimer. The other interpretation—the electronic transitions around 13,000 and 15,000 cm^{-1} belong to Γ_R and Γ_N, respectively—contradicts the intensities of Γ_R and Γ_N derived from the energy diagrams.

The optical density from 8000 to 10,000 cm^{-1} is weaker than that around 13,000 cm^{-1} but comparable to that in the midinfrared transition [49]. The electronic spectra in this region are composed of two electronic transitions, around 8000 cm^{-1} and 10,000 cm^{-1} [49].

The assignment of these electronic transitions on the basis of the tight dimer mechanism and the combination mechanism is different from that based on the loose and tight dimers mechanism. The charge-rich dimer in the former mechanisms exhibits a HL inversion, but that in the latter mechanism does not. From the viewpoints of the tight dimer mechanism and the combination mechanism, electronic transitions around 8000 cm^{-1} are assigned to Δ and those around 10,000 cm^{-1} to Γ_R. The opposite assignments are applied to the CO state due to the loose and tight dimers mechanism. According to the conductivity spectra of β'-$Me_4N[Pt(dmit)_2]_2$, the intensity of the electronic transition around 10,000 cm^{-1} is more than that around 8000 cm^{-1} [8, 56]. These results might be consistent with the results of b-22Sb: the electronic transition around 8000 cm^{-1} is assigned to Δ and that around 10,000 cm^{-1} to Γ_R. However, the intensities around 8000 cm^{-1} of β'-$Me_4P[Pt(dmit)_2]_2$ and β'-$Me_4N[Pt(dmit)_2]_2$ are not always remarkably weak as compared with those of b-22Sb and b-Cs [8, 9, 35, 49, 56]. Whether or not the HL inversion is canceled in the CO transition is an interesting subject from the general viewpoint in the physics and chemistry of molecular solids. Another interpretation is that Δ is not always forbidden when molecular distortion in a TDM induces the mixture of HOMO and LUMO at a monomer as predicted by some theoretical studies [14, 16, 17, 63]. In this sense, the detailed assignment of Δ in a series of $X[Pt(dmit)_2]_2$ salts is a task for the future.

2.6 Method for Analyzing Intermolecular Interaction and Charge Separation on the Basis of Vibrational Spectroscopy Focused on the C=C Stretching Modes

The CO states of molecular conductors containing BEDT-TTF have been experimentally analyzed in terms of vibrational spectroscopy, nuclear magnetic resonance, X-ray diffraction method, etc. On the other hand, the methodology for analyzing the CO state of $X[M(dmit)_2]_2$ has not been established. The present author has accumulated the vibrational spectra of several $X[M(dmit)_2]_2$ since 2008. By comparing the experimental results previously published, the author has established a methodology [19–21, 23, 36]. In this section, the author introduces four vibrational modes (mode sensitive to the molecular charge, mode sensitive to the dimer charge, mode sensitive to the intradimer interaction, and mode sensitive to the interdimer interaction) and the minimal repeating unit.

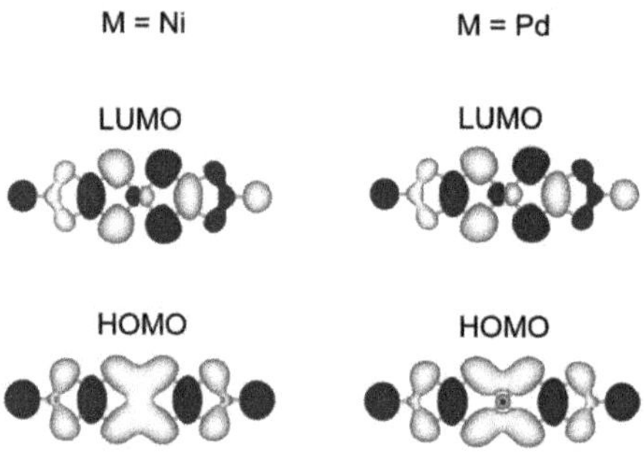

Figure 2.10 HOMO and LUMO of $[Ni(dmit)_2]$ and $[Pd(dmit)_2]$ obtained from B3LYP/6-311G* and B3LYP/LANL2DZ.

The carbon-carbon double bond (C=C bond) is constructed by the σ and π bonds, the latter of which contributes to the conducting, magnetic, and optical properties. Figure 2.10 shows schematic views of the HOMO and LUMO of $[Ni(dmit)_2]$ and $[Pd(dmit)_2]$ obtained from B3LYP/6-311G* and B3LYP/LANL2DZ, respectively [19]. The C=C bonds in the HOMO and LUMO of a monomer exhibit a bonding interaction. We had also obtained the HOMO and LUMO of neutral dimers, $[Ni(dmit)_2]_2$ and $[Pd(dmit)_2]_2$. The C=C bonds in LDM1–LDM4, TDM1–TDM4, TM1–TM8, and OM1–OM16 show a bonding interaction. Therefore, the frequencies of the C=C stretching modes are sensitive to a subtle change in the electrical density in a $M(dmit)_2$ molecule. Furthermore, the C=C bonds between neighboring molecules in a dimer form a bonding interaction in TDM2, TDMr2, and TDMp2. This result means that the frequencies are also sensitive to the magnitude of the intermolecular CT. Indeed, the electron-molecular vibrational (e-mv) coupling constant (g) is large for the symmetric C=C stretching mode, that is, A_g ν_1 mode [63, 64]. On the contrary, the electrical densities at the carbon atoms in the C=S bonds are remarkably smaller than those at the carbon atoms in the C=C bonds and those at the sulfur atoms in the C=S bonds. This result is significantly different from MOs obtained from the extended Hückel method. This condition indicates that the frequencies of the C=S stretching modes are not always sensitive to the change in the electrical density in a $M(dmit)_2$ molecule. Indeed, the Raman spectra revealed that the frequency of the symmetric C=S stretching mode, that is $A_g(2)$, is insensitive to the fractional charge [63]. Hereafter, we focused on the C=C stretching modes.

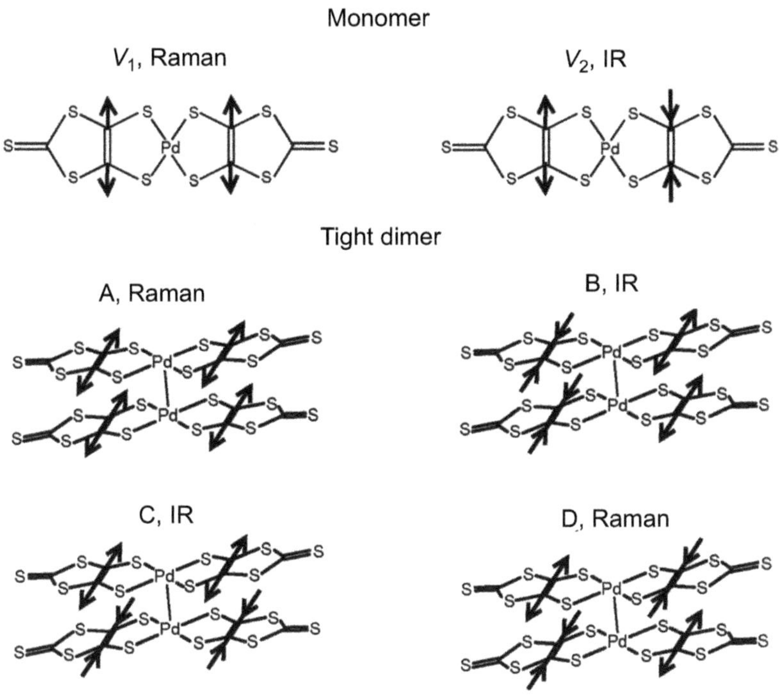

Figure 2.11 C=C stretching modes of a monomer and a tight dimer.

Figure 2.11 shows the two C=C stretching modes of a monomer and four C=C stretching modes of a TDM. Two C=C bonds of a monomer vibrate in phase and out of phase in the ν_1 and ν_2 modes, respectively [63]. Note that the out-of-phase mode in this review is denoted as ν_2. When a monomer has the center of the inversion symmetry, the ν_1 mode belongs to the A_g mode. The ν_1 modes of monomers constitute the A and C modes at a TDM, and the ν_2 modes constitute the B and D modes. Figures 2.12 and 2.13 show the Raman and IR spectra of b-22Sb, t-13P, m-13P, b-04P, and b-13Sb at 300 K [19–21, 23, 36]. We assume that any charge separation is negligible at this temperature and that dimers are regularly arranged in the 2D

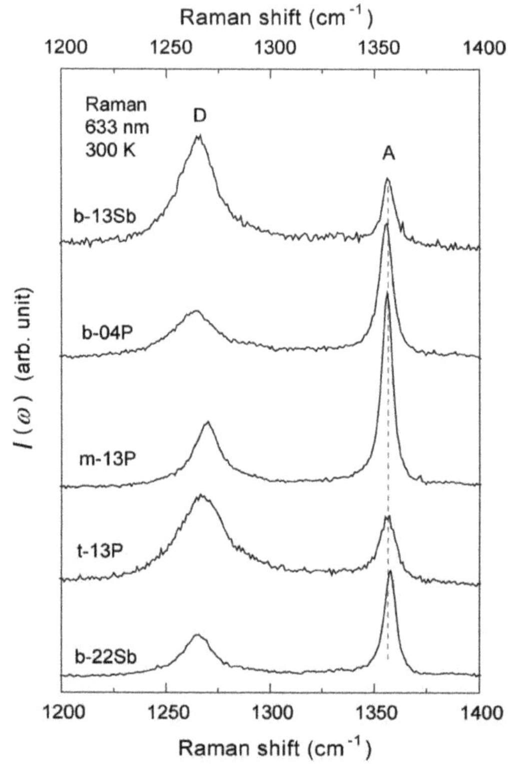

Figure 2.12 Raman spectra of b-13Sb, b-04P, m-13P, t-13P, and b-22Sb at 300 K.

layer. Two vibrational modes are observed in the Raman spectra, and two modes are observed in the IR spectra. Because the frequencies of two Raman modes are significantly different from those of the two IR modes, the four vibrational modes are independent of each other. The frequency of the C mode is lower than that of the A mode, which is ascribed to the e-mv interaction. Interestingly, the frequency of the D mode is significantly lower than that of the B mode. By assuming a LDM, on the contrary, the frequency of the B mode should be identical to that of the D mode because the asymmetric vibrational mode at a monomer is inherently free from the e-mv interaction. The remarkable difference in the frequencies of B and D modes in our spectra is characteristic of the tight dimerization. Concerning the B mode, the nearest C=C bonds between the neighboring monomers in a TDM vibrate in phase whereas two C=C bonds in a monomer vibrate out of phase. In this condition, intermolecular CT is not induced whereas intramolecular CT is induced. Concerning the D mode, on the other hand, the nearest C=C bonds between the neighboring monomers in a TDM vibrate out of phase and two C=C bonds in a monomer also vibrate out of phase. Both intermolecular and intramolecular CTs are induced by the D mode, which leads to a large perturbation in the frequency of the D mode. Because the degree of perturbation depends on the magnitude of the CT, the frequencies of A–D modes depend on the parities of the vibrational modes. The frequencies of A, B, C, and D modes become the highest, the next highest, the next lowest, and the lowest.

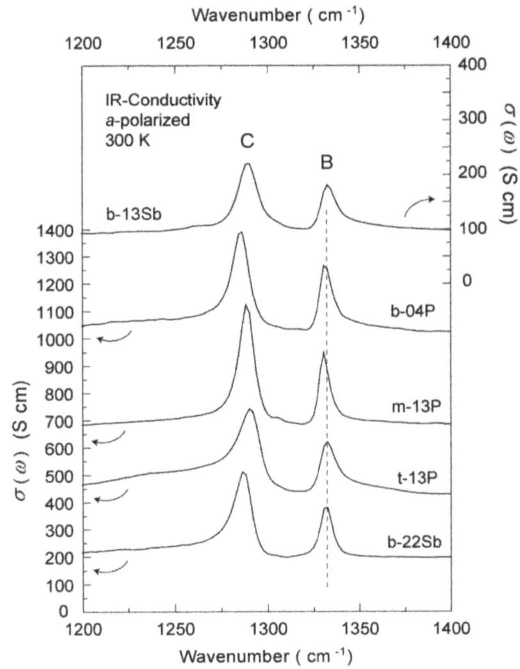

Figure 2.13 IR conductivity spectra of b-13Sb, b-04P, m-13P, t-13P, and b-22Sb at 300 K. The spectra of t-13P, m-13P, and b-04P are offset by 200, 500, and 800 S·cm.

According to the calculation done by Ramakumar et al., the e-mv coupling constant (g) of the ν_1 mode was estimated to be 0.1 eV [64]. This value is comparable to that of the ν_3 mode in BEDT-TTF and significantly larger than that of the ν_2 mode in BEDT-TTF [65]. The author examined the coupling constant of the ν_1 mode from the experimental point of view [19]. The difference in the frequencies of the A and C modes is reproduced under the assumption of g = 0.1 eV and T_{dimer} = 0.45 eV, where T_{dimer} denotes the intradimer transfer integral. However, T_{dimer} obtained from the extended Hückel calculation, ca. 0.45 eV, is smaller than that estimated from the reflectance spectra, ca. 0.7 eV [30]. Nevertheless, it is confirmed that the frequency of the C mode is perturbed by both a large g and a large T_{dimer}. Furthermore, the frequency is also perturbed by the interdimer interaction because of the large g. The C mode is not always useful for analyzing the interdimer and intradimer interactions because the contribution of the interdimer interaction is not always distinguished from that of the intradimer interaction. The frequency of the A mode is also perturbed by the e-mv interaction. Owing to the in-phase vibration, the degree of perturbation is smaller than that of the C mode. The A mode tends to exhibit the frequency corresponding to the formal charge, particularly, the formal charge at a dimer. When the 2D layer does not exhibit the CO state, the frequency of the A mode corresponds to that of $[Pd(dmit)_2]^{0.5-}$. Because of the large coupling constant, the frequency of the A mode in the CO state is not always proportional to the molecular charge. Even when the charge-rich and charge-poor monomers constitute the TDM, the frequency is almost identical to or slightly higher than that of $[Pd(dmit)_2]^{0.5-}$ because the monomer charge is averaged by the CT [19]. When a dimer is composed of two charge-rich monomers, on the other hand, the frequency is higher than that of $[Pd(dmit)_2]^{0.5-}$. Similarly, the frequency is lower than that of $[Pd(dmit)_2]^{0.5-}$ when two charge-poor monomers constitute the dimer. These phenomena are ascribed to the property that the frequency tends to be reflected in the formal charge at a dimer. As shown in Fig. 2.14, the dimer charges in an OM can be analyzed because ([A+A]+[A+A]) is independent of ([A–A]–[A–A]): one belongs to the charge-rich dimer and the other to the charge-poor dimer. Therefore, the dimer charge can be analyzed from the frequency of the A mode in the Raman spectra. The interdimer charge separations in the CO states of b-Cs and b-22Sb are analyzed from the behavior of the A modes in the Raman spectra [19, 23]. The frequency of the A mode is also useful in the analysis of the interdimer interaction when the interdimer interaction shows the alternation, which is described later in this section.

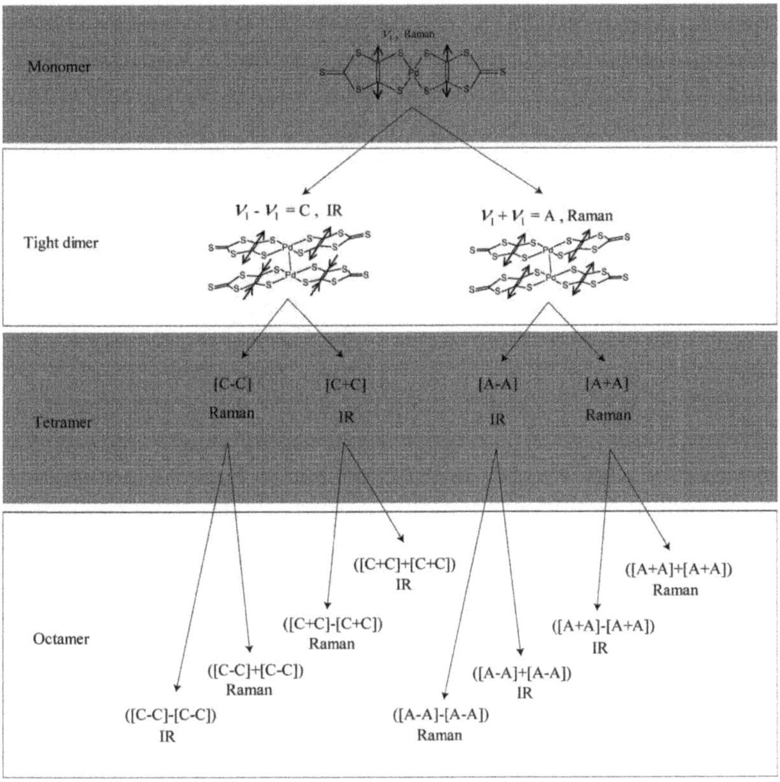

Figure 2.14 Correlation diagram of the C=C stretching modes derived from the ν_1 mode.

The B mode belongs to the ν_2 mode and is free from intermolecular interaction. Because the B mode is less perturbed by the CT interaction, the frequency is sensitive to the monomer charge. This property is consistent with that of the ν_{27} mode of BEDT-TTF, which is also the out-of-phase vibration in a monomer [66–68]. Concerning the BEDT-TTF salts, the frequency shows a linear relationship with the monomer charge as long as the BEDT-TTF molecules retain planar structures [68]. However, the monomer in a TDM of $X[M(dmit)_2]_2$ (M = Pt and Pd) exhibits a boat structure. The molecule in the single crystal of BEDT-TTF^0 also exhibits a boat structure. The frequency of the boat BEDT-TTF^0 is lower than that of the planar structure, and the difference is estimated to be 25 cm^{-1}, which is obtained from the calculation and experiment [68]. The degree of molecular distortion in $X[M(dmit)_2]_2$ depends on the degree of dimerization: the molecular distortion is enhanced for the neutral-like molecule rather than the anionic molecule [28]. These facts indicate that the frequency of the B mode, ν, shows a nonlinear relationship with the monomer charge, ρ. Unfortunately, the exact value of $d^2\nu/d\rho^2$ cannot be obtained from the experimental data available. We tentatively estimate the molecular charge by the use of three kinds of linear relationships: $d\nu/d\rho$ of $-0.5 < \rho \leq -1$, $d\nu/d\rho$ of $0 \leq \rho < -0.5$, and $d\nu/d\rho$ around $\rho = -0.5$ [23]. $d\nu/d\rho$ of $-0.5 \leq \rho \leq -1$ is smaller than that of $0 \leq \rho \leq -0.5$ ($|d\nu/d\rho|$ of $-0.5 \leq \rho \leq -1$ is larger than that of $0 \leq \rho \leq -0.5$). Hereafter, the relationship of $-0.5 \leq \rho \leq -1$ is denoted as $(d\nu/d\rho)_A$ and that of $0 \leq \rho \leq -0.5$ is denoted as $(d\nu/d\rho)_N$, where the subscripts A and N signify anionic and neutral-like species, respectively. $(d\nu/d\rho)_A$ is obtained from the frequency of the dimer at $\rho = -0.5$ (1329 cm^{-1}) and that of the planar monomer at $\rho = -1$ (1375 cm^{-1}) [23]. $(d\nu/d\rho)_A$ is estimated to be = -92 cm^{-1}/electron. Because there is no data concerning the frequency at $\rho = 0$, $(d\nu/d\rho)_N$ is indirectly estimated from the frequency of the charge-poor molecule in the CO state and the frequency of the dimer at $\rho = -0.5$. From the frequency of $\rho = -0.25$ in the CO state of t-13P, 1317 cm^{-1}, the relationship is estimated to be $(d\nu/d\rho)_N = -48$ cm^{-1}/electron [21, 23]. The linear relationship around $\rho = -0.5$ is denoted as $(d\nu/d\rho)_{0.5}$. By averaging the linear relationships of $(d\nu/d\rho)_A$ and $(d\nu/d\rho)_N$, $(d\nu/d\rho)_{0.5}$ is estimated to be -70 cm^{-1}/electron [23].

The B mode induces intramolecular CT because of the out-of-phase vibration in a monomer [19, 66, 68]. Because the transition dipole moment is induced along the long axis of a molecule, the B mode is polarized along the molecular long axis. In general, the long axis of the molecule in the conducting $[Pd(dmit)_2]_2$ salts is almost perpendicular to the large plane of the crystal. Therefore, when the direction of polarized light is parallel to the interplane direction, the relative intensity of the B mode is large [19–21, 23, 68]. On the other hand, the relative intensity becomes small when the incident lights are irradiated onto the largest plane of the single crystal [19–21, 23, 68]. This polarization dependence is also consistent with the ν_{27} mode in the CT salts containing BEDT-TTF.

Let us examine the B modes of a TM and an OM by using the correlation diagram in Fig. 2.15. The vibration at the inner molecules in the TM is independent of that at the outer molecules. In this condition, either [B+B] or [B–B] belongs to the inner molecules and the other belongs to the outer molecules. Therefore, the molecular charge is estimated from the frequency of the B mode. When the repeating unit is the symmetric OM shown in Fig. 2.15, the number of the independent B mode becomes four. Indeed, four vibrational modes belonging to the B mode are observed in the CO states of b-Cs and b-22Sb [19, 23].

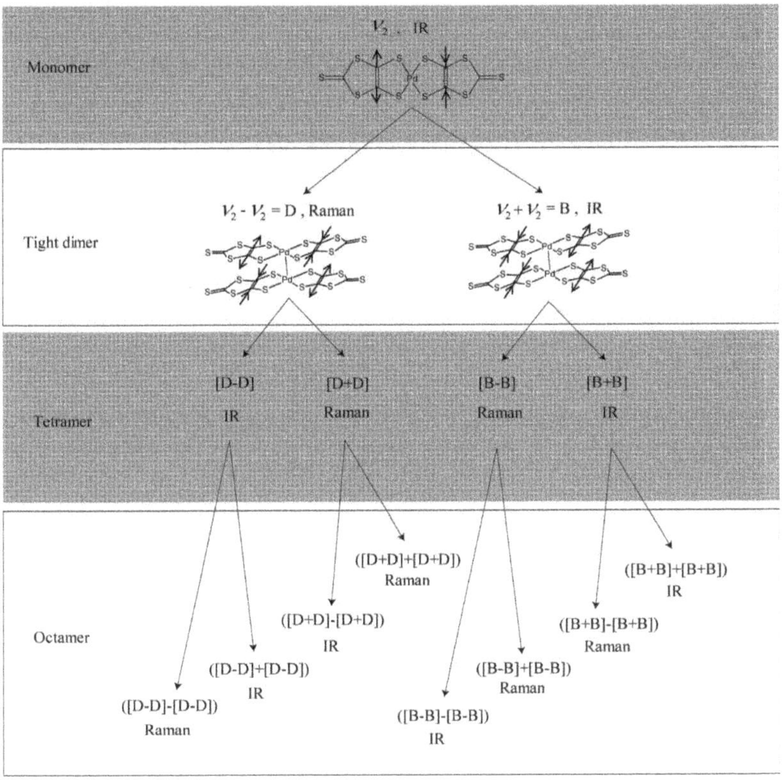

Figure 2.15 Correlation diagram of the C=C stretching modes derived from the ν_2 mode.

Let us discuss interdimer interaction. The number of the C=C stretching mode depends on the number of molecules in the repeating unit. Figures 2.14 and 2.15 show the correlation diagram of the C=C stretching modes derived from the ν_1 and ν_2 modes when the repeating unit is a monomer, a TDM, a symmetric TM consisting of two dimers, and a symmetric OM consisting of four TDMs. The brackets and parentheses denote the TM and the OM. The TDM in a TM and an OM is no longer symmetric. The vibrations of the monomers in the same dimer become inequivalent. Concerning the TM, the inner molecules are equivalent and the outer molecules are also equivalent—when the outer molecules make a large contribution to [A+A] (or [A–A]), the inner molecules make a large contribution to [A–A] (or [A+A]). Because of a large coupling constant (g = 0.1 eV), the frequencies of both [A+A] and [A–A] are perturbed by the e-mv interaction. When [A–A] is affected by the transfer integral between the inner molecules, [A+A] is affected by that between the outer molecules. The former is significantly perturbed by the e-mv interaction, and the latter is reflected in the dimer charge. Therefore, the difference in the frequencies of [A+A] and [A–A] is convenient for estimating the degree of interdimer bond alternation. The TMs of m-13P and t-13P can be analyzed from the behavior of the A modes [20, 21].

Similarly, the frequencies of four A modes in an OM are affected by the interdimer bond alternations, particularly the intertetramer bond alternation. It should be noted that the A mode is also sensitive to the dimer charge. When the OM exhibits both inter- and intradimer charge separations, the frequencies of the A modes are affected by the dimer charge and the intertetramer bond alternation. In the CO state accompanying the OM, the total number of A modes becomes four: two Raman and two IR modes. The two Raman modes shown in Fig. 2.14 are sensitive to the dimer charges because the intertetramer CT is less induced. On the other hand, the neighboring dimers show the out-of-phase vibration in the IR modes, which induces the intertetramer CT. These behaviors suggest that the intertetramer interaction can be analyzed from the differences in the frequencies between Raman and IR modes. Indeed, the intertetramer interactions in the CO states of b-Cs and b-22Sb can be analyzed from the behaviors of the A modes [19, 23].

The frequencies of C and D modes in a TM and an OM are also affected by the intermolecular interactions. However, the frequencies of the C modes in a TM and an OM are significantly perturbed by not only the interdimer interaction but also the intradimer interaction. When the frequency is significantly perturbed by the e-mv interaction, the line width becomes broad. Owing to the broad line width, it is difficult to give the assignment of the multiple peaks belonging to the C modes. The quantitative analyses of the inter- and intradimer bond alternations are not always applicable to the C mode. Nevertheless, the frequencies of some C modes in an OM can be significantly lower than those of other C=C stretching modes because the intradimer, interdimer, and intertetramer interactions all participate in the perturbation due to the e-mv interaction. In the CO state of b-Cs, two C=C stretching modes belonging to the C mode take the lowest frequency of all C=C stretching modes [23]. In this case, the assignments are easy, which allows us to investigate the interdimer interaction in an OM by using the C mode [23].

The D mode induces intradimer CT, which suggests that the frequency of the D mode is useful in analyzing the intradimer interaction. When the intradimer interaction shows alternation, the frequency of the D mode for an enhanced intradimer interaction is higher than that for a reduced intradimer interaction. As shown in Fig. 2.14, the dimer charges in an OM can be analyzed because ([A+A]+[A+A]) is independent of ([A–A]–[A–A]). In the same manner, as shown in Fig. 2.15, ([D+D]+[D+D]) is independent of ([D–D]–[D–D]). Either of the two modes belongs to the outer dimers and the other to the inner dimers. These facts indicate the frequencies of the two Raman modes belonging to the D mode assigned as the D modes for the outer and inner dimers, whose intradimer transfer integrals are different from each other. Significant differences in the frequencies of the D modes are observed in the CO state of b-Cs and b-22Sb [19, 23].

The D mode belongs to the A_g mode. The symmetric vibration indicates that the e-mv interaction accompanied by an interdimer CT is nonnegligible. The additional modes, which cannot be assigned to the C mode, are observed in the IR spectra of t-13P, m-13P, and b-22P salts [20, 21, 23]. These phenomena suggest that the additional modes belong to the D modes. The difference in the frequencies of the D modes in the IR and Raman spectra, that is, [D+D] and [D–D], is affected by the interdimer CT. Similar to the A mode, the interdimer interaction can be analyzed from the behavior of the D mode. The D mode is a characteristic of tight dimerization; however, the e-mv coupling constant of the D mode depends on the degree of dimerization. Therefore, quantitative analysis can be applied when the degrees of dimerization are almost unchanged.

In summary, the B and A modes are sensitive to the molecular charge and dimer charge, respectively. The A mode is also sensitive to the interdimer interaction whereas the D mode to the intradimer interaction.

2.7 C=C Stretching Modes of X[M(dmit)$_2$]$_2$ Salts

2.7.1 *Triclinic*-EtMe$_3$P[Pd(dmit)$_2$]$_2$ [21]

The behaviors of the C=C stretching modes in several X[M(dmit)$_2$]$_2$ salts are introduced. Intermolecular interactions and distributions of molecular charges in the 2D layer are also described on the basis of our experimental results. In this section, we introduce the intradimer CO state of *triclinic*-EtMe$_3$P[Pd(dmit)$_2$]$_2$, which is abbreviated to t-13P in Table 2.1 [21]. The 3D crystal structure belongs to the non-solid-crossing structure (Fig. 2.3), where the S, diagonal, and transverse directions in a 2D layer are identical to those in the nearest neighboring 2D layers separated by the countercation layers [21]. The temperature dependence of the magnetic susceptibility revealed the spin-singlet state below 50 K [21]. The TM was observed by an X-ray structural analysis at 10 K [21]. These results indicate the CO state shown by Fig. 2.8b. However, no positive result suggesting the CO state was obtained from the bond length analysis on the basis of the structural analysis [21].

Figures 2.16a and 2.16b show the IR conductivity spectra obtained from the reflectance spectra polarized in the interlayer (IL) direction [21]. As shown in the insets of these figures, the incident lights are irradiated on the crystal edge of the platelike crystal. In this configuration, the relative intensity of the B mode becomes large compared with that of the C mode. The reflectance spectra were obtained at the Institute for Molecular Science (IMS) and at a synchrotron radiation facility, Spring-8 (BL-43IR). Because the intensity of the incident light at BL-43IR is significantly larger than that of the infrared radiation source in the commercially available spectrometer, the number of scans required is remarkably smaller than that by the commercially available spectrometer. This property is suitable for observing the temperature dependence of the reflectance spectra in the small area. However, it is not easy to obtain the absolute reflectivity because the microgoniometer is not equipped with the cryostat at Spring-8. On the other hand, the absolute reflectivity can be obtained at the Institute for the Molecular Science.

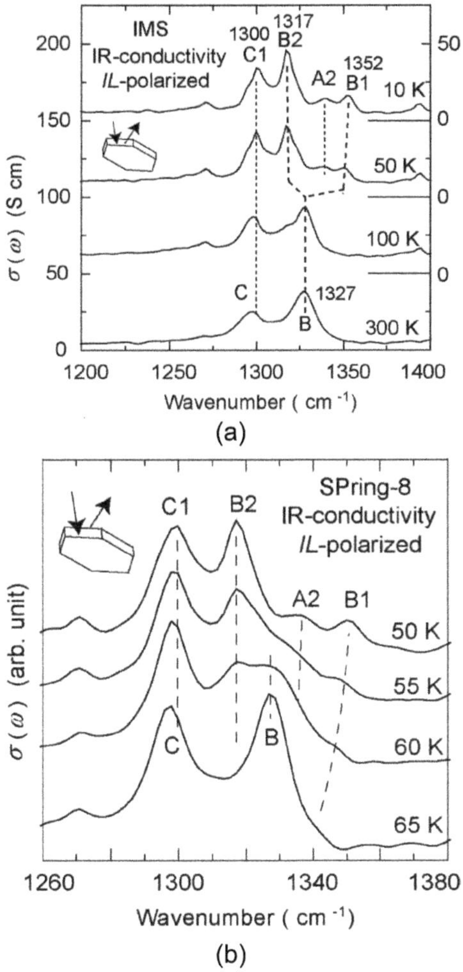

Figure 2.16 IR conductivity spectra of t-13P. The incident light was polarized in the interlayer (IL) direction of the single crystals.

As shown in Fig. 2.16a, the conductivity spectra at 300 and 100 K show two peaks at 1327 and 1300 cm^{-1}. The former is assigned to the B mode and the latter to the C mode. The intensity of the B mode at 1327 cm^{-1} decreases at 50 and 10 K, whereas the frequency of the C mode remains unchanged. By comparing the conductivity spectra between 100 and 50 K, it is suggested that the B mode becomes B1 and B2 at a low temperature. The peak separation is confirmed from the temperature dependence observed at BL43-IR in Spring-8, shown in Fig. 2.16. The B mode observed at 65 K becomes suppressed with a decreasing temperature. On the other hand, B1 and B2 modes are developed. These results indicate that the molecular charge changes from $[Pd(dmit)_2]^{0.5-}$ to $[Pd(dmit)_2]^{(0.5+\delta)-}$ and $[Pd(dmit)_2]^{(0.5-\delta)-}$ in the low-temperature phase. Either B1 or B2 belongs to [B+B] in Fig. 2.15 and the other to [B–B]. From the viewpoint of Fig. 2.15, [B–B] is the symmetric vibration. Nevertheless both B1 and B2 are observed in the IR spectra. One of the reasons is that the 2D layers separated by the countercation layer are crystallographically independent. The averaged frequency of B1 and B2, 1334.5 cm^{-1}, is higher than the frequency of the B mode in the high-temperature phase, 1327 cm^{-1}. This observation is consistent with the relationship between molecular distortion and fractional charge: the degree of distortion is enhanced when the molecule changes from radical anion to neutral species. The molecular charges are estimated to be $[Pd(dmit)_2]^{0.75-}$ and $[Pd(dmit)_2]^{0.25-}$ for the charge-rich and charge-poor molecules, respectively.

Figure 2.17 shows the Raman and the IR conductivity spectra in the CO state [21]. The polarization directions are the S, the perpendicular-to-the-stacking (P), and the IL directions. The IL-polarized spectra are the same as that of Fig. 2.16. Interestingly, the relative intensity of the B2 mode in the IL-polarized spectra is larger than that of the B1 mode. On the other hand, the relative intensity of the B2 mode in the P- and S-polarized spectra is smaller than that of the B1 mode. This polarization dependence is ascribed to the following facts: [B+B] induces intramolecular CT, but [B–B] induces interdimer CT as well as intramolecular CT. Because the transition dipole moment of [B+B] is induced along the long axis of the molecule, the intensity of the B2 mode is enhanced in the IL-polarized spectra. On the other hand, B1 (B1 = [B–B]) is enhanced in the P- and S-polarized spectra. Under the assumption that a unit cell contains one 2D layer, [B–B] is inherently symmetric vibration, which means that [B–B] is silent in the IR spectra. Nevertheless, the neighboring 2D layers separated by a countercation layer are independent for t-13P. Therefore, [B–B] can be observed in the IR spectra. Because the $[Pd(dmit)_2]$ molecule is not normal to the conducting plane, the resultant dipole moment is oriented along any direction within the conducting plane. The same polarization dependencies are also observed in the low-temperature phase of m-13P [20].

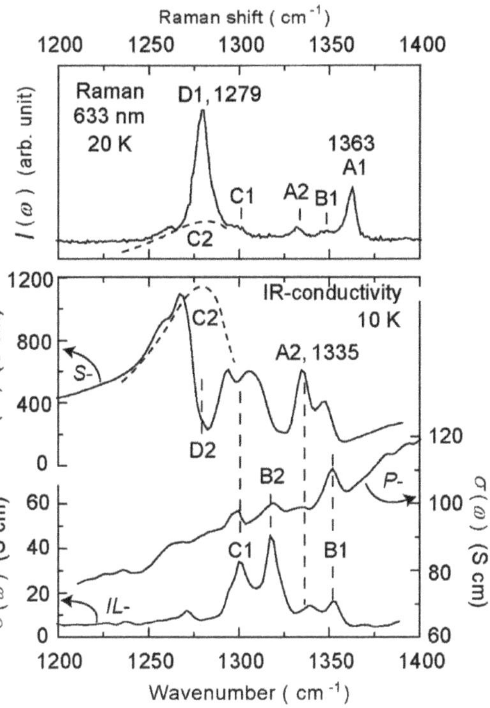

Figure 2.17 Raman and IR conductivity spectra in the CO state of t-13P.

A weak but non-negligible peak is observed between B1 and B2 modes in the low-temperature phase, which is denoted as A2. This vibrational mode has polarization dependence. Although the relative intensity is weak in the IL- and P-polarized spectra, that in the S-polarized spectra is significantly large. This polarization dependence is inconsistent with the property of the B1 and B2 mode and indicates that the A2 mode is relevant to the interdimer interaction along the S direction. The A2 mode is assigned to the [A–A] mode in Fig. 2.14. This polarization dependence supports 1D TM in Fig. 2.8b, where TMs are oriented along the S direction. The counterpart of the A2 mode is the A1 mode in the Raman spectra. The relative intensity of the A1 mode as well as the D1 mode is large in the Raman spectra. The A1 mode is assigned to [A+A] in Fig. 2.14 and the D1 to [D+D] in Fig. 2.15. The difference in the frequencies of A1 and A2 modes is affected by the difference in the interdimer transfer integrals: the transfer integral between inner molecules, t_{inner}, and that between outer molecules, t_{outer} (1D TM in Fig. 2.8b). The differences in the frequencies are reflected in the magnitude of the bond alternation in the TM mechanism. The differences are larger than those of other $X[Pd(dmit)_2]_2$ salts. The details of the bond alternation are discussed in a later section.

Because the frequency of the A1 mode is comparable to that of $[Pd(dmit)_2]_2^-$, the 2D layer is comprised of dimers exhibiting intradimer charge separation instead of interdimer charge separations comprised of charge-rich and charge-poor dimers. In the frequency region from 1250 cm^{-1} to 1300 cm^{-1} in the S-polarized spectra, the C2 and D2 modes overlap each other. The spectral shape in this region is interpreted as an antiresonance behavior due to the Fermi resonance. The observations of additional C and D modes in the IR and Raman spectra are consistent with the TM as a minimal unit. C2 corresponds to [C–C]. Because the 2D layers separated by the countercation layer are crystallographically independent, the D2 in the S-polarized spectra is comprised of [D+D] and [D–D]. Accurate frequency of the D2 mode cannot be obtained from the S-polarized spectra. The C1 mode in the IL-polarized spectra has a weak shoulder around 1295 cm^{-1}. This result suggests that the frequency of [D–D] might be ca. 1295 cm^{-1}. The frequencies of C1 and C2 are different from each other, and the frequencies of [D+D] and [D–D] are also different from each other. Similar to the A mode, these phenomena are ascribed to interdimer bond alternation.

Figure 2.18 shows the Raman and polarized IR conductivity spectra above the temperatures of the CO transition [21]. The frequencies of the A and D modes in the Raman spectra are slightly lower than those at 10 K, which is ascribed to thermal expansion. The line shape of the B mode in the S-polarized spectra at 100 K is asymmetric compared with that at 300 K, which means that the A2 mode is non-negligible at 100 K. This result indicates the fluctuation of TM. Furthermore, the line width of the D mode is slightly larger than that of the m-13P and b-22P [20, 21, 23]. Because the frequency of the D mode depends on the degree of dimerization, the broad line width indicates the fluctuation in the degree of dimerization. In addition, the C mode in the S-polarized spectra has shoulders in the low-frequency region, which suggests that the additional vibrational modes belonging to the C2 and/or D2 modes are non-negligible [21]. These results support the fact that there is competition between the TM mechanism and the combination mechanism.

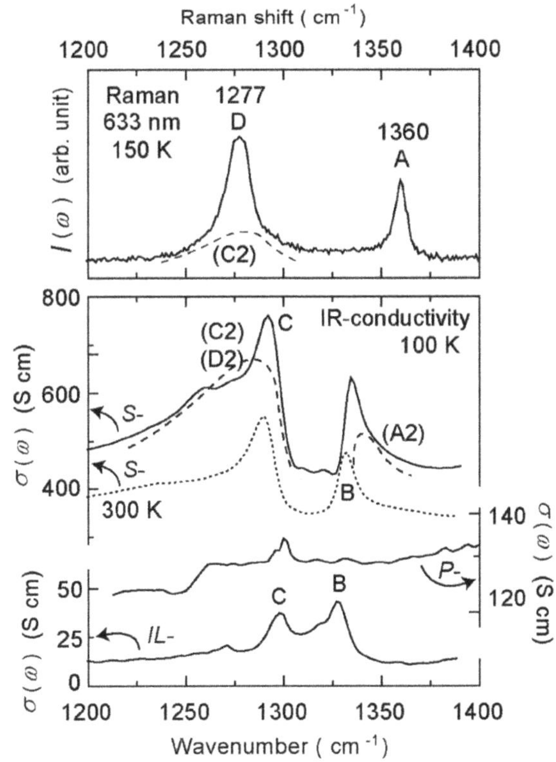

Figure 2.18 Raman and IR conductivity spectra of t-13P. These spectra were obtained at temperatures above the CO transition temperature.

As shown in Fig. 2.17, the A2 mode is observed in the Raman spectra although the relative intensity is weak. The observation of the [A–A] in the Raman spectra might be consistent with the fact that the 2D layers separated by the countercation layer are crystallographically independent. However, there is no corresponding vibrational mode in the Raman spectra of m-13P, which is shown in the next section. The observation of [A–A] in the Raman spectra indicates that the mutual exclusion rule cannot be exactly applicable to both IR and Raman spectra. This phenomenon is consistent with the competition between the TM mechanism and the combination mechanism because the competition leads to domain formation.

2.7.2 *Monoclinic*-$EtMe_3P[Pd(dmit)_2]_2$

In this section, we introduce the tetramerization of *monoclinic*-$EtMe_3P[Pd(dmit)_2]_2$, which is abbreviated to m-13P in Table 2.1 [20]. The 3D crystal structure belongs to a non-solid-crossing structure [39]. As is often the case with the $X[Pd(dmit)_2]_2$ salts, the temperature dependence of the electrical resistivity exhibits an insulating behavior under ambient pressure [39]. The SC phase transition is observed by applying a pressure as low as 2 kbar (T_c = 5.5 K) [39, 40]. The 2D layer is thought to be close to the equilateral triangular lattice [39, 55]. Nevertheless, the magnetic susceptibility exhibits none of the tendency suggesting a SL or AF state. On the contrary, nonmagnetic (NM) behavior is observed below 25 K [38]. Magnetic susceptibility under ambient pressure reveals a direct phase transition from a NM state to a SC state [38]. The alternation of the interdimer interaction is derived from an X-ray structural analysis at a low temperature [39]. This compound is of considerable interest from the following viewpoints: the absence of a SL state and the direct phase transition from the NM to the SC state [33, 38].

Figure 2.19 shows the IL-, *a*-, and *c*-polarized IR conductivity spectra and the Raman spectra obtained from the 633 nm laser [20]. The *a* and *c* directions correspond to the S and the P directions in the conducting layer. The A1 and D1 modes in the Raman spectra are assigned to [A+A] and [D+D] in Figs. 2.14 and 2.15, respectively. The A2 mode is observed in the *a*-polarized spectra at 5 K, which is assigned to [A–A]. On the other hand, no trace of the A2 mode is observed in the *a*-polarized spectra at 100 K, which is significantly different from the behavior of t-13P. This result means that the fluctuation of TM is negligible above the NM transition temperature. In the low-frequency region near the C1 mode in the *a*-polarized spectra, an additional vibrational mode, denoted as "dc2," is observed at 5 K, whereas it is not observed above 100 K. In the high-frequency region near the D1 mode in the Raman spectra, an additional mode, denoted as "dc1," is observed. These vibrational modes are assigned to [D–D] and [C–C]. A2 and dc2 modes are also observed in the *c*-polarized spectra, although the intensities are small. Similar to the A2 modes, the dc1 and dc2 modes are ascribed to tetramerization in the NM phase.

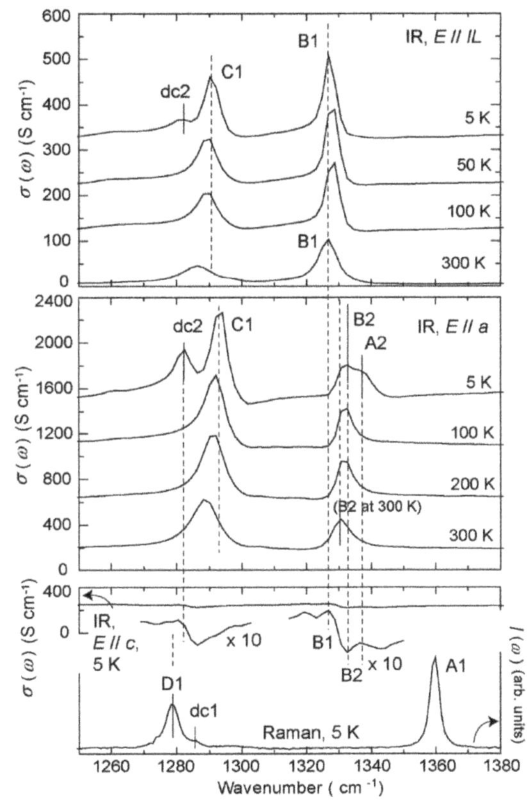

Figure 2.19 IR conductivity and Raman spectra of m-13P. The upper panel shows the IL-polarized conductivity spectra. The middle panel shows the *a*-polarized spectra. The bottom panel shows the *c*-polarized spectra at 5 K and the Raman spectra obtained by a 633 nm laser.

The B mode is observed in the IR spectra, but no remarkable peak suggesting [B–B] is observed in the Raman spectra. The frequency of the B2 mode in the *a*-polarized spectra monotonically increases with a decreasing temperature. Interestingly, the frequency of the B1 mode in the IL-polarized spectra exhibits the different temperature dependence from that of the B2 mode in the *a*-polarized spectra. These results mean that the B1 mode in the IL-polarized spectra is different from the B2 mode in the *a*-polarized spectra. This polarization dependence is ascribed to the same mechanism as that in the CO state of t-13P [21]. Because the transition dipole moment is induced along the molecular long axis for [B+B], the resultant moments are aligned in the IL direction. On the other hand, [B–B] is inherently symmetric because the transition dipole moment is induced in the conducting plane. However, the neighboring 2D layers are independent. As a result, [B–B] is observed in the a- and *c*-polarized spectra. Therefore, it is concluded that a charge-sensitive mode exhibits a peak separation. However, the difference in the frequencies, ca. 6 cm^{-1}, is significantly smaller than that of t-13P, ca. 35 cm^{-1} [21]. Δ of $[Pd(dmit)_2]^{(0.5+\delta/2)-}$ and $[Pd(dmit)_2]^{(0.5-\delta/2)-}$ is estimated as $0 < \delta \leq 0.1$.

The difference in the frequencies of A1 and A2 mode is smaller than that in the CO state of t-13P, which is described in Section 2.8 [21]. The relative intensity of the A2 mode in the *a*-polarized spectra is slightly smaller than that in the CO state of t-13P [21]. The relative intensity of the dc2 in the *a*-polarized spectra is comparable to the corresponding mode of t-13P at 100 K [21]. The transition temperature, ca. 20 K, is lower than that of t-13P, ca. 50 K [21, 38, 39]. These results as well as the small δ are ascribed to the weakly bound TM due to the isotropic 2D layer, which is shown as "2D-tetramer" in Fig. 2.8b. However, the polarization dependence of the A2 mode indicates that the 2D layer is not exactly identical to the equilateral triangular lattice. A similar result is pointed out by thermal compression [69]. According to the X-ray structural analysis, however, the alternation in the interdimer distance in the diagonal and transverse directions is non-negligible [46]. The polarization dependence of the plasma frequency indicates that the 2D layer is slightly anisotropic [55]. The robustness of the TM is ascribed to the 2D TM, where the TM mechanism operates in at least two of the three directions in the 2D layer. The puzzling phenomenon—the SL state is absent in this triangular lattice—is attributed to electron pairing due to the cooperation between VBO and intersite coulomb repulsions in the different orbitals, which is characteristic of the interchange of MO levels (HL inversion) [20, 33]. It is noteworthy that m-13P shows a small but non-negligible δ, as observed in the β''-type BEDT-TTF slats exhibiting the SC transition [36, 37]. This consistency indicates that not only magnetic fluctuation but also fluctuations of charge and VBO contribute to the electron pairing in the SC transition of $X[Pd(dmit)_2]_2$ salts.

2.7.3 β'-$Et_2Me_2Sb[Pd(dmit)_2]_2$

In this section, we introduce the OM in the CO state of β'-$Et_2Me_2Sb[Pd(dmit)_2]_2$, which is abbreviated to b-22Sb [19, 23]. The 3D crystal structure belongs to the solid-crossing structure [28]. The temperature dependence of the electrical resistivity exhibits an insulating behavior [34]. The compound b-22Sb is one of the two compounds whose 2D layer at room temperature is close to the equilateral triangular lattice [28]. The other is m-13P, described in the previous section [39]. Nevertheless, the magnetic susceptibility exhibits none of the tendency suggesting a SL or AF state. On the contrary, NM behavior is observed below 70 K [34, 35]. An X-ray structural analysis reveals that the 2D layer consists of two kinds of TDMs whose intradimer distances are significantly different from each other [28]. The NIR spectra show the peak separation in the CT transition at the TDM. The analyses of the transition energy and intensity reveal that the intradimer distance in the charge-rich dimer is larger than that in the charge-poor dimer [35]. A pair of charge-rich dimers and a pair of charge-poor dimers are alternately arranged in the S direction [28]. However, the difference in the charges of the two molecules in a dimer cannot be detected from the structural analysis although these molecules are crystallographically independent. This compound is mysterious from the following three viewpoints: why the ground state is the CO state instead of the SL state, why the 2D layer does not exhibit a checkerboard pattern, and why two molecules in a dimer become independent. These problems were clarified on the basis of the results of vibrational spectroscopy [19, 23].

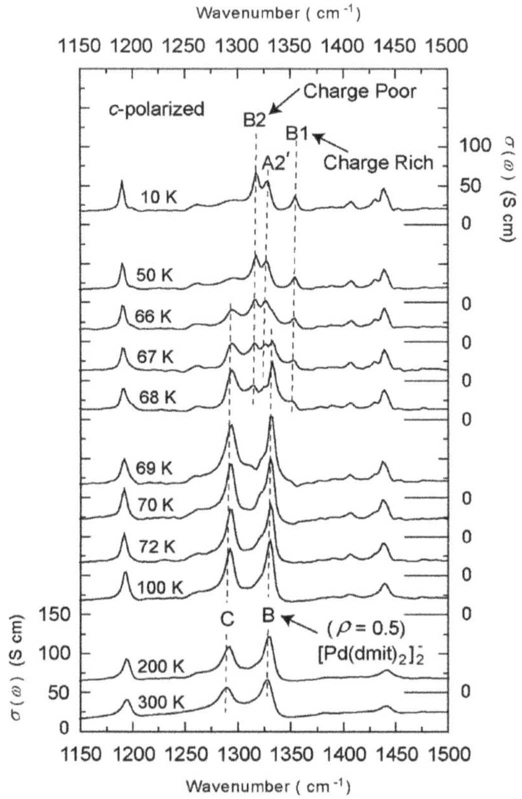

Figure 2.20 Temperature dependence of the IR conductivity spectra of b-22Sb. The incident lights polarized along the *c* axis (IL direction) were irradiated onto the edge of the single crystal.

Figure. 2.20 shows the temperature dependence of *c*-polarized conductivity spectra [19]. The *c* axis corresponds to the IL direction. In the high-temperature phase, above 69 K, one B and one C mode are observed. These vibrational modes are suppressed when the temperature decreases from 69 K to 50 K. Additional vibrational modes, denoted as B1, A2′, and B2, are developed. The average frequency between B1 and B2 modes, 1336 cm^{-1}, is close to but slightly higher than the frequency of the B mode in the high-temperature phase, 1333 cm^{-1}. As described in the previous sections, this phenomenon is ascribed to the fact that the neutral-like molecule is more distorted than the ionic molecule. Either of ([B+B]+[B+B]) or ([B–B]–[B–B]) in Fig. 2.15 is assigned to B1 and the other to B2. Figure 2.21 shows the Raman and IR conductivity spectra in the CO state [19]. The frequencies of B1′ and B2′ in the Raman spectra are close to but slightly different from those of B1 and B2, respectively. One of ([B+B]–[B+B]) and ([B–B]+[B-B]) in Fig. 2.15 is assigned to B1′ and the other to B2′. Observation of four independent B modes means that the symmetric OM is composed of four kinds of molecules whose fractional charges are different from each other. These four vibrational modes belong to pairs [M1 and M8], [M2 and M7], [M3 and M6], and [M4 and M5] in Fig. 2.7b. Applying the tentative relationships between fractional charge and the frequency of the B mode, the molecular charges are estimated to be -0.7_7, -0.7_1, -0.3_5, and -0.2_5. This result evidences intradimer charge separation. The intradimer separation is superimposed on the interdimer separation.

The A2′ mode is observed between B1 and B2 modes in the IR spectra. This behavior is consistent with that in the CO state of t-13P [21]. This vibrational mode belongs to the A mode rather than the B mode. In a manner that is the same as the B mode, the number of the A mode is four per OM. A1, A1′, A2, and A2′ are assigned to ([A+A]+[A+A]), ([A+A]–[A+A]), ([A–A]–[A–A]), and ([A–A]+[A–A]) in Fig. 2.14, respectively. Concerning the A2 and A2′ modes, the mutual exclusion rule is applicable to both IR and Raman spectra. The A1′ mode, which is inherently observed in the IR spectra, is also observed in the Raman spectra. Because the A1′ mode is selectively enhanced by the 514 nm laser, this result is ascribed to the resonant effect. The A1 and A1′ modes belong to the charge-rich dimer, and the A2 and A2′ modes belong to the charge-poor dimer. Because the frequency between A1 and A2 mode, ca. 36 cm^{-1}, is comparable to that between B1 and B2, the dimers are composed of two charge-rich molecules and two charge-poor molecules rather than one charge-rich and one charge-poor molecule. Therefore, $[Pd(dmit)_2]^{0.77-}$ and $[Pd(dmit)_2]^{0.71}$ constitute the charge-rich dimer. $[Pd(dmit)_2]^{0.35-}$ and $[Pd(dmit)_2]^{0.25-}$ constitute the charge-poor dimer. The difference in the frequencies of A1 and A1′ modes is significantly larger than that of A2 and A2′ modes. This result indicates that the interdimer transfer integral between charge-rich dimers is larger than that between charge-poor dimers, which evidences intertetramer bond alternation.

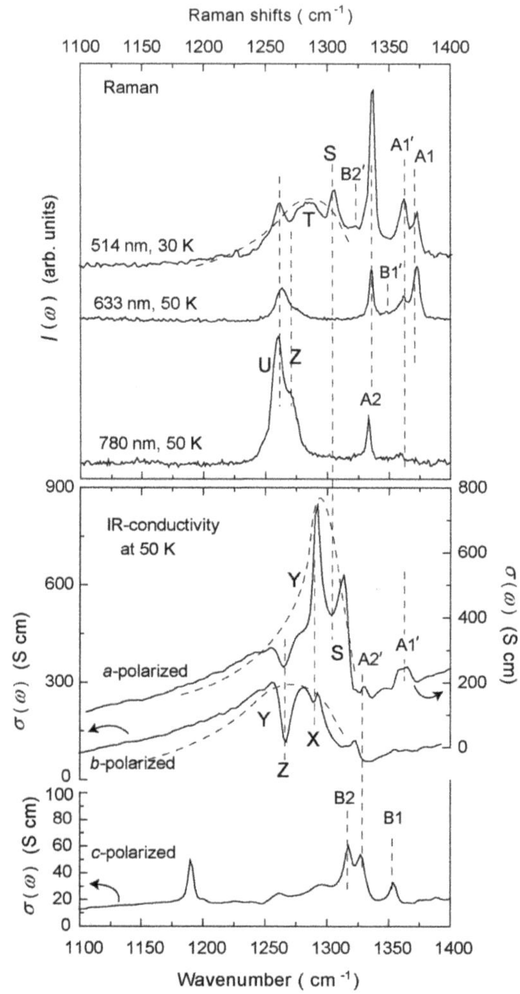

Figure 2.21 Raman and IR conductivity spectra in the CO state of b-22Sb. The *c*-polarized (IL-polarized) spectra are identical to that in Fig. 2.20.

In the frequency region from 1200 to 1300 cm^{-1}, multiple vibrational modes are observed. These vibrational modes belong to the C and D modes, those that are denoted as S, U, and Z modes (W and V modes are not shown in these spectra). The line width of the T mode in the Raman spectra is larger than the line widths of the S, U, and Z modes. The T mode belongs to the C mode because the large line width is ascribed to the e-mv interaction. The *a*- and *b*-polarized IR conductivity spectra exhibit the Fermi resonance. The S, X, and Z modes overlap with the Y mode, whose line width is remarkably large. The Y mode also belongs to the C mode. The S mode is enhanced by the 514 nm laser. This behavior is consistent with those of the A modes belonging to the charge-rich dimers, that is, A1 and A1′. Because the frequency of the D mode increases with a decreasing intradimer transfer integral, the S mode belongs to the D mode of the charge-rich dimer, whose intradimer distance is larger than that of the charge-poor dimer. The X mode is a counterpart of the S mode. The U and Z modes belong to the D modes of the charge-poor dimer. From the viewpoint of Fig. 2.15, the S and U modes belong to ([D+D]+[D+D]) and ([D−D]−[D−D]) and the X and Z modes belong to ([D+D]−[D+D]) and ([D−D]+[D−D]). The Z mode is inherently observed in the IR conductivity spectra. The Z mode as well as the U mode are enhanced by the 780 nm laser. This phenomenon is ascribed to the resonance effect. The transition energy of Γ_N is comparable to the 780 nm laser. Γ_N in Fig. 2.9b is assigned to the electronic transitions of OM1,2→OM11,12 and OM5,6→OM15,16 at the charge-poor dimers in an OM. The difference in the frequencies of S and X modes is affected by the interdimer interaction between the charge-rich dimers. The differences between Z and U are affected by the interdimer interaction between charge-poor dimers. The difference between S and X is larger than that between Z and U, which is in agreement with the behavior of the A1, A1′, A2, and A2′ modes.

It is not easy to assign the C mode because the line widths of the Y and T modes are large and since the multiple vibrational modes overlap each other. Thus, the Y and T modes cannot be separated into ([C+C]+[C+C]), ([C−C]−[C−C]), ([C+C]−[C+C]), and ([C−C]+[C−C]) in Fig. 2.14. The C mode is sensitive to both intradimer and interdimer interactions. Owing to sensitivity to intermolecular interactions, the C mode is not necessarily useful for analyzing intermolecular interactions. When frequencies of several C modes are remarkably different from each other, however, the interdimer interaction might be analyzed from the C mode, which is described in the next sections.

The puzzling phenomena—the CO state does not show a checkerboard pattern, the intradimer charge separation is superimposed on the interdimer charge separation, and the ground state is not SL—are resolved from the viewpoint of the combination mechanism shown in Fig. 2.7b and the left panel of Fig. 2.8c. Because of the cooperation between VBO and the intersite coulomb repulsion in the different orbitals, the charge-rich dimers neighbor each other. The cooperative interaction also induces bond formation between charge-rich and charge-poor dimers. The interdimer cooperative interaction induces intradimer charge separation in both charge-rich and charge-poor dimers. The OM is more stable than the spin frustration because not only intradimer cooperative interaction but also interdimer cooperative interaction favors the ordered state. As described in the previous section, both cooperative interactions cannot be applied to all of the S, diagonal, and transverse directions. The A1′ mode is enhanced in the *a*-polarized spectra, which suggests that interdimer bond alternations operate in the S or diagonal direction. Thus, the charge distribution corresponds to the left and right panels of Fig. 2.8c. The orange rectangle in the left panel of Fig. 2.8c denotes the OM where the interdimer bond alterations operate in the S direction. On the other hand, the charge distribution in the cyan rectangle is different from that in the orange rectangle. The cyan rectangle does not correspond to an OM because the interdimer bond alternation due to the combination mechanism does not operate. According to the X-ray structural analysis, alternation in the interdimer interaction is observed in the S direction. Both experimental results indicate the charge distribution shown in the left panel of Fig. 2.8c. It should be noted that the 2D layers shown in Fig. 2.8c deviate from the equilateral triangular lattice although the 2D layer at room temperature is close to the equilateral triangular lattice. Thus, the reduction of the SL state is ascribed to the fact that the combination mechanism enhances deviation from the equilateral triangular lattice.

2.7.4 β'-Cs[Pd(dmit)$_2$]$_2$

In this section, we introduce the OM in the CO state of β'-Cs[Pd(dmit)$_2$]$_2$, which is abbreviated to b-Cs [23]. The 3D crystal structure belongs to the solid-crossing structure [9]. The temperature dependence of the electrical resistivity exhibits a metallic behavior from room temperature to 56 K, whereas most of X[Pd(dmit)$_2$]$_2$ salts exhibit an insulating behavior below room temperature [9]. The metal-insulator transition is observed around 56 K [9]. The magnetic susceptibility exhibits NM behavior below the transition temperature [9]. Conductivity spectra obtained from the polarized reflectance spectra were reported by Underhill et al. [9], where the NIR transition exhibits peak separation below the transition temperature. This peak separation is ascribed to the interdimer charge separation [34, 35]. An X-ray structural analysis reveals that the structure of the 2D layer in the insulator phase is identical to that of b-22Sb in the CO state [28]. Two charge-rich dimers and two charge-poor dimers are alternately arranged in the S and diagonal directions [28]. Therefore, interdimer charge separation is confirmed from the previous studies. As far as two molecules in the charge-rich dimers are crystallographically independent, the intradimer charge separations should be observed using a certain method. Similarly, the intradimer charge separation in the charge-poor dimer should be observed. However, bond length analysis based on an X-ray structural analysis is not always useful for estimation of the molecular charge. We have studied intradimer charge separation from the viewpoint of vibrational spectroscopy [23]. In addition to intradimer charge separation, there are unresolved phenomena as follows: (i) the 2D layer does not exhibit a checkerboard pattern, (ii) the electrical resistivity exhibits a metallic behavior, and (iii) the polarization dependence of the midinfrared transition is changed with a decreasing temperature [9]. These puzzling phenomena are also examined by using vibrational spectroscopy.

Figure 2.22 shows the Raman and IR conductivity spectra in the insulator phase [23]. In the same way as the CO state of b-22Sb, the number of the independent B mode is four: B_W–B_Z. Furthermore, the number of the independent A mode is also four: A_W–A_Z. Observation of the A_X in the Raman spectra is ascribed to the resonant effect because this mode is enhanced by the 514 nm laser. These results indicate that the minimal unit is the symmetric OM where the charge-rich dimers and charge-poor dimers exhibit intradimer charge separation. The correlation diagrams of the symmetric OM shown in Figs. 2.14 and 2.15 are applicable to the vibrational modes in Fig. 2.22. Under the assumption that A_W belongs to ([A+A]+[A+A]), A_X, A_Y, and A_Z belong to ([A+A]–[A+A]), ([A–A]–[A–A]), and ([A–A]+[A–A]), respectively. Under the assumption that B_W belongs to ([B+B]+[B+B]), B_X, B_Y, and B_Z belong to ([B+B]–[B+B]), ([B–B]+[B–B]), and ([B–B]–[B–B]), respectively.

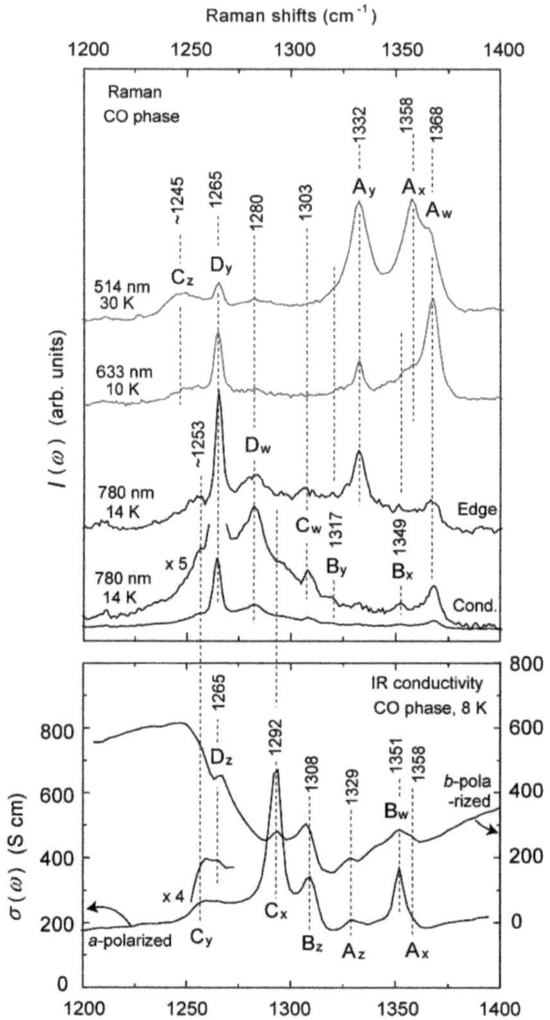

Figure 2.22 Raman and IR conductivity spectra in the CO state of b-Cs.

Applying the tentative relationships between the molecular charge and the frequency of the B mode, the molecular charges are estimated to be -0.7_8, -0.7_4, -0.4_2, and -0.0_6. The averaged frequency between B_w and B_x corresponds to that of the B mode as a charge-rich dimer, and the averaged frequency between B_y and B_z corresponds to that of the B mode as a charge-poor dimer. The difference in the averaged frequencies, 38 cm^{-1}, is in good agreement with the difference in the frequencies of A_w and A_y. This result means that two charge-rich molecules constitute a charge-rich dimer and two charge-poor molecules constitute a charge-poor dimer. It should be noted that intradimer charge separation in the charge-poor dimer is more enhanced than that in the CO state of b-22Sb. This result indicates that the interdimer bond alternation is more enhanced than that in the CO state of b-22Sb. The enhancement of the bond alternation can be analyzed by using the A and C modes. The difference in the frequencies of A_w and A_x modes is affected by the interdimer interaction between the charge-rich dimers and that of A_y and A_z by the interdimer interaction between charge-poor dimers. The difference in the former, 10 cm^{-1}, is larger than that of the latter, 3 cm^{-1}. These values are almost comparable to those in the CO state of b-22Sb, respectively. These results mean that the intradimer interaction between charge-rich dimers of b-Cs is almost identical to that of b-22Sb and the intradimer interactions between charge-poor dimers are almost identical to each other. These results indicate that any remarkable difference in the CO states of b-Cs and b-22Sb lies in the interaction between charge-rich and charge-poor dimers. However, the interdimer interaction between charge-rich and charge-poor dimers cannot be analyzed from the behavior of the A mode because the difference in the frequencies of A_w and A_y (and also A_x and A_z) modes is affected by the difference in the dimer charges rather than by the interdimer interaction.

The interdimer interaction between charge-rich and charge-poor dimers can be indirectly analyzed from the behavior of the C mode. The C_z mode does not overlap with any other C=C stretching modes because the frequency of the C_z mode is the lowest due to the large perturbation. The C_z mode belongs to ([C–C]–[C–C]) in Fig. 2.14, where all molecules in an OM exhibit out-of-phase vibration. Not only intradimer interaction but also interdimer interaction plays an important role in the large perturbation of the C_z mode. The frequency of the C_z mode is almost identical to the CII mode in the high-temperature phase, which is shown in Fig. 2.23. The behavior of the C=C stretching modes in the high-temperature phase is explained from the TM whose details are described in the next paragraphs. The large perturbation in the C_{II} mode is ascribed to the bond alternation due to the TM. The consistency in the frequencies between C_z and C_{II}

indicates that the two TMs in the high-temperature phase become an OM in the low-temperature phase. The pair of TMs shown in Fig. 2.7a become the OM in Fig. 2.7b. The interdimer transfer integral between the inner molecules in a TM, t_{inner}, is almost identical to that between the charge-rich and charge-poor dimers in the OM, t_{23} and t_{67}, in Figs. 2.8c. Therefore, not only t_{inner} but also t_{23} and t_{67} are related to the difference in the frequencies of A_I and A_{II} in the high-temperature phase. The difference in the frequencies, 29 cm^{-1}, is comparable to that of t-13P and significantly larger than A_w–A_x and A_y–A_z in the CO state of b-Cs, 10 and 3 cm^{-1}. This result means that the interdimer bond formation in the CO state is enhanced between charge-rich and charge-poor dimers, that is, t_{23} and t_{67}. Owing to the enhancement of t_{23} and t_{67}, CT from a charge-rich dimer to a charge-poor dimer is induced, resulting in the enhancement of the charge density at M2 and M7 in the charge-poor dimers shown in Fig. 2.7b. Therefore, the remarkable intradimer charge separation in the charge-poor dimer is ascribed to the interdimer cooperative interaction shown in Fig. 2.7b.

Intradimer bond alternation was observed by an X-ray structural analysis [28]. It can also be observed by vibrational spectroscopy. As shown in Fig. 2.22, two D modes, that is, D_w and D_y, are observed in the Raman spectra, those that belong to ([D+D]+[D+D]) and ([D–D]–[D–D]) in Fig. 2.15. The observation of two D modes in the Raman spectra indicates that the intradimer interaction between the charge-poor dimers is more enhanced than that between charge-rich dimers. The D_w and D_y modes belong to the charge-rich and charge-poor dimers, respectively. The frequency of the D_w mode is significantly lower than that of the corresponding mode of b-22Sb. This phenomenon is explained from the viewpoint of the interdimer interaction at t_{23} and t_{67}. When t_{23} and t_{67} are enhanced, CT is induced from the charge-rich dimer to the charge-poor dimer, which contributes to a reduction of the intermolecular coulomb repulsion within the charge-rich dimer. Owing to the reduction of the electrical density in the charge-rich dimer, the expansion of the charge-rich dimer due to the coulomb repulsion is suppressed. As a result, the intradimer transfer integral at the charge-rich dimer, T_R shown in Fig. 2.8c, remains large even in the CO state. Indeed, the frequency of the D_w is close to that of D_I in the high-temperature phase. T_R estimated from the extended Hückel calculation and the result of the X-ray structural analysis are larger than that for b-22Sb [28]. This phenomenon is also ascribed to the cooperation between VBO and coulomb repulsion in the different orbitals at dimers. Furthermore, this phenomenon is linked to the enhancement of t_{23} and t_{67}: the cooperative interaction within a TM. Therefore, the CO state of b-Cs is attributed to both dimer and TM mechanisms, that is, the combination mechanism. One of the puzzling phenomena—the 2D layer does not exhibit a checkerboard pattern—is ascribed to the combination mechanism.

The line width of the D_W is larger than that of D_y, which indicates inhomogeneous dimerization in the charge-rich dimer. Such inhomogeneity is also observed in both high- and low-temperature phases of t-13P [21]. The inhomogeneous dimerization in the insulating state is ascribed to the competition between the TM mechanism and the combination mechanism. Figure 2.23 shows the Raman and IR conductivity spectra in the high-temperature phase at 100 K. These spectra are explained from the viewpoint of the symmetric TM exhibiting an inhomogeneous molecular charge, as shown in Fig. 2.8b. A small but non-negligible difference between the frequencies of the B_I and B_{II} modes is ascribed to the inhomogeneous charges due to the TM mechanism. Applying the tentative relationship between the frequency and the molecular charge around $\rho = -0.5$, the molecular charges are estimated to be -0.5_6 and -0.4_4. On the other hand, the line width of the D_I remains broad in the high-temperature phase, which indicates fluctuation in the magnitude of dimerization. These observations indicate that the 2D layer is not comprised of the static TM. The TMs and the OMs dynamically exchange themselves in the 2D layers. This dynamical exchange induces interdimer CT because the dimer charge in a TM is different from that in an OM. The metallic behavior above the temperature of the CO transition is ascribed to the CT due to the dynamical exchange between TMs and OMs in a 2D layer. The broad line width in the insulating state, on the other hand, is ascribed to the static and spatial inhomogeneity, where the minor domains composed of TMs are intercalated into the major domains composed of OMs. The behavior of the vibrational mode in the CO state of t-13P also suggests spatial inhomogeneity, but the major domain is composed of TMs [21].

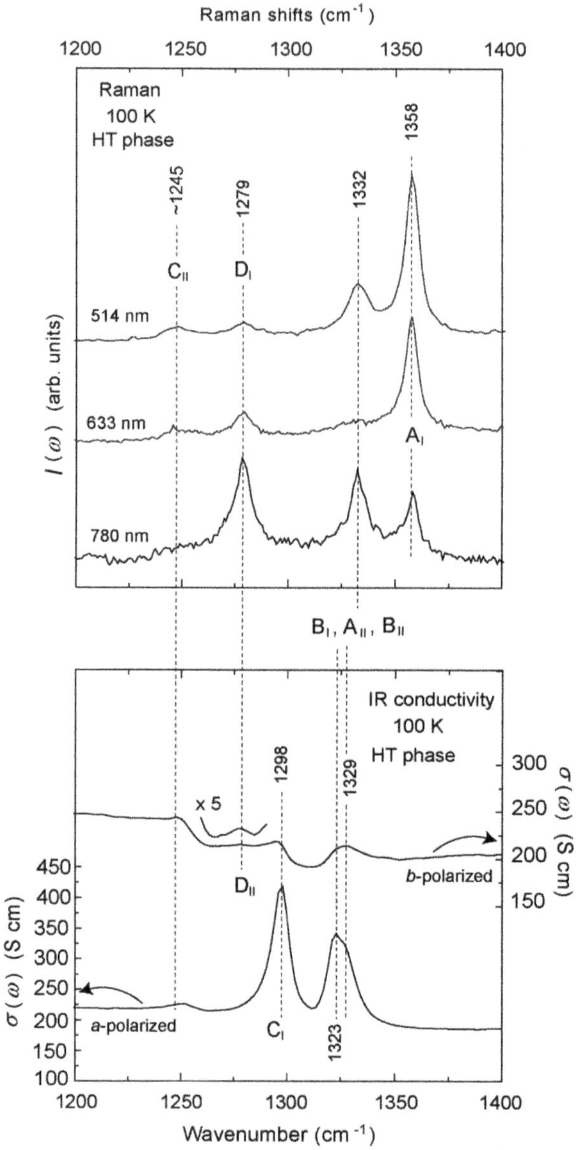

Figure 2.23 Raman and IR conductivity spectra above the transition temperature of b-Cs.

Conductivity spectra were reported by Underhill et al., where the polarization dependencies in the IR region at 200, 80, and 50 K are different from each other [9]. This phenomenon is in agreement with the competition between the TM and combination mechanisms because the direction of the most enhanced interdimer interaction in the 2D TM is not always identical to that in the OM. As shown in Table 2.2 and Fig. 2.4, the 2D layer of b-Cs is remarkably deviated from the equilateral triangular lattice. The 2D layers in the CO states shown by Figs. 2.8b and 2.8c belong to the isosceles or scalene triangular lattices rather than the equilateral triangular lattice. It should be noted that the competition is observed although the 2D layer deviates from the equilateral triangular lattice. This result suggests that the SL state, instead of the CO state, is realized when the 2D layer at a very low temperature satisfies the condition of the competition. Interestingly, anomalous phenomena suggesting subtle entropy rereading were reported by the specific heat measurement, thermal conductivity, nuclear magnetic resonance, and magnetic torque of b-13Sb [41, 43, 44, 71–73]. It is worthwhile to examine the mechanism of the SL state in b-13Sb from the viewpoint of the competition accompanied by the different charge distributions because the TM and combination mechanisms contribute to entropy release.

2.8 Ground States and Bond Alternations

In this section, we summarize the diversity in the ground state of the $X[Pd(dmit)_2]_2$ salts from the viewpoint of vibrational spectroscopy [23]. In the previous sections, we have introduced two kinds of bond alternations, interdimer bond alternation and intradimer bond alternation. The former induces the TM and the intradimer charge separation, and the latter induces the interdimer charge separation. The combination of inter- and intradimer bond alternations induces the OM accompanied by both intra- and interdimer charge separations. The intradimer bond alternation is quantified by the difference in the frequencies of the D modes at the charge-rich and charge-poor dimers, or the line width of the D mode, which is defined as ΔD [23]. The interdimer bond alternation in the TM mechanism is experimentally quantified by the difference in the frequencies of [A+A] and [A–A] in Fig. 2.14, which is defined as ΔA [20, 21, 23]. The interdimer bond alternations in the combination mechanism were indirectly estimated from the behaviors of the A and C modes. Because the OM is comprised of two TMs, ΔA in the CO state of b-Cs was indirectly estimated from the difference in the frequencies of [A+A] and [A–A] in the high-temperature phase at 100 K. This estimation is based on the consistency of the interdimer interactions in a TM below and above the CO transition temperature, which is supported from the fact that the frequency of the C_z mode in the CO state is almost identical to that of the C_{II} mode at 100 K. It is not easy to estimate ΔA in the CO state of b-22Sb because the C_z mode is overlapped with other C=C stretching modes and since there is no tendency to form TM in the high-temperature phase. We have tentatively estimate the maximum value of ΔA as the line width of the T mode in Fig. 2.21, ca. 20 cm^{-1}, and estimate the minimum value as the difference in the frequencies of A_w and A_x, ca. 11 cm^{-1}. ΔA obtained from the line width of the T mode might be overestimated because the intradimer interaction as well as the interdimer interaction participate in the perturbation of the C mode. On the other hand, ΔA obtained from $|A_w - A_x|$ might be underestimated because this value is reflected in the interdimer interaction between charge-rich dimers rather than the interaction between charge-rich and charge-poor dimers. The estimated values of ΔD and ΔA are plotted in Fig. 2.24.

The ΔA's in the CO states of m-13P, t-13P, b-Cs, and b-22Sb are non-negligibly large. On the other hand, ΔD's of these compounds are different from each other. These compounds are classified into two regions: region II and region III. The former corresponds to the TM due to the TM mechanism and the latter to the OM due to the combination mechanism. The CO transition temperature increases with increasing ΔD, that is, from the TM to the OM. This phenomenon is consistent with the viewpoint that the TM mechanism and the tight dimer mechanism both enhance the charge localization due to the cooperation between VBO and the intersite coulomb repulsions. The open circle denotes the b-Cs at 100 K, where the TM is more enhanced than the OM. The closeness of the filled circle of t-13P to the open circle of b-Cs indicates that the boundary between the TM and the OM lies between the filled circles of t-13P and b-Cs.

$X[Pd(dmit)_2]_2$ salts can take other ground states, that is, AF and SL states [3, 4, 31, 41, 43, 44]. According to the vibrational spectra of b-22P, whose ground state is the AF state, no remarkable change was observed in the temperature dependence of the A–D modes [19]. This result is consistent with the speculation that b-22P can be regarded as a Mott insulator. However, the line shape of the C mode becomes asymmetric at a low temperature [19, 23]. This observation suggests that any additional vibrational mode is hidden even in the AF state. The curve fitting the conductivity spectra reveals that the [D–D] mode is hidden in the low-frequency region of the C mode. The difference between the frequencies of [D–D] and [D+D], which is defined as δD, indicates the interdimer bond alternation. However, the difference in the frequencies is significantly lower than that between the ΔA's in the CO states of m-13P, t-13P, b-Cs, and b-22Sb [19–21, 23, 36]. Furthermore, any remarkable vibrational mode suggesting the [A–A] mode is not observed in the IR conductivity spectra of b-22P. These results indicate that the magnitude of the bond alternation in b-22P is significantly weak and the bond alternation exhibits a fluctuation. Therefore, the AF state lies in region I in Fig. 2.24, where both ΔA and ΔD are small.

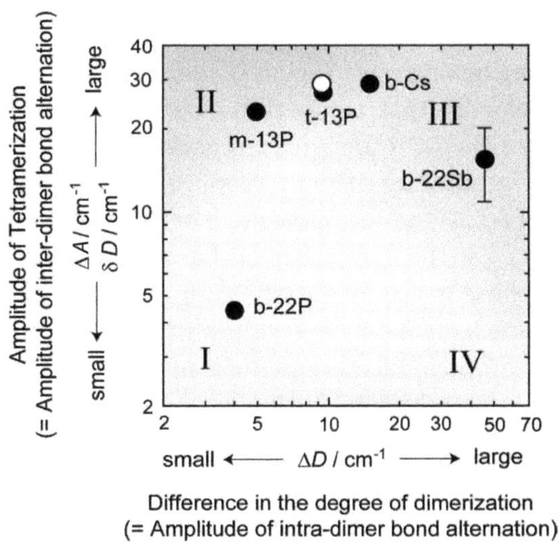

Figure 2.24 Amplitude of the interdimer and intradimer bond alternation estimated from the frequencies of the C=C stretching modes.

In the bottom-right area, region IV, interdimer cooperative interaction is reduced, whereas intradimer cooperative interaction is enhanced. The CO state in region IV is ascribed to the tight dimer mechanism in Fig. 2.7c. Under the assumption that there is no additional interaction except for the Madelung interaction between charge-rich dimers, the 2D layer exhibits one of the distributions shown in Fig. 2.8a. The CO states of β'-$Me_4P[Pt(dmit)_2]_2$ are the candidates belonging to region IV because the charge distribution estimated from the X-ray structural analysis is consistent with the left panel in Fig. 2.8a [49]. The Pt–Pt distance in a dimer at room temperature, that is, 3.31 Å, is larger than the Pd–Pd distances of b-22Sb and b-Cs, that is, 3.14 Å and 3.17 Å, respectively. In addition, the averaged distance in the CO state, that is, 3.17 Å, is a little larger than the averaged Pd–Pd distances in the CO states of b-22Sb and b-Cs, 3.13 Å and 3.15 Å, respectively. These results suggest that the large intradimer distance favors the tight dimer mechanism rather than the combination mechanism because the tight dimer mechanism is free from any additional contraction due to the interdimer distance. To confirm the mechanism of the CO transition of β'-$Me_4P[Pt(dmit)_2]_2$, further experiments need to be conducted, related to vibrational and electronic spectra of several $X[Pt(dmit)_2]_2$ salts. To our best knowledge, on the other hand, there is no report on the CO states belonging to region IV in the $X[Pd(dmit)_2]_2$ salts.

It is noteworthy that competition between TMs and OMs is observed above the CO transition temperatures of t-13P and b-Cs. The boundary separating region II from region III runs around the filled circles of t-13P and the open circle of b-Cs. As described in the last of Section 2.7.4, the fluctuation of charge and lattice in the high-temperature phase is ascribed to the competition between TM and combination mechanisms. This result indicates that the absence in any ordering can be induced regardless of whether the 2D layer is close to the equilateral triangular lattice. Furthermore, it is expected that the AF state can participate in the competition in the central area of Fig. 2.24. Because any ordering is suppressed in the central area, the fluctuation remains even in the low temperature. Indeed, the absence of any ordering is reported for mixed crystals whose parent materials are b-13Sb and b-04Sb [4, 53]. The same behavior is also observed in the mixed crystal consisting of b-04Sb and b-22Sb [4, 53]. These results indicate that the SL state is allowed in the central area of Fig. 2.24, where both ΔA and ΔD have intermediate values. This result also indicates that the SL state is suppressed even in the equilateral triangular lattice when competition between the different CO states and the AF state is forbidden. Indeed, any behavior suggesting a competition was not observed in the C=C stretching mode in the high-temperature phases of m-13P and b-22Sb [19, 20]. Although the 2D layers are close to the equilateral triangular lattice, the ground states are the CO states. These results suggest that the absence of competition leads to an ordered state and any competition plays a non-negligible role in the SL state. It is a task for the future to analyze ΔA and ΔD in the SL state of b-13Sb.

In the last of this review, we comment on the pressure-induced SC transition from the viewpoint of the C=C stretching modes. The pressure inducing the SC transition is the lowest for m-13P compared with other $X[Pd(dmit)_2]_2$ salts exhibiting the AF ground state. As shown in Fig. 2.24, ΔD's of both m-13P and b-22P are small. On the other hand, ΔA of m-13P is larger than that of b-22P. Because the pressure decreases the intermolecular distance, the interdimer and/or intradimer bond alternation are enhanced. These behaviors suggest that the SC transitions of these compounds are induced by the electron pairing due to the interdimer and/or intradimer bond alternations cooperating with the intersite coulomb repulsions. Inhomogeneous molecular charges of the superconductor are also observed in the other molecular conductors belonging to the β''-type BEDT-TTF salts. Furthermore, inhomogeneous charges are observed in the κ-type BEDT-TTF salts [73]. Similar to $X[Pd(dmit)_2]_2$ salts, one electron is formally accommodated into a dimer of κ-type BEDT-TTF salts. Although the β''- and κ-type BEDT-TTF salts do not belong to the HL inversion system, the roles of an electron pairing in the SL and SC states are examined on the basis of the theoretical model, including the electron-electron and electron-phonon interactions [26, 27]. It is interesting and a task for the future to apply a similar calculation to the HL inversion system.

Acknowledgments

The author is grateful to Prof. R. Kato, Prof. K. Yakushi, Prof. Y. Nakazawa, Dr. S. Yamashita, Prof. M. Tamura, and Prof. T. Naito for collaboration and advice. The author expresses his thanks to Dr. M. Uruichi, of the Institute for Molecular Science, and Dr. Y. Ikemoto and Dr. T. Moriwaki, of JASRI/SPring-8, for help with the experimental setup. The author would like to thank to Dr. T. Ishikawa, Prof. K.

Onda, Dr. Y. Okimorto, and Prof. S. Koshihara for useful discussions on optical spectra. Some of the experiments were conducted at the Instrumental Center at the Institute for Molecular Science, partly supported by the Nanotechnology Platform Program (Molecule and Material Synthesis) of the Ministry of Education, Culture, Sports, Science and Technology (MEXT), Japan. Part of the analysis was conducted using the facilities of the Advanced Center for Computing and Communication, RIKEN. This research was supported by a Grant-in-Aid for Scientific Research (No. 24750127, 20850024, 16H06346, and 15K05478) from JSPS, by the Morino Fund for Molecular Science, and by the Research Unit for Materials Science under Ultra-high Pressure, Ehime University.

References

1. Canadell, E. (1999). Coord. *Chem. Rev.*, 185, p. 629.
2. Cassoux, P., Valade, L., Kobayashi, H., Kobayashi, A., Clark, R. and Underhill, A. E. (1991). Coord. *Chem. Rev.*, 110, p. 115.
3. Kato, R. (2004). *Chem. Rev.*, 104, p. 5319.
4. Kato, R. (2014). Bull. *Chem. Soc. Jpn.*, 87, p. 355.
5. Pullen, A. E. and Olk, R.-M. (1999). Coord. *Chem. Rev.*, 188, p. 211.
6. Robertson N. and Cronin, L. (2002). Coord. *Chem. Rev.*, 227, p. 93.
7. Canadell, E., Rachidi, I.-I., Ravy, S., Pouget, J. P., Brossard, L. and Legros, J. (1989). *J. Phys.*, 50, p. 2967.
8. Tajima, H., Naito, T., Tamura, M., Kobayashi, A., Kuroda, H., Kato, R., Kobayashi, H., Clark, R. and Underhill, A. (1991). *Solid State Commun.*, 79, p. 337.
9. Underhill, A., et al. (1991). *J. Phys.: Condens. Matter*, 3, p. 933.
10. Kobayashi, A., Fujiwara, E. and Kobayashi, H. (2004). *Chem. Rev.*, 104, p. 5243.
11. Tanaka, H., Okano, Y., Kobayashi, H., Suzuki, W. and Kobayashi, A. (2001). *Science*, 291, p. 285.
12. Ishibashi, S., Terakura, K. and Kobayashi, A. (2008). *J. Phys. Soc. Jpn.*, 77, p. 024702.
13. Seo, H., Ishibashi, S., Okano, Y., Kobayashi, H., Kobayashi, A., Fukuyama, H. and Terakura, K. (2008). *J. Phys. Soc. Jpn.*, 77, p. 023714.
14. Bonnet, M.-L., Robert, V., Tsuchiizu, M., Omori, Y. and Suzumura, Y. (2010). *J. Chem. Phys.*, 132, p. 214705.

15. Omori, Y., Tsuchiizu, M. and Suzumura, Y. (2011). *J. Phys. Soc. Jpn.*, 80, p. 024707.

16. Tsuchiizu, M., Omori, Y., Suzumura, Y., Bonnet, M.-L. and Robert, V. (2012). *J. Chem. Phys.*, 136, p. 044519.

17. Tsuchiizu, M., Omori, Y., Suzumura, Y., Bonnet, M.-L., Robert, V., Ishibashi, S. and Seo, H. (2010). *J. Phys. Soc. Jpn.*, 80, p. 013703.

18. Canadell, E., Ravy, S., Pouget, J. and Brossard, L. (1990). *Solid State Commun.*, 75, p. 633.

19. Yamamoto, T., Nakazawa, Y., Tamura, M., Fukunaga, T., Kato, R. and Yakushi, K. (2011). *J. Phys. Soc. Jpn.*, 80, p. 074717.

20. Yamamoto, T., Nakazawa, Y., Tamura, M., Nakao, A., Fukaya, A., Kato, R. and Yakushi, K. (2014). *J. Phys. Soc. Jpn.*, 83, p. 053703.

21. Yamamoto, T., Nakazawa, Y., Tamura, M., Nakao, A., Ikemoto, Y., Moriwaki, T., Fukaya, A., Kato, R. and Yakushi, K. (2011). *J. Phys. Soc. Jpn.*, 80, p. 123709.

22. Yamamoto, T., Nakazawa, Y., Tamura, M., Yakushi, K. and Kato, R. (2012). *Phys. Status Solidi B*, 249, p. 971.

23. Yamamoto, T., Tamura, M., Yakushi, K. and Kato, R. (2016). *J. Phys. Soc. Jpn.*, 85, p. 104711.

24. Kobayashi, H., Bun, K., Naito, T., Kato, R. and Kobayashi, A. (1992). *Chem. Lett.*, 21, p. 1909.

25. Tajima, A., Nakao, A. and Kato, R. (2005). *J. Phys. Soc. Jpn.*, 74, p. 412.

26. Clay, R., Mazumdar, S. and Campbell, D. (2003). *Phys. Rev. B*, 67, p. 115121.

27. Dayal, S., Clay, R., Li, H. and Mazumdar, S. (2011). *Phys. Rev. B*, 83, p. 245106.

28. Nakao, A. and Kato, R. (2005). *J. Phys. Soc. Jpn.*, 74, p. 2754.

29. Tamura, M. and Kato, R. (2004). *J. Phys. Soc. Jpn.*, 73, p. 3108.

30. Tamura, M. and Kato, R. (2004). *Chem. Phys. Lett.*, 387, p. 448.

31. Tamura, M. and Kato, R. (2005). *Polyhedron*, 24, p. 2817.

32. Tamura, M. and Kato, R. (2009). *Sci. Technol. Adv. Mater.*, 10, p. 024304.

33. Tamura, M., Nakao, A. and Kato, R. (2006). *J. Phys. Soc. Jpn.*, 75, p. 093701.

34. Tamura, M., Tajima, A. and Kato, R. (2005). *Synth. Met.*, 152, p. 397.

35. Tamura, M., Takenaka, K., Takagi, H., Sugai, S., Tajima, A. and Kato, R. (2005). *Chem. Phys. Lett.*, 411, p. 133.

36. Yamamoto, T. (2012). *Crystals*, 2, p. 893.

37. Yamamoto, T., et al. (2008). *Phys. Rev. B*, 77, p. 205120.

38. Ishii, Y., Tamura, M. and Kato, R. (2007). *J. Phys. Soc. Jpn.*, 76, p. 033704.

39. Kato, R., Tajima, A., Nakao, A. and Tamura, M. (2006). *J. Am. Chem. Soc.*, 128, p. 10016.

40. Shimizu, Y., Akimoto, H., Tsujii, H., Tajima, A. and Kato, R. (2007). *Phys. Rev. Lett.*, 99, p. 256403.

41. Itou, T., Oyamada, A., Maegawa, S., Tamura, M. and Kato, R. (2008). *Phys. Rev. B*, 77, p. 104413.

42. Tamura, M. and Kato, R. (2002). *J. Phys.: Condens. Matter*, 14, p. L729.

43. Yamashita, M., Nakata, N., Senshu, Y., Nagata, M., Yamamoto, H. M., Kato, R., Shibauchi, T. and Matsuda, Y. (2010). *Science*, 328, p. 1246.

44. Yamashita, S., Yamamoto, T., Nakazawa, Y., Tamura, M. and Kato, R. (2011). *Nat. Commun.*, 2, p. 275.

45. Kato, R., Fukunaga, T., Yamamoto, H. M., Ueda, K. and Hengbo, C. (2012). *Phys. Status Solidi B*, 249, p. 999.

46. Kato, R. and Hengbo, C. (2012). *Crystals*, 2, p. 861.

47. Faulmann, C., Legros, J.-P., Cassoux, P., Cornelissen, J., Brossard, L., Inokuchi, M., Tajima, H. and Tokumoto, M. (1994). *J. Chem. Soc., Dalton Trans.*, p. 249.

48. Aonuma, S., Sawa, H. and Kato, R. (1997). *Synth. Met.*, 86, p. 1881.

49. Ishikawa, T., et al. (2015). *Science*, 350, p. 1501.

50. Rouziere, S., Yamaura, J.-I. and Kato, R. (1999). *Phys. Rev. B*, 60, p. 3113.

51. Yamaura, J.-I., Nakao, A. and Kato, R. (2004). *J. Phys. Soc. Jpn.*, 73, p. 976.

52. Mori, T., Mori, H. and Tanaka, S. (1999). *Bull. Chem. Soc. Jpn.*, 72, p. 179.

53. Kanoda, K. and Kato, R. (2011). *Annu. Rev. Condens. Matter Phys.*, 2, p. 167.

54. Yoshida, Y., et al. (2015). *Nat. Phys.*, 11, p. 679.

55. Tamura, M., Yamamoto, T., Kato, R. and Yakushi, K. (2009). Proc. 8th Int. Symposium on Crystalline Organic Metals, Superconductors and Ferromagnets, Niseko, p. 183.

56. Tajima, H., et al. (1991). *Synth. Met.*, 42, p. 2417.

57. Kakiuchi, T., Wakabayashi, Y., Sawa, H., Itou, T. and Kanoda, K. (2007). *Phys. Rev. Lett.*, 98, p. 066402.

58. Yamamoto, K., Yamamoto, T., Yakushi, K., Pecile, C. and Meneghetti, M. (2005). Phys. Rev. B, 71, p. 045118. (Although the CO state exhibiting the tetrameriation is emphasized in this article, another charge distribution cannot be excluded. Although most of the C=C and C=N

stretching modes follow the mutual exclusion rule, some vibrational modes around 1200 ~ 1300 cm^{-1} are observed in both IR and Raman spectra, which suggests the coexistence of another charge distribution with the symmetric tetramer).

59. Yamamoto, T., Uruichi, M., Yakushi, K. and Kawamoto, A. (2006). *Phys. Rev. B*, 73, p. 125116.
60. Yamamoto, T., Uruichi, M., Yakushi, K., Yamaura, J.-I. and Tajima, H. (2004). *Phys. Rev. B*, 70, p. 125102.
61. Li, H., Clay, R. and Mazumdar, S. (2010). *J. Phys.: Condens. Matter*, 22, p. 272201.
62. Ihara, Y., Seki, H. and Kawamoto, A. (2013). *J. Phys. Soc. Jpn.*, 82, p. 083701.
63. Pokhodnya, K., Faulmann, C., Malfant, I., Andreu-Solano, R., Cassoux, P., Mlayah, A., Smirnov, D. and Leotin, J. (1999). Synth. Met., 103, p. 2016.
64. Ramakumar, R., Tanaka, Y. and Yamaji, K. (1997). *Phys. Rev. B*, 56, p. 795.
65. Kozlov, M., Pokhodnia, K. and Yurchenko, A. (1989). *Spectrochim. Acta, Part A*, 45, p. 437.
66. Świetlik, R. and Garrigou-Lagrange, C. (1993). Synth. Met., 56, p. 2217.
67. Świetlik, R., Kushch, N., Kushch, L. and Yagubskii, E. (1994). *Phys. Status Solidi B*, 181, p. 499.
68. Yamamoto, T., Uruichi, M., Yamamoto, K., Yakushi, K., Kawamoto, A. and Taniguchi, H. (2005). *J. Phys. Chem. B*, 109, p. 15226.
69. Manna, R. S., de Souza, M., Kato, R. and Lang, M. (2014). *Phys. Rev. B*, 89, p. 045113.
70. Itou, T., Oyamada, A., Maegawa, S. and Kato, R. (2010). *Nat. Phys.*, 6, p. 673.
71. Watanabe, D., et al. (2012). *Nat. Commun.*, 3, p. 1090.
72. Yamashita, S., Yoshimoto, R., Fukuoka, S., Nakazawa, Y. and Kato, R. (2015). *Quantum Matter*, 4, p. 314.
73. Löhle, A., Rose, E., Singh, S., Beyer, R., Tafra, E., Ivek, T., Zhilyaeva, E. I., Lyubovskaya, R. N. and Dressel, M. (2016). *J. Phys.: Condensed Matter*, **29**(5), p. 055601.

Chapter 3

Study of High-Temperature Oxidation Behavior of Antimony and Bismuth Tellurides by Thermopiezic Analysis and Powder X-Ray Diffraction: A Case Study of Thermochemistry

Makio Kurisu
Department of Physics, Graduate School of Science and Engineering, Ehime University, 2-5 Bunkyo-cho, Matsuyama, Ehime 790-8577, Japan
kurisu@ehime-u.ac.jp, makiokurisu@gmail.com

This chapter exemplifies a thermochemical study in detail. High-temperature tolerance of antimony and bismuth tellurides as thermoelectric materials, Sb_2Te_3, $(Bi_{0.5}Sb_{0.5})_2Te_3$ and $(Bi_{0.9}Sb_{0.1})_2Te_3$, has been investigated under argon, nitrogen, or oxygen atmospheres at 0.2 MPa and 547°C. An apparatus for thermopiezic analysis (TPA) has been developed to examine the reactions of various gases with thermoelectric materials as functions of temperature and time, and its instrumental performance is presented. It is demonstrated that TPA is a powerful means for monitoring the reaction of an

Functional Materials: Advances and Applications in Energy Storage and Conversion
Edited by Toshio Naito

ISBN 978-981-4800-09-9 (Hardcover), 978-0-429-46813-1 (eBook)
www.panstanford.com

introduced gas with the test solid materials and desorption of an unknown gas from them. The reaction products have been examined by microscope observation and powder X-ray diffraction with the aid of Rietveld refinement. It is found that at temperatures up to 547°C, the three types of solid thermoelectric materials do not react with Ar or N_2 gas. On the other hand, all the compounds begin to be oxidized at 400°C. Sb_2Te_3 is oxidized initially to form $SbO_{1.5}$ and precipitate Te and by further oxidation reactions to form Sb_2O_4 and TeO_2. More interesting is the anisotropic lattice expansion found in the mother compound Sb_2Te_3; when the relative number of moles of reacted O_2 gas amounts to 5.3, the lattice parameter *a* is increased by 2.4%, while the *c* spacing remains unchanged by oxidation in the hexagonal unit cell. $(Bi_{0.5}Sb_{0.5})_2Te_3$ is most oxidation resistant across the series. None of the antimony and tellurium oxides, such as $SbO_{1.5}$, Sb_2O_4, and TeO_2, are produced; however, anisotropic lattice expansion is found in the mother compound; the *c* plane expands by 0.5%, whereas along the *c* axis the lattice spacing remains unchanged. The degree of expansion is totally dependent on the exposure temperature. In $(Bi_{0.9}Sb_{0.1})_2Te_3$, a bismuth tellurite Bi_2TeO_5 is produced and there is no trace of antimony and tellurium oxides formed. An expansion of 0.7% is found in the *c* plane, while shrinking is as much as 1.3% along the *c* axis, so the mother phase cell volume remains unchanged. These observed modifications in the unit cell structure of the mother phase in the pseudobinary system of Sb_2Te_3-Bi_2Te_3 are attributed to the rearrangement of constituent elements in the unit cell; the Sb and Te site occupancies in the stacked layers are appreciably modified by oxidation. It is proposed that by gas reaction with solid thermoelectric materials of the $(Bi, Sb)_2Te_3$ pseudobinary alloys, tuning of their optimum thermoelectric performance is possible.

3.1 Introduction

Thermoelectric materials have recently attracted a lot of attention since they are capable of generating electric power from waste-heat resources available in a wide range and on a large scale [1]. Thermoelectric devices work without causing environmental pollution, since they can directly convert heat energy to electric one, neither producing any waste materials such as exhaust gas nor emitting noise. More attention has been directed to the optimization of the thermoelectric performance of existing well-known materials and devices and to search for novel materials possessing a larger thermoelectric figure merit for various thermoelectric applications. In general, in using a thermoelectric device for thermoelectric generation, setting one of its ends (junctions of p- and n-type of thermoelectric materials) at a temperature as high as possible is very favorable for optimizing the performance of the thermoelectric generator since the conversion efficiency is given by the product of the Carnot efficiency and the device efficiency [2].

Further aspects one should be concerned with should be the tolerance of the materials and their deterioration in terms of thermoelectric as well as mechanical properties when the materials are placed under extreme conditions: high temperatures, various gas atmospheres and vacuum state, strong acids, or strong oxidizing/reducing gases. The formation of oxides or nitrides on the solid surface should be dependent on the thermodynamic activities of constituent elements in the samples under given conditions of temperature and oxygen/nitrogen gas pressure. A fundamental understanding of oxidation and nitridization behavior in thermoelectric materials contributes to the progress of application as well as basic study of thermoelectricity.

In this chapter, we concern ourselves with the high temperature tolerance of the pseudobinary alloy series of Sb_2Te_3-Bi_2Te_3 in gas atmospheres of Ar, N_2, or O_2. This system has been known for a long time as a family of excellent thermoelectric materials for thermoelectric generators in the temperature range between room temperature and 300°C [3]. From the metallurgical point of view, this alloy system is most suitable for the study of oxidation rates across the alloys since Sb_2Te_3 forms a complete solid solution series with Bi_2Te_3 [4, 5]. High-quality single crystals of hexagonal $(Bi, Sb)_2Te_3$ compounds are relatively easily grown, and abundant data have been accumulated for both structural and thermoelectric properties [3, 6]. Although $(Bi, Sb)_2Te_3$ compounds have been described as having a rhombohedral structure, for easier representation of symmetry, we treat them as having a hexagonal one. We can intuitively guess some of the structural and thermoelectric properties of oxidized/nitridized alloys when their properties are not significantly different from the original ones.

To examine gas reaction processes with $(Bi, Sb)_2Te_3$ compounds at a high temperature as functions of temperature and time, we designed and built a thermopiezic analysis (TPA) apparatus according to the descriptions of the pioneer development of the apparatus by Ryan and Coey [7]. The performance of the TPA apparatus designed and assembled for the study is presented in the next section.

The desorption as well as absorption behaviors of the $(Bi, Sb)_2Te_3$ compounds at temperatures up to 547°C are shown to demonstrate how the TPA apparatus is sensitive and precise in operation for examining the high-temperature tolerance of thermoelectric materials in a specific gas atmosphere. The reaction volume of the gas and the reaction temperature are readily determined by monitoring the pressure versus temperature and/or time. The net amount of gas that has reacted with the test compounds is evaluated after completion of the repeating TPA scans and compared with that estimated by the changes in mass of the recovered sample exposed in the gas atmosphere at any given temperature.

Optical microscope observation shows that all the samples are colored brown after they are exposed in an O_2 atmosphere at temperatures above 421°C, indicating that oxidation takes place on the surface of the sample. The oxidization products are identified by analyzing the powder X-ray diffraction (XRD) patterns by the aid of RIETAN-FP software [8] and observed oxidization processes are discussed.

3.2 Experimental Details

In this section, the experimental details are described in three parts: (i) design of the TPA apparatus and its instrumental performance, (ii) sample preparation, and (iii) powder XRD.

3.2.1 Thermopiezic Analysis Apparatus

The TPA is one of the thermal analysis techniques in which some physical properties of a specimen and/or its gas reaction products are measured by controlling the pressure and temperature of the introduced gas confined to a constant volume. Usually the properties of interest are monitored, displayed, and recorded as functions of temperature and time. This thermopiezic (thermomanometric) method is superior to the thermogravimetric analytic one [9] in terms of higher sensitivity to observing the light gas reaction with the solid specimen. A pioneer apparatus was already developed and investigations were accomplished dealing with hydrogen gas absorption and desorption to small amounts (mg) of samples [10, 11]. We designed our TPA apparatus with better temperature homogeneity around the sample position by preparing a larger-sized furnace and a relatively smaller blank volume to gain the required pressure sensitivity during the absorption/desorption processes.

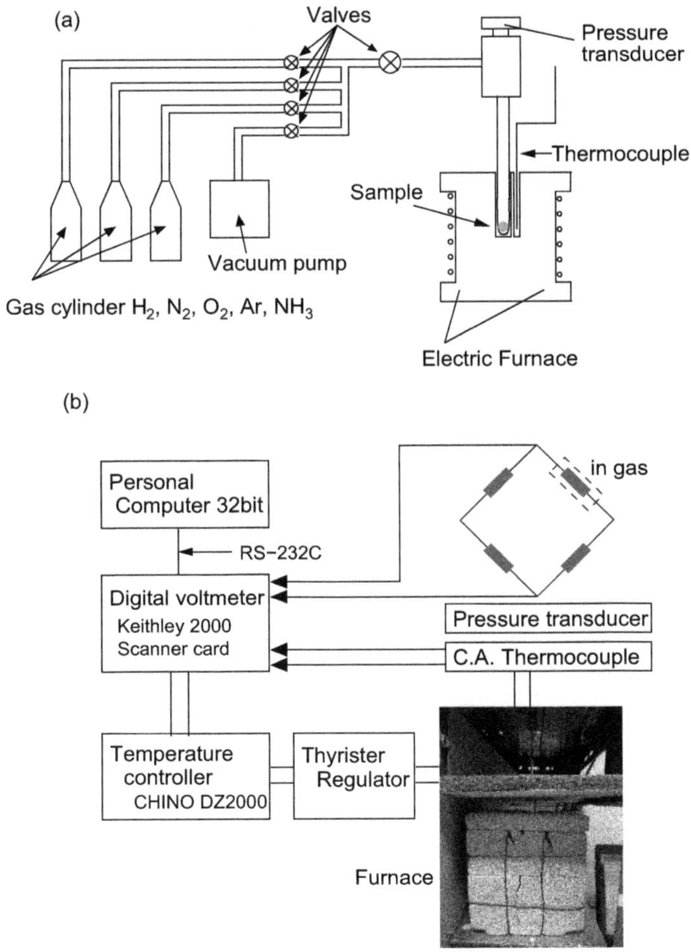

Figure 3.1 Schematic diagram of the thermopiezic analysis apparatus: (a) mechanical construction and (b) electrical connections.

The details of the designed and built thermopiezic analyzer are presented in the following order: structural construction (Fig. 3.1a) and electrical connections (Fig. 3.1b), a picture in its entirety (Fig. 3.2), components, an example of TPA scans (Figs. 3.3 and 3.4), and total performance.

Figure 3.1a shows a schematic diagram for the structural construction of the TPA apparatus. One sets the sample grains at the bottom of the quartz glass tube and evacuates and subsequently fills the tube at the desired gas pressure by handling the valves. Then, the gas pressure and sample temperature are monitored and recorded as a function of time by controlling the electric furnace in order to raise/reduce the temperature at the sample position at a constant rate.

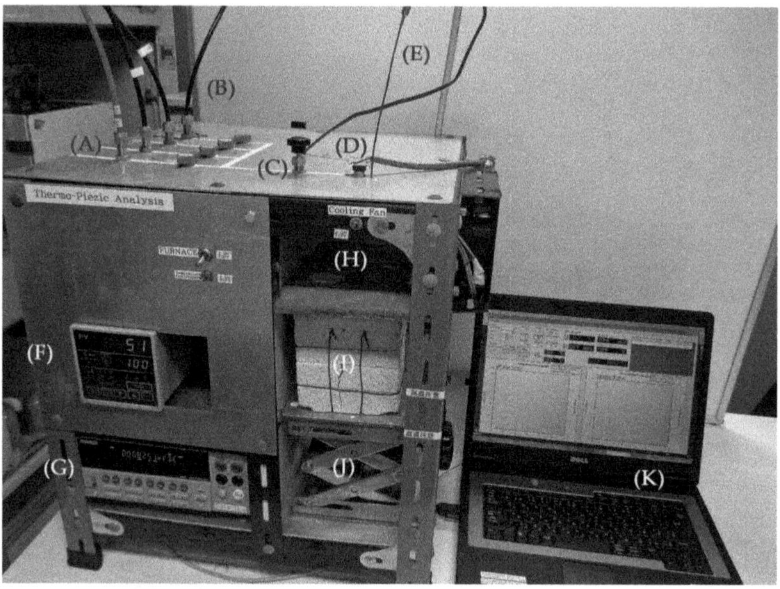

Figure 3.2 Thermopiezic analysis apparatus. Upper surface: Three gas transfer lines (A) and an evacuation line (B), vacuum valves (C), a pressure transducer (D), and a thermocouple (E). Front panel: A temperature controller (F) and a digital voltmeter (G). On the right: A quartz tube (not shown) (H) fitting the electric furnace (I) on a manual lift device (J) and a laptop computer (K) on the working table.

Figure 3.1b displays the electrical connections. The signals from the pressure transducer and the thermocouple are measured by a digital voltmeter with a scanner function. A laptop computer is in charge of digital data acquisition from the digital voltmeter and displaying all the current and accumulated data. A commercially available conventional temperature controller with a proportional-integral-derivative (PID) control function is equipped to operate the electric furnace.

Figure 3.2 shows a picture of the TPA apparatus in its entirety on the working desk. See the upper surface for gas transfer lines and an evacuation line, vacuum valves, a pressure transducer, and a thermocouple. On the front panel, one finds a temperature controller and a digital voltmeter. In the right part, there is a quartz glass tube (not shown) connected to the pressure transducer and an electric furnace on a manual lift device. A conventional laptop computer displays numeric data with full graphic capabilities.

3.2.1.1 Sample room

The sample room consists of a transparent quartz glass tube having a fine inner diameter of 2.1 mm with its tip end sealed, a pressure transducer, a needle valve, and a stainless-steel pipe connecting these parts. It has a net inner capacity of 990 mm^3, capable of about 1 millimole of gas at 0.2 MPa; half the net is from the quartz glass tube.

Small grains of solid samples are accommodated in the bottom part of the quartz glass tube, which fits into the electric furnace core. Although only a volume of 121 mm^3, 12% of the net volume, is placed in the high-temperature thermal environment, this gives an extra rise in sample pressure, for example, by 0.022 MPa at 600°C when the initial pressure is set at 0.2 MPa. A way of subtraction of this thermal background is described in section 3.2.1.5. To start the analysis, a set of evacuation by the oil-sealed rotary vacuum pump and subsequent filling of test gas in the sample room is repeated several times to reduce contamination of air as much as possible.

3.2.1.2 Pressure transducer

Requirements for a fast and accurate response to the fluctuation of pressure that neither greatly influences other properties nor impedes the miniaturization of the sample room are fulfilled by adapting a piezo-resistive transducer, a model FPM-15PAR from Fujikura Ltd.

A DC current is supplied to the transducer from the constant-voltage power source composed of a three-terminal voltage regulator. The unbalanced voltage in the bridge circuit in the transducer is amplified with a low-drift and ultra-high-precision operational amplifier device, enabling precise and reproducible measurement of the pressure sensed. A drift of the output voltage from the transducer by the change of room temperature is found to be totally canceled. Careful pressure calibration of the transducer shows that the measurable pressure range is from 0.013 to 0.243 MPa, for which the linearity is confirmed between the output voltage of the transducer and the applied pressure, at 94.6 Pa/mV.

3.2.1.3 Electric furnace

A miniature cylindrical electric furnace is designed to improve the temperature uniformity along the central vertical axis. A stainless round steel is machined to a cylindrical bobbin of 64 mm outer diameter and 80 mm length. Two cylindrical holes 40 mm in length are drilled in close proximity to one another in the bobbin, one for the sample tube and the other for the thermocouple. Kanthal® resistive wire of 14 SWG is wound onto the bobbin and fixed with cement. For thermal insulation, glass wool is packed between the bobbin and the stainless-steel cover tube whose outside is covered with firebricks, as shown in Figs. 3.1b and 3.2.

3.2.1.4 Temperature control and data acquisition

The sample temperature is measured with a sheathed thermocouple of type K, chromel-almel, set in another hole 5 mm away from the sample. The thermoelectric voltage is fed to a digital voltmeter, K2000 from Keithley, for the temperature reading and to the temperature controller, a model DZ 2000 from CHINO, for programmed temperature control of the furnace. The controlled output voltage from the temperature controller is supplied to the thyristor regulator to heat the furnace at required temperatures. Full controlling of the sample temperature is achieved both in increasing and decreasing runs at the largest rate of 10°C/min. up to 900°C.

A 32-/64-bit Microsoft Windows laptop is used to plot the gas pressure versus temperature data in real time and to record them for further manipulation in off-line mode. The Visual Basic 2010 Express toolkit from Microsoft is installed to build the measuring system.

The computer communicates with the K2000 digital voltmeter through a USB-RS-232C converter. Both time and temperature intervals for data acquisition are arbitrarily set even during running the program. A 10-channel scanner card is installed in K2000, and it provides a 2-channel selection of 2-pole signals, respectively, from thermocouple and pressure transducer circuit output.

The current temperature and pressure are displayed as a function of time on the screen at assigned intervals, and all the stored data are easily retrieved on the screen table and graphs.

3.2.1.5 Operating and system performance

Figure 3.3 provides an example of a thermopiezic scan for no sample set in the sample room. The temperature control program is set so as to give a relatively rapid increase in temperature at a constant rate of 3°C/min., to regulate the temperature at 514°C for 60 min., and then to lower the temperature at −3°C/min. The recorded dependence of temperature on time, $T(t)$, shows high actual temperature followability to a programmed target temperature in the whole temperature range except for a descending temperature run below 200°C where the furnace is cooled naturally. The transitional overshoot is reduced to less than 0.5°C, and 10 s are needed to stabilize the sample temperature in the ascending temperature scan.

The gas pressure sensed at the pressure transducer shows a concave curve as a function of temperature. There are two points to be noted for the $p(T)$ and $p(t)$ curves: the increasing blank pressure with temperature and the pressure leak being proportional to time.

The increment in pressure at the highest temperature, 0.02 MPa, is accounted for by assuming two isolated rooms with different volumes (respectively 0.88 and 0.12 of net volume) and their respective temperatures (30°C and 514°C). A blank signal in pressure is brought about by heating the test gas in the bottom part of the quartz glass tube touching the inner wall of the furnace core. It is not the absolute volumes but the relative volumes of the two parts of the tube that determine the relative magnitude of the blank signal; the upper part of the tube is kept at room temperature, while the bottom part is accommodated in the furnace core. The more important issue is how to keep the room temperature constant during the actual measurement time of 10 h. A change in room temperature by 1°C brings about a relative pressure change of 0.33% in the sample room in our TPA system according to Boyle–Charles's law.

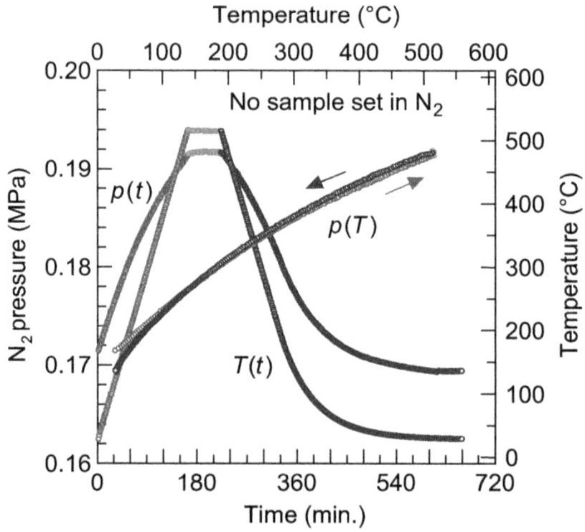

Figure 3.3 Thermopiezic analysis scan with no sample set in a N_2 gas atmosphere. N_2 gas pressure p and temperature T as functions of time t and N_2 pressure as a function of temperature $p(T)$ for increasing, constant, and decreasing scans are depicted by marks of different colors.

Gas leak occurrence is detected in the following two ways: one is by monitoring the pressure at room temperature regulated at a constant within 0.1°C, and the other is by comparing initial and final pressures after completion of the scan. The checking of the data in Fig. 3.3 gives (0.1694–0.1716) MPa/660 min. = -3.3×10^{-6} MPa/min. = -2.00×10^{-4} MPa/h for the initial pressure of 0.1716 MPa at room temperature, indicating that the leakage of test gas is negligibly small compared with the rates of test gas reactions with the test specimens in 1 min., which we will see in the following sections.

In Fig. 3.4, observed $p(T)$ curves are shown for increasing and decreasing runs, respectively. The descending process is not in accordance with the ascending one. Firstly, it is due to a technical thermal hysteresis brought about by a finite temperature gradient between the sample and the thermocouple. Secondly, the amount of gas leakage is proportional to the measuring time as described above. In the actual calibration of pressure as a function of temperature, subtraction of the blank $p(T)$ curves is made, respectively, for temperature increasing and decreasing runs.

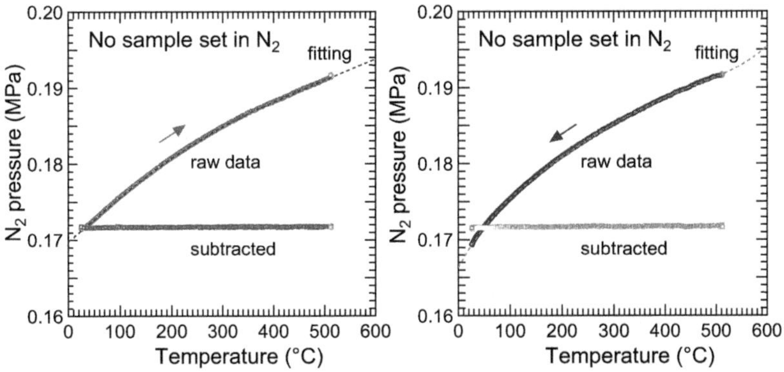

Figure 3.4 Fitting of curves of N_2 pressure as a function of temperature for temperature increasing (left) and decreasing (right) thermopiezic analysis scans with no sample set in a N_2 gas atmosphere. Subtracted curves are also depicted for increasing and decreasing temperature scans.

The weight of the reacted gas is determined by subtracting the initial total weight from that after the reaction. Here the total weight means the weight of the quartz glass tube plus that of the raw sample. The reaction products often stick to the inner surface of the quartz glass tube. The measured increase in the sample mass is in good agreement with that estimated from the pressure change in the TPA scan within 10%.

The system performance of the TPA apparatus is summarized in Table 3.1. See the text for details.

Table 3.1 Instrumental performance of TPA

Item	Performance
Temperature range	RT to 900°C
Heating rate	0°C to 10°C/min.
Temperature lag at 3°C/min.	<+/−1°C for increasing/decreasing
Measurement interval	>2 s
Pressure range	0.013 to 0.243 MPa
Resolution	2×10^{-8} moles
Leakage	-2.0×10^{-4} MPa/h for an initial pressure of 0.17 MPa

3.2.2 Sample Preparation

Single crystalline ingots of Sb_2Te_3, $(Bi_{0.5}Sb_{0.5})_2Te_3$, and $(Bi_{0.9}Sb_{0.1})_2Te_3$ were grown by using the Bridgman method in a gradient freeze (GF) furnace. Firstly, starting pure elements of Sb, Bi, and Te, with a purity of 99.9999%, from Osaka Asahi Metal Mfg. Co., Ltd., were sealed into a transparent quartz glass tube at a pressure of 10^{-3} Pa and melted at 810°C for 12 h. Then the single crystal was grown by giving a temperature gradient of 30°C at the solidification interface at a constant growth rate of 0.3 mm/h by controlling two independent electric heaters in the vertical GF furnace [6]. Thinly sliced disks were obtained from the central parts of the single crystalline ingots by using a spark erosion wire cutting machine, and they were crushed by hands to small grains for the TPA measurements and structural analyses. Cleavage planes are parallel to the grown axis in the ingots; for the cleaved planes, Sb_2Te_3 is more brittle than Bi_2Te_3 alloyed crystals.

3.2.3 Powder X-Ray Diffraction

A RIGAKU SmartLab® diffractometer with a parallel Cu K_α beam (45 kV, 100 mA) was used for collecting the reflections from a small amount of the powder sample. A structural analysis was done by using RIETAN-FP Rietveld refinement software [8] and the current ICDD database.

The grains of samples examined thermopiezically, typically 30 mg in weight, were ground for the powder XRD by using agate mortar and pestle. Since crystals of Sb_2Te_3 and its solid solutions with Bi_2Te_3 are easily cleaved, even powders tend to align to the pressed directions when pressed into the sample holder plate made of glass [12]. The focus was on the following: The powder was deposited in a shallow well of a glass sample holder and lightly pressed down by using a flat glass plate to pack it into the well to help minimize the preferred orientation. 20 mg powder of Sb_2Te_3 formed nearly a disk shape 10 mm in diameter on the glass sample holder. However, careful tuning of the "preferred-orientation parameter" was necessary to fit the reflection patterns in the Rietveld analysis, leading to a reduction in the values of the "reliability indices" and the "goodness-of-fit indicator."

3.3 Experimental Results and Discussion

In this section, the results of TPA measurements, microscope observations, and powder XRD measurements are presented for three (Bi, Sb)$_2$Te$_3$ alloy compounds and oxidization reaction processes are discussed on the basis of them.

3.3.1 Thermopiezic Analysis

Three (Bi, Sb)$_2$Te$_3$ alloy compounds were examined thermopiezically in Ar, N_2, or O_2 atmospheres. The data were collected as functions of time and temperature under the following conditions: starting pressures of 0.15–0.18 MPa as functions of time and temperature, by setting the heating curve at rates of +/−3°C/min. and at holding temperatures of 246°C–547°C and for holding time ranging from 0.5 to 6 h.

3.3.1.1 TPA for Sb_2Te_3 in Ar and N_2

The two graphs in Fig. 3.5 provide, respectively, the initial (#01) and subsequent (#02) TPA scans for Sb_2Te_3 with an initial mass of 22.80 mg in an Ar atmosphere. In #01 of Fig. 3.5, desorption behavior is found at temperatures above 250°C when the temperature increases, while when the temperature lowers, the gas pressure is constant below 400°C. The increase in gas pressure at room temperature after the initial scan (Fig. 3.5, #01), 0.1698 – 0.1680 = 0.0018 MPa, means that 0.02 moles of gas are desorbed from 1 mole of the Sb_2Te_3 sample. The second run (#02 of Fig. 3.5) shows less active desorption behavior than initial run #01. These two sets of scans indicate that desorption behavior in Sb_2Te_3 is certainly temperature and time dependent; however, its volume is negligibly small compared to the oxidation volume shown in the next.

The TPA of Sb_2Te_3 in a N_2 gas atmosphere also exhibits desorption behavior at temperatures higher than 300°C, as shown in Fig. 3.6. The increment of gas pressure at room temperature after the scan, 0.1697 – 0.1679 = 0.0018 MPa, gives the same amount of gas desorption from the tested Sb_2Te_3 as in Fig. 3.5, #01, in Ar. These increases in gas pressure above certain high temperatures in atmospheres of N_2

as well as Ar gas indicate that both Ar and N_2 gases don't react with Sb_2Te_3 in the present conditions of gas pressure, temperature, and heating time.

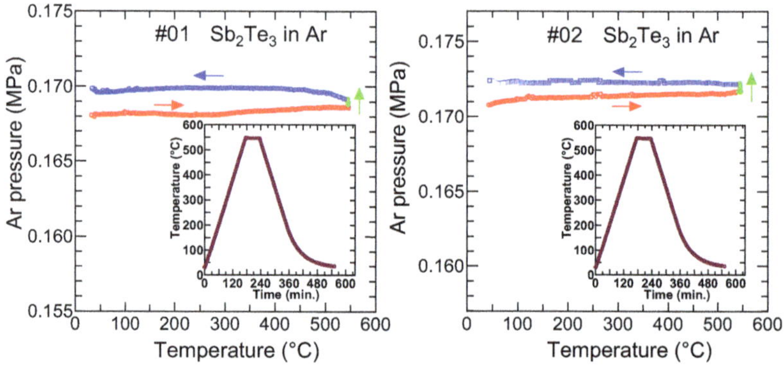

Figure 3.5 Thermopiezic analysis scans for Sb_2Te_3 in an Ar gas atmosphere; gas pressure versus temperature; insets show the heating curves as a function of time. Red, green, and blue arrows/points are, respectively, for increasing, constant, and decreasing temperature runs (at +3.0°C/min., 546°C for 60 min., and –3.0°C/min. down to 200°C and natural cooling to room temperature).

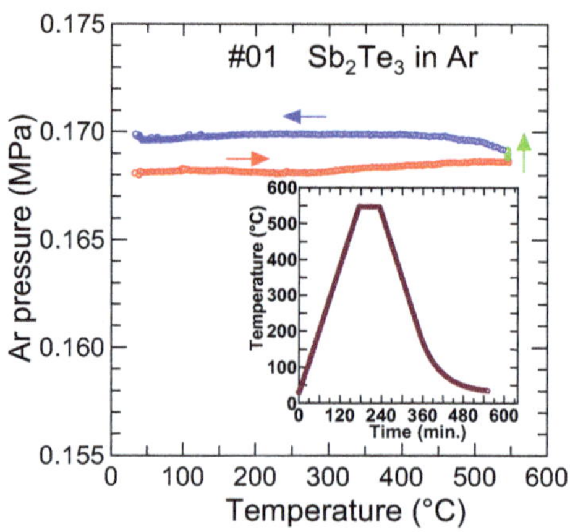

Figure 3.6 Thermopiezic analysis scan for Sb_2Te_3 in N_2 gas; gas pressure versus temperature; inset for the heating curve as a function of time. Red, green, and blue arrows/points are, respectively, for increasing, constant, and decreasing temperature runs (at +3.0°C/min., 546°C for 60 min., and –3.0°C/min. down to 200°C and natural cooling to room temperature).

3.3.1.2 TPA for Sb_2Te_3 in O_2

A set of consecutive TPA scans are made for two samples consisting of grains of single crystalline Sb_2Te_3; four scans (#01–#04) are made for a sample consisting of 9 relatively larger grains with a net weight of 50.03 mg, and eight scans (#11–#18) are done for 10 smaller grains with a net weight of 21.03 mg. It is stressed that the two sets of sample grains of different sizes are prepared for TPA scans; the mass of grains for the latter sample is reduced by as much as a factor of 3, ensuring the reaction(s) are improved more rapidly. The holding time for the samples being exposed in O_2 gas at the highest temperature is different in each scan; however, the net exposure time for the samples at temperatures above 400°C is set so as to be not very different between the two sets of samples. For the larger samples, the holding times in the four scans are set to be longer: 120, 120, 300, and 300 min.; for the smaller samples it is 60 min. for each scan.

Figure 3.7 displays the four consecutive TPA scans, #01–#04, for smaller size sample grains. In the virgin scan, #01, one observes a drastic absorption of O_2 gas above 420°C. 78% of the total absorption volume is observed for 40 min. in the run from 420°C to the holding temperature at 536°C, while at the holding temperature, 22% is absorbed for 120 min. It is noted that the relative number of moles of reacted O_2 gas in #01 is 0.74 to 1 mole of the Sb_2Te_3 sample. It should be also noted that there is another larger change in the reaction rate at 480°C, indicative of the existence of a complex reaction in Sb_2Te_3.

In the subsequent scans, #02–#04, one finds that the oxidation reaction rate is getting smaller even at temperatures above 400°C. In #03 and #04, where the sample is kept at 536°C for 300 min., oxidation takes place at 400°C but its oxidation rate with time is small.

Plots of pressure decrease with oxidation time at 536°C are presented in Fig. 3.8a for #01–#04. In this work, our TPA apparatus was not capable of instantaneous heating of the sample to 500°C from room temperature. We raised the sample temperature at a moderate rate of 3°C/min. from room temperature. Isothermal exposure examinations starting at a given high temperature were not performed; therefore, no exact information is available for the dependence of oxidation volume on time at a high temperature.

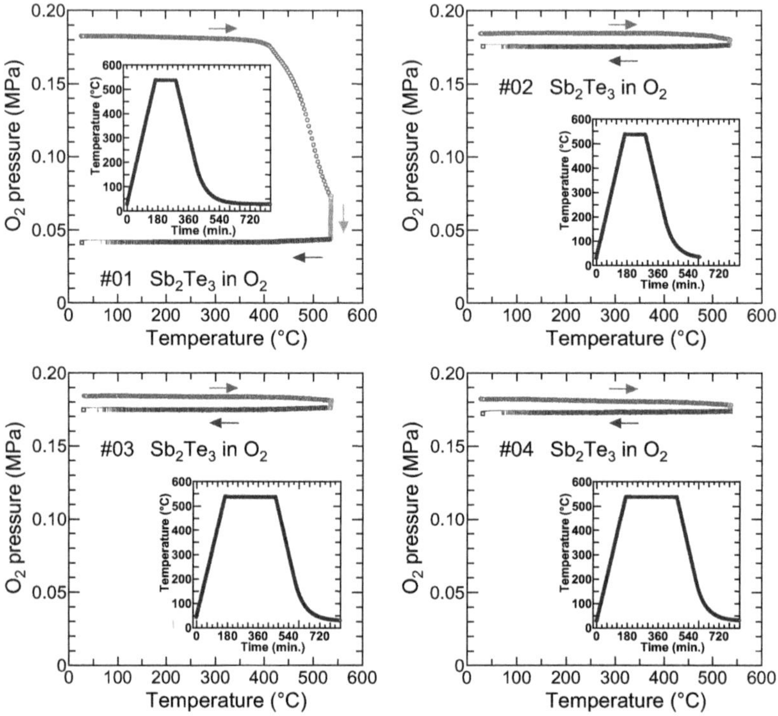

Figure 3.7 Thermopiezic analysis scans, #01–#04, for Sb_2Te_3 in O_2 gas; gas pressure versus temperature; insets show the heating curves with different holding times at a holding temperature of 536°C: 120, 120, 300, and 300 min. in order. Note that the scaling of the vertical line for pressure is altered from previous graphs for Ar and N_2 in Figs. 3.5 and 3.6.

However, the pressure variations over time at temperatures above 400°C and the isothermal oxidation at 536°C in #01 and #02 are suggestive of the oxidation kinetics obeying a parabolic law [13]. Various empirical rate laws have been observed for oxidations of many metals and alloys; a parabolic law is one of them; the increase in oxidation volume is proportional to the square root of exposure time. Wagner's parabolic model assumes that the growth of the oxide scale is controlled by metal and/or oxygen ions. For reactions at the holding temperature in #03 and #04, a parabolic law is not applicable; however, the logarithm of pressure-versus-time plot gives the relaxation time for oxidation reaction, which is greater than 10,000 min. in the scans of #03 and #04. These indicate that the oxidation reaction is not further accelerated after scan #01. A summation of reacted oxygen gas volume over the whole scans gives 0.89 moles of reacted O_2 gas to 1 mole of Sb_2Te_3. A comparison of the O_2 reaction volume is made later, with the case of smaller sample grains, in scans #11–#18.

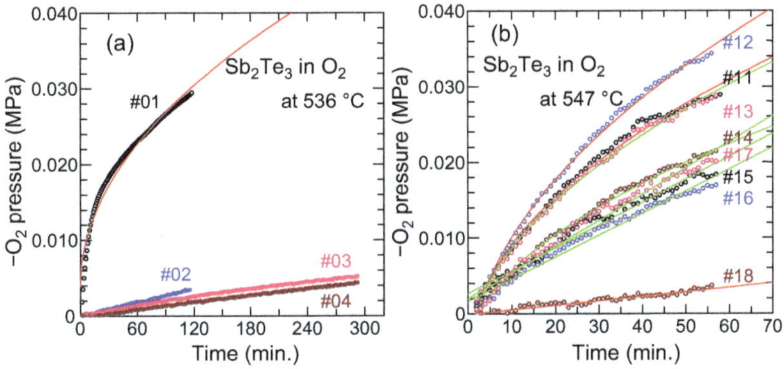

Figure 3.8 Pressure change data during isothermal oxidation of Sb_2Te_3; (a) For TPA scans #01–#04 at 536°C and (b) for scans #11–#18 at 547°C. See Figs. 3.7, 3.9, and 3.10 for starting pressures for each TPA run. Data points are least-squares fitted with the square root of exposure time and the fitted curves are shown by red solid lines.

Figures 3.9 and 3.10 show eight consecutive TPA scans for smaller-sized grains under the condition of a holding time being 60 min. at 547°C. In the virgin scan, #11, the largest amount of oxygen gas reacted with the sample among the series of scans. A marked change in the slope of $p(T)$ curve is found at 400°C, above which the reaction rate is larger. The estimated relative reaction volume of O_2 in #11 is 1.05, which is comparable to that in #01 and the overall volume for #01 to #04 for a larger grain sample. The setting of two steps at 246°C in the heating curve does not bring about any appreciable change in observed O_2 pressure in the vicinity of this temperature, supporting the fact that oxidation takes place in Sb_2Te_3 apparently at a higher temperature than 400°C.

It is noted that oxidation in the subsequent scan, #12, is comparably active to that in the virgin scan, #11. The reaction volume of O_2 is estimated to be 1.03 moles to 1 mole of Sb_2Te_3 in #12. Scans are repeated another six times, until oxidation is almost saturated in scan #18. Figure 3.8b displays that the reactions obey a parabolic law for scans up to #17, whereas in #18, the change in pressure is fairly small and shows a rather linear dependence on the exposure time.

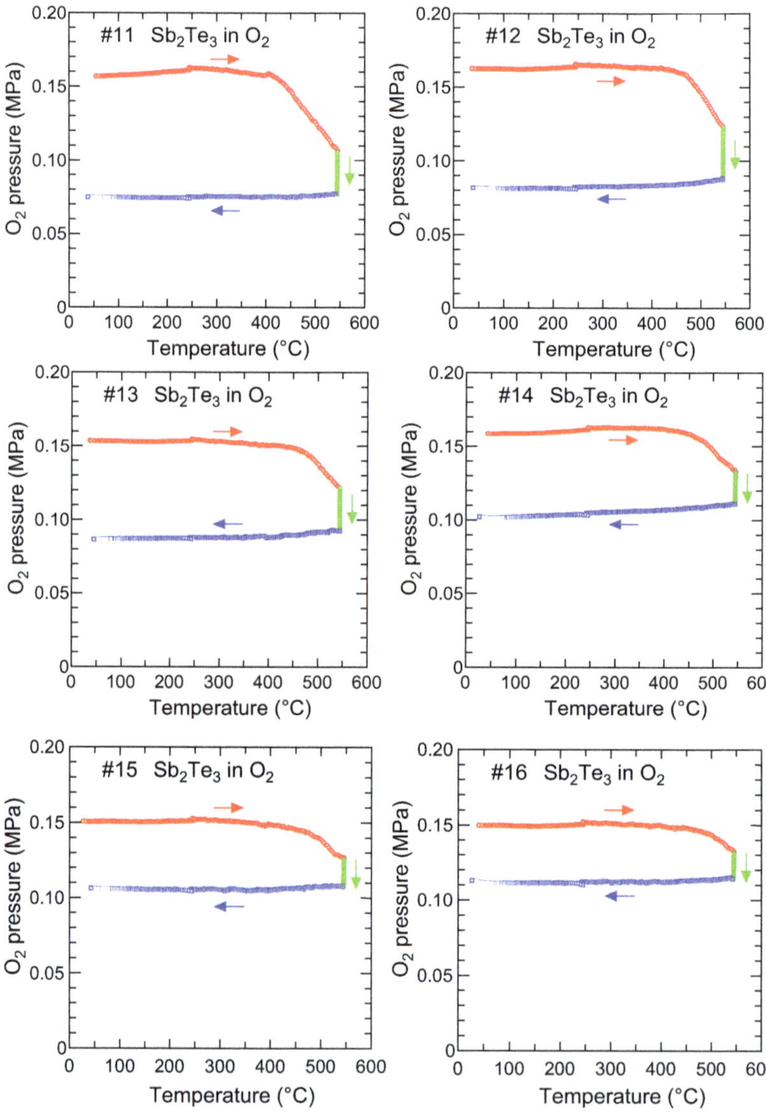

Figure 3.9 Thermopiezic analysis scans, #11–#16, for Sb_2Te_3 in O_2 gas; gas pressure versus temperature; the heating curve is depicted in Fig. 3.10. Red, green, and blue points are, respectively, for increasing, constant, and decreasing temperature runs (at +3.0°C/min., 547°C for 60 min., and –3.0°C/min. down to 200°C and natural cooling to room temperature).

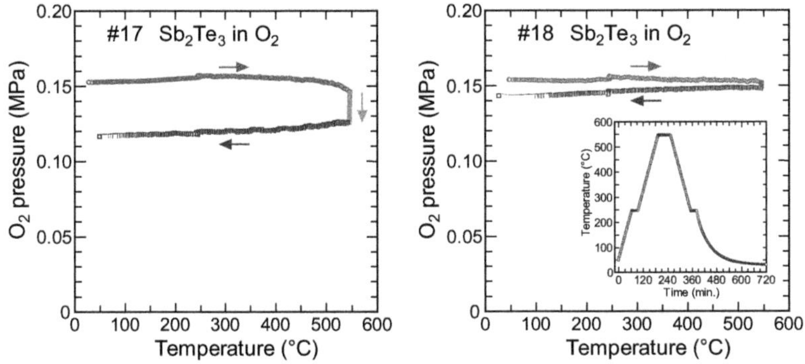

Figure 3.10 Subsequent thermopiezic analysis scans of #17 and #18 after #16 in Fig. 3.9, for Sb_2Te_3 in O_2 gas; gas pressure versus temperature; inset for the heating curve common to scans #11–#16 represented in Fig. 3.9; three steps: at 246°C for 30 min., 546°C for 60 min., and 246°C for 30 min.

The total relative mole number of reacted O_2 gas in these eight scans, #11–#18, amounts to 5.3 for these smaller sample grains; it is 6 times larger than that of the larger-sized grains. It should be also noted that the oxidation start temperature is 470°C for #12 to #18, suggesting a process to the next stage of reaction.

Later we will learn from the XRD analysis that reaction products in these two sets of experiments are different from each other: formation of $SbO_{1.5}$ and segregation of Te for the relatively larger grains and, on the other hand, complex reaction products of Sb_2O_4 and TeO_2 for the smaller grains. Totally different reaction volumes of O_2 gas observed between the two series of TPA scans may reflect the complex reaction in Sb_2Te_3 under an O_2 gas atmosphere at high temperatures. Observed different oxidation temperatures between #11 and the subsequent scans for the smaller-sized sample are also associated with the existence of a complex reaction leading to the formation of antimony tetroxides from trioxides. The former oxides, cervantite, are usually synthesized by heating the latter in O_2 [14–16].

3.3.1.3 TPA for $(Bi_{0.5}Sb_{0.5})_2Te_3$ in Ar and N_2

Figure 3.11 provides two TPA scans for $(Bi_{0.5}Sb_{0.5})_2Te_3$, one done in an Ar gas atmosphere and the other in a N_2 gas atmosphere. Note that the *Y* axis's scale is changed so that one can see easily the small variation of pressure with temperature. No distinct changes in the slope of the gas-pressure-versus-temperature curves are found, and the discrepancies in the room temperature pressure after the initial scan are about one-third of those for Sb_2Te_3. Therefore, we may conclude that desorption and absorption hardly take place in $(Bi_{0.5}Sb_{0.5})_2Te_3$ in Ar and N_2 gas atmospheres at temperatures up to 547°C.

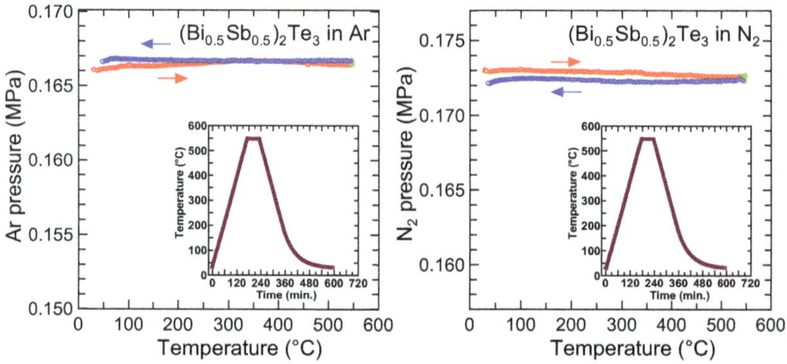

Figure 3.11 Thermopiezic analysis scans for $(Bi_{0.5}Sb_{0.5})_2Te_3$ in Ar (left) and N_2 (right) gases; gas pressure versus temperature; insets show the heating curves. Red, green, and blue points are, respectively, for increasing, constant, and decreasing temperature runs (at +3.0°C/min., 547°C for 60 min., and −3.0°C/min. down to 200°C and natural cooling to room temperature).

3.3.1.4 TPA for $(Bi_{0.5}Sb_{0.5})_2Te_3$ in O_2

TPA scans were made in O_2 gas atmospheres for two samples of 8 to 10 grains of $(Bi_{0.5}Sb_{0.5})_2Te_3$ single crystals, uniform in size, at holding temperatures of 421°C and 527°C, respectively.

Figure 3.12 shows four TPA scans for $(Bi_{0.5}Sb_{0.5})_2Te_3$ performed at a lower temperature, 421°C. One finds that temperature-activated oxidation is dominant in the temperature range from 375°C to the holding temperature of 421°C. In the holding temperature stage at 421°C, one finds a parabolic decrease in pressure for #01 and #02; on the other hand, there is an exponential decrease in pressure for #03 and #04, as shown in Fig. 3.13a, suggesting that the reaction rate is getting smaller in subsequent scans. The total amount of O_2 gas reaction is estimated to be 0.72 moles to 1 mole of $(Bi_{0.5}Sb_{0.5})_2Te_3$.

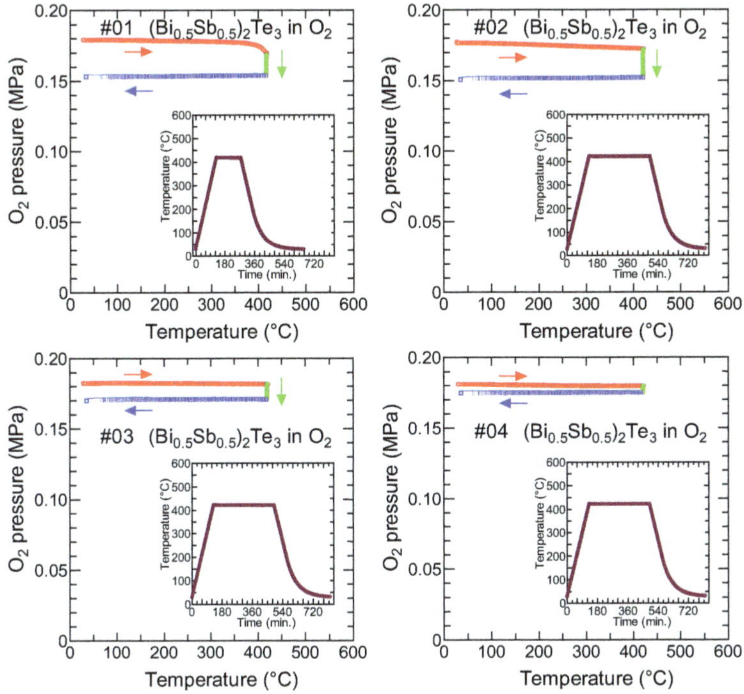

Figure 3.12 Thermopiezic analysis scans, #01–#04, for $(Bi_{0.5}Sb_{0.5})_2Te_3$ in O_2 gas; gas pressure versus temperature; insets show the heating curves. Red, green, and blue arrows/points are, respectively, for increasing, constant, and decreasing temperature runs (at +3.0°C/min.; 421°C for 150, 360, 360, and 360 min.; and –3.0°C/min. down to 200°C and natural cooling to room temperature).

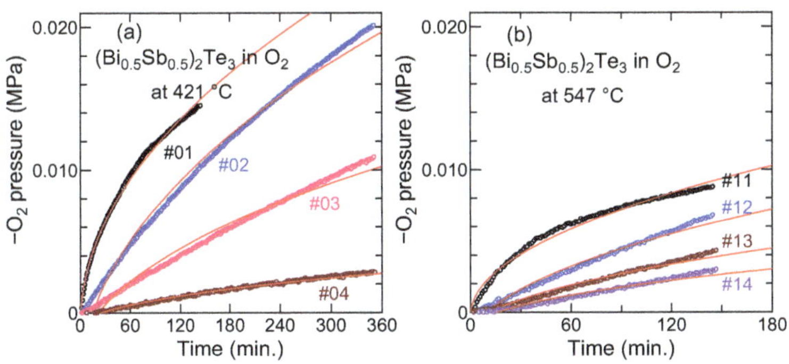

Figure 3.13 Pressure change data during isothermal oxidation of $(Bi_{0.50}Sb_{0.50})_2Te_3$; (a) for TPA scans #01–#04 at 421°C and (b) for scans #11–#14 at 547°C. See Figs. 3.12 and 3.14 for starting pressures for each TPA run. Data points are least-squares fitted with the square root of exposure time and the fitting curves are shown by red solid lines.

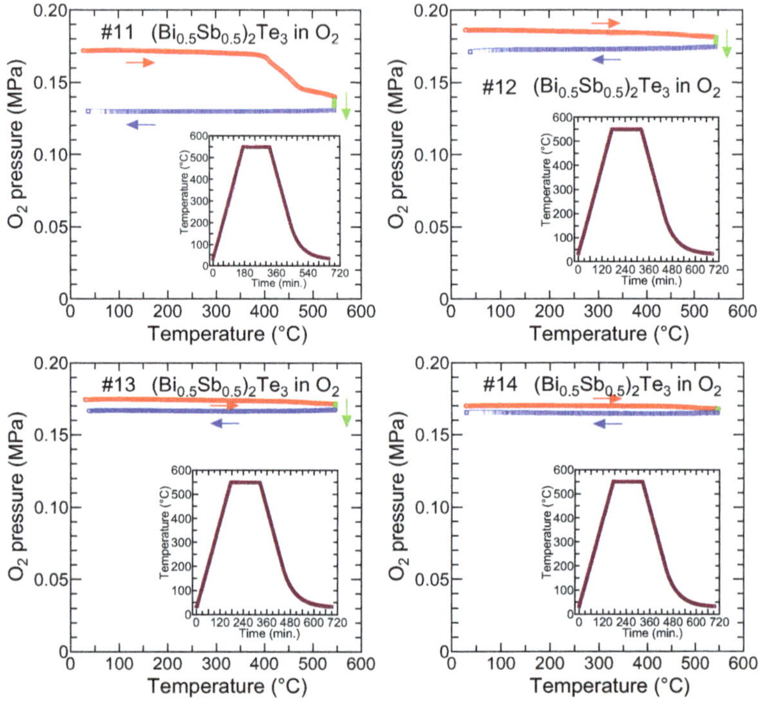

Figure 3.14 Thermopiezic analysis scans, #11–#14, for $(Bi_{0.5}Sb_{0.5})_2Te_3$ in O_2 gas; gas pressure versus temperature; insets show the heating curves. Red, green, and blue arrows/points are, respectively, for increasing, constant, and decreasing temperature runs (at +3.0°C/min., 547°C for 150 min., and –3.0°C/min. down to 200°C and natural cooling to room temperature).

Figure 3.14 shows the results of higher-temperature TPA scans #11–#14. In scan #11, oxidation reaction starts at 400°C and its reaction rate becomes small at 470°C with further increasing temperature. It is also noted that the reaction-volume-of-O_2-gas-versus-time curve follows a parabolic law in scan #11 (Fig. 3.13b). The subsequent scans, #12 and #13, show that oxidation becomes less active even at the holding temperature stage, following an exponential decrease with the exposure time. The net amount of reacted O_2 gas by the four scans is 0.75 moles to 1 mole of $(Bi_{0.5}Sb_{0.5})_2Te_3$. These results suggest that exposure of the sample at a higher temperature results in faster saturation behavior of oxidation, with the resultant net oxidation reaction volume being almost the same as that exposed at a lower temperature. The net exposure time for the two sets of TPA scans above 421°C is not so different for the two sets of scans done at different holding temperatures, one at 694 K (421°C) and the other at 824 K (547°C). However, as we will learn later from the XRD analysis, the reaction products are different in the degree of lattice expansion between the two sets of TPA scans conducted at different temperatures. Here, we note that the reactions taking place are highly dependent on the exposure temperature in $(Bi_{0.5}Sb_{0.5})_2Te_3$.

3.3.1.5 TPA for $(Bi_{0.9}Sb_{0.1})_2Te_3$ in Ar and N_2

Figure 3.15 provides two TPA scans for $(Bi_{0.9}Sb_{0.1})_2Te_3$, one conducted in an Ar gas atmosphere and the other in a N_2 gas atmosphere. Note that the *Y* axis scale is rescaled to show clearly the small variation in pressure with temperature. Under the Ar gas atmosphere, desorption is found in the $p(T)$ curve in the temperature range below 350°C, above which absorption is likely to take place to show a gradual decrease in pressure up to the holding temperature of 547°C. The curve attains the initial point at room temperature in the lowering run.

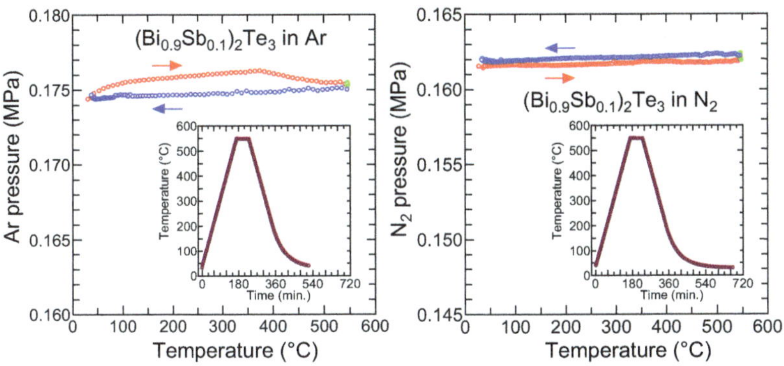

Figure 3.15 Thermopiezic analysis scans for $(Bi_{0.9}Sb_{0.1})_2Te_3$ in Ar (left) and N_2 (right) gases; gas pressure versus temperature; insets show the heating curves. Red, green, and blue points are, respectively, for increasing, constant, and decreasing temperature runs (at +3.0°C/min., 547°C for 60 min., and −3.0°C/min. down to 200°C and natural cooling to room temperature).

Exposure to the N_2 gas atmosphere shows neither desorption nor any appreciable reaction with the sample in the experimental conditions, as shown in Fig. 3.15 (right).

3.3.1.6 TPA for $(Bi_{0.9}Sb_{0.1})_2Te_3$ in O_2

A set of TPA scans was made at a holding temperature of 547°C for a sample of 15 grains (net mass = 19.22 mg) of $(Bi_{0.9}Sb_{0.1})_2Te_3$ single crystals of uniform size.

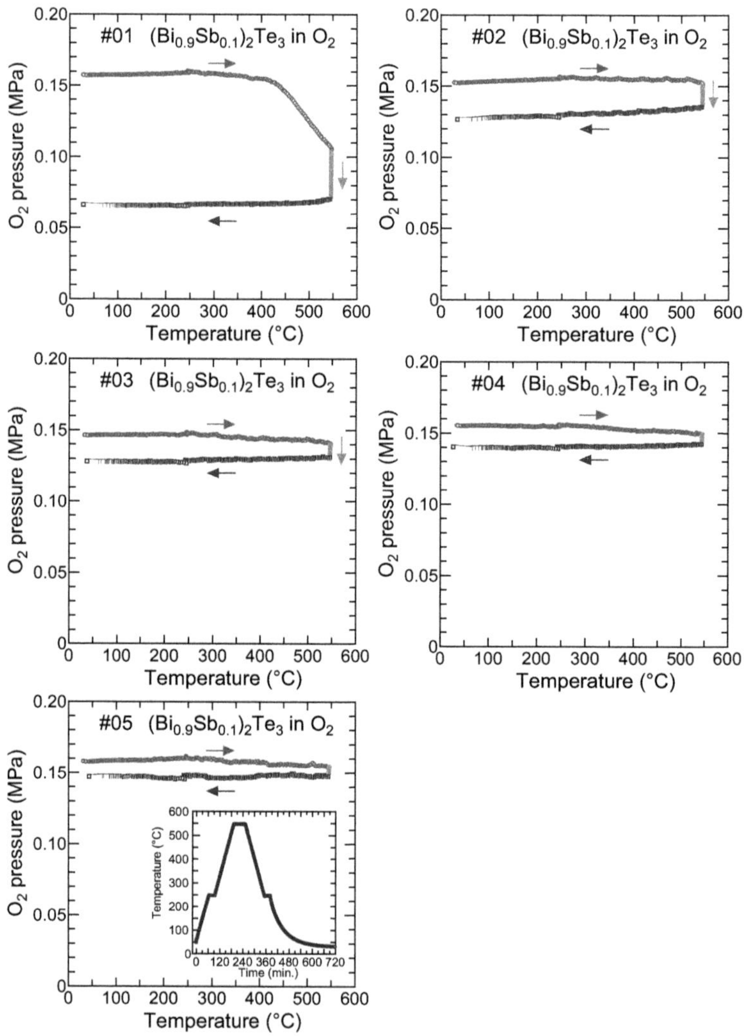

Figure 3.16 Thermopiezic analysis scans, #01–#05, for $(Bi_{0.9}Sb_{0.1})_2Te_3$ in O_2 gas; gas pressure versus temperature; inset shows the heating curve common to the scans of this sample; three steps: at 246°C for 30 min., at 547°C for 60 min., and at 246°C for 30 min.

Figure 3.16 displays five consecutive TPA scans for $(Bi_{0.9}Sb_{0.1})_2Te_3$ in an O_2 gas atmosphere. In scan #01, one finds that an oxidation reaction takes place at 420°C and proceeds with further increasing temperature up to the holding temperature of 547°C. In the holding temperature run, a parabolic plot with respect to the exposure time is possible (#01 in Fig. 3.17).

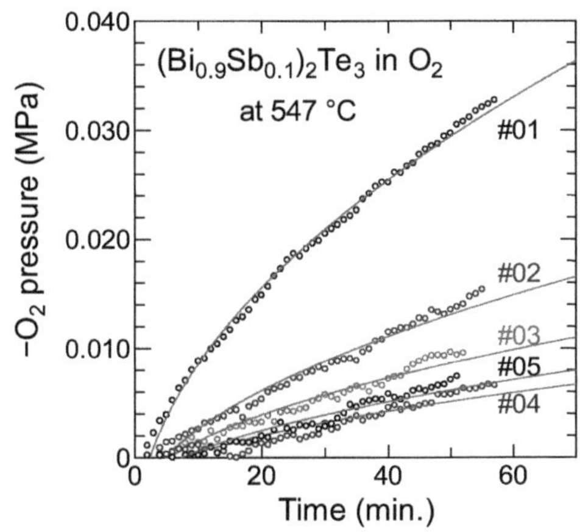

Figure 3.17 Pressure change data during isothermal oxidation of $(Bi_{0.9}Sb_{0.1})_2Te_3$; TPA scans #01–#05 at 547°C. See Fig. 3.16 for starting pressures for each TPA run. Data points are least-squares fitted with the square root of exposure time and the fitting curves are shown by red solid lines.

Subsequent scans show that the oxidation becomes less active and reaches saturation after the completion of #05. The technical saturation of the reaction is suggested by the increasing time constants for an exponential decrease in pressure with time: 1130 to 957 min. for #02 to #05, respectively. The net amount of the O_2 gas reaction is estimated to be 2.9 moles to 1 mole of $(Bi_{0.9}Sb_{0.1})_2Te_3$.

The relative amounts of reacted O_2 gas are compared among the $(Bi, Sb)_2Te_3$ alloys examined in this work. The smallest O_2 mole number, 0.7, is obtained for the reaction with $(Bi_{0.5}Sb_{0.5})_2Te_3$, indicating that the intermediate composition alloy of $(Bi_{0.5}Sb_{0.5})_2Te_3$ is most oxidation resistant in the pseudobinary alloys of $(Bi, Sb)_2Te_3$. Two end members of Sb_2Te_3 and $(Bi_{0.9}Sb_{0.1})_2Te_3$ exhibit fairly larger oxidation volumes, respectively 5.3 and 2.9, than the intermediate composition does. Although the reaction rates are dependent on primarily what the reaction products are and secondarily what the reaction rates for relevant complex reactions are under given temperature and exposure time conditions, the determined relative number of moles of reacted O_2 gas to Sb_2Te_3, 5.3, is explained when we consider the reaction product formulas that we will discuss later in the XRD analysis.

3.3.2 Microscope Observation

Figure 3.18 shows the microscope pictures for Sb_2Te_3 sample grains, respectively, taken before and after a series of TPA scan heatings in an O_2 atmosphere (#01 to #04). Grains of a raw Sb_2Te_3 sample typically 2 mm in length are prepared by crushing a bulk single crystal; the cleavage planes are shining brilliantly. After four sets of TPA scans in an O_2 gas atmosphere, whose $p(T)$ curves are shown in Fig. 3.7, most parts of surface are found to be colored brown with oxidation. Light-brown parts are also visible to indicate that there are multiple reaction products on the surface of the grains.

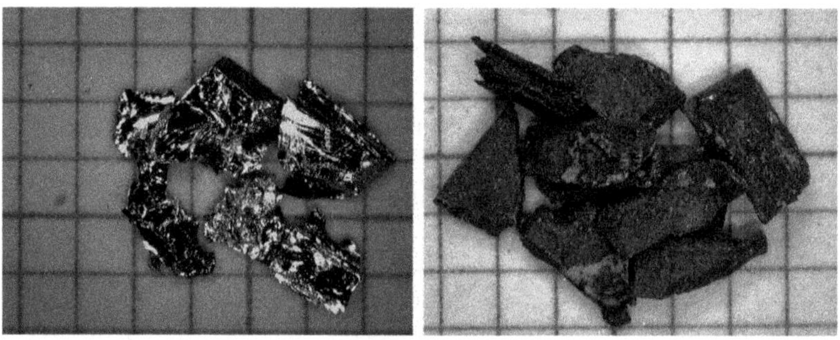

Figure 3.18 Microscope observation of raw (left) and oxidized (right) Sb_2Te_3 exposed in O_2 gas through #01–#04 at 536°C, on section paper with squares (side = 1 mm). Refer to text in Section 3.3.3.1 and Figs. 3.20 and 3.21.

Figure 3.19 shows microscope pictures of $(Bi_{0.5}Sb_{0.5})_2Te_3$ for raw and two oxidized samples exposed, respectively, at temperatures of 420°C and 547°C. Cleavage planes, hexagonal c planes, of the raw compound shine brilliantly. It is very interesting to note that the two samples exposed in O_2 at different temperatures exhibit totally different reaction products with each other. The surfaces of the sample oxidized at lower temperature, 421°C, are found to be covered with white-colored precipitates. On the other hand, oxidation at a higher temperature gives the grains the appearance of being semitransparent. The majority of the grains look dark brown, with a small but very appreciable surface area being colored light brown. We will learn later, from the powder XRD analysis in Section 3.3.3.2, that reaction products crystallize in the original structure with larger unit cells.

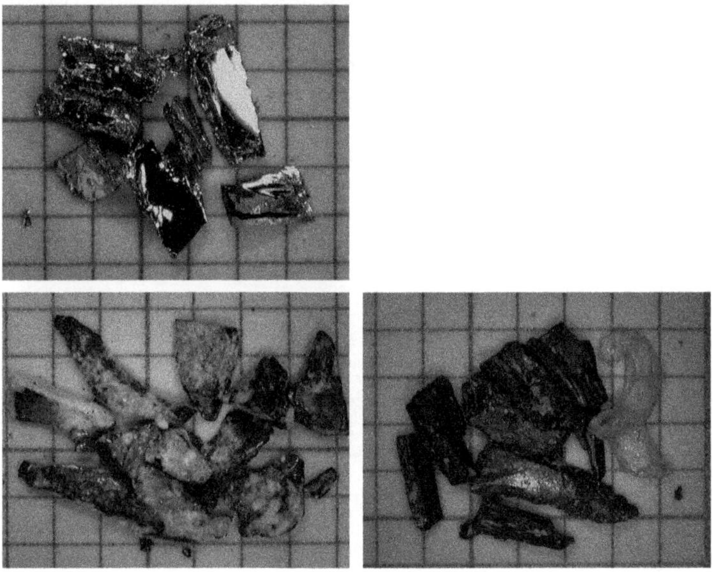

Figure 3.19 Microscope observation of raw (top) and two oxidized $(Bi_{0.5}Sb_{0.5})_2Te_3$ samples treated in different thermal conditions; lower (bottom left) and higher holding temperatures (bottom right). See text for details.

3.3.3 X-Ray Diffraction

3.3.3.1 XRD for Sb_2Te_3 in O_2

The structural properties of the two Sb_2Te_3 samples oxidized in different exposure temperatures and times, whose estimated relative number of moles of O_2 gas reacted with them is totally different from each other (respectively, 0.89 and 5.3 estimated by TPA), are compared.

Figure 3.20 shows the powder XRD pattern of a raw Sb_2Te_3 sample at 300 K. As a reference, International Tables for Crystallography,

Vol. A [17] has been used in association with RIETAN-FP. When we have assumed a hexagonal symmetry ($R\bar{3}m$, No. A-166) for Sb_2Te_3, the refinement has resulted in a reasonably good fit with the value of the reliability factor of the weighted profile, R_{wp} = 7.9%. The obtained structural parameters for the raw compound together with those of its oxidation products are summarized in Table 3.2. The lattice parameters of raw Sb_2Te_3 are in good agreement with those published [3, 5]. As noted in Section 3.2.3, the preferred orientation parameters were carefully tuned to give 0.4, 0.533, 0.6, 1.134, 0, and 0, respectively, for $r1, f1, f2, r2, f3$, and $r3$ for a best fit.

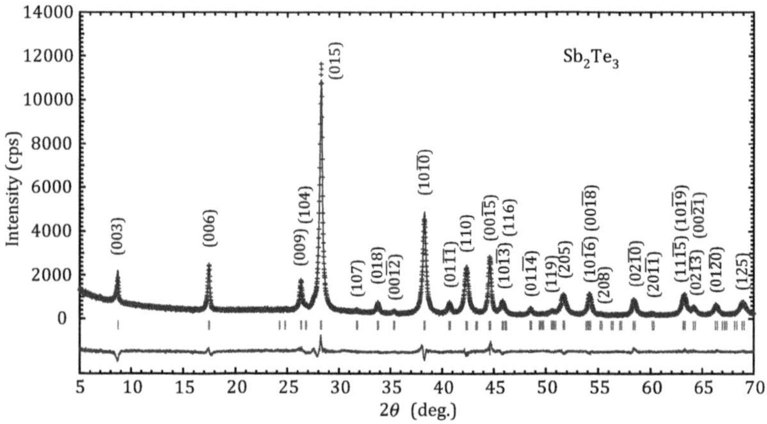

Figure 3.20 Powder X-ray diffraction pattern of raw Sb_2Te_3 at 300 K. The observed and calculated patterns are given by the plus signs and the solid line, respectively. The difference between observed and calculated intensities is indicated by the curve drawn at the bottom. Short vertical lines below the pattern indicate positions of possible reflections of Sb_2Te_3 in the space group $R\bar{3}m$ (No. A-166).

Figure 3.21 displays the XRD pattern of oxidized Sb_2Te_3 with a smaller O_2 reaction volume in the TPA scans of #01–#04 shown in Fig. 3.7. The refinement is converged with R_{wp} = 6.8%. One finds that the whole volume of the sample is not oxidized; the reflections from the mother phase remain with relatively stronger intensities than those from foreign phases appeared. Reaction products are assigned as a mixture of $SbO_{1.5}$ and Te when we assume cubic ($Fd\bar{3}m$, No. A-227) and trigonal ($P3_121$, No. A-152) symmetries for them, respectively. The corrected mole fractions are 0.484, 0.164, and 0.352, respectively, for mother, $SbO_{1.5}$, and Te constituents; the formation of $SbO_{1.5}$ oxide and the segregation of Te take place by exposing the sample in an O_2 atmosphere at 536°C for a total of 14 h.

It should be also noted for the comparison of mother phase modifications by oxidation in the pseudobinary alloy series of Sb_2Te_3-Bi_2Te_3 that the crystal unit cell of Sb_2Te_3 is not modified substantially when exposed in O_2 under the condition at 536°C; the cell volume remains almost constant. Refer to the lattice constants of raw Sb_2Te_3 listed in Table 3.2.

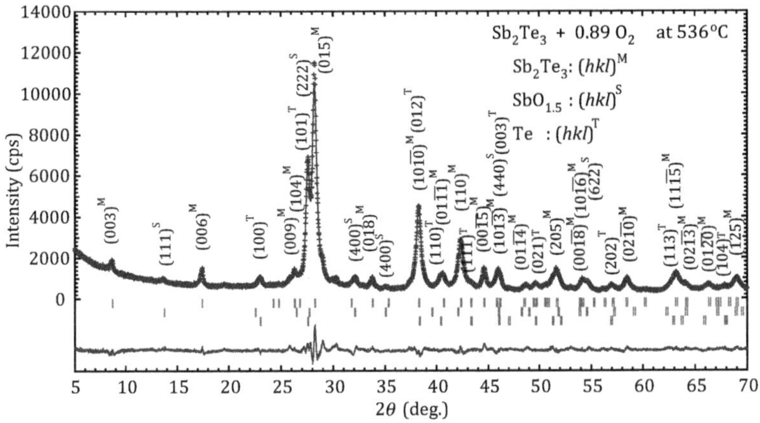

Figure 3.21 Powder X-ray diffraction pattern of Sb_2Te_3 after O_2 gas TPA scans #01–#04, whose oxidization TPA curves are shown in Fig. 3.7. The majority phase is the mother compound of Sb_2Te_3. Reaction products are $SbO_{1.5}$ and Te, respectively, with relative molar fractions of 0.16 and 0.35 to Sb_2Te_3. The strongest reflection is from Sb_2Te_3 at $2\theta = 28.223$ degrees, indexed as $(015)^M$, and the reflections appearing at its lower-angle side are, respectively, $SbO_{1.5}$ $(222)^S$ and Te $(101)^T$. See the starting-phase XRD pattern in Fig. 3.20 for comparison.

Figure 3.22 shows the XRD pattern of the sample oxidized with a larger resultant O_2 reaction volume in the TPA scans #11–#18 shown in Figs. 3.9 and 3.10. The refinement is converged with R_{wp} = 6.5%. We see that reflections of the mother Sb_2Te_3 compound are even further suppressed and peaks from reaction products are dominant in the profile. It should be noted that reaction products are different from those when less amount of O_2 gas is reacted; reaction products are assigned as a mixture of Sb_2O_4 and TeO_2 when we assume orthorhombic ($Pna2_1$, No. A-33) and tetragonal ($P4_12_12$, No. A-92) symmetries for them, respectively.

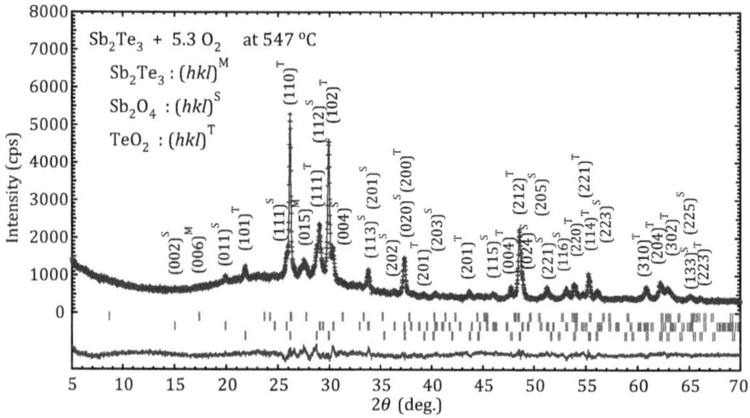

Figure 3.22 Powder X-ray diffraction pattern of Sb_2Te_3 after O_2 gas TPA scans #11–#18, whose oxidization TPA curves are shown in Figs. 3.9 and 3.10. The reaction products are Sb_2O_4 and TeO_2, respectively, with relative molar fractions of 0.25 and 0.71 to Sb_2Te_3. The most intense reflection is indexed as TeO_2 (110) at 2θ = 26.167 degrees. See the pattern of the starting phase in Fig. 3.20 and that of the less oxidized sample in Fig. 3.21 for comparison.

Estimations of corrected stoichiometries of the products are made by fixing each occupancy of the sites by constituent elements in assumed crystal unit cells at unity. The obtained mole fractions are 0.04, 0.25, and 0.71, respectively, for mother, Sb_2O_4, and TeO_2 components. This compositional analysis is consistent with the observed reflection from the mother phase assigned as $(015)^M$, which is the strongest one in the profile for raw Sb_2Te_3; see the pattern for raw Sb_2Te_3 shown in Fig. 3.20 for comparison. Most parts of the mother grains are decomposed into antimony and tellurium oxides; TeO_2 occupies the greater part of the product at a mole percentage of 71% to the composite product.

More interesting is the larger expansion of the crystal cell of the mother phase whose lattice constants are listed in Table 3.2; the expansion by 2.4% is proved in the *c* plane of the hexagonal crystal, while the lattice constant *c* remains almost constant, resulting in a cell volume expansion by 5.2%. The degree of cell expansion observed in oxidized Sb_2Te_3 will be compared with that in its solid solutions with Bi_2Te_3 in the following sections.

Table 3.2 Structural parameters of Sb_2Te_3 compound and its reaction products in an O_2 gas atmosphere refined by the Rietveld profile fitting

Sample R_{wp}, S	Product SG	w	X	a (Å)	b (Å)	c (Å)	V (Å^3)
Sb_2Te_3 R_{wp} = 7.878 S = 1.6853	Sb_2Te_3 $R\bar{3}m$			4.26869 (0.00014)	4.26869 (0.00014)	30.47911 (0.00070)	480.9747 (0.0248)
Sb_2Te_3 536°C	Sb_2Te_3 (Z=3) $R\bar{3}m$	0.8153	0.4844	4.26633 (0.00025)	4.26633 (0.00025)	30.47171 (0.00217)	480.326 (0.0526)
R_{wp} = 6.795 S = 1.9447	$SbO_{1.5}$ (Z=32) $Fd\bar{3}m$	0.0641	0.1635	11.15399 (0.00001)	11.15399 (0.00001)	11.15399 (0.00001)	1387.6832 (0.0025)
	Te (Z=3) $P3_121$	0.1207	0.3520	4.46101 (0.00069)	4.46101 (0.00069)	5.92205 (0.00194)	101.8797 (0.0324)
Sb_2Te_3 547°C	Sb_2Te_3 (Z=3) $R\bar{3}m$	0.1150	0.0395	4.37051 (0.00612)	4.37051 (0.00612)	30.58074 (0.06913)	505.8744 (1.5200)
R_{wp} = 6.487 S = 1.5459	Sb_2O_4 (Z=4) $Pna2_1$	0.3559	0.2486	5.43963 (0.00115)	4.81537 (0.00088)	11.7746 (0.00155)	308.4223 (0.0951)
	TeO_2 (Z=4) $P4_12_12$	0.5289	0.7119	4.81224 (0.00012)	4.81224 (0.00012)	7.61549(0.00037)	176.3571 (0.0107)

Note: Experimental conditions; reliability factor of weighted profile R_{wp}; reliability factor of goodness of fit S; reaction products and each crystal space group; mass/mole fractions corrected for microabsorption w/X; lattice constants a, b, and c; and unit cell volume V with standard deviation (error) in parentheses

Structural parameters of Sb_2Te_3 compound and its reaction products in an O_2 gas atmosphere refined by the Rietveld profile fitting are summarized in Table 3.2.

3.3.3.2 XRD for $(Bi_{0.5}Sb_{0.5})_2Te_3$ in O_2

The oxidation process is compared between the two samples of $(Bi_{0.5}Sb_{0.5})_2Te_3$ oxidized at different exposure temperatures, whose resultant reaction volumes of O_2 determined by the TPA are very close to each other: 0.72 for low-temperature exposure and 0.75 for high-temperature exposure.

Figure 3.23 shows the powder XRD pattern of a raw $(Bi_{0.5}Sb_{0.5})_2Te_3$ sample at 300 K. As a reference, space group No. A-166 ($R\bar{3}m$) in International Tables for Crystallography, Vol. A [17] has been used for $(Bi_{0.5}Sb_{0.5})_2Te_3$ structural analysis in association with RIETAN-FP. A reasonably good fit is obtained with the value of the reliability factor of the weighted profile R_{wp} = 7.7%. The lattice parameters of raw $(Bi_{0.5}Sb_{0.5})_2Te_3$ are in good agreement with those published [5]. Obtained parameters for the raw compound together with those of its oxidation products are summarized in Table 3.3.

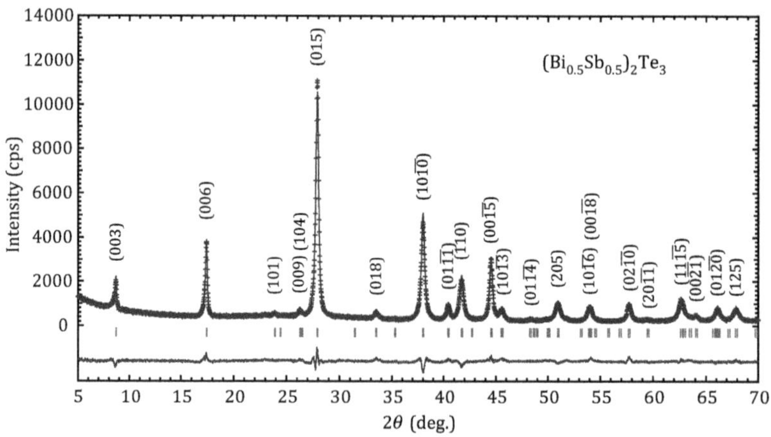

Figure 3.23 Powder X-ray diffraction pattern of raw $(Bi_{0.5}Sb_{0.5})_2Te_3$ at 300 K. The observed and calculated patterns are given by the plus signs and the solid line, respectively. The difference between observed and calculated intensities is indicated by the bottom curve. Short vertical lines below the pattern indicate positions of possible reflections of $(Bi_{0.5}Sb_{0.5})_2Te_3$ in space group $R\bar{3}m$ (No. A-166).

Figure 3.24 shows the powder XRD pattern of $(Bi_{0.5}Sb_{0.5})_2Te_3$ exposed at 421°C in the TPA scans #01–#04, shown in Fig. 3.12. Of particular interest is that all the reflections are indexed by assuming the original mother crystal symmetry with modified lattice parameters with R_{wp} = 8.5%. There is no trace of oxide compounds in the XRD pattern.

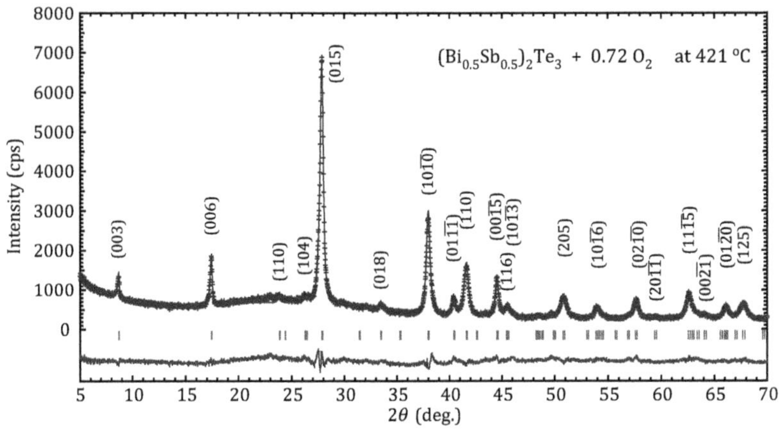

Figure 3.24 Powder X-ray diffraction pattern after TPA scanned $(Bi_{0.5}Sb_{0.5})_2Te_3$ at 421°C. The observed and calculated patterns are given by the plus signs and the solid line, respectively. The difference between observed and calculated intensities is indicated by the bottom curve. Short vertical lines below the pattern indicate positions of possible reflections of $(Bi_{0.5}Sb_{0.5})_2Te_3$ in the space group $R\bar{3}m$ (No. A-166).

One finds anisotropic cell expansion and shrinkage: an increase by 0.22% in the lattice constant a and a decrease by 0.07% along the c direction, with a resultant expansion by 0.37% in the hexagonal unit cell.

Figure 3.25 shows the XRD pattern of the $(Bi_{0.5}Sb_{0.5})_2Te_3$ sample exposed at 547°C. All the reflections are indexed with a hexagonal symmetry with R_{wp} = 8.1% by referring to No. A-166 ($R\bar{3}m$) in International Tables for Crystallography, Vol. A [17]. It should also be noted that no antimony and tellurium oxides are formed in this higher-temperature reaction in an O_2 atmosphere.

Table 3.3 Structural parameters of $(Bi_{0.5}Sb_{0.5})_2Te_3$ compound and its reaction products in an O_2 atmosphere refined by the Rietveld profile fitting

Sample R_{wp}, S	Product SG	a (Å)	b (Å)	c (Å)	V (Å³)
$(Bi_{0.5}Sb_{0.5})_2Te_3$ R_{wp} = 7.739 S = 1.7081	$(Bi_{0.5}Sb_{0.5})_2Te_3$ $R\bar{3}m$	4.32970 (0.00015)	4.32970 (0.00015)	30.51944 (0.00078)	495.4765 (0.0279)
$(Bi_{0.5}Sb_{0.5})_2Te_3$ 421°C R_{wp} = 8.538 S = 1.9289	$(Bi_{0.5}Sb_{0.5})_2Te_3$ $R\bar{3}m$	4.33905 (0.00027)	4.33905 (0.00027)	30.49763 (0.00163)	497.2948 (0.0507)
$(Bi_{0.5}Sb_{0.5})_2Te_3$ 547°C R_{wp} = 8.069 S = 1.7516	$(Bi_{0.5}Sb_{0.5})_2Te_3$ $R\bar{3}m$	4.35140 (0.00028)	4.35140 (0.00028)	30.49856 (0.00208)	500.1135 (0.0567)

Note: Experimental conditions; reliability factor of weighted profile R_{wp}; reliability factor of goodness of fit S; reaction products and each crystal space group; mass/mole fractions corrected for microabsorption w/X; lattice constants a, b, and c; and unit cell volume V with standard deviation (error) in parentheses

Compared to the $(Bi_{0.5}Sb_{0.5})_2Te_3$ exposed at the lower temperature, the sample exposed at the higher temperature shows larger cell expansion: an increase in the lattice parameter a by 0.50%; on the other hand a decrease of 0.07% in c results in a volume expansion by 0.94%. It is an extremely interesting finding that we have succeeded in synthesizing solid solutions of $(Bi, Sb)_2Te_3$ with interatomic distances expanded in the hexagonal c plane while along the c axis the distances between stacked c planes are shortened by exposing the starting materials in O_2 gas at temperatures higher than 400°C.

Table 3.3 lists the structural parameters for $(Bi_{0.5}Sb_{0.5})_2Te_3$ compound and its reaction products in an O_2 atmosphere refined by the Rietveld profile fitting.

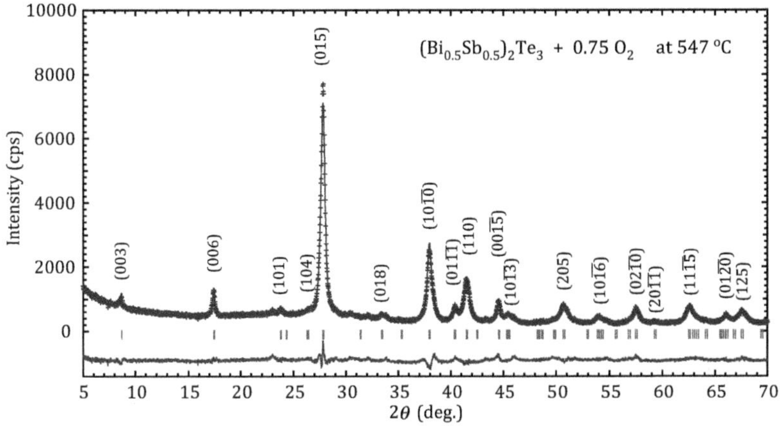

Figure 3.25 Powder X-ray diffraction pattern after TPA scanned $(Bi_{0.5}Sb_{0.5})_2Te_3$ at 547°C. The observed and calculated patterns are given by the plus signs and the solid line, respectively. The difference between observed and calculated intensities is indicated by the bottom curve. Short vertical lines below the pattern indicate expected positions of reflections of $(Bi_{0.5}Sb_{0.5})_2Te_3$ in space group $R\bar{3}m$ (No. A-166).

3.3.3.3 XRD for $(Bi_{0.9}Sb_{0.1})_2Te_3$ in O_2

Oxidation of $(Bi_{0.9}Sb_{0.1})_2Te_3$ does not bring about formation of antimony or tellurium oxides such as Bi_2O_3, $SbO_{1.5}$, and Sb_2O_4 but exhibits the formation of multiple oxides of Bi_2TeO_5. In this section, the powder patterns of raw and oxidized $(Bi_{0.9}Sb_{0.1})_2Te_3$ are shown and the formation of a modified mother phase compound with larger lattice parameters are presented.

Figure 3.26 shows the powder XRD pattern of raw $(Bi_{0.9}Sb_{0.1})_2Te_3$ at 300 K. We have assumed a hexagonal symmetry ($R\bar{3}m$, No. A-166) for $(Bi_{0.9}Sb_{0.1})_2Te_3$. Rietveld refining confirms the hexagonal structure of crystals with R_{wp} = 8.9%. See the lattice parameters, listed in Table 3.4, which are in good agreement with the literature [5, 6].

Figure 3.27 displays the XRD pattern of the $(Bi_{0.9}Sb_{0.1})_2Te_3$ sample exposed at 547°C. It shows not only hexagonal peaks but also comparably strong peaks from the reaction product of Bi_2TeO_5. All the reflections are indexed and refined with R_{wp} = 12.9% with the hexagonal ($R\bar{3}m$) for $(Bi_{0.9}Sb_{0.1})_2Te_3$ and orthorhombic (No. A-39, *Aem*2) for Bi_2TeO_5.

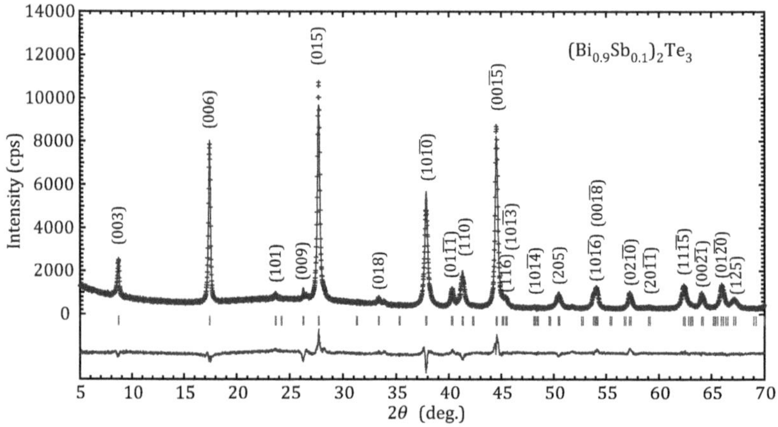

Figure 3.26 Powder X-ray diffraction pattern of raw $(Bi_{0.9}Sb_{0.1})_2Te_3$ at 300 K. The observed and calculated patterns are given by the plus signs and the solid line, respectively. The difference between observed and calculated intensities is indicated by the bottom curve, indicating less amount of the mother phase. Short vertical lines below the pattern indicate positions of possible reflections of $(Bi_{0.9}Sb_{0.1})_2Te_3$ with the space group $R\bar{3}m$ (No. A-166).

The lattice parameter *a* is increased by as much as 0.86%, while *c* is decreased by 1.4%, resulting in the cell volume remaining almost unchanged in dimension in the form of $(Bi_{0.9}Sb_{0.1})_2Te_3$. At this point, we may conclude that this anisotropic cell volume expansion of the mother phase brought about by oxidation is a common feature of pseudobinary $(Bi, Sb)_2Te_3$ compounds. More interestingly, we find the appearance of orthorhombic Bi_2TeO_5 compound in the oxidation of $(Bi_{0.9}Sb_{0.1})_2Te_3$, in contrast to the oxidation in $(Bi_{0.5}Sb_{0.5})_2Te_3$, where no antimony tellurites are produced. The determined lattice parameters of Bi_2TeO_5, listed in Table 3.4, are in fairly good agreement with the literature [18–22]. Previous studies to synthesize Bi_2TeO_5 were carried out by the solid-state reaction of a mixture of Bi_2O_3 and TeO_2 at 690°C [20], 650°C–750°C [21], and 650°C–850°C [23].

Table 3.4 Structural parameters of $(Bi_{0.9}Sb_{0.1})_2Te_3$ compound and its reaction products in an O_2 atmosphere refined by the Rietveld profile fitting

Sample R_{wp}, S	Product SG	w	X	a (Å)	b (Å)	c (Å)	V (Å^3)
$(Bi_{0.9}Sb_{0.1})_2Te_3$ R_{wp} = 8.862 S = 2.1449	$(Bi_{0.9}Sb_{0.1})_2Te_3$ $R\bar{3}m$			4.37396 (0.00019)	4.37396 (0.00019)	30.5182 (0.00053)	505.6375 (0.0330)
$(Bi_{0.9}Sb_{0.1})_2Te_3$ 547°C	$(Bi_{0.9}Sb_{0.1})_2Te_3$ (Z = 3) $R\bar{3}m$	0.6629	0.7076	4.41190 (0.00149)	4.41190 (0.00149)	30.09183 (0.01402)	507.2588 (0.3385)
R_{wp} = 12.872 S = 2.6848	$Bi_4Te_2O_{10}$ (Z = 4) *Aem*2	0.3923	0.2924	11.59344 (0.00075)	16.44408 (0.00167)	5.52076 (0.00049)	1052.497 (0.1574)

Note: Products are listed with their space group (SG); mass/mole fractions corrected for microabsorption w and X; lattice constants a, b, and c; and the unit cell volume V. R_{wp} and S are, respectively, the factor of weighted profile and the factor of goodness of fit.

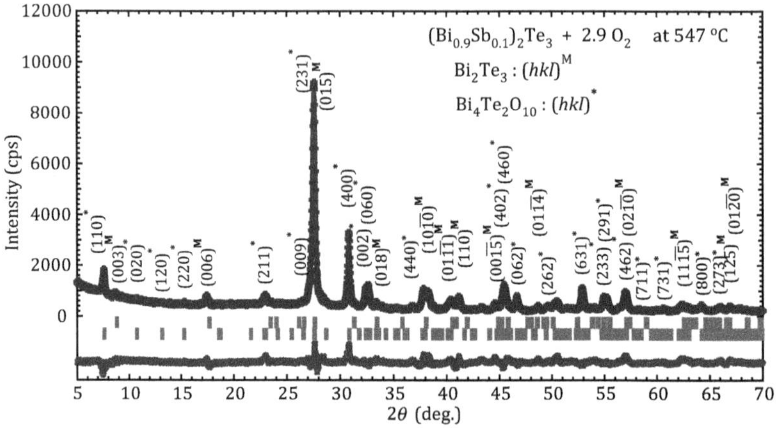

Figure 3.27 Powder X-ray diffraction pattern of $(Bi_{0.9}Sb_{0.1})_2Te_3$ after O_2 gas exposure in the subsequent TPA scans #01–#05, whose oxidization curves are shown in Fig. 3.16. The observed and calculated patterns assumed for the mixture of $(Bi_{0.9}Sb_{0.1})_2Te$ and Bi_2TeO_5 compounds are given by the plus signs and the solid line, respectively. The difference between observed and calculated intensities is indicated by the bottom curve, indicating less amount of the mother phase. Short vertical lines below the pattern indicate positions of possible reflections of $(Bi_{0.9}Sb_{0.1})_2Te_3$ with the space group $R\bar{3}m$ (No. A-166) and Bi_2TeO_5 with the space group *Abm*2 (No. A-39).

3.4 Discussions

In this section, comparisons of the reaction volume of O_2 gas, reaction products, and resultant anisotropic lattice expansions in the mother phase are made for Sb_2Te_3, $(Bi_{0.5}Sb_{0.5})_2Te_3$, and $(Bi_{0.9}Sb_{0.1})_2Te_3$, each representative of an edge, centered, and another nearly edge in the pseudobinary alloy series of Sb_2Te_3-Bi_2Te_3. A possible tuning of an optimum thermoelectric performance by oxidation of the (Bi, Sb)$_2$Te$_3$ pseudobinary system is proposed.

3.4.1 Oxidation Resistance

TPA measurements are found to be a powerful means for getting quantitative information about the intensity of the reaction temperature and the reacted volume of the gas atmosphere with the solid thermoelectric materials of concerns. It has been demonstrated in Sb_2Te_3, $(Bi_{0.5}Sb_{0.5})_2Te_3$, and $(Bi_{0.9}Sb_{0.1})_2Te_3$ that the oxidation reaction rate is dependent on temperature and time, as usually observed in metal oxidation. The TPA curves observed for the three alloys indicate a rapid decrease in O_2 gas pressure at 400°C above which the initial oxidation process for respective constituents takes place. It is stressed again for the case of Sb_2Te_3 that the oxidation of constituent elements is a complex reaction and reaction volume of O_2 gas is relatively larger than those of others. It should be also remarked that the reaction rate is very much dependent on the sample grain size, which is discussed below.

For Sb_2Te_3 samples, a large amount of O_2 gas, as much as 5.3 moles, is reacted with the smaller grains, while only 0.9 moles of O_2 is reacted with relatively larger grains. For the nearly edge alloy of $(Bi_{0.9}Sb_{0.1})_2Te_3$, its reaction volume is 2.9 moles. On the other hand, in the intermediate alloy $(Bi_{0.5}Sb_{0.5})_2Te_3$ across the pseudobinary series of alloys, one finds that the smallest amount of O_2 gas, 0.7 moles, is reacted with the sample grains in the two sets of TPA experiments exposed, respectively, at 421°C and 547°C. It follows that an intermediate composition alloy is most oxidation resistant in the pseudobinary alloy series of Sb_2Te_3-Bi_2Te_3 under the present experimental conditions.

The largest relative number of moles of O_2 gas reacted with Sb_2Te_3, 5.3, is accounted for when we assume the following reaction:

$$Sb_2Te_3 + 5\ O_2 \rightarrow Sb_2O_4 + 3\ TeO_2$$

The consumptions of 2 and 3 moles of O_2 account for the formation of antimony tetroxide and tellurium dioxides, respectively. The mole ratio of these two oxides in the formula above is in accordance with the mole fraction determined by the XRD analysis; see Table 3.2.

3.4.2 Reaction Products

Reaction products are examined by microscope observation and powder XRD with the aid of Rietveld refinement. The results elucidate that (i) Sb_2Te_3 is oxidized initially to form $SbO_{1.5}$ and a Te precipitate and further oxidation reactions produce Sb_2O_4 and TeO_2,

(ii) $(Bi_{0.5}Sb_{0.5})_2Te_3$ is oxidized to form a compound with a larger unit cell, and (iii), whereas in $(Bi_{0.9}Sb_{0.1})_2Te_3$, bismuth tellurite Bi_2TeO_5 is the main product, and there is no trace of antimony and tellurium oxides, such as $SbO_{1.5}$, Sb_2O_4, and TeO_2, detected.

3.4.2.1 Sb_2Te_3

The formation of oxides on the solid surface or the incorporation of oxygen atoms into the solid should be dependent primarily on the thermodynamic activities of constituent elements of the samples under given conditions of temperature and oxygen pressure. The difference in products between the two series of TPA scans for Sb_2Te_3 oxidation is attributed to the complex reaction of O_2 with Sb_2Te_3; only antimony trioxides are formed in the TPA scans #01–#04 at 536°C, while in the other set of scans, #11–#18, at 547°C the antimony trioxide and Te are oxidized, respectively, to antimony tetroxide and tellurium dioxide. Actually, the two sets of TPA experiments are indicative of two different oxidation start temperatures in Sb_2Te_3: 400°C for scans #01 (Fig. 3.7) and #11 (Fig. 3.9) and 470°C for scans #12–#18 (Figs. 3.9 and 3.10). It may be concluded that the reaction starting at 470°C is associated with the formation of antimony tetroxide ($Sb_2O_3 + 0.5\ O_2 \rightarrow Sb_2O_4$). The formation of an Sb_2O_4 crystal by oxidation in air of Sb_2O_3 was reported [14–16], where the calcination temperatures are higher than the oxidation temperature of the bulk Sb_2Te_3 in the present work: 750°C [15] and three successive heatings at 690°C, 800°C, and 950°C [16]. It is also noted that synthesis of antimony tellurites, such as Sb_2TeO_7 or Sb_2TeO_5, were reported by calcinating a mixture of TeO_2 and either Sb_2O_3 or Sb_2O_4 above 500°C [24].

In the two sets of TPA scans, their O_2 exposure time and temperature are not different from each other. The total exposure time for temperatures higher than 400°C is approximately 20 h for both series. The holding time temperatures, respectively 536°C and 547°C, for the two sets of TPA scans are apparently lower than the melting point of the starting material Sb_2Te_3, at 617°C [3]. Therefore, we can conclude that the difference of 11°C in the exposing temperatures does not appear to affect the reaction of O_2 with Sb_2Te_3. Rather, we may reason that large difference in total relative mole number of reacted O_2 gas in the two sets of TPA scans for Sb_2Te_3 is due to the difference in grain size of the samples.

Different specific surface areas of sample grains employed for the two sets of TPA for Sb_2Te_3 can possibly be an important factor to control the reaction rate in this alloy. The solid sample oxidizes at different rates on different areas/surfaces of the grain. There may be formation of protective scales on some places, and at the same time its breakdown takes place upon further exposure in an O_2 gas atmosphere. In this context, smaller grains favor a faster reaction of O_2 gas on solid surfaces.

3.4.2.2 $(Bi_{0.5}Sb_{0.5})_2Te_3$

In the case of $(Bi_{0.5}Sb_{0.5})_2Te_3$, the oxidation products in the two sets of TPA measurements are found to be hexagonal $(Bi_{0.5}Sb_{0.5})_2Te_3$ with larger unit cells whose lattice parameters are listed in Table 3.3. Any forms of oxides are not detected in the XRD; it is partly because of smaller volume of produced oxides.

A different degree of cell volume expansion between the two $(Bi_{0.5}Sb_{0.5})_2Te_3$ samples should come from the reactions of O_2 gas with the samples at different exposure temperatures, 421°C and 547°C, respectively. Neither antimony nor tellurium oxides are formed as $SbO_{1.5}$, Sb_2O_4, TeO_2, or $(Sb, Te)_mO_n$ in $(Bi_{0.5}Sb_{0.5})_2Te_3$. We assume that the preferential reaction of constituent elements of the mother phase compound with O_2 gas would lead to a local deviation in the composition of the mother phase, resulting in a different crystalline arrangement in the unit cell where the Sb and Te site occupancies in the stacked layers are appreciably modified. It follows that the reaction rate of Sb to form Sb_2O_3 or Sb_2O_4 is larger than those of Bi and Te in the reactant; consequently, the formation of a Bi-rich phase is favored.

The lattice parameters data has been accumulated for the Bi_2Te_3-Sb_2Te_3 pseudobinary system [3, 5]. The lattice parameter a is an almost linearly increasing function of Bi_2Te_3 contents; it increases in dimension from Sb_2Te_3 toward Bi_2Te_3; on the other hand, the parameter c does not change very much with the composition across the system. In this sense, we may attribute observed cell expansion in the c plane with oxidation to the composition modification in the $(Bi_{0.5}Sb_{0.5})_2Te_3$ mother phase. Following the phase diagram for the Sb_2Te_3-Bi_2Te_3 system of the solid solution type [5], one can estimate the composition of the products as follows: $(Bi_{0.57}Sb_{0.43})_2Te_3$ and

$(Bi_{0.68}Sb_{0.32})_2Te_3$ for lower- and higher-temperature exposure samples, respectively.

3.4.2.3 $(Bi_{0.9}Sb_{0.1})_2Te_3$

It should be stressed that we demonstrated for the first time the synthesis of Bi_2TeO_5 crystals by oxidation of crystals of bismuth tellurium compounds. A well-known preparation method for bismuth tellurium oxides is to mix the required amounts of Bi_2O_3 with TeO_2 and then heat them at an elevated temperature for a period until the appropriate reaction occurs [18–25].

It has been clearly shown that by the oxidation the lattice parameters of intermediate $(Bi_{0.5}Sb_{0.5})_2Te_3$ and nearly edge $(Bi_{0.9}Sb_{0.1})_2Te_3$ tend to increase toward the edge alloy Bi_2Te_3 in the phase diagram. Given the atomic rearrangement in the *c* plane of mother-phase crystal associated with oxidation resulting from the formation of bismuth tellurite Bi_2TeO_5, one can estimate from the increase in the lattice parameter *a* that the modified composition of the mother phase is $(Bi_2Te_3)_{0.93}(Sb_2Te_3)_{0.07}$ rather than $(Bi_2Te_3)_{0.90}(Sb_2Te_3)_{0.10}$.

When we restrict ourselves to the thermoelectric property of a partially oxidized $(Bi_{0.9}Sb_{0.1})_2Te_3$ crystal, it is most likely that the thermoelectric property is representative of a composition modified $(Bi_2Te_3)_{0.9+m}(Sb_2Te_3)_{0.1-m}$ compound whose bulk surface is covered with dielectric Bi_2TeO_5 scales. The electronic structure should also be modified appreciably to be sensed by the usual thermoelectric property measurement. It can be a subject of study to characterize or optimize the thermoelectric performance of the modified composition sample by oxidation.

3.4.3 Anisotropic Lattice Expansion in Modified Mother Cells

Anisotropic cell expansion is found by oxidation in these hexagonal mother materials; the lattice parameter *a* is largely increased, while the *c* plane spacing remains almost unchanged. These behaviors are attributed to the rearrangement of constituent elements in the unit cell where the Sb and Te site occupancies in the stacked layers are appreciably modified by oxidation, as discussed above. Rearrangement of constituent elements in the crystal cell should accompany the modification of the composition as well as the lattice parameters. These consequences suggest a possible way of tuning of an optimum thermoelectric performance by gas reaction with solid thermoelectric materials of the (Bi, Sb)$_2$Te$_3$ pseudobinary system.

To provide a detailed explanation for the anisotropic expansion of the cell by oxidation in these class of alloys, a more detailed structural examination is necessary. Neutron diffraction [26] or X-ray absorption fine structure (XAFS) is a candidate to deal with the light element of oxygen, which may be involved in the interstitial sites of the stacked *c* plane layers or chemically coupled with Sb and/or Te atoms of the mother lattice.

We may conclude that the expansion in the *c* plane is a common feature brought about by oxidation in the pseudobinary alloy series of Sb_2Te_3-Bi_2Te_3. When we presume that Sb atoms are preferably separated from the mother compound by being exposed in O_2 at a high temperature, the modified local atomic arrangement favors reformation of a Bi-rich (Sb, Bi)$_2$Te$_3$ alloy, resulting in a larger lattice parameter of *a* in the hexagonal cell. For Sb_2Te_3, where Bi atoms are absent, random occupation of Sb or Te appears to take place in the sublattices of the hexagonal cell.

3.5 Conclusions

In this study, high-temperature tolerance has been investigated for antimony and bismuth telluride thermoelectric materials, Sb_2Te_3, $(Bi_{0.5}Sb_{0.5})_2Te_3$, and $(Bi_{0.9}Sb_{0.1})_2Te_3$, under argon, nitrogen, or oxygen gas atmospheres at 0.2 MPa and 547°C. For this purpose, an apparatus for TPA has been developed to examine the reaction of various gases with thermoelectric materials as functions of temperature and time, and its instrumental details and performance are presented. TPA is found to be a highly powerful means for monitoring the reaction of an introduced gas with the test solid materials and desorption of an unknown gas from them. The reaction products have been examined by microscope observation and powder XRD with the aid of Rietveld refinement.

It is confirmed that neither Ar nor N_2 gas reacts with Sb_2Te_3, $(Bi_{0.5}Sb_{0.5})_2Te_3$, and $(Bi_{0.9}Sb_{0.1})_2Te_3$ at temperatures up to 547°C. On the other hand, all the compounds start to oxidize in O_2 gas at 400°C.

Sb_2Te_3 is oxidized initially to produce $SbO_{1.5}$ oxide and a Te precipitate. By further oxidation reactions, crystals of Sb_2O_4 and TeO_2 oxides are synthesized. More interesting is the anisotropic lattice expansion in the mother compound Sb_2Te_3; when the relative number of moles of reacted O_2 gas amounts to 5.3, the lattice parameter a is increased by 2.4% while the c spacing remains unchanged by oxidation in the hexagonal unit cell.

For $(Bi_{0.5}Sb_{0.5})_2Te_3$, none of the antimony and tellurium oxides, such as $SbO_{1.5}$, Sb_2O_4, and TeO_2, are produced. The relative number of moles of reacted O_2 gas is 0.7 irrespective of exposure temperatures in the two sets of TPA scans, demonstrating that intermediate $(Bi_{0.5}Sb_{0.5})_2Te_3$ is most oxidation resistant across the series. Anisotropic lattice expansion is also found in the mother crystals. However, the degree of expansion is totally dependent on the exposure temperature. The c plane is expanded by 0.5%, whereas along the c axis the lattice spacing remains unchanged for exposure at 547°C.

For the near-edge compound of $(Bi_{0.9}Sb_{0.1})_2Te_3$ in the pseudobinary series of $(Bi, Sb)_2Te_3$, it is demonstrated that bismuth tellurium oxide (Bi_2TeO_5) rather than antimony and tellurium oxides is produced by oxidation of crystal of bismuth tellurium compound. The expansion is found to be by 0.86% in the c plane, while the shrinking is by as much as 1.4% along the c axis, so the mother-phase cell volume remains unchanged.

These observed modifications by oxidation in the unit cell structure of the mother phase in the pseudobinary system of Sb_2Te_3-Bi_2Te_3 are attributed to the rearrangement of constituent elements in the unit cell; the Sb and Te site occupancies in the stacked layers are appreciably modified by oxidation. The increased lattice parameter a and the almost constant c parameter are in accordance with the change in lattice parameters across the Sb_2Te_3-Bi_2Te_3 pseudobinary system of the perfect solid solution, indicating that the composition in the mother phase crystal is appreciably modified so as to shift toward the edge Bi_2Te_3. These statements suggest a possible way of tuning of an optimum thermoelectric performance by gas reaction with solid thermoelectric materials of the $(Bi, Sb)_2Te_3$ pseudobinary system.

Acknowledgments

I am grateful to Prof. J. M. D. Coey for his having introduced the interesting absorption and desorption behavior of various gases on solid materials to me during my stay in Dublin. I would also like to take this opportunity to express my sincere thanks to Dr. G. Nakamoto, Dr. T. Nobata, and Mr. G. Ito for their help in the construction of the TPA apparatus; to Mr. S. Noguchi for his help in Rietveld refinements of XRD profiles of TPA samples; and to Mr. I. Shimizu for his help in part of TPA data acquisition for Sb_2Te_3. This work has been supported financially by the Adaptable and Seamless Technology Transfer Program from Japan Science and Technology Agency, JST, and a Grant-in-Aid for Research Promotion from Ehime University. I thank the financial support by JSPS KAKENHI Grant Numbers 11640344 and 13650716 in designing and building the thermopiezic apparatus at an early stage of the research.

References

1. Goldsmit, H. J. (2016). *Introduction to Thermoelectricity*, 2nd Ed., Springer-Verlag, Berlin Hidelberg, pp. 1–278.
2. Harman, T. C. and Honig, J. M. (1967). Elementary theory of the efficiency of thermoelectric devices, in *Thermoelectric and Thermomagnetic Effects and Applications*, McGraw-Hill, New York, pp. 273–310.
3. Scherrer, H. and Scherrer, S. (1995). Bismuth telluride, antimony telluride, and their solid solutions, in *CRC Handbook of Thermoelectrics*, Rowe, D. M. (ed.), CRC Press, New York, pp. 211–237.
4. Hansen, M. and Anderko, K. (1988). *Constitution of Binary Alloys*, 2nd Ed., Genium Pub. Corp., p. 1232.
5. Smith, M. J., Knight, R. J. and Spencer, C. W. (1962). Properties of Bi_2Te_3-Sb_2Te_3 alloys, *J. Appl., Phys.*, **33**, pp. 2186–2190.
6. Huong, N. T., Setou, Y., Nakamoto, G., Kurisu, M., Kajihara, T., Mizukami, H. and Sano, S. (2004). High thermoelectric performance at low temperature of p-$Bi_{1.8}Sb_{0.2}Te_{3.0}$ grown by the gradient freeze method from Te rich melt, *J. Alloys Compd.*, **368**, pp. 44–50.
7. Ryan, D. H. and Coey, J. M. D. (1986). Thermopiezic analysis: gas absorption and desorption studies on milligram samples, *J. Phys. E: Sci. Instrum.*, **19**, pp. 693–694.

8. Izumi, F. and Momma, K. (2007). Three-dimensional visualization in powder diffraction, *Solid State Phenom.*, **130**, pp. 15–20.
9. Skoog, D. A., Holler, F. J. and Crouch, S. R. (2007). Thermal method, in *Principles of Instrumental Analysis*, 6th Ed., Thomson Brooks/Cole, USA, pp. 894–908.
10. Ryan, D. H. and Coey, J. M. D. (1986). An automatic thermomanometer for hydrogen absorption and desorption studies on milligram samples, *Surf. Coat. Technol.*, **28**, pp. 383–386.
11. Coey, J. M. D. and Wang, X. Z. (1988). Thermopiezic analysis of rare-earth cuprate superconductors and related oxides, *J. Solid State Chem.*, **77**, pp. 204–209.
12. Fukuda, K., Imaizumi, H., Ishii, T., Toyoda, F., Yamanashi, M. and Kibayashi, Y. (1996). Orientational distribution and procedure parameters in hot pressed $(Bi_2Te_3)_{0.90}(Bi_2Se_3)_{0.10}$, *Proceedings of 15th International Conference on Thermoelectrics*, pp. 37–41.
13. Wagner, C. (1952). Theoretical analysis of the diffusion processes determining the oxidation rate of alloys, *J. Electrochem. Soc.*, **99**, pp. 369–380.
14. Gopalakrishnan, P. S. and Manohar, H., (1975). Cervantite, α-Sb_2O_4, *Cryst. Struct. Commun.*, **4**, pp. 203–206.
15. Thornton, G. (1977). A neutron diffraction study of α-Sb_2O_4, *Acta Cryst.*, **B33**, pp. 1271–1273.
16. Amador, J., Gutierrez Puebla, E., Monge, M. A., Rasines, I. and Ruiz Valero, C. (1988). Diantimony tetraoxides revisited, *Inorg. Chem.*, **27**, pp. 1367–1370.
17. Hahn, Th., Ed. (2002). *International Tables for Crystallography, Vol. A, Space-Group Symmetry*, 5th ed., Kluwer Academic.
18. Domoratsky, K. V., Dudnik, E. F., Katkov, V. F. and Sadovskaya, L. Ya. (1999). X-ray investigations of phase transition in Bi_2TeO_5 single crystals, *Condens. Matter Phys.*, **2**, pp. 591–594.
19. Földvári, I., Péter, Á., Voszka, R. and Kappers, L. A. (1990). Growth and properties of Bi_2TeO_5 single crystals, *J. Cryst. Growth*, **100**, pp. 75–77.
20. Ok, K. M., Bhuvanesh, N. S. P. and Halasyamani, P. S. (2001). Bi_2TeO_5: synthesis, structure, and powder second harmonic generation properties, *Inorg. Chem.*, **40**, pp. 1978–1980.
21. Carvalho, J. F., Fabris, Z. V., de Oliveira, I. and Frejlich, J. (2014). Crystal growth of Bi_2TeO_5 by a double crucible Czochralski method, *J. Cryst. Growth*, **401**, pp. 795–797.

22. Yadav, Y. K., Sahoo, M. P. K. and Choudhary, R. N. P. (2010). Electrical properties of Bi_2TeO_5 ceramic, *J. Alloys Compd.*, **490**, pp.589–593.
23. Kumaragurubaran, S., Krishnamurthy, D., Subramanian, C. and Ramasamy, P. (1999). Investigations on the growth of Bi_2TeO_5 and TeO_2 crystals, *J. Cryst. Growth*, **197**, pp. 210–215.
24. Evans, W. P. (1985). Tellurium-containing cathodes for nonaqueous cells, US Patent 4,536,456.
25. Chagraoui, A., Yakine, I., Tairi, A., Moussaoui, A., Talbi, M. and Naji, M. (2011). Glasses formation, characterization, and crystal-structure determination in the Bi_2O_3-Sb_2O_3-TeO_2 system prepared in an air, *J. Mater. Sci.*, **46**, pp. 5439–5446.
26. Nobata, T., Nakamoto, G., Kurisu, M., Makihara, Y., Ohyama, K. and Ohashi, M., (2002). Neutron diffraction study on the Heusler compound $Co_{1.50}TiSn$ and its nitrogenation products, *J. Alloys Compd.*, **347**, pp. 86–90.

Chapter 4

Organic Rechargeable Batteries

Yohji Misaki
Department of Applied Chemistry, Graduate School of Science and Engineering, Ehime University, and Research Unit for Power Generation and Storage Materials, Ehime University, 3 Bunkyo-cho, Matsuyama, Ehime 790-8577, Japan
misaki.yohji.mx@ehime-u.ac.jp

Chapter 4 deals with organic positive-electrode materials for rechargeable batteries. Recently, organic materials, which can act as positive-electrode materials, have attracted attention as alternatives to inorganic metal oxides. A large number of organic molecules have been examined and have been newly synthesized so far. Among them, molecular materials are expected to exhibit a high capacity and accordingly a high energy density compared with polymer materials. This chapter reviews molecular positive-electrode materials for organic rechargeable batteries.

4.1 Introduction

The effective use of electric energy is quite important to save energy as well as reduce carbon dioxide emission [1]. Electric storage

Functional Materials: Advances and Applications in Energy Storage and Conversion
Edited by Toshio Naito

ISBN 978-981-4800-09-9 (Hardcover), 978-0-429-46813-1 (eBook)
www.panstanford.com

devices represented as rechargeable batteries attract considerable attention. Among various rechargeable batteries, lithium ion batteries (LIBs) play a central role. Metal oxides, including minor metals such as Ni, Co, and Mn, are often used for the active materials of positive electrodes in LIBs. However, such inorganic materials have problems. Mn induces environmental problem, and Ni and Co are expensive. In this context, the development of new positive-electrode materials without any minor metals is desirable. Recently, iron-based materials such as $LiFePO_4$ [2], Li_2FePO_4F [3], and Li_2FeSiO_4 [4] have been developed. However, the examples are still limited and their cell performance is inferior to nickel- and cobalt-based materials. Thus, the development of new active electrode materials is an urgent task to act as substitutes or to compensate for the present inorganic materials, considering demands for LIBs will explosively increase in the future.

Organic materials that do not include any heavy metals have several advantages as active electrode materials for LIBs, that is, they are abundant in nature and have less environmental load. Furthermore, various organic molecules can be easily designed and synthesized conveniently compared to inorganic materials. Thus, redox-active organic materials are one of the possible candidates for positive-electrode materials of the next generation.

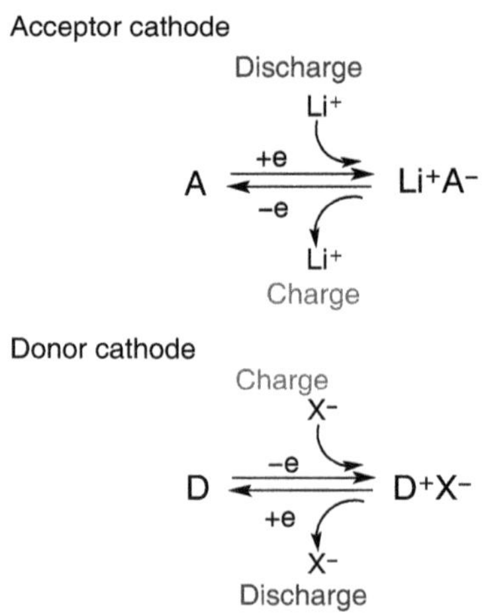

Figure 4.1 Redox reactions at the positive electrode in a LIB.

As for redox-active organic materials, different types of counterions participate in the charge-discharge processes of donor and acceptor molecules. The neutral states of acceptor molecules correspond to the charged states, and their reduced states correspond to the discharged states, as shown in Fig. 4.1. In this case, cations in the electrolyte are involved in discharged processes. This behavior is the same as that of an ordinary LIB. In contrast, the neutral states of donor molecules correspond to the discharged states and oxidized states correspond to the charged states. In this case, anions of the electrolyte are involved in charged processes.

4.2 Design of Organic Positive-Electrode Materials

To design organic molecules that can be utilized as positive-electrode materials for rechargeable batteries with high performance, the following requirements should be satisfied:

- The molecules should exhibit reversible redox processes.
- Electrons per weight participating in redox reactions need to be increased to increase the electric capacity:

$$\text{capacity [mAh/g]} = \left[\frac{\text{n} \times 96500}{\text{Mw}}\right] \times \left[{}^{1000}\!/\!{}_{3600}\right]$$

- Redox potentials should be high within the range to avoid decomposition of the electrolyte solutions.
- There should be enough stability in both oxidation and reduction for a long period.
- The material should have low solubility in the ordinary organic solvents in the electrolyte solution to achieve a high cycle performance.

Various organic materials have been examined so far. However, most materials do not satisfy all of the requirements above. In particular, the suppression of solubility in electrolyte solutions is an important but demanding problem yet to be solved, because organic molecules usually have a strong affinity for organic solvents. Utilization of polymerized organic materials is one of the solutions. However, insertion of a linkage group is usually required to polymerize monomeric organic molecules, which results in a decrease in the theoretical capacity. Organic molecules that are hardly soluble in ordinary organic solvents might be the candidates for new electrode materials for rechargeable batteries.

4.3 Representative Organic Materials for Positive-Electrode Materials

This section briefly surveys representative organic molecules used for molecular positive-electrode materials (Fig. 4.2) [5–30]. They are classified into three groups, namely acceptors, donors, and dual redox systems.

4.3.1 Benzoquinone Derivatives

As for acceptor molecules, *p*-benzoquinone (BQ) derivatives are most extensively investigated [5–7]. BQ is a two-electron electron system, and both mono- and dianion states are stabilized by a significant contribution from resonance structures with an aromatic benzene structure (Fig. 4.3). The theoretical capacity for utilizing two-electron redox reaches 496 mAh g^{-1}. However, BQ itself is difficult to use as a positive-electrode material because of its strong tendency toward sublimation and high solubility in organic solvents.

Yao and coworkers reported that 2,5-dimethoxybenzoquinone (DMBQ) exhibits a large capacity (312 mAh g^{-1}) and a resultant high energy density (810 mWh g^{-1}) [5] (see Fig. 4.4). A conventional cycle-life test reveals that the DMBQ electrode shows a relatively good cycle-life performance; the discharge capacity is still 255 mAh g^{-1} after 10 cycles (82% of the original discharge capacity).

Condensation of aromatic rings is a promising strategy to decrease solubility in the electrolyte solution and the resultant enhancement of the cycle performance (Fig. 4.5). 9,10-Anthraquinone (AQ) actually works as a positive-electrode material for rechargeable batteries, although the AQ electrode tends to degrade upon cycling owing to the dissolution of the AQ molecule in the electrolyte solution during cycling [9, 10].

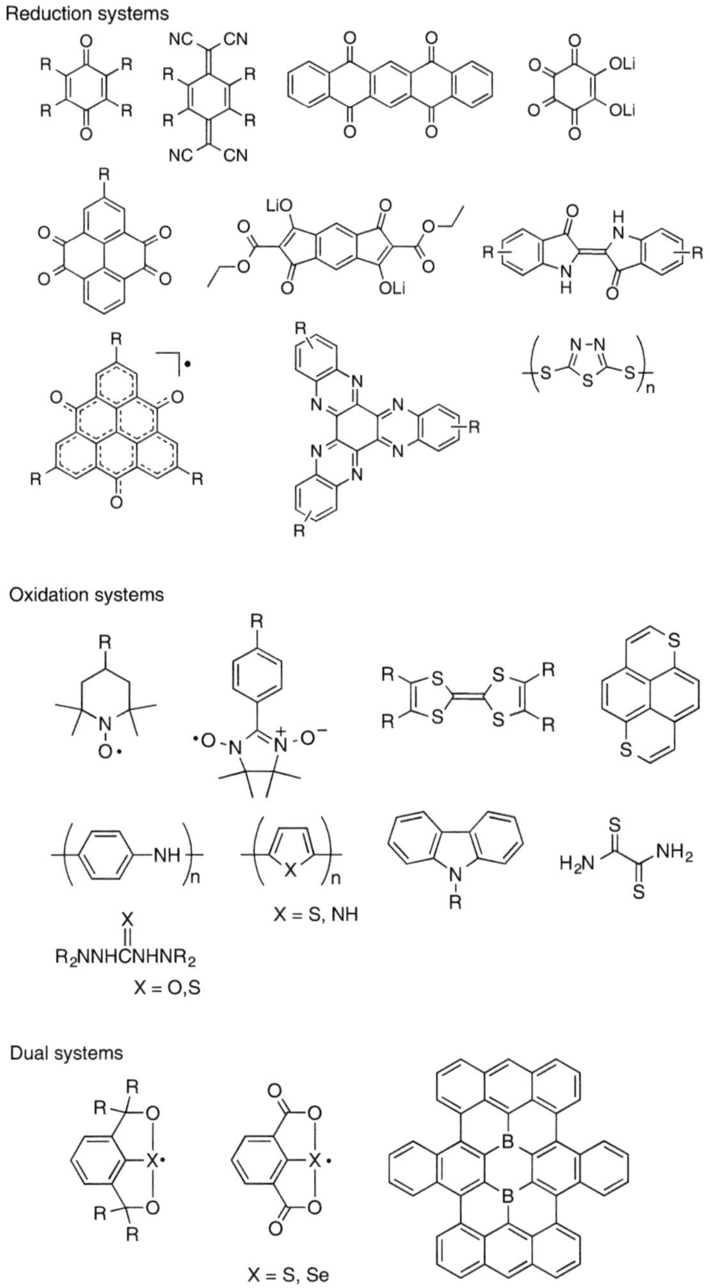

Figure 4.2 Representative acceptor molecules examined as positive-electrode materials.

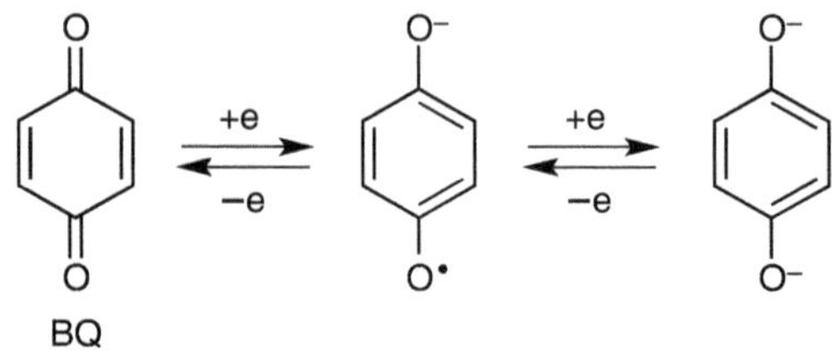

Figure 4.3 Redox reactions of BQ.

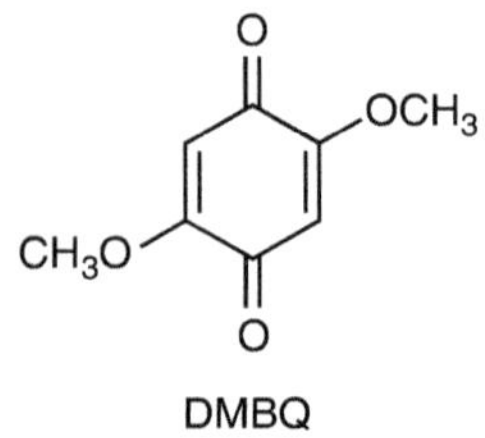

Figure 4.4 Chemical structure of DMBQ.

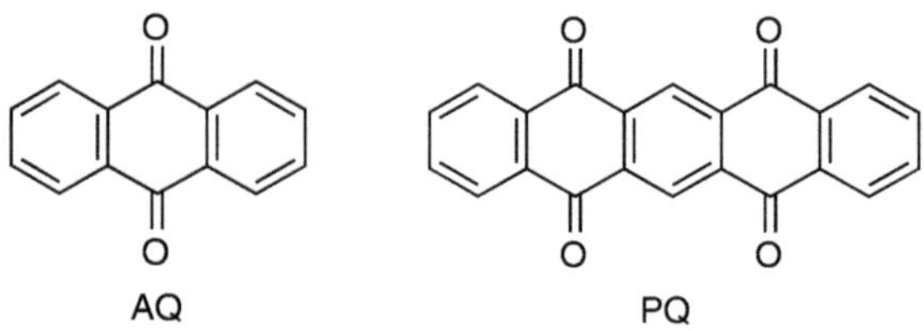

Figure 4.5 Chemical structures of AQ and PQ.

Yao and coworkers reported on 5,7,12,14-pentacenetetrone (PQ) [10] (Fig. 4.5). It is a four-electron redox system whose cyclic voltammogram shows three pairs of redox waves, at –0.9 ($1e^-$), –1.3 ($1e^-$), and –1.9 V ($2e^-$) (V versus ferrocene/ferrocenium [Fc/Fc^+]). The numbers in parentheses denote the number of electrons participating in the redox processes. The practical discharge capacities of AQ/Li and PQ/Li are 217 and 236 mA h g^{-1}, which are 84% and 74% of the theoretical values (257 mA h g^{-1} for AQ and 317 mA h g^{-1} for PQ, respectively), as shown in Fig. 4.6. A conventional cycle-life test reveals that the capacity of the AQ/Li cell rapidly decreased to 49 mAh g^{-1} after 100 cycles (23% of the initial discharge capacity), as shown in Fig. 4.7. In contrast, the PQ/Li cell shows a relatively good cycle-life performance; the discharge capacity after 100 cycles is maintained at 183 mAh g^{-1} (78% of the initial discharge capacity).

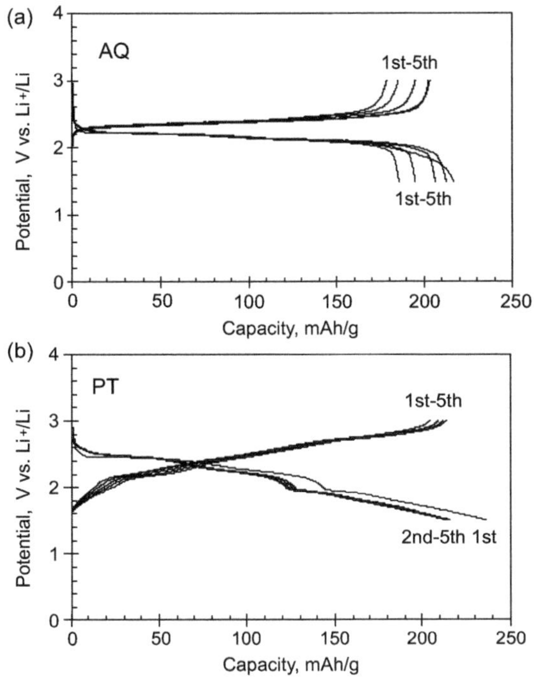

Figure 4.6 Charge-discharge curves for (a) PT/Li and (b) AQ/Li cells. Reprinted from Ref. [10], Copyright (2012), with permission from Elsevier.

Matsubara and coworkers reported on dimeric-BQ systems, NDQ and BBQ [31] (Fig. 4.8). Both materials belong to the four-electron redox systems, and the theoretical capacity of NDQ reaches 570 mA h g^{-1} if four electrons can be utilized for the charge-discharge process. The practical initial discharge capacities of NDQ/Li and BBQ/Li cells are 347 and 326 mA h g^{-1}, respectively, which are 61% and 65% of their theoretical values, respectively. The degradation might be caused by the dissolution of the positive-electrode active materials into the electrolyte solution before the charge-discharge tests. Anyway, the practical initial discharge capacities of NDQ/Li and BBQ/Li cells are very high; they are more than twice the discharge capacities of $LiCoO_2$. However, after many cycles the performance of these cells is not so good; the discharge capacities after 20 cycles are 19% and 52% of their initial discharge capacities.

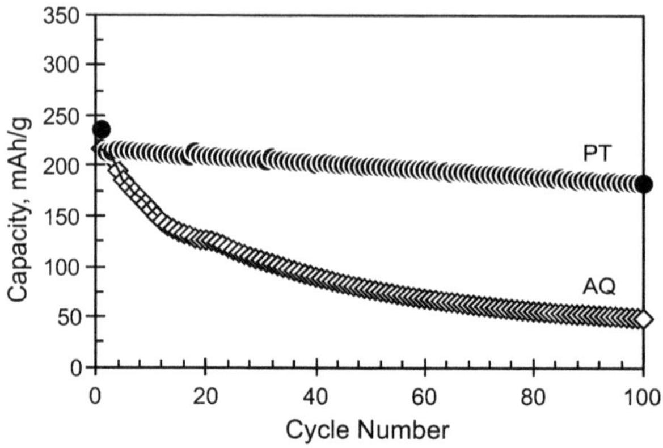

Figure 4.7 Cycle-life performances for PT/Li and AQ/Li cells. Reprinted from Ref. [10], Copyright (2012), with permission from Elsevier.

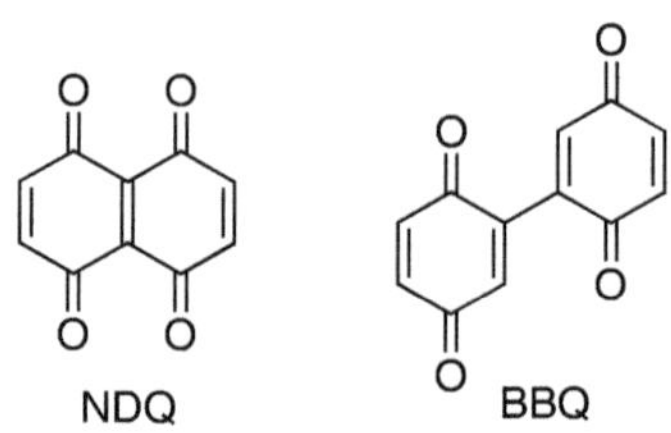

Figure 4.8 Chemical structures of NDQ and BBQ.

4.3.2 Pyrene-4,5,9,10-Tetraone

Yoshida and coworkers reported on pyrene-4,5,9,10-tetraone (PYT) derivatives (Fig. 4.9) [13, 14]. PYT is a four-electron redox system (Fig. 4.10) whose theoretical capacity utilizing four-electron redox is 409 mA h g^{-1}. A PYT/Li cell shows an initial discharge capacity of 320 mA h g^{-1}; however, the capacity drops considerably during charge-discharge cycles; the discharge capacity after 20 cycles (76 mA h g^{-1}) is 25% of the initial capacity. The cycle-life performance is significantly improved by substitution of carboxylate groups or by binding to polymethacrylate [13]. The discharge capacity of LCPYT/Li, where LCPYT is 2,7-bis(lithiooxycarbonyl)pyrene-4,5,9,10-tetraone, after 20 cycles (187 mA h g^{-1}) is maintained at 86% of the initial discharge (217 mA h g^{-1}). On the other hand, PPYT/Li, where PPYT is polymer-bound pyrene-4,5,9,10-tetraone, shows an excellent cycle performance; the discharge capacity after 500 cycles is maintained at 83% of the initial capacity [14]. It is noted that PPYT exhibits a fast charge-discharge ability; the capacity even at 30 C (2 min. for full discharge) is about 90% of that at the 1 C rate (1 hour for full discharge), as shown in Fig. 4.11 [13]. This behavior might be attributed to the physical flexibility and the affinity to Li ions of the methacrylate polymer backbone, although the details are not clear at present.

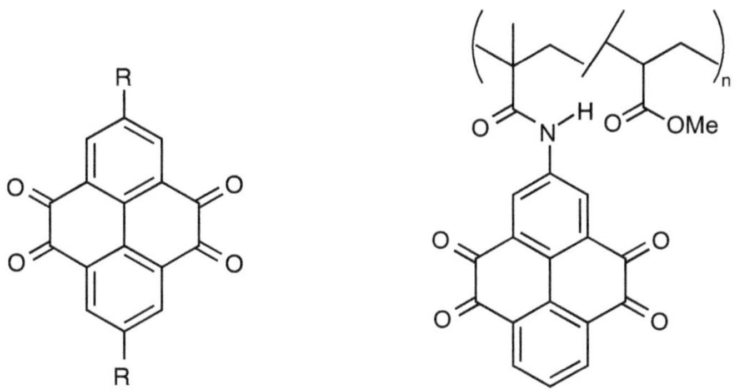

R = H, PYT
R = CO_2 Li, LCPYT

PPYT

Figure 4.9 Chemical structures of PYT and PPYT.

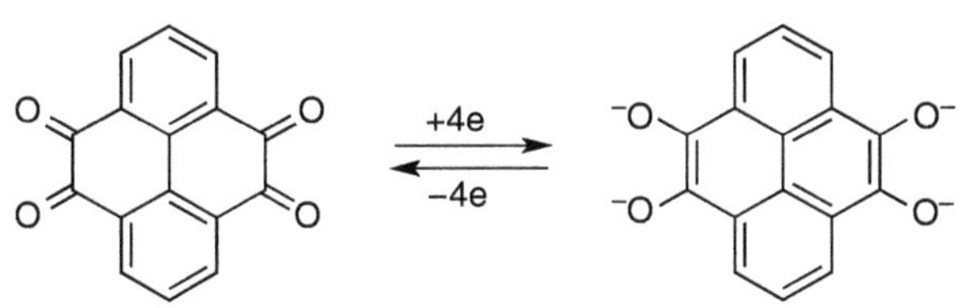

Figure 4.10 Redox process of PYT. Reprinted with permission from Ref. [13]. Copyright (2012) American Chemical Society.

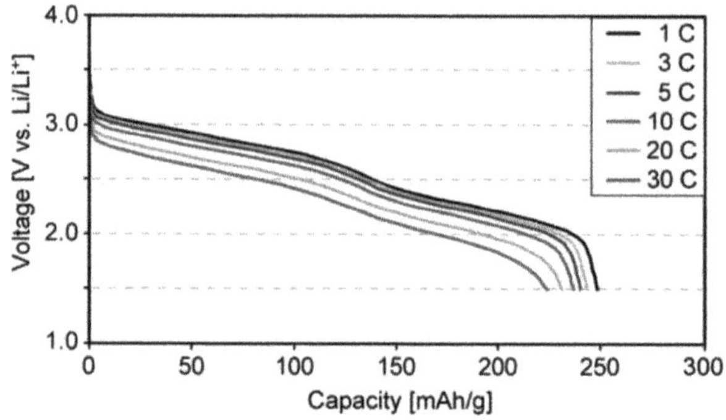

Figure 4.11 Discharge curves at various rates (1–30 C) of PPYT/Li using $LiNTf_2$/tetraglyme as an electrolyte solution at 45°C. Reprinted with permission from Ref. [13]. Copyright (2012) American Chemical Society.

4.3.3 Indigo Carmine

Indigo has been widely used for dyeing and printing with dark blue. For example, it is utilized as a dye for blue jeans. On the other hand, indigo belongs to a two-electron redox system (Fig. 4.12). Therefore, indigo and its derivatives are candidates for positive-electrode materials that can be obtained at a reasonable manufacturing cost. Yao and coworkers reported on indigo carmine (5,5'-indigodisulfonic acid sodium salt) [16].

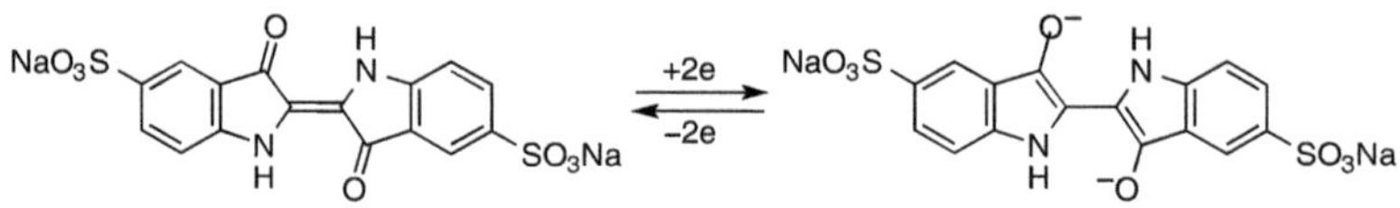

Figure 4.12 Redox process of indigo carmine. Adapted from Ref. [16] with permission from the Chemical Society of Japan.

The indigo carmine electrode shows an initial discharge capacity of 110 mA h g^{-1}, which is close to the theoretical capacity for a two-electron redox process (115 mA h g^{-1}). The average potential at the first discharge is 2.20 V (versus Li/Li^+). It is noteworthy that it shows a good cycle-life performance; the cell shows an almost constant value after a slight drop at the second cycle (Fig. 4.13). The discharge capacity after 100 cycles is maintained at 87 mA h g^{-1}.

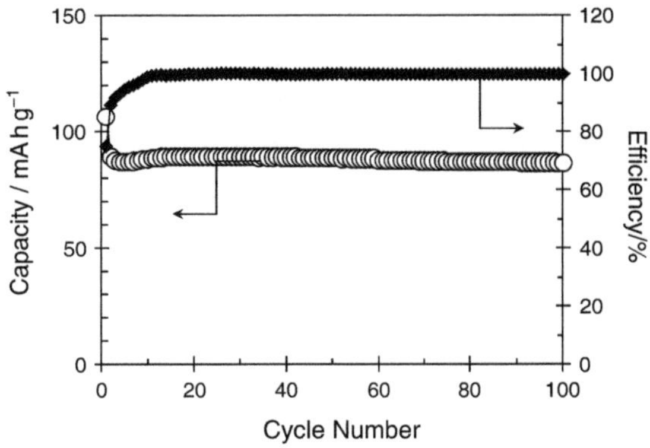

Figure 4.13 Cycle-life performance of a rechargeable battery using indigo carmine as the positive-electrode material (current density: 20 mA g^{-1}; potential range: 1.50–3.00 V versus Li/Li^+). Adapted from Ref. [16] with permission from the Chemical Society of Japan.

4.3.4 Application to Sodium and Magnesium Batteries

Lithium is an indispensable element as the carrier ion in a LIB. However, lithium resources are unevenly distributed on earth, rich in South America, particularly in Chile. In this connection, the development of other battery systems using a different carrier ion instead of the lithium ion is an urgent issue. Sodium and magnesium ions are promising substitutes for lithium because they are more widely distributed and more abundant. Furthermore, magnesium-based batteries are expected to be safer and more reliable than LIBs because magnesium metal is less reactive toward water and oxygen in an ambient atmosphere than lithium metal. However, variations in inorganic materials for sodium- or magnesium-based electrodes are limited, probably owing to the size effect of the metal ions. The large ionic sizes of sodium and magnesium ions would hinder their movement in the crystals.

Yao and coworkers reported on sodium ion batteries using indigo carmine [32] and on magnesium batteries using DMBQ as the positive-electrode materials [33], both of which are introduced in the previous sections (4.3.3 and 4.3.1, respectively). Both sodium-based cells using the indigo carmine electrodes and magnesium-based cells using the DMBQ electrodes exhibit charge-discharge curves and discharge capacities similar to the corresponding LIBs. Energy-dispersive X-ray spectroscopy (EDX) measurement reveals that the Mg concentration in the electrode increases during the first discharge process and decreases during the subsequent charge process (Fig. 4.14) [33].

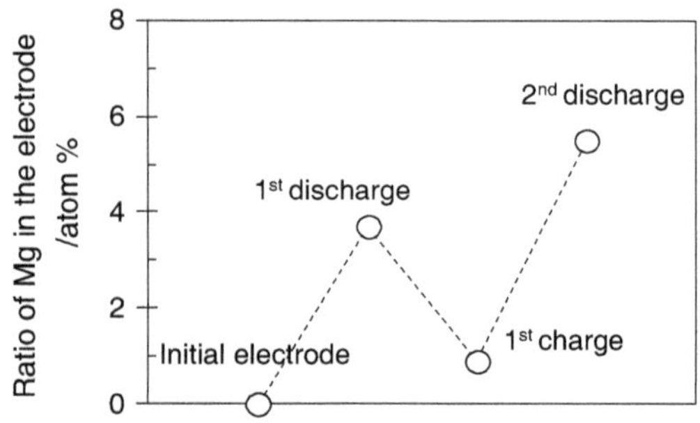

Figure 4.14 Atomic content ratio of Mg to Mg+C+O+F in the DMBQ electrode during cycling obtained by EDX measurement. Adapted from Ref. [33] with permission from the Chemical Society of Japan.

4.3.5 Radical Compounds

Morita and coworkers reported on derivatives of 6-oxophenalenoxyl (6OPO) and trioxotriangulene (TOT) (Fig. 4.15) [17]. The initial discharge capacity of the 6OPO/Li cell is 152 mA h g^{-1}, which is close to the theoretical value for two-electron redox of (6OPO)/$(6OPO)^{2-}$ (147 mA h g^{-1}). A conventional cycle-life test reveals that the discharge capacity after 100 cycles falls to 33 mA h g^{-1} (22% of the initial capacity), probably owing to the dissolution of 6OPO into the electrolyte solution. On the other hand, cyclic voltammograms of TOT derivatives show four pairs of one-electron redox waves, which correspond to the reduction processes from TOT to $(TOT)^{3-}$. The cycle-life performance of the $(Br)_3TOT$/Li cell is fairly good; the discharge capacity after 100 cycles is 71% of the initial capacity while the discharge capacity of $(t\text{-}Bu)_3TOT$/Li after 100 cycles is only 17% of the initial capacity.

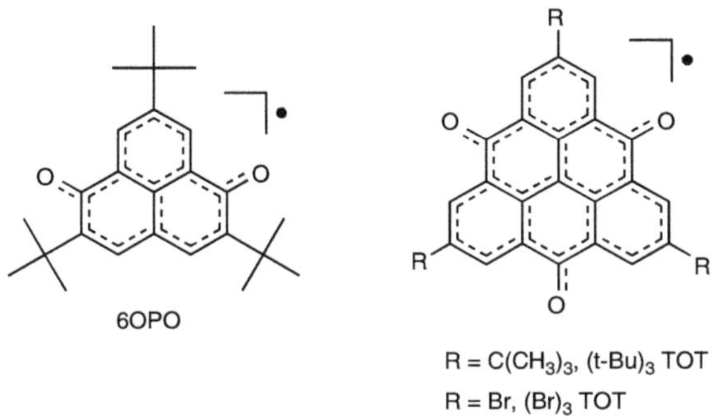

Figure 4.15 Chemical structures of OPO and TOT derivatives.

4.3.6 Tetrathiafulvalene Derivatives

Tetrathiafulvalene (TTF) is a well-known π-electron donor. Its oxidized species are stabilized by the aromatic character of the 1,3-dithiolylium ion, with six π-electrons, as shown in Fig. 4.16 [34, 35]. It exhibits two successive one-electron reversible redox processes. The redox potentials of TTF are 3.10 and 3.50 V versus Li/Li^+, in propylene carbonate containing 1 M $LiBF_4$ as an electrolyte. The radical cation is stabilized by the delocalization of a positive charge between two 1,3-dithiole rings, in addition to aromatic stabilization. Aromatic stabilization in $(TTF)^{2+}$ is larger than that in $(TTF)^{\bullet+}$ because $(TTF)^{2+}$ has two 1,3-dithiolylium rings. However, on-site coulomb repulsion between two positive charges significantly destabilizes the dicationic state. In other words, the second oxidation potential (E_2^{ox}) is much more positive than the first oxidation potential (E_1^{ox}). As a result, the two-electron oxidized state of TTF is less stable than the one-electron oxidized state. The TTF and its derivatives have been utilized as components for metallic and superconducting molecular conductors, and the TTF derivatives have actually produced a great number of molecular metals and superconductors [36–39].

The theoretical capacity of TTF assuming two-electron redox is 262 mAh/g, which is considerably larger than the theoretical capacities of active positive-electrode materials for commercially available LIBs. On the other hand, redox potentials of TTF (3.10 and

3.50 V versus Li^+/Li) are in the same range as that of the LIB (3.00–4.00 V). These properties can be important advantages in an active electrode material. However, TTF itself dissolves easily in organic solvents for electrolyte solutions, as many organic molecules do.

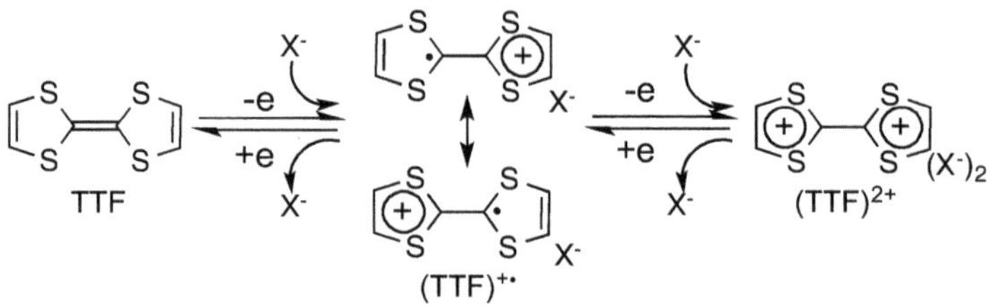

Figure 4.16 Redox process of TTF.

Introduction of rigid substituents on TTF is a promising strategy for the utilization of active positive-electrode materials. Inatomi and coworkers reported that bis(ethylenedithio)tetrathiafulvalene (BEDT-TTF) (Fig. 4.17) works as a positive-electrode material for rechargeable batteries [23]. However, the theoretical capacity of BEDT-TTF falls to 141 mAh/g due to the presence of ethylenedithio groups. Morita and coworkers reported charge-discharge properties of a dibenzotetrathiafulvalene (DB-TTF)/Li battery (Fig. 4.17) [8]. The initial discharge capacity of the DB-TTF cell was 167 mAh/g. This value is close to the theoretical value (176 mAh/g).

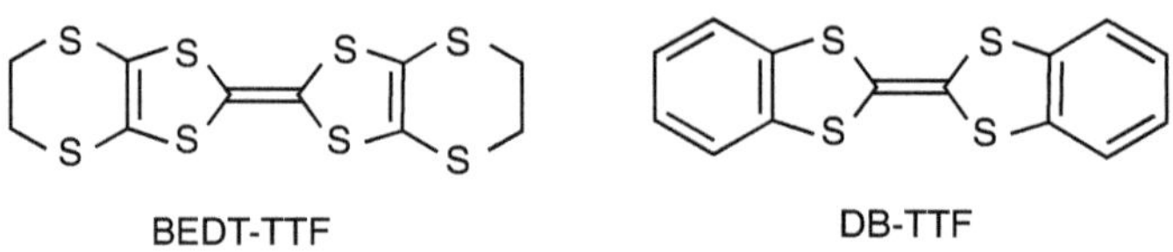

Figure 4.17 Chemical structures of BEDT-TTF and DB-TTF.

4.4 Fused TTF Systems

4.4.1 Fused TTF Dimers and Trimers

Bis- and tris-fused TTFs, 2,5-bis(1,3-dithiol-2-ylidene)-1,3,4,6-tetrathiapentalene (TTP) [40] and 2,2'-bi[5-(1,3-dihthiol-2-

ylidene)-1,3,4,6-tetrathiapentanylidene] (TTPY) [41], and their derivatives [42–44] were originally developed as donor components of molecular conductors exhibiting metallic conductivity down to low temperatures [36, 37]. They have a strong tendency to construct two-dimensional conduction networks in the solid states of charge transfer (CT) salts because they have ladder-like arrays of sulfur atoms [44–46]. Because of this network, TTP and TTPY might aggregate themselves closer than TTF in the oxidized states. Therefore, TTP and TTPY are less soluble than the dimeric and trimeric TTFs linked by various spacers [47–51]. For example, carbon disulfide has to be used as the eluent for column chromatography. Furthermore, TTPY is scarcely soluble even in carbon disulfide. TTP and TTPY have four and six redox-active 1,3-dithiol-2-ylidene sites, respectively. Thus, TTP and TTPY are expected to exhibit charge-discharge properties with four and six electrons participating in redox reactions. Their theoretical capacities assuming full-electron redox is involved reach 282 and 289 mAh g^{-1} for TTP and TTPY, respectively. Insertion of any linkage group is not required for the fused TTF molecules, unlike ordinary oligomer. Thus, the theoretical capacities of fused TTF systems rather increase with an increasing number of fused TTF moieties. Inatomi, the present author, and coworkers examined TTPP and TTPY electrodes (Fig. 4.18) [52].

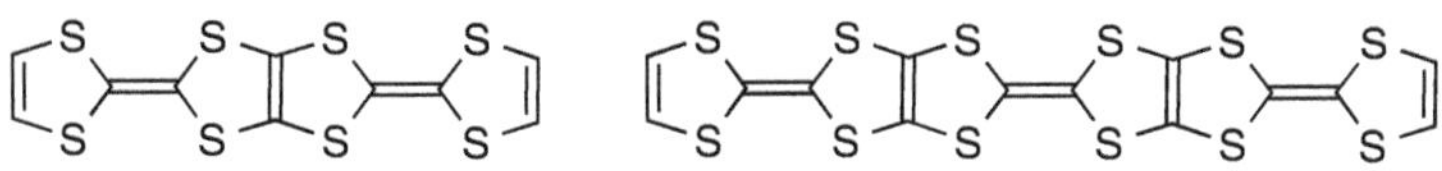

Figure 4.18 Chemical structures of TTP and TTPY.

Cyclic voltammetry of TTP and TTPY in the solid state was carried out by using a TTP or TTPY electrode as the working electrode in the mixed solution of ethylene carbonate and diethyl carbonate (1:5, v/v) containing 1 M $LiBF_4$ as an electrolyte solution. Whereas TTP in a benzonitrile solution exhibits four pairs of one-electron redox waves, at –0.01, 0.22, 0.65, and 0.83 V (versus Fc/Fc^+), respectively, the TTP electrode exhibits three oxidation peaks, observed at 3.17, 3.70, and 4.09 V (versus Li/Li^+), respectively, as shown in Fig. 4.19. The oxidation peak at 3.70 V in the solid electrode might correspond to the redox waves at –0.01 and 0.22 V in the TTP solution and that at 4.09 V corresponds to 0.65 and 0.83 V waves. Coalescence of the waves and appearance of the unexpected wave at 3.00 V are probably induced by a solid-state effect. The numbers of participating electrons in the three redox processes were estimated to be 0.5, 1.5, and 2.0 electrons, respectively, by analysis of the voltammogram and elemental analyses of the electrode in each oxidation state.

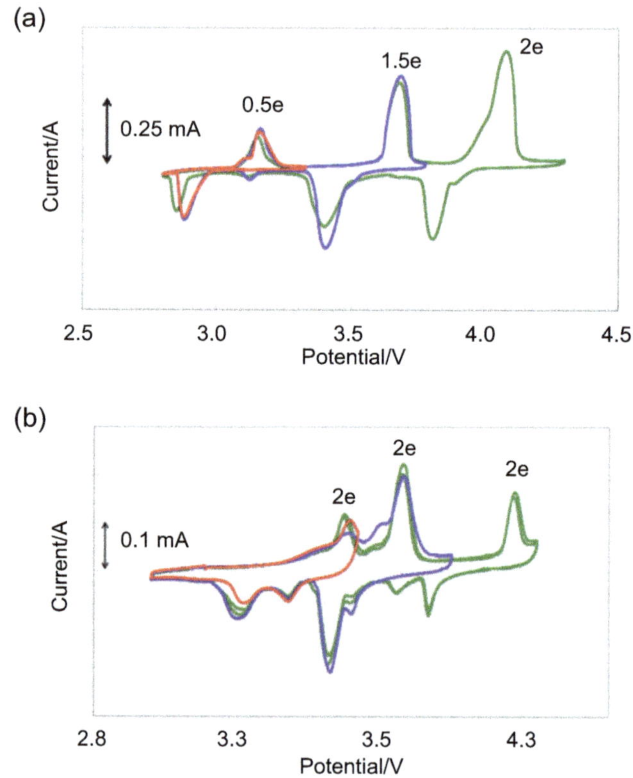

Figure 4.19 Cyclic voltammograms of a TTP electrode with different potential windows of 2.70–4.30 V (green), 2.70–3.80 V (blue), and 2.70–3.40 (red) (V versus Li/Li$^+$, scan rate 0.05 mV/sec.). Adapted from Ref. [52] with permission from John Wiley and Sons.

The dissolution of TTP and TTPY from the electrode into the electrolyte solution was observed in the highest oxidation state (+4 state for TTP and +6 state for TTPY), although it was not observed in the lower oxidation states (+0.5 and +2 states for TTP and +2 and +4 states for TTPY). The dissolution of TTP and TTPY in the highest oxidation states can be understood by considering intermolecular interactions and intermolecular coulomb repulsion. The dication and tetracation of TTPY in the solid states are stabilized by intermolecular CT interaction in addition to the van der Waals force (Fig. 4.20). As a result, percolation of the solvent molecules into the solid states of active materials can be avoided. In contrast, stabilization by CT interaction is not anticipated in the highest oxidation state of TTPY. In this case, the solvent molecules may percolate into the solids so as to relax the strong intermolecular coulomb repulsion, which results in the dissolution of the active materials (Fig. 4.20). In a similar manner, $(TTP)^{4+}$ may dissolve in the solvent, while $(TTP)^{2+}$ may hardly dissolve in the solvent.

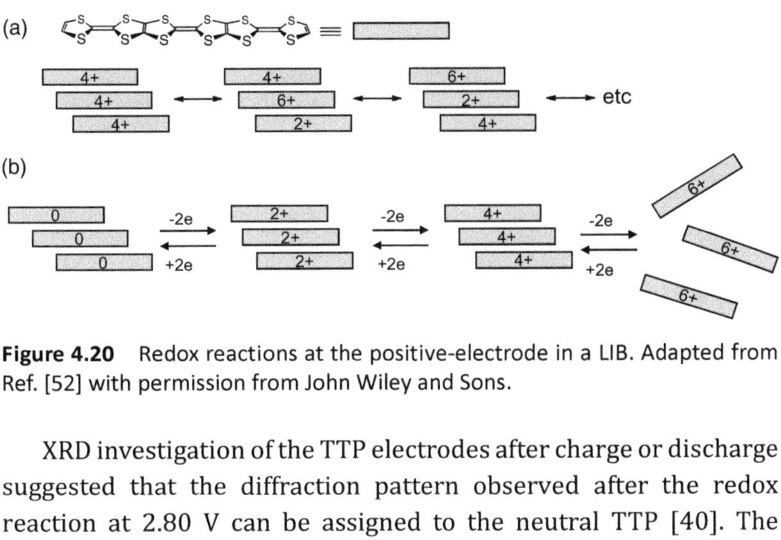

Figure 4.20 Redox reactions at the positive-electrode in a LIB. Adapted from Ref. [52] with permission from John Wiley and Sons.

XRD investigation of the TTP electrodes after charge or discharge suggested that the diffraction pattern observed after the redox reaction at 2.80 V can be assigned to the neutral TTP [40]. The diffraction pattern after the redox reaction at 3.20 V can be assigned to $(TTP)_2X$ ($X = BF_4$ and ClO_4) [45]. These findings indicate that TTP molecules in the electrode (solid state) do not occur as independent molecules but are aggregated, as in their crystalline states.

The TTP cells were charged and discharged in the voltage range of 3.00–3.80 V and of 3.00–4.30 V. The oxidation states of the turning back voltages at 3.80 and 4.30 V correspond to dicationic and tetracationic states, respectively. The discharge capacity of the TTP cell in the 3.00–3.80 V range is 99.8 mAh/g and that in the 3.00–4.30 V range is 198 mAh g^{-1}. They are 70.7 and 70.2% of the theoretical values (141 or 282 mAh g^{-1} for two-electron or four-electron redox processes, respectively). The TTP cell operated in the 3.00–3.80 V range exhibits a better cycle performance than that operated in the 3.00–4.30 V range. That is, 99.7% of the discharge capacity of the initial run is retained after 30 cycles (see Fig. 4.21) operated in the 3.00–3.80 V range. When the cell is operated in the 3.00-4.30 V range, the capacity of the cell after several cycles increases from the initial capacity but it decreases significantly after 10 or more cycles. This seems to correspond to the dissolution of TTP molecules contained in the electrode in the solution.

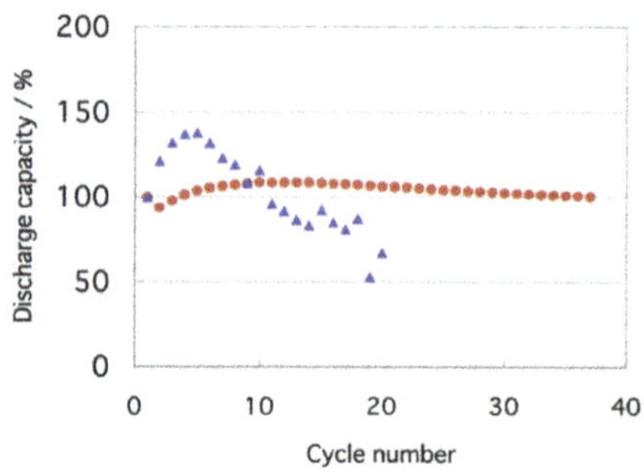

Figure 4.21 Charge-discharge coulombic efficiencies of a TTP/Li cell as a function of cycle number. The cell was cycled between 0 and (a) 3.80 (red) and (b) 4.30 (blue) V at a rate of 1/5 C respectively. The discharge capacity determined in the first cycle discharge was assumed to be 100%. Discharge capacities were 198 mAh g^{-1} in (a) and 99 mAh g^{-1} in (b) at the first cycles. Adapted from Ref. [52] with permission from John Wiley and Sons.

The TTPY electrode by utilizing four-electron redox exhibited the best performance from the viewpoint of compatibility of high capacity and multicycle characteristics. Therefore, further investigation of battery performance was carried out for TTPY in the voltage range of 3.00–4.05 V in order to utilize the four-electron redox process. The initial discharge capacity of the TTPY cell in the voltage range of 3.00–4.05 V was 168 mAh g^{-1} (87% of the theoretical capacity), which is comparable to those of the positive-electrode active materials for commercially available LIBs (150–170 mAh g^{-1}). The capacity deceased gradually as the cycle number increased, as shown in Fig. 4.22. The discharge capacity in the voltage range of 3.00–4.05 V after 20 cycles was 90% of the initial capacity. The capacity decreased more slowly after 20 cycles than the initial stages. It is noteworthy that the capacity of 138 mAh g^{-1} (84% of the initial capacity) was retained even after 100 cycles.

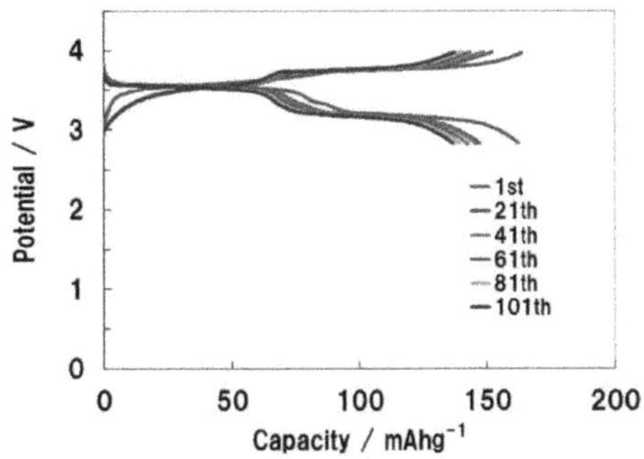

Figure 4.22 Galvanostatic charge-discharge curves for TTPY measured at 298 K under a 1/5 C rate. The inset shows the discharge capacities under various current densities under identical charging conditions. Adapted from Ref. [52] with permission from John Wiley and Sons.

4.4.2 Fused TTF Systems Containing Vinyl-Extended TTFs

The positive electrodes incorporating TTP and TTPY as active materials exhibit a good cycle performance by controlling the number of electrons participating in the redox reaction, as described in the section above. In contrast, they dissolve in the electrolyte solutions in the highest oxidation state (+4 for TTP and +6 for TTPY), as mentioned above. Possible molecular modifications for TTPY to reduce solubility in electrolyte solvents are as follows:

- Rigid substituents such as ethylenedithio groups
- Rigid extended-TTF units
- Increase in the number of (extended) TTF units

As for the rigid extended-TTF units, insertion of a π-spacer is sometimes effective. A vinylogous TTF (EBDT, Fig. 4.23) [53–56] is known to exhibit lower solubility in ordinary organic solvents than TTF. Thus, we examined vinyl-extended TTPY analogs (**1a–c** in Ref.

[57]). The tetrakis(*n*-hexylthio) derivative of **1** (**1c**) exhibits four pairs of redox waves (Fig. 4.24). A comparison of peak currents of the redox waves indicates that the first two pairs of redox waves correspond to two-electron redox waves and the last two pairs of redox waves correspond to a one-electron transfer process.

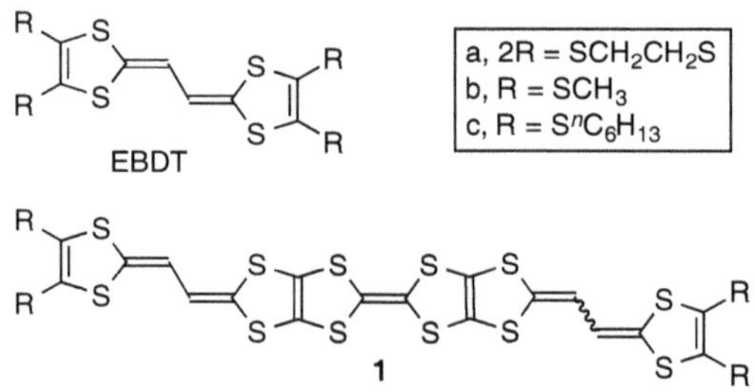

Figure 4.23 Chemical structures of EBDT and **1**.

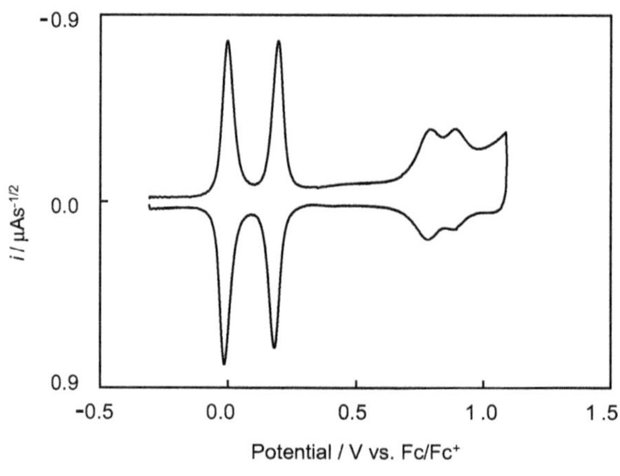

Figure 4.24 Deconvoluted cyclic voltammogram of **1c**; see Ref. [57].

Figure 4.25 shows the first five charge-discharge curves of **1a**/Li and the cycle-life performance of **1a**/Li and **1b**/Li, respectively. No distinct plateau is observed in both charge and discharge processes, while well-separated redox waves are observed in a solution. This is possibly because of the apparent overlap of the redox processes in the solid state. The initial discharge capacities of **1a**/Li and **1b**/Li cells are 157 and 168 mAh g^{-1}, respectively. They correspond to 93% and

99% of the theoretical capacities of the five-electron redox of **1a** and **1b**, respectively. Their initial discharge capacities are comparable to those of the positive-electrode materials for commercially available LIBs (150–170 mAh g^{-1}). The discharge capacities after 40 cycles are 86% and 73% and 74% of the first discharge capacities for **1a**/Li and **1b**/Li cells, respectively. The result that the **1a** cell exhibits a higher cycle performance than the **5c** cell is consistent with the lower solubility of **1a** with rigid ethylenedithio substituents rather than **1b** with flexible methylthio substituents.

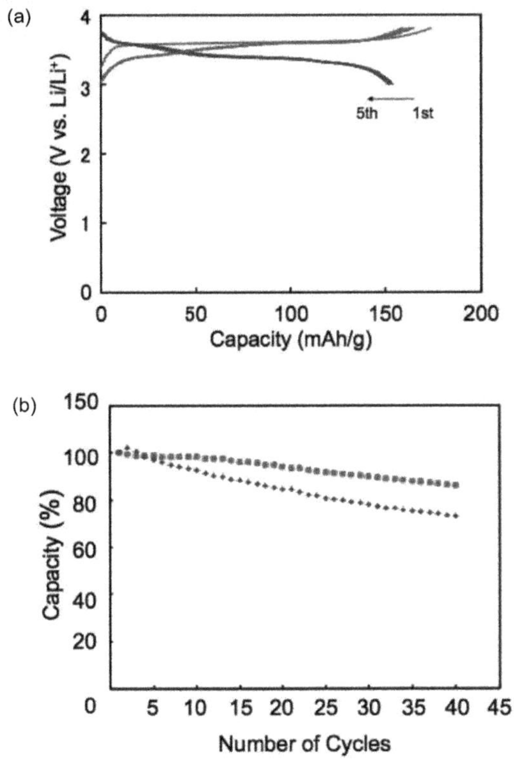

Figure 4.25 (a) Galvanostatic charge-discharge curves of **1a**/Li cell and (b) cycle-life performances for **1a**/Li and **1b**/Li cells; see Ref. [57].

As described above, the six-electron redox system, TTPY, is scarcely soluble in organic solvents in the oxidation state of up to +4, while it is soluble in the highest oxidation state, of +6. Thus, TTPY can be utilized as a positive-electrode material for a rechargeable battery by utilizing the four-electron redox. Increase of TTF units might be the best way to reduce solubility. As for such higher homologs of multifused TTFs composed of more than three TTF units, utilization of maximum electrons might be expected. Otherwise, the utilization rate should be enhanced. For example, pentakis-fused TTF derivatives can cause redox reactions involving 10-electron processes and thus they achieve the theoretical maximum utilization rate of 8/10 = 80% within their cycle performances. In the meantime, the TTPY derivatives, belonging to the six-electron redox systems, achieve the theoretical maximum utilization rate of 4/6 = 67% within their cycle performances. Accordingly, the pentakis-fused TTF derivatives exhibit a higher rate than the rest. However, such higher homologs of fused TTF systems have several problems. Firstly, the synthesis of such large molecules is usually difficult because an increase in the dimensions of the precursors should lower the solubility of the precursors in solvents used in syntheses. Furthermore, their maximum oxidation potentials should shift toward more positive voltages because of enhanced on-site coulomb repulsion. In that case, the multielectron oxidation potentials might reach the potential region where the electrolyte solution begins to decompose.

The thiophene-inserted TTF analog (**2**, Fig. 4.26) [58–60] is known to be more soluble than TTF itself because of flexibility of the 1,3-dithiole rings around the aromatic thiophene ring. In addition, the second oxidation potential of **2** is considerably lower than that of TTF. Thus, insertion of thiophene spacers might be a promising strategy for the synthesis of highly fused TTF oligomers. We synthesized a pentakis-fused TTF analog extended by the insertion of two thiophene rings (**3** and **4**) [57]. A cyclic voltammogram of **4** shows six pairs of redox waves (Fig. 4.27). On the basis of the comparison of peak currents of the redox waves and considering the presence of 10 redox-active 1,3-dithiol-2-ylidene sites in **4**, it is indicated that the last two pairs of redox waves of **4** correspond to one-electron transfer processes, while the others correspond to two-electron transfer processes.

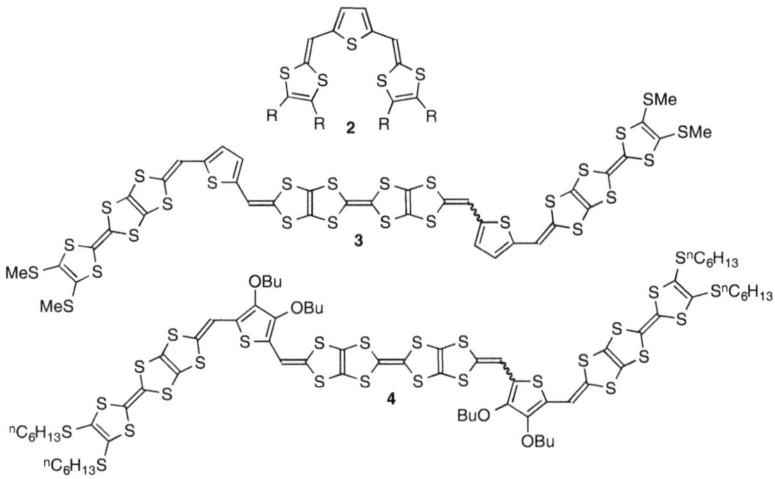

Figure 4.26 Chemical structures of **2**, **3**, and **4**.

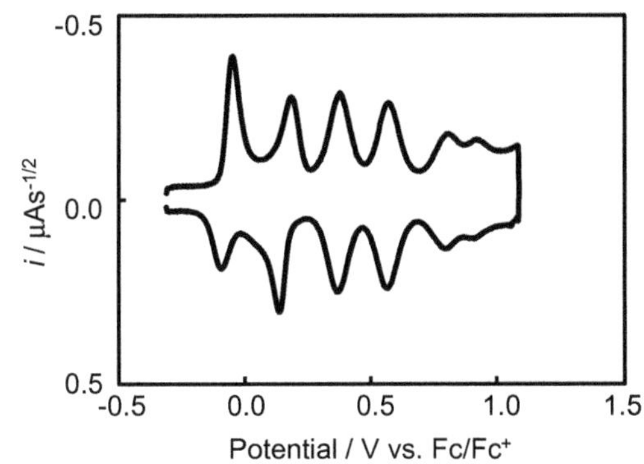

Figure 4.27 Deconvoluted cyclic voltammogram of **4**; see Ref. [57].

Figure 4.28 shows the first five charge-discharge curves of a **3**/Li cell and its cycle performance up to 40 cycles. The initial discharge capacity is 190 mAh g^{-1}, which corresponds to 93% of the theoretical capacity for the 10-electron process (205 mAh g^{-1}). This result strongly indicates that 10-electron redox per molecule takes place in the charge and discharge processes. The initial discharge capacity of the **3**/Li cell (190 mAh g^{-1}) is larger than the initial discharge capacities of the positive-electrode materials for commercially available LIBs. The discharge capacity after 40 cycles (121 mAh g^{-1}) is 64% of the initial discharge capacity. The high cycle-life performance in spite of utilization of the highest oxidation state of +10 is probably attributable to the strong van der Waals force between the large π-electron framework of the thiophene-inserted pentakis-fused TTF.

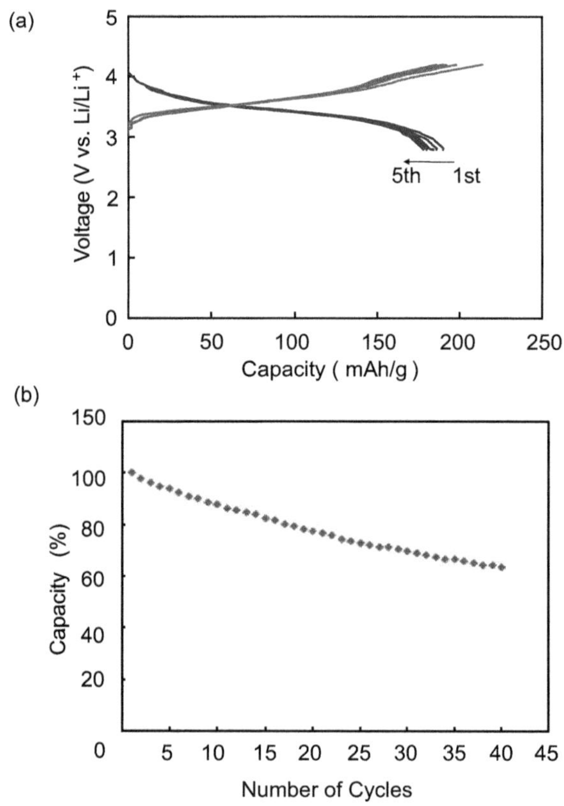

Figure 4.28 (a) Galvanostatic charge-discharge curves of **3**/Li and (b) cycle-life performances; see Ref. [57].

4.4.3 Fused TTF Systems Containing Cyclohexene-1,4-diylidene

Utilization of an extended TTF analog with a cyclohexene-1,4-diylidene (**5**) [61] as a unit of fused TTF molecules has several advantages for the development of positive-electrode materials for rechargeable batteries. First, it can be synthesized by using the retro Diels–Alder reaction of a highly soluble cyclopentadiene adduct (Fig. 4.29). Such a strategy is very useful for the synthesis of higher homologs of fused TTF oligomers, that is, the large precursors are expected to be soluble for the common solvents. Next, **5** is less soluble in organic solvents than TTF because of the presence of a rigid cyclohexene-1,4-diylidene moiety. Thus, tris-fused and pentakis-fused TTF derivatives possessing one and two cyclohexene-1,4-diylidene-inserted TTF units (**6**, **7**) were investigated [62, 63] (Fig. 4.30).

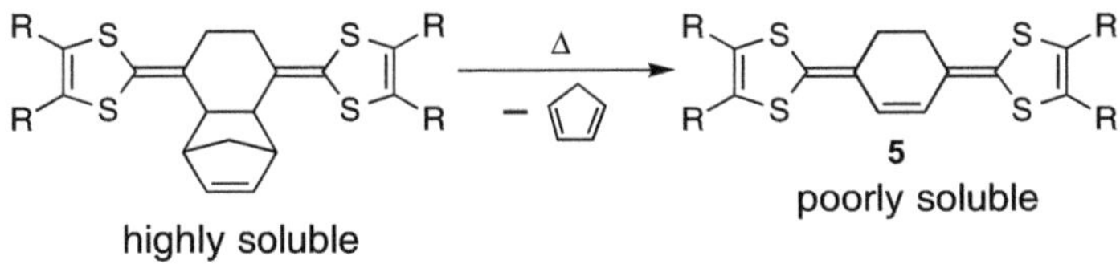

Figure 4.29 Synthesis of **5**.

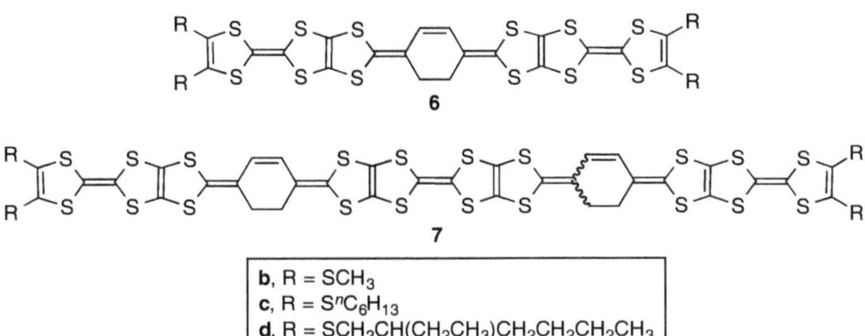

Figure 4.30 Chemical structures of **5**, **6**, and **7**.

Syntheses of **6** and **7** were carried out according to Fig. 4.31. The reaction of **8** with 2.2 equivalent moles of **9b,c** [64] in the presence of BuLi in tetrahydrofuran (THF) at −78°C gave bisadducts **10b** and **10c**. When an excess of the diketone 8 was used, the monoadduct **11** was obtained as the main product. The reaction of **11b,d** with a bisphosphonate with a TTF core (**12**) in the presence of *n*-BuLi afforded **13b,d**. The precursors **10b** and **13b** were soluble enough in carbon disulfide to be purified by column chromatography.

Finally, the target molecules **6b,c** and **7b,d** were obtained by heating of **10b,c** and **13b,d** in the solid state under a reduced pressure. The compound **6c** is scarcely soluble in carbon disulfide, and a sample of high purity can be obtained by washing it thoroughly with carbon disulfide. Solubility of **6c** in carbon disulfide (0.11 mg/mL at 23°C) was considerably lower than that of the TTC$_6$-TTPY (3.9 mg/mL) (Fig. 4.32) [43]. Thus, solubility in the neutral state is reduced by the insertion of the cyclohexene-1,4-diylidene moiety, as expected.

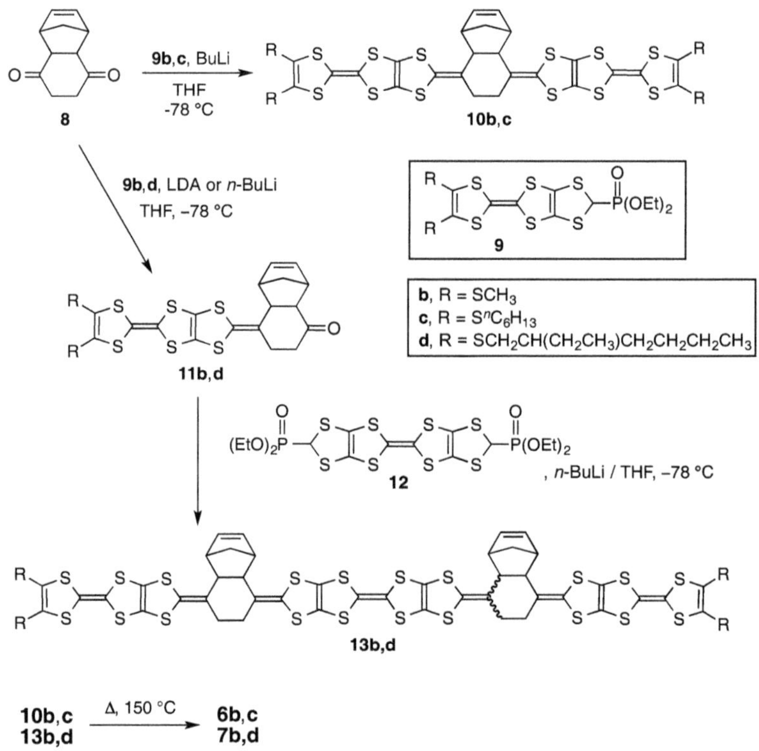

Figure 4.31 Synthetic schemes of **6** and **7**.

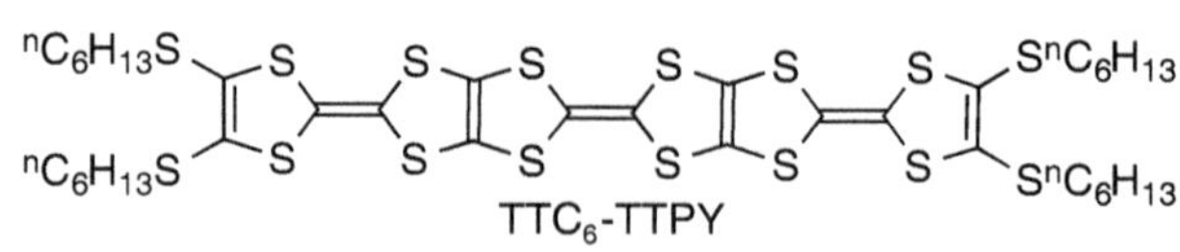

Figure 4.32 Chemical structure of TTC$_6$-TTPY.

The cyclic voltammogram of **6c** measured in a carbon disulfide-benzonitrile solution (2:1, v/v) consists of three pairs of two-electron redox waves, at +0.05, +0.15, and +0.58 (V versus Fc/Fc) [62]. On the other hand, **7d** exhibited five pairs of redox waves, at +0.05 ($4e^-$), +0.15 ($2e^-$), +0.36 ($2e^-$), +0.58 ($1e^-$), and +0.72 ($1e^-$) V (V versus Fc/Fc^+) (Fig. 4.33). The parentheses after the potentials are the number of electrons participating in the redox process, which were estimated by a comparison of the peak current of each redox wave. The E_{10} of **7d** (+0.72 V) is more negative by 0.16 V than the six-electron oxidation potential of TTC_6-TTPY (+0.88 V) [43]. This suggests 10 electrons may be utilized for rechargeable batteries using **7** as a positive-electrode material within the appropriate voltage at which the electrolyte solution does not decompose.

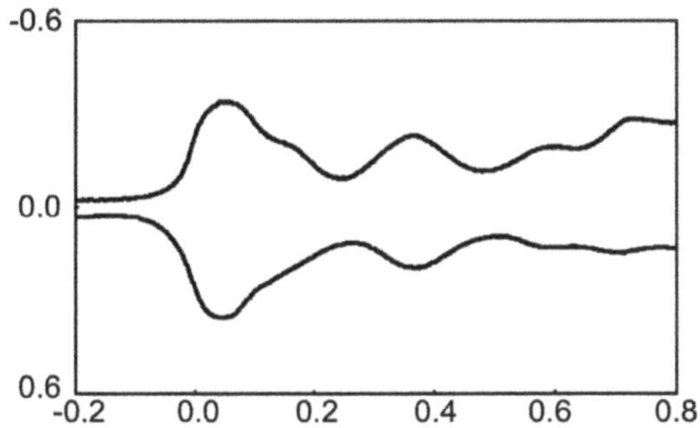

Figure 4.33 Deconvoluted cyclic voltammogram of **7d** in carbon disulfide-benzonitrile (2:1, v/v); see Ref. [63].

Figures 4.34 shows the first five charge-discharge curves of the prepared **6a**/Li cell cycled between 2.50 and 4.00 V at room temperature. The initial discharge capacity is 174 mAh g^{-1}, which is 89% of the theoretical capacity of the six-electron process (196 mAh g^{-1}) [62]. This result strongly indicates that six-electron redox per molecule takes place in the charge-discharge process. The discharge capacity after 20 cycles is 78% of the first run. Such good cycle performance in spite of utilization of the highest oxidation state of +6 might be because the van der Waals force has been overcome compared with the intermolecular coulomb repulsion between adjacent $\mathbf{6a}^{6+}$ molecules.

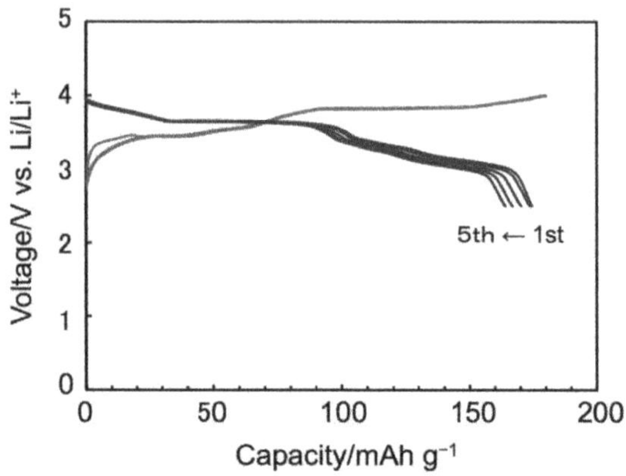

Figure 4.34 Galvanostatic charge-discharge curves of **6a**/Li. Adapted from Ref. [62] with permission from the Chemical Society of Japan.

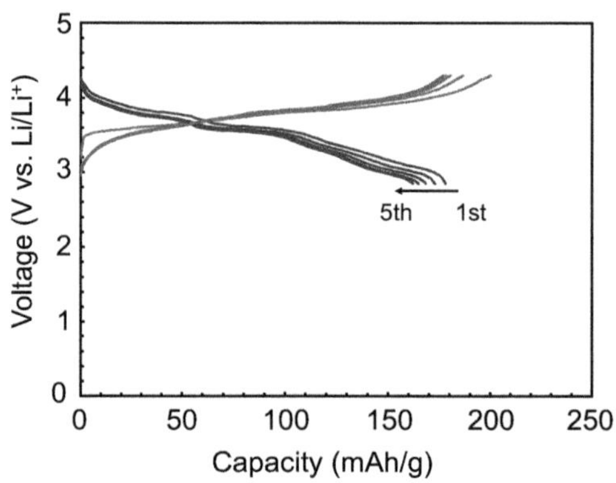

Figure 4.35 (a) Galvanostatic charge-discharge curves of **7a**/Li; see Ref. [63].

On the other hand, the initial charge capacity of the **7a**/Li cell was 196 mAh/g. This value corresponds to 92% of the theoretical value for the 10-electron redox process (214 mAh g^{-1}), indicating 10 electrons per molecule involved in the charge-discharge process of the **7a** electrode (Fig. 4.35). The average voltage of the first discharge is 3.56 V, which is higher by 1.00–1.20 V than the average voltages of organic materials with a high capacity, tribromotriquinoxalinylene (Br_3TQX, Fig. 4.36) (2.34 V) [18] and PPVT (2.54 V) (Fig. 4.9) [13].

Realization of a higher average voltage of **7a** compared with Br_3TQX and PPVT may be explained as follows: Multifused TTF systems such as TTPY and **7a** are multielectron donors, while Br_3TQX and PPVT are multielectron acceptors. The maximum oxidation potentials of the donors shift to a more positive voltage region as the number of electrons increases because of intramolecular coulomb repulsion among positive charges. In contrast, the maximum reduction potentials of acceptors must shift to more negative voltage regions as the number of electron increases due to intramolecular coulomb repulsion among negative charges.

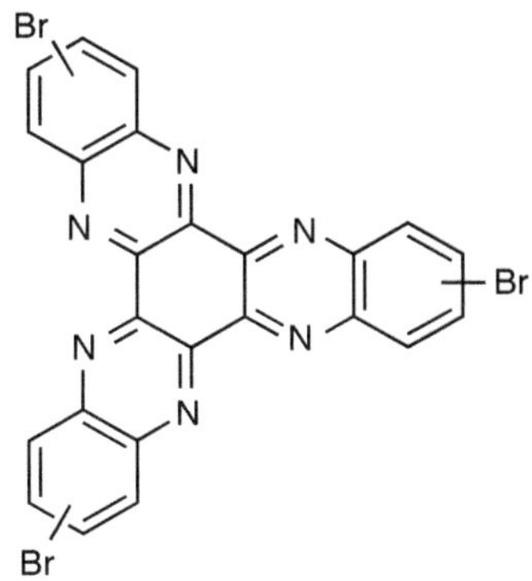

Figure 4.36 Molecular structure of Br_3TQX.

The energy density of **7a** at the first discharge reaches 700 mWh g^{-1}. The energy density of **7a** is superior to the energy densities of most inorganic positive-electrode materials for LIBs [65, 66], although it is inferior to those of Li_2MnO_3-$LiMO_2$ (M = Mn, Co, and Ni) systems (~1000 mWh g^{-1}) [67–69]. The cycle-life performance test revealed that 72% of the initial capacity is maintained after 30 cycles in spite of utilization of the highest oxidation state of +10 (Fig. 4.37).

Figure 4.38 compares the discharge curves of the **7a**/Li cell at the current densities from 0.5 to 200 C rates. The discharge capacity at 100 C shows 64% of the obtained capacity at the low current density of 0.5 C, although it decreases as the current density increases.

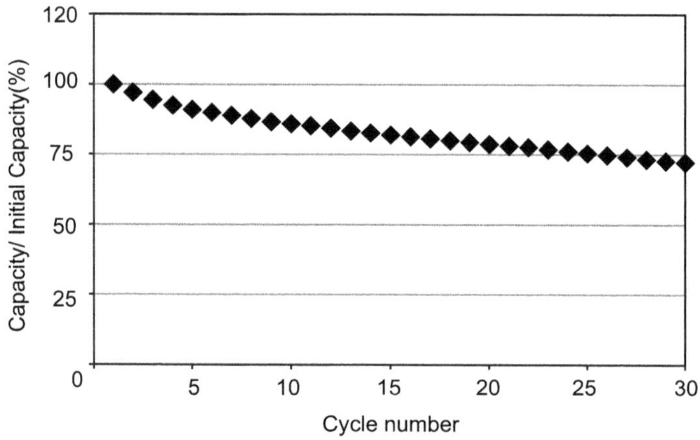

Figure 4.37 Cycle-life performance of **7a**/Li; see Ref. [63].

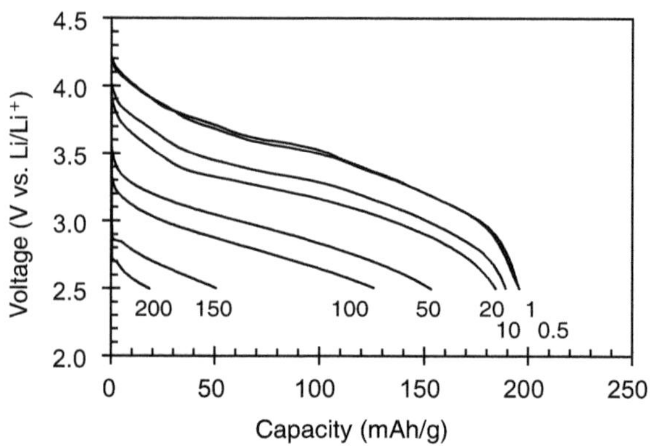

Figure 4.38 Discharge curves of the **7a**/Li cell at various current densities. The inset numbers designate the C rates. (Voltage range 2.50–4.30 V versus Li^+/Li, 30°C); see Ref. [63].

Figure 4.39 shows current–voltage (*I*–*V*) plots obtained by a pulse-type high-rate discharge test (5 second discharge). A power maximum (69 W g^{-1}) was obtained at the current density of around 160 C. This power output performance is comparable to that of an electrode incorporating a commercially available carbon-layer-deposited $LiFePO_4$ that has been specially designed for high-power usage. High-rate performances of the **7a**/Li cell might be attributed to possible high conductivity in the oxidized state of **7a**.

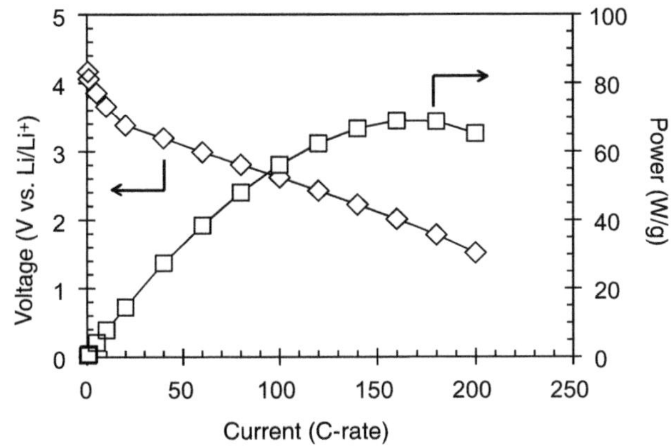

Figure 4.39 Current-voltage (*I*-*V*) and current-power (*I*-*P*) plots of the **7a**/Li cell. The voltages after a 5-second discharge were plotted and used for power calculation. The state of charge (SOC) was controlled to be constant (~100%) by charging the discharged capacity at each step with the current density of 0.5 C; see Ref. [63].

4.4.4 Fused TTF Systems Extended with an Anthraquinoid Spacer

A TTF analog with an anthraquinoid spacer (TTFAQ, Fig. 4.40) is known to adopt a nonplanar structure in both neutral and oxidized states because the planar structure is significantly destabilized by steric hindrance between the sulfur atoms in the 1,3-dithiole rings and the hydrogen atoms at the peri-positions [70–73]. $(TTFAQ)^{2+}$ has a conformation with the orthogonally twisted structure between anthracene and two 1,3-dithiolium rings (T form) [72, 74, 75], while neutral TTFAQ adopts a folded saddle-like conformation (S form), as shown in Fig. 4.41 [75–77]. A theoretical calculation suggests that one-electron oxidized TTFAQ also adopts S form [74], in which a positive charge is hardly delocalized. As a result, the dicationic state with a planar and aromatic anthracene moiety is more stable than the monocationic state, although the dicationic state of most TTF-type molecules is more unstable than the monocationic state because of on-site coulomb repulsion between two positive charges.

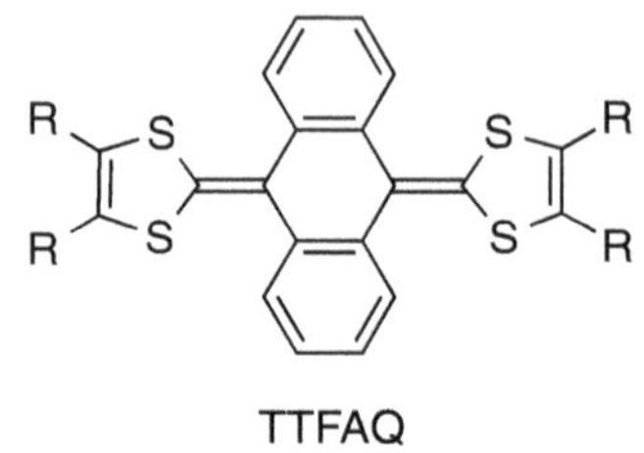

Figure 4.40 Chemical structures of TTFAQ.

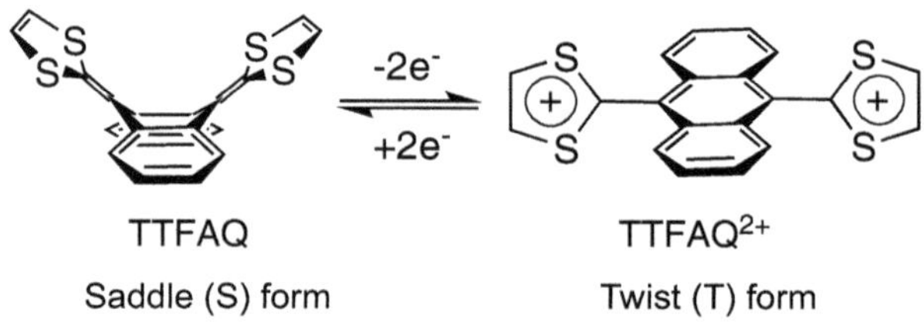

Figure 4.41 Redox process–accompanied schematic structures of two conformations of TTFAQ.

Fused TTF systems, including one or more TTFAQ units [78, 79], might have advantages as positive-electrode materials, as follows. Firstly, the nonplanar structure of the TTFAQ unit in the neutral state might enhance the solubility of the precursors containing TTFAQ units, which will facilitate the synthesis of highly fused TTF systems. Secondly, solubility of the molecule in the highest oxidation state might be suppressed because the whole molecule should be partially packed so as to avoid intermolecular steric repulsion (Fig. 4.42). As a result, intermolecular coulomb repulsion will be reduced compared with the highly planar molecules.

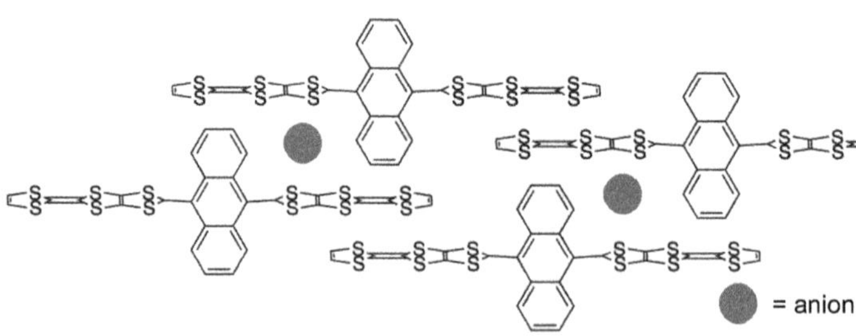

Figure 4.42 Schematic molecular packing of oxidized TTPYAQ.

The present author and coworkers have synthesized derivatives of a tris-fused donor incorporated with an anthraquinoid spacer (TTPYAQ, **14**, Fig. 4.43) [79]. X-ray structure analysis of **14b** reveals that the molecule adopts a saddle structure, as is observed in most TTFAQ derivatives (Fig. 4.44). Cyclic voltammograms of all the derivatives contain three pairs of two-electron redox waves. The ultraviolet-visual-near-infrared (UV-vis-NIR) spectrum of **14c**$^{2+}$ suggests that two positive charges formed by the first two-electron oxidation process are mainly located on the central TTFAQ moiety. The positive charges formed by the second and third redox processes are probably distributed over two TTF moieties at both ends.

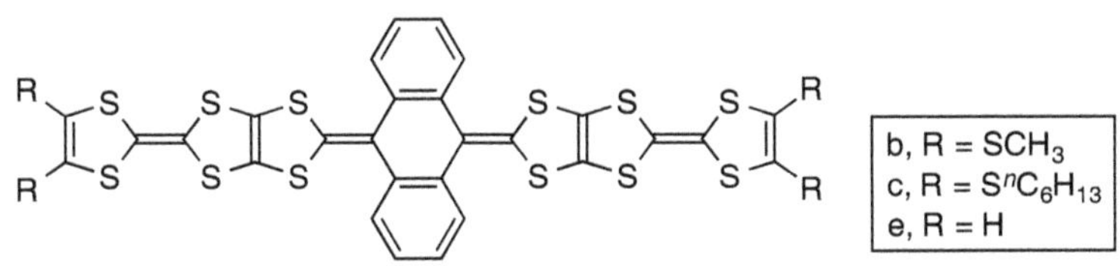

14 (TTPYAQ)

Figure 4.43 Chemical structures of TTPYAQ (**14**).

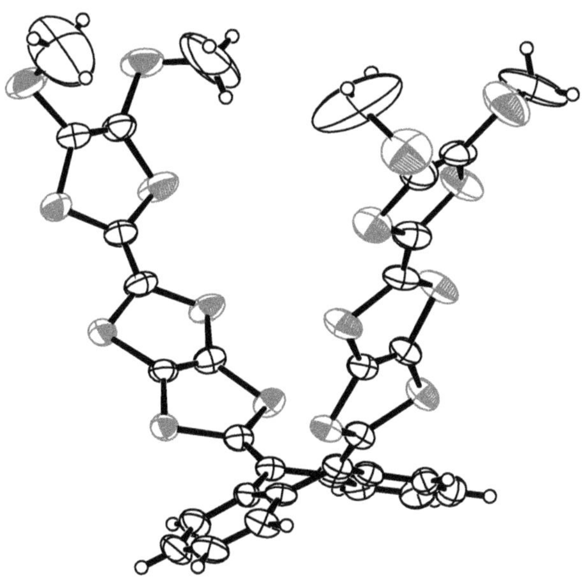

Figure 4.44 Molecular structure determined by the X-ray structure analysis of **14b** [79].

The charge-discharge property of a rechargeable battery using the unsubstituted derivative **14e** as the active material was examined. Figure 4.45 shows the first five charge and discharge curves of the prepared **14e**/Li cell. The initial discharge capacity is 192 mAh g^{-1}, which is 85% of the theoretical capacity of six-electron utilization (220 mAh g^{-1}). The discharge capacity after 100 cycles is 38% of the first one. This result strongly indicates that six-electron redox per molecule participates in the charge-discharge process, although the cycle performance is not so good. This result suggests the nonplanar hexacation of **14e**, with the T forms weakly interacting with each other in the solid state probably due to reduced intermolecular coulomb repulsion.

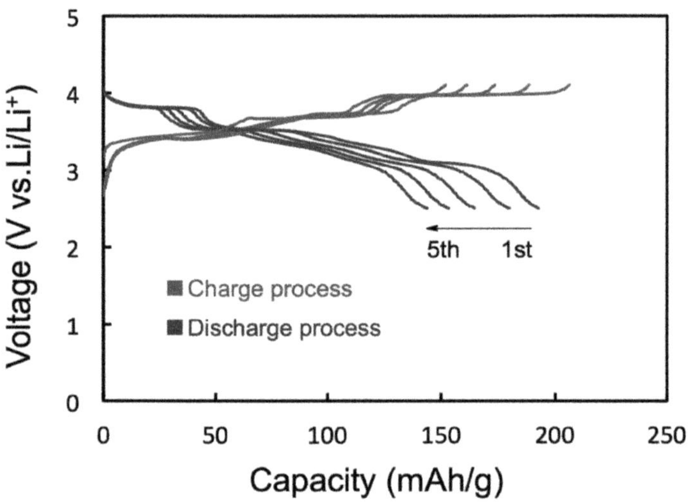

Figure 4.45 Galvanostatic charge-discharge curves of **14e**/Li [79].

4.4.5 Fused Donor and Acceptor Triads Composed of TTF and *p*-Benzoquinones

The fused TTF-*p*-benzoquinone (BQ) systems [80–82] are expected to exhibit higher capacities and energy densities derived from the low molecular weight of the strongly electron-accepting BQ unit. In addition, a long cycle-life would be also expected because of their low solubility in the solvents, attributed to the intermolecular charge-transfer interaction.

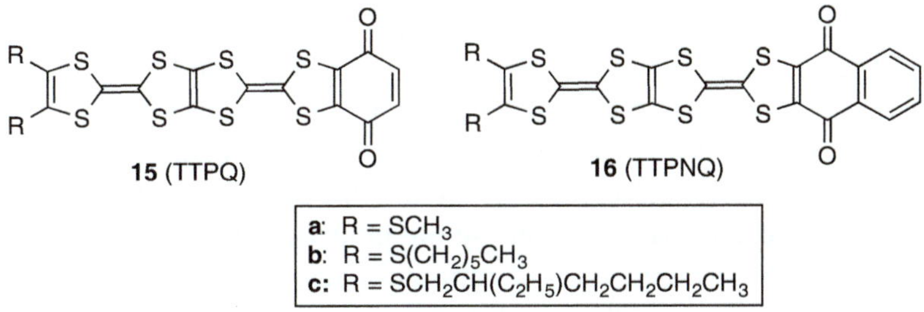

Figure 4.46 Chemical structures of TTPQ and TTPNQ.

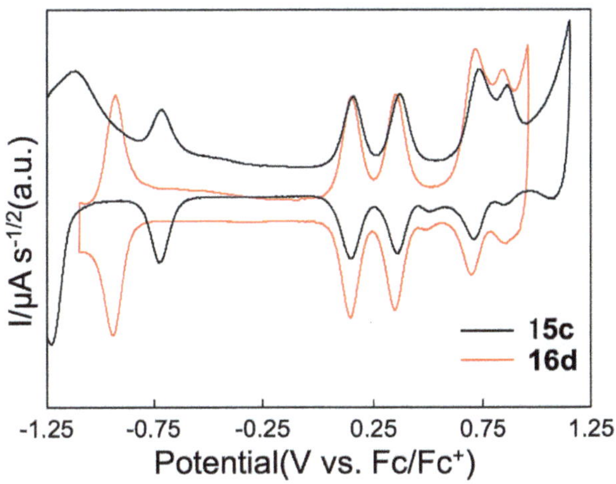

Figure 4.47 Deconvoluted cyclic voltammograms (CVs) of **15c** (black) and **16d** (red) in benzonitrile-carbon disulfide (1:1, v/v) containing 0.1 M Bu_4NPF_6; the scan rate is 10 mV s^{-1} (**15c**) and 50 mV s^{-1} (**16d**). Adapted from Ref. [82] with permission from the Chemical Society of Japan.

The author and coworkers have synthesized derivatives of a fused TTF-TTF-BQ triad, **15,** and its naphthoquinone (NQ) analog, **16** (Fig. 4.46) [82]. The tetrakis(*n*-hexylthio) derivative **15c** exhibits six pairs of one-electron redox waves at −1.18, −0.72, +0.15, +0.37, +0.73, and +0.87 V (versus Fc/Fc^+), as shown in Fig. 4.47. A comparison of each of the redox potentials reveals that the redox waves at −1.18 and −0.72 V are assigned to the BQ moiety while the other redox waves are ascribed to the oxidation of the TTP moiety of **15c** (Table 4.1). The possible redox process of **15** is shown in Fig. 4.48. The potentials E_1^{ox}–E_4^{ox} of **15c** are substantially higher than those of TTM-TTP [42] because the cationic species formed by oxidation of **15c** are destabilized by the presence of electron-withdrawing carbonyl groups. On the other hand, the potentials corresponding to the reduction of the benzoquinone moiety are also higher than those of BQ. This result indicates that the anionic species formed by the reduction of **15c** are stabilized by the inductive effect of the sulfur atoms. Such destabilization of cationic species and stabilization of the anionic species might raise the operation voltage of rechargeable batteries using derivatives of **15**.

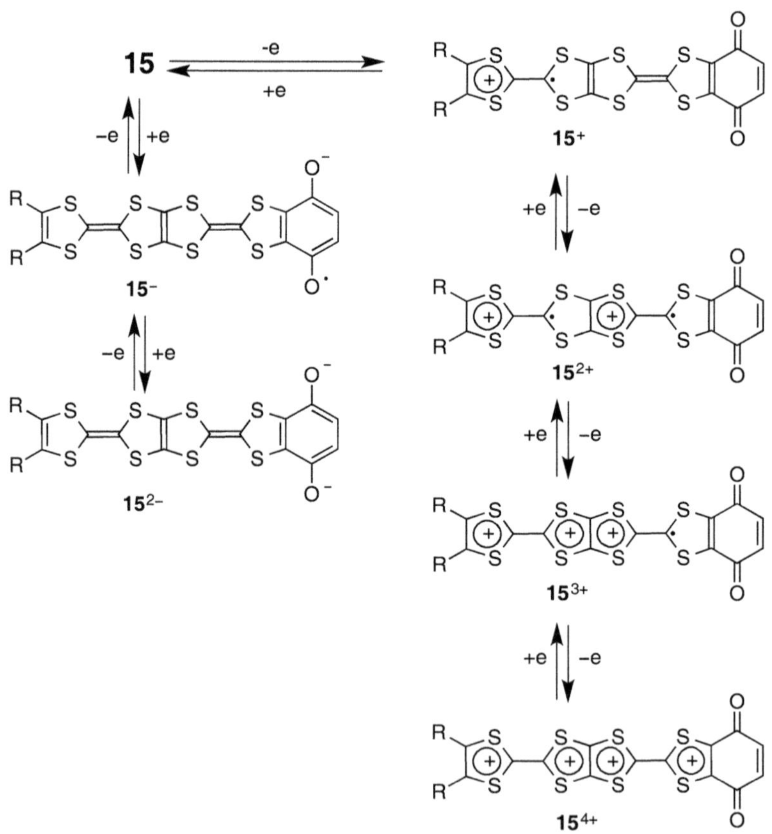

Figure 4.48 Possible redox scheme of **15**. Adapted from Ref. [82] with permission from the Chemical Society of Japan.

Table 4.1 Redox potentials of **15c** and **16d** and their related compounds (V versus Fc/Fc^+)[a]. The chemical structures are given below the table.

Compound	Solvent	E_2^{red}	E_1^{red}	E_1^{ox}	E_2^{ox}	E_3^{ox}	E_4^{ox}
15c	CS_2-C_6H_5CN (1:1, v/v)	−1.18	−0.72	+0.15	+0.37	+0.73	+0.87
16d			−0.93	+0.15	+0.35	+0.72	+0.84
TTM-TTP				+0.11	+0.29	+0.56[b]	+0.68[b]
NQ			−1.17				
BQ	CH_2Cl_2		−0.98				
NQ			−1.18				

[a]Measured in a solution containing 0.1 M $^{n}Bu_4NPF_6$ at 25°C, with Pt working electrodes and counterelectrodes; the scan rate is 50 mV s^{-1}, except **15c**, for which the rate is 10 mV s^{-1}.
[b]Irreversible step. Anodic peak potentials.
Source: Adapted from Ref. [82] with permission from the Chemical Society of Japan.

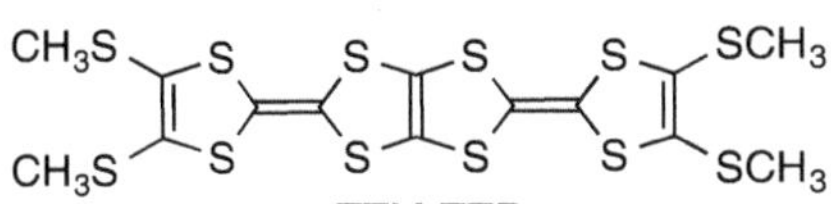

TTM-TTP

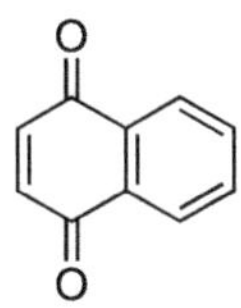

NQ

Figure 4.49 shows the charge and discharge curves of the initial five cycles of the **15b**/Li and **16b**/Li cells. The first discharge process corresponds to the two-electron reduction (i.e., Li^+ insertion) of the BQ or NQ moieties. The two-electron oxidation (Li^+ release) of the quinone moiety is followed by the four-electron oxidation (PF_6^- insertion) of the bis-fused TTF part during the subsequent charge process. The discharge capacities at the second cycle are 270 and 266 mA h g^{-1} for **15b**/Li and **16b**/Li, respectively, which are 93% and 100% of the theoretical capacities (291 and 267 mA h g^{-1} for **15b**/Li and **16b**/Li, respectively) for a six-electron redox reaction. These values are larger by more than 1.5 times than the initial discharge capacity of **6b** (174 mA h g^{-1}), which is also a six-electron redox system. The average discharge potentials are 3.04 and 2.94 V (versus Li/Li^+) for **15b** and **16b**, respectively, which are transformed into the energy densities of 810 and 768 mWh g^{-1}, respectively. These energy densities are superior to those of the conventional positive-electrode materials, $LiCoO_2$, that is, 540–600 mWh g^{-1}. After 25 cycles, these cells maintain the discharge capacities of 231 (**15b**) and 204 (**16b**) mA h g^{-1}, which correspond to 86% and 77% of the second discharge capacities.

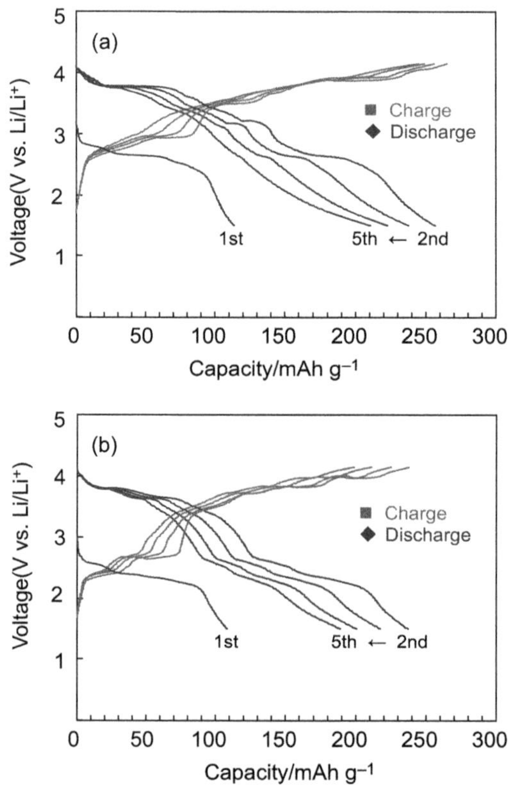

Figure 4.49 (a) Galvanostatic charge-discharge curves of (a) the **15b**/Li cell and (b) the **16b**/Li cell at the current densities of 40 mA g^{-1} (charge) and 100 mA g^{-1} (discharge). Adapted from Ref. [82] with permission from the Chemical Society of Japan.

The charge and discharge potential profiles of the **16b**/Na cell during the initial five cycles are shown in Fig. 4.50. The discharge capacity at the second cycle is 250 mAh g^{-1}, which is 94% of the theoretical value, suggesting that TTPNQ also undergoes the six-electron transfer in the sodium ion battery, accompanied by the two Na^+ release and four PF_6^- insertion. The average discharge potential is 2.60 V (versus Na/Na^+), which is lower by 0.34 V than that of the **16b**/Li cell. The cycle performance of the **16b**/Na cell is inferior to that of the **16b**/Li cell; the discharge capacity drops to 155 mA h g^{-1} after 25 cycles (62% of the second discharge capacity).

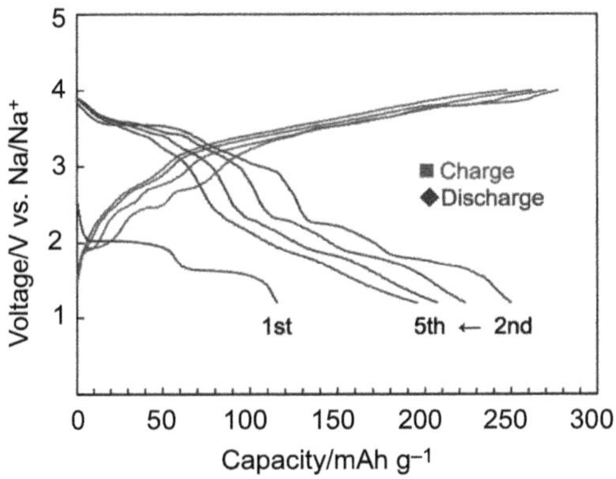

Figure 4.50 Galvanostatic charge-discharge curves of the **16b**/Na cell at the current densities of 40 mA g^{-1} (charge) and 100 mA g^{-1} (discharge). Adapted from Ref. [82] with permission from the Chemical Society of Japan.

4.5 Summary and Outlook

In this chapter, organic molecular positive-electrode materials have been overviewed. Various organic molecules have been examined so far as positive-electrode materials for rechargeable batteries are concerned. However, the molecules that are scarcely soluble in the electrolyte solution during the charge-discharge process are still limited.

Fused TTF materials explained in this chapter are promising as organic positive-electrode materials. However, there are a number of problems that should be overcome in terms of both performance and cost. In terms of the manufacturing cost, the following points should be considered: the present fused TTF materials are synthesized by multistep reactions from commercially available precursors; purification by column chromatography is required. However, expensive reagents, such as transition metal catalysts, are not required for the synthesis of fused TTF molecules. Therefore, an inexpensive synthetic method might be established by a synthetic route to a different target molecule with fewer steps or a synthetic route to the same target molecule with a different synthetic methodology.

In the rechargeable batteries using electron donors such as fused TTF systems as positive-electrode materials, the counteranions in the electrolyte are inserted during the charge process. In other words, expensive lithium ions are not required for the electrolyte. Therefore, reduction of the total production cost of the rechargeable batteries might be expected if inexpensive and widely distributed sodium and/or magnesium ions can be used.

As for cell performance, fused TTF systems such as **7b** and **15b** exhibit considerably high energy densities, of 700 and 810 mAh g^{-1}. However, improvement of cycle-life performance is still required because dissolution in the maximum oxidized state cannot be completely suppressed even in a fused-TTF pentamer **7b**. In addition, increase of the ratio of the positive-electrode material is indispensable for improvement of the energy density of the batteries themselves. Improvement of the molecular skeleton, such as increasing the number of cyclohexene-1,4-diylidene units, increasing the number of donor units, and oligomerization of fused-TTF molecules, might be also useful. On the other hand, dissolution of positive-electrode molecules can be suppressed by using a solid or pseudosolid electrolyte. Low-weight molecular electrode materials such as TTF dissolve slightly in a pseudosolid electrolyte [83]. Therefore, TTPY and unsubstituted **15** might be promising for realization of both high energy density and long cycle-life performance. To enhance capacities, design of new molecules in which BQ units are incorporated might be promising.

Although organic molecules have several problems to be solved, there is infinite degree of freedom to design and modify them. It is expected that further syntheses of new molecules designed by adopting advantages of existing inorganic and organic materials will realize higher cell performance suitable for practical use in the future.

Acknowledgments

This work was partially supported by JSPS KAKENHI Grant Numbers JP23550155, JP26410095, and JP15H00948; by MEXT KAKENHI Grant Number JP15H03798; and by Grant-in-Aid for Research Promotion, Ehime University, to the Research Unit for Power Generation and Storage Materials (PGeS).

References

1. Armand, M. and Tarascon, J.-M. (2008). Building better batteries, *Nature*, **451**, pp. 652–657.
2. Padhi, A. K., Nanjundaswamy, K. S. and Goodenough, J. B. (1997). Phospho-olivines as positive-electrode materials for rechargeable lithium batteries, *J. Electrochem. Soc.*, **144**, pp. 1188–1194.
3. Khasanova, N. R., Drozhzhin, O. A., Storozhilova, D. A., Delmas, C. and Antipov, E. V. (2012). New form of Li_2FePO_4F as cathode material for Li-ion batteries, *Chem. Mater.*, **24**, pp. 4271–4273.
4. Nytén, A., Abouimrane, A., Armand, M., Gustafsson, T. and Thomas, J. O. (2005). Electrochemical performance of Li_2FeSiO_4 as a new Li-battery cathode material,*Electrochem. Commun.*, **7**, pp. 156–160.
5. Yao, M., Senoh, H., Yamazaki, S., Siroma, Z., Sakai, T. and Yasuda, K. (2010). High-capacity organic positive-electrode material based on a benzoquinone derivative for use in rechargeable lithium batteries, *J. Power Sources*, **195**, pp. 8336–8340.
6. Yokoji, T., Matsubara, H. and Satoh, M. (2014). Rechargeable organic lithium-ion batteries using electron-deficient benzoquinones as positive-electrode materials with high discharge voltages, *J. Mater. Chem. A*, **2**, pp. 19347–19354.
7. Yokoji, T., Kameyama, Y., Sakaida, S., Maruyama, N., Satoh, M. and Matsubara, H. (2015). Steric effects on the cyclability of benzoquinone-type organic cathode active materials for rechargeable batteries, *Chem. Lett.*, **44**, pp. 1726–1728.
8. Nishida, S., Yamamoto, Y., Takui, T. and Morita, Y. (2013). Organic rechargeable batteries with tailored voltage and cycle performance, *ChemSusChem*, **16**, pp. 794–797.
9. Song, Z., Zhan, H. and Zhou, Y. (2009). Anthraquinone based polymer as high performance cathode material for rechargeable lithium batteries, *Chem. Commun.*, pp. 448–450.

10. Yao, M., Yamazaki, S., Senoh, H., Sakai, T. and Kiyobayashi, T. (2012). Crystalline polycyclic quinone derivatives as organic positive-electrode materials for use in rechargeable lithium batteries, *Mater. Sci. Eng. B*, **177**, pp. 483–487.
11. Chen, H., Armand, Demailly, G., Dolhem, F., Poizot, P. and Tarascon, J.-M. (2008). From biomass to a renewable LixC_6O_6 organic electrode for sustainable Li-ion batteries, *ChemSucChem*, **1**, pp. 348–355.
12. Chen, H., Armand, M., Courty, M., Jiang, M., Grey C. P., Dolhem, F., Tarascon, J.-M. and Poizot, P. (2009). Lithium salt of tetrahydroxybenzoquinone: toward the development of a sustainable Li-ion battery, *J. Am. Chem. Soc.*, **131**, pp. 8984–8988.
13. Nokami, T., Matsuo, T., Inatomi, Y., Hojo, N., Tsukagoshi, T., Yoshizawa, H., Shimizu, A., Kuramoto, H., Komae, K., Tsuyama, H. and Yoshida, J. (2012). Polymer-bound pyrene-4,5,9,10-tetraone for fast-charge and -discharge lithium-ion batteries with high capacity, *J. Am. Chem. Soc.*, **134**, pp. 19694–19700.
14. Shimizu, A., Kuramoto, H., Tsujii, Y. Nokami, T., Inatomi, Y., Hojo, N., Suzuki, H. and Yoshida, J. (2014). Introduction of two lithiooxycarbonyl groups enhances cyclability of lithium batteries with organic cathode materials, *J. Power Sources*, **260**, pp. 211–217.
15. Walker, W., Grugeon, S., Mentre, O., Laruelle, S., Tarascon, J.-M. and Wudl, F. (2010). Ethoxycarbonyl-based organic electrode for Li-batteries, *J. Am. Chem. Soc.*, **132**, pp. 6517–6523.
16. Yao, M., Araki, M., Senoh, H., Yamazaki, S., Sakai, T. and Yasuda, K. (2010). Indigo dye as a positive-electrode material for rechargeable lithium batteries, *Chem. Lett.*, **39**, pp. 950–952.
17. Morita, Y., Nishida, S., Murata, T., Moriguchi, M., Ueda, A., Satoh, M., Arifuku, K., Sato, K. and Takui, T. (2011). Organic tailored batteries materials using stable open-shell molecules with degenerate frontier orbital, *Nat. Mater.*, **10**, pp. 947–950.
18. Matsunaga, T., Kuboia, T., Sugimoto, T. and Satoh, M. (2013). High-performance lithium secondary batteries using cathode active materials of triquinoxalinylenes exhibiting six electron migration, *Chem. Lett.*, **40**, pp. 750–752.
19. Nova'k, P., Müller, K., Santhanam, K. S. V. and Hass, O. (1997). Electrochemically active polymers for rechargeable batteries, *Chem. Rev.*, **97**, pp. 207–281 and references therein.
20. Nakahara, K., Iwasa, S., Satoh, M., Morioka, Y., Iriyama, J., Suguro, M. and Hasegawa, E. (2002). Rechargeable batteries with organic radical cathodes, *Chem. Phys. Lett.*, **359**, pp. 351–354.

21. Oyaizu, K., Suga, T., Yoshimura, K. and Nishide, H. (2008). Synthesis and characterization of radical-bearing polyethers as an electrode-active material for organic secondary batteries, *Macromolecules*, **41**, pp. 6646–6652.

22. Oyaizu, K., Sukegawa, T. and Nishide, H. (2011). Dual dopable poly(phenylacetylene) with nitronyl nitroxide pendants for reversible ambipolar charging and discharging, *Chem. Lett.*, **40**, pp. 184–185.

23. Inatomi, Y., Hojo, N., Yamamoto, T., Shimada, M. and Watanabe, S. (2008). *ECS Meeting Abstracts*, *MA2008-1*, Abstract #167, 213th ECS meeting, Phoenix, presentation No. 167.

24. Tang, J., Song, Z.-P., Shan, N., Zhan, L.-Z., Zhang, J.-Y., Zhan, H., Zhou, Y.-H. and Zhan, C.-M. (2008). Poly[3,4-(ethylenedithio)thiophene]: high specific capacity cathode active material for lithium rechargeable batteries, *J. Power Sources*, **185**, pp. 1434–1438.

25. Liu, M., Visco, S. J. and Jonghe, L. C. D. (1991). Novel solid redox polymerization electrodes electrochemical properties, *J. Electrochem. Soc.*, **138**, pp. 1896–1901.

26. MacDiarmid, A. G., Yang, L. S., Huang, W. S. and Humphrey, B. D. (1987). Polyaniline: electrochemistry and application to rechargeable batteries, *Synth. Met.*, **18**, pp. 393–398.

27. Yao, M., Senoh, H., Sakai, T. and Kiyobayashi, T. (2012). Redox active poly(N-vinylcarbazole) for use in rechargeable lithium batteries, *J. Power Sources*, **202**, pp. 364–368.

28. Sugimoto, T. and Senoo, K. (2012). Thiocarbohydrazides and oxalyl dihydrazides exhibiting two- to four-electron redoxes for positive electrode active materials in organic rechargeable lithium batteries, *Chem. Lett.*, **41**, pp. 1475–1477.

29. Imada, Y., Nakano, H., Furukawa, K., Kishi, R., Nakano, M., Maruyama, H., Nakamoto, M., Sekiguchi, A., Ogawa, M., Ohta, T. and Yamamoto, Y. (2016). Isolation of hypervalent group-16 radicals and their application in organic-radical batteries, *J. Am. Chem. Soc.*, **138**, pp. 479–482.

30. Osumi, S., Saito, S., Dou, C., Matsuo, K., Kume, K., Yoshikawa, H., Awaga, K. and Yamaguchi, S. (2016). Boron-doped nanographene: Lewis acidity, redox properties, and battery electrode performance, *Chem. Sci.*, **7**, pp. 219–227.

31. Yokoji, T., Kameyama, Y., Maruyama, N. and Matsubara, H. (2016). High-capacity organic cathode active materials of 2,2'-bis-*p*-benzoquinone derivatives for rechargeable batteries, *J. Mater. Chem. A*, **4**, pp. 5456–5466.

32. Yao, M., Kuratani, K., Kojima, T., Takeichi, N., Senoh, H. and Kiyobayashi, T. (2014). Indigo carmine: an organic crystal as a positive-electrode material for rechargeable sodium batteries, *Sci. Rep.*, **4**, pp. 3650 (6 pages).
33. Sano, H., Senoh, H., Yao, M., Sakaebe, H. and Kiyobayashi, T. (2012). Mg^{2+} storage in organic positive-electrode active material based on 2,5-dimethoxy-1,4-benzoquinone, *Chem. Lett.*, **41**, pp. 1594–1596.
34. Yamada, J. and Sugimoto, T. (eds.) (2004). *TTF Chemistry–Fundamental and Applications of Tetrathiafulvalene*, Kodansha-Springer, Tokyo.
35. Canevet, D., Sallé, M., Zhang, G., Zhang, D. and Zhu, D. (2009). Tetrathiafulvalene (TTF) derivatives: key building-blocks for switchable processes, *Chem. Commun.*, pp. 2245–2269.
36. Ishiguro, T., Yamaji, K. and Saito, G. (1998). *Organic Superconductors*, 2nd Ed., Springer Series in Solid-State Sciences, Vol. 88, Springer, Berlin.
37. Williams, J. M., Ferraro, J. R., Thorn, R. J., Carlson, K. D., Geiser, U., Wang, H. H., Kini, A. M. and Whangbo, M.-H. (1992). *Organic Superconductors*, Prentice Hall, New Jersey.
38. Saito, G. and Yoshida, Y. (2007). Development of conductive organic molecular assemblies: organic metals, superconductors, and exotic functional materials, *Bull. Chem. Soc. Jpn.*, **80**, pp. 1–137.
39. Mori, H. (2006). Materials viewpoint of organic superconductors, *J. Phys. Soc. Jpn.*, **75**, pp. 051003 (15 pages).
40. Misaki, Y., Matsui, T., Kawakami, K., Nishikawa, H., Yamabe, T. and Shiro, M. (1993). 2,5-Bis(1',3'-dithiol-2'-ylidene)-1,3,4,6-tetrathiapentalene and its related unsymmetrical donors, *Chem. Lett.*, **22**, pp. 1337–1340.
41. Misaki, Y., Kasakami, K., Higuchi, N., Nishikawa, H., Miura, T. and Yamabe, T. (1996). Synthesis and properties of higher homologues of bis-fused TTF, *Mol. Cryst. Liq. Cryst.*, **284**, pp. 337–344.
42. Misaki, Y., Nishikawa, H., Kawakami, K., Koyanagi, S., Yamabe, T. and Shiro, M. (1992). Methylthio substituted 2,5-bis(1',3'-dithiol-2'-ylidene)-1,3,4,6-tetrathiapentalene. A bis-fused tetrathiafulvalene, *Chem. Lett.*, **21**, pp. 2321–2324.
43. Nishikawa, H., Kawauchi, S., Misaki, Y. and Yamabe, T. (1996). Synthesis and properties of alkylthio-substituted tris-fused tetrathiafulvalenes, *Chem. Lett.*, **25**, pp. 43–44.
44. Misaki, Y. (2009). Tetrathiapentalene-based organic conductors, *Sci. Technol. Adv. Mater.*, **10**, pp. 024301 (22 pages).

45. Misaki, Y., Fujiwara, H. T., Yamabe, T., Mori, T., Mori, H. and Tanaka, S. (1994). Structure and conducting properties of BDT-TTP salts, *Chem. Lett.*, **23**, pp. 1653–1656.

46. Mori, T., Misaki, Y., Fujiwara, H. and Yamabe, T. (1994). Crystal structure and physical properties of $(BDT\text{-}TTP)_2ClO_4$, *Bull. Chem. Soc. Jpn.*, **67**, pp. 2685–2689.

47. Hasegawa, M., Daigoku, K., Hashimoto, K., Nishikawa, H. and Iyoda, M. (2012). Face-to-face dimeric tetrathiafulvalenes and their cation radical and dication species as models of mixed valence and π-dimer states, *Bull. Chem. Soc. Jpn.*, **85**, pp. 51–60.

48. Iyoda, M., Hasegawa, M. and Miyake, Y. (2004). Bi-TTF, bis-TTF, and related TTF oligomers, *Chem. Rev.*, **104**, pp. 5085–5113.

49. Otsubo, T., Aso, Y. and Talkimiya, K. (1996). Dimeric tetrathiafulvalenes: new electron donors, *Adv. Mater.*, **8**, pp. 203–211.

50. Hasegawa, M., Nakamura, K., Tokunaga, S., Baba, Y., Shiba, R., Shirahata, T., Mazaki, Y. and Misaki, Y. (2016). Synthesis, structure, optical, and electrochemical properties of triple- and quadruple-decker co-facial tetrathiafulvalene arrays, *Chem. Eur. J.*, **22**, pp. 10090–10101.

51. Nakamura, K., Hashimoto, T., Shirahata, Hino, S. T., Hasegawa, M., Mazaki, Y. and Misaki, Y. (2011). Synthesis and properties of new trimeric and tetrameric tetrathiafulvalenes with alternate links, *Chem. Lett.*, **40**, pp. 883–885.

52. Inatomi, Y., Hojo, N., Yamamoto, T., Watanabe, S. and Misaki, Y. (2012). Construction of rechargeable batteries using multifused tetrathiafulvalene systems as cathode materials, *ChemPlusChem*, **77**, pp. 973–976.

53. Yoshida, Z., Kawase, T., Awaji, H., Sugimoto, I., Sugimoto T. and Yoneda, S. (1983). Synthesis and properties of ethanediylidene-2,2′-bis(1,3-dithiole), *Tetrahedron Lett.*, **24**, pp. 3469–3472.

54. Sugimoto, T., Awaji, H., Sugimoto, I., Misaki, Y., Kawase, T., Yoneda, S., Yoshida, Z., Kobayashi, T. and Anzai, H. (1989). Ethylene analogues of tetrathiafulvalene and tetraselenafulvalene: new donors for organic metals, *Chem. Mater.*, **1**, pp. 535–547.

55. Moore, A. J., Bryce, M. R., Ando D. J. and Hursthouse, M. B. (1991). New bis(ethylenedithio)tetrathiafulvalene derivatives with low oxidation potentials, *J. Chem. Soc., Chem. Commun.*, pp. 320–321.

56. Moore, A. J. and Bryce, M. R. (1992). New vinylogous tetrathiafulvalene (TTF) π-electron donors. *Tetrahedron Lett.*, **33**, pp. 1373–1376.

57. Iwamoto, S., Inatomi, Y., Ogi, D., Shibayama, S., Murakami, Y., Kato, M., Takahashi, K., Tanaka, K., Hojo, N. and Misaki, Y. (2015). New tris- and pentakis-fused donors containing extended tetrathiafulvalenes: new positive electrode materials for rechargeable batteries, *Beilstein J. Org. Chem.*, **11**, pp. 1136–1147.

58. Takahashi, K., Nihira, T., Yoshifuji, M. and Tomitani, K. (1993). Conjugation-extended tetrathiafulvalene analogues involving a central aromatic 5-membered heterocyclic linking group, *Bull. Chem. Soc. Jpn.*, **66**, pp. 2330–2335.

59. Hansen, T. K., Lakshmikantham, M. V., Cava, M. P., Niziurski-Mann, R. E., Jensen, F. and Becher, J. (1992). New extended π-electron donors. Tetrathiafulvalene systems with heterocyclic spacer groups, *J. Am. Chem. Soc.*, **114**, pp. 5035–5039.

60. Benahmed-Gasmi, A. S., Frère, P., Garrigues, B., Gorgues, A., Jubault, M., Carlier, R. and Texier, F. (1992). 2,5-bis(1,4-dithiafulven-6-yl) furans, thiophenes and N-methyl pyrroles as extended analogues of tetrathiafulvalene, *Tetrahedron Lett.*, **33**, pp. 6457–6460.

61. Yamashita, Y., Tomura, M., Uruichi, M. and Yakushi, K. (2002). Synthesis and properties of π-extended TTF analogues and their cation radical and dication salts, *Mol. Cryst. Liq. Cryst.*, **376**, pp. 19–24.

62. Kato, M., Ogi, D., Yao, M. and Misaki, Y. (2013). A tris-fused tetrathiafulvalene extended with cyclohexene-1,4-diylidene: a new positive electrode material for organic rechargeable batteries, *Chem. Lett.*, **42**, pp. 1556–1558.

63. Kato, M., Senoo, M., Yao, M. and Misaki, Y. (2014). A pentakis-fused tetrathiafulvalene system extended by cyclohexene-1,4-diylidenes: a new positive electrode material for rechargeable batteries utilizing ten electron redox, *J. Mater. Chem. A*, **2**, pp. 6747–6754.

64. Takahashi, K., Tanioka, H., Fueno, H., Misaki, Y. and Tanaka, K. (2002). Preparation and characterization of novel aromatic-inserted tris-fused tetrathiafulvalenes, *Chem. Lett.*, **43**, pp. 1002–1003.

65. Aravindan, V., Gnanaraj, J., Lee, Y.-S. and Madhavi, S. (2013). $LiMnPO_4$ – A next generation cathode material for lithium-ion batteries, *J. Mater. Chem. A*, **1**, pp. 3518–3539.

66. Thackeray, M. M., Wolverton, C. and Isaacs, E. D. (2012). Electrical energy storage for transportation—approaching the limits of, and going beyond, lithium-ion batteries, *Energy Environ. Sci.*, **5**, pp. 7854–7863.

67. Numata, K., Sakai, C. and Yamanaka, S. (1997). Synthesis of solid solutions in a system of $LiCoO_2$-Li_2MnO_3 for cathode materials of secondary lithium batteries, *Chem. Lett.*, **26**, pp. 725–726.

68. Thackeray, M. M., Kang, S.-H., Johnson, C. S., Vaughey, J. T., Benedek, R. and Hackney, S. A. (2007). Li_2MnO_3-stabilized $LiMO_2$ (M = Mn, Ni, Co) electrodes for lithium-ion batteries, *J. Mater. Chem.*, **17**, pp. 3112–3125.

69. Johnson, C. S., Li, N., Lefief, C. and Thackeray, M. M. (2007). Anomalous capacity and cycling stability of $x Li_2MnO_3 \cdot (1-x)LiMO_2$ electrodes (M = Mn, Ni, Co) in lithium batteries at 50 °C, *Electrochem. Commun.*, **9**, pp. 787–795.

70. Akiba, K., Ishikawa, K. and Inamoto, N. (1978). Synthesis of 1,4-dithiafulvenes and 1,4-dithiafulvalenes by carbonyl olefination using 2-dimethoxyphosphinyl-1,3-benzodithiole, *Bull. Chem. Soc. Jpn.*, **51**, pp. 2674–2683.

71. Bryce, M. R. and Moore, A. J. (1988). A new highly-conjugated TTF analogue: synthesis, electrochemistry and a conducting TCNQ complex of 9,10-anthracenediylidene-2,2′-bis(4,5-dimethyl-1,3-dithiole), *Synth. Met.*, **25**, pp. 203–205.

72. Bryce, M. R., Moore, A. J., Hasan, M., Ashwell, G. J., Fraser, A. T., Clegg, W., Hursthouse, M. B. and Karaulov, A. I. (1990). Electrical and magnetic properties and X-ray structure of a highly conductive 4:1 complex of tetracyanoquinodimethane and a tetrathiafulvalene derivative, *Angew. Chem. Int. Ed.*, **29**, pp. 1450–1452.

73. Cerrada, E., Bryce, M. R. and Moore, A. J. (1993). Highly functionalised analogues of tetrathiafulvalene: new 9,10-bis(1,3-dithiol-2-ylidene)-9.10-dihydroanthracene donors, *J. Chem. Soc., Perkin Trans. 1*, pp. 537–538.

74. Gruhn, N. E., Macías-Ruvalcaba, N. A. and Evans, D. H. (2006). Studies of potential inversion in an extended tetrathiafulvalene, *Langmuir*, **22**, pp. 10683–10688.

75. Jones, A. E., Christensen, C. A., Perepichka, D. F., Batsanov, A. S., Beeby, A., Low, P. J., Bryce, M. R. and Parker, A. W. (2001). Photochemistry of the π-extended 9,10-bis(1,3-dithiol-2-ylidene)-9,10-dihydroanthracene system: generation and characterisation of the radical cation, dication, and derived products, *Chem. Eur. J.*, **7**, pp. 973–978.

76. Bryce, M. R., Finn, T., Moore, A. J., Batsanov, A. S. and Howard, J. A. K. (2000). Syntheses and X-ray crystal structures of functionalised 9,10-bis(1,3-dithiol-2-ylidene)-9,10-dihydroanthracene derivatives, *Eur. J. Org. Chem.*, pp. 51–60.

77. Christensen, C. A., Batsanov, A. S. and Bryce, M. R. (2007). Thiolated π-extended tetrathiafulvalenes: versatile multifunctional π-systems, *J. Org. Chem.*, **72**, pp. 1301–1308.
78. Ogi, D., Fujita, Y., Mori, S., Shirahata, T. and Misaki, Y. (2016). Bis- and tris-fused tetrathiafulvalenes extended with anthracene-9,10-diylidene, *Org. Lett.*, **18**, pp. 5868–5871.
79. Ogi, D., Fujita, Y., Kato, M., Tahara, A., Mori, S., Shirahata, T., Yao, M. and Misaki, Y. (in preparation).
80. Sun, D., Krawiec, M. and Watson, W. H. (1997). Benzoquinone derived 1,3-dithiole-2-ones and thiones, *J. Chem. Crystallogr.*, **27**, pp. 515–526.
81. Dumur, F., Gautier, N., Gallego-Planas, N., Sahin, Y., Levillain, E., Mercier, N., Hudhomme, P., Masino, M., Girlando, A., Lloveras, V., Vidal-Gancedo, J., Veciana, J. and Rovira, C. (2004). Novel fused D-A dyad and A-D-A triad incorporating tetrathiafulvalene and *p*-benzoquinone, *J. Org. Chem.*, **69**, pp. 2164–2177.
82. Kato, M., Noda, S., Kiyobayashi, T., Yao, M. and Misaki, Y. (2017). Fused donor–donor–acceptor triads composed of tetrathiafulvalene and benzoquinone derivatives as the positive electrode materials for rechargeable lithium and sodium batteries, *Chem. Lett.*, **46**, pp. 368–370.
83. Hanyu, Y. and Honma, I. (2012). Rechargeable quasi-solid state lithium battery with organic crystalline cathode, *Sci. Rep.*, **2**, pp. 453 (6 pages).

Chapter 5

Development of Purely Organic Superconductors

Takashi Shirahata
Department of Applied Chemistry, Graduate School of Science and Engineering, Ehime University, and Research Unit for Development of Organic Superconductors, Ehime University, 3 Bunkyo-cho, Matsuyama, Ehime 790-8577, Japan
shirahata.takashi.mj@ehime-u.ac.jp

Chapter 5 deals with purely organic superconductors. Materials of this kind have attracted much attention from the viewpoint of not only scientific interest but also practical applications in organic electronic devices. Quite a few researchers have devoted a lot effort to achieving superconductivity in purely organic materials. Despite many attempts, there are only six superconductors consisting of purely organic molecules. This chapter reviews the six kinds of purely organic superconductors as well as the efforts on for the development of related materials.

Functional Materials: Advances and Applications in Energy Storage and Conversion
Edited by Toshio Naito

ISBN 978-981-4800-09-9 (Hardcover), 978-0-429-46813-1 (eBook)
www.panstanford.com

5.1 Introduction

5.1.1 Molecular Conductors Based on Closed-Shell Molecules

It is well known that common organic materials are insulating. The highest occupied molecular orbital (HOMO) of a closed-shell molecule is fully occupied. Therefore, the valence band is completely filled with electrons and a conduction band is vacant in the solid state. The large energy gap (E_g) between the valence and conduction bands, which is associated with the energy gap (ΔE) between the HOMO and lowest unoccupied molecular orbital (LUMO) levels, also gives rise to the insulating nature of organic materials (Fig. 5.1a). To produce organic conducting materials, the E_g should be reduced. A small HOMO-LUMO gap (ΔE) and an effective overlap of molecular orbitals (MOs) are essential requirements to reduce the E_g. On the basis of the strategy, transition metal complexes with tetrathiafulvalene (TTF, **1**) dithiolato ligands were synthesized [1–13]. The first single-component molecular metal, [Ni(tmdt)$_2$] (**2**, Fig. 5.2), under ambient pressure, was developed by Kobayashi et al. [1]. The ΔE of the [Ni(tmdt)$_2$] molecule was estimated to be 0.10 eV by the extended Hückel MO calculation, and the small ΔE was consistent with the result of ab initio calculation. The transfer integrals (23–75 meV) were calculated to be relatively large because the HOMO and LUMO of [Ni(tmdt)$_2$] overlapped effectively. As a result, the calculated Fermi surfaces of [Ni(tmdt)$_2$] were composed of holes and electrons. That is, the HOMO band overlaps with the LUMO band and partially filled semimetallic bands are formed (Fig. 5.1b). The existence of Fermi surfaces was elucidated by detecting the quantum oscillation (de Haas–van Alphen (dHvA) effect) [14]. Recently, superconductivity of the single-component molecule was achieved by [Ni(hfdt)$_2$] (**3**) under an applied high pressure (T_C = 5.5 K at 8.1 GPa) [15]. However, these molecules contain transition metals and, therefore, they are not purely organic molecules.

On the other hand, the development of single-component molecular metals and superconductors using closed-shell organic compounds has been also reported. Figure 5.3 shows TTF derivatives (**4–8**) containing electron acceptor units, where they are designed to reduce the HOMO-LUMO gap (ΔE) [16–21]. A series of donor-acceptor (D-A) systems afforded organic semiconductors; however, organic metals and superconductors were not obtained in the system.

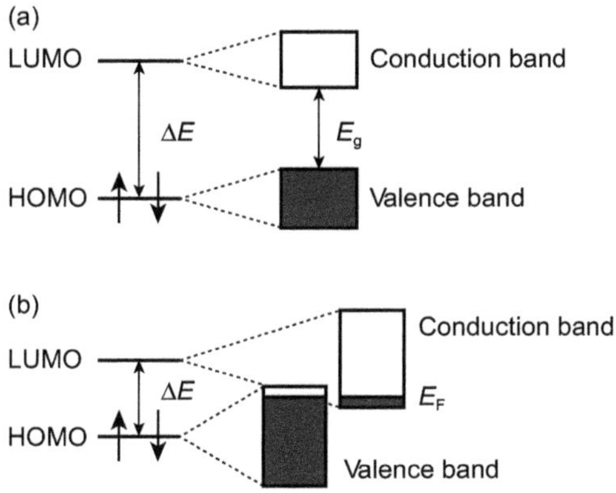

Figure 5.1 Schematic illustration of the band structure of a closed-shell organic molecule with a HOMO-LUMO gap (ΔE); (a) a large ΔE and a small intermolecular overlap of MOs form an insulating band with an energy gap (E_g); (b) a small ΔE and an effective overlap of MOs form partially filled semimetallic bands.

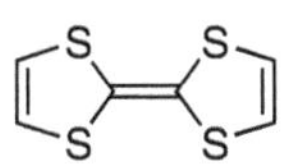

1 (TTF)

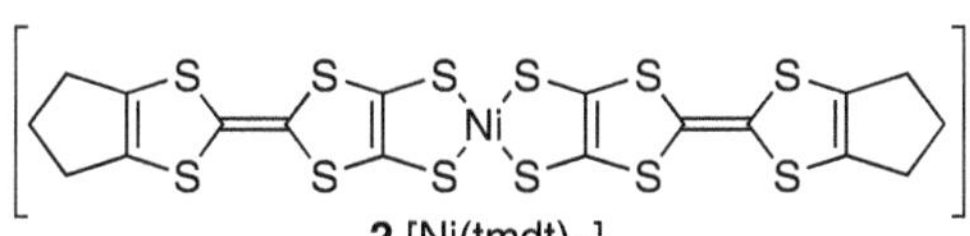

2 [Ni(tmdt)$_2$]

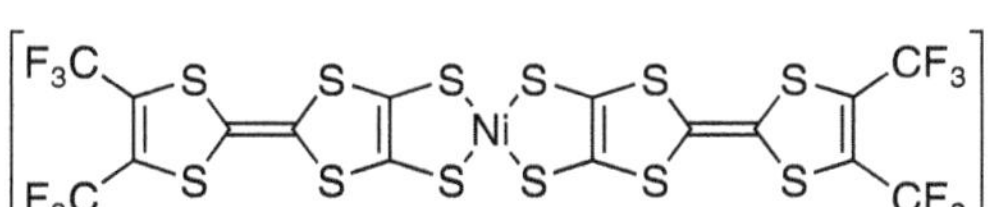

3 [Ni(hfdt)$_2$]

Figure 5.2 Structure of TTF (**1**) and transition metal complexes with extended-TTF dithiolato ligands. [Ni(tmdt)$_2$] (**2**) is the first single-component molecular metal under ambient pressure, and [Ni(hfdt)$_2$] (**3**) is a single-component molecular superconductor.

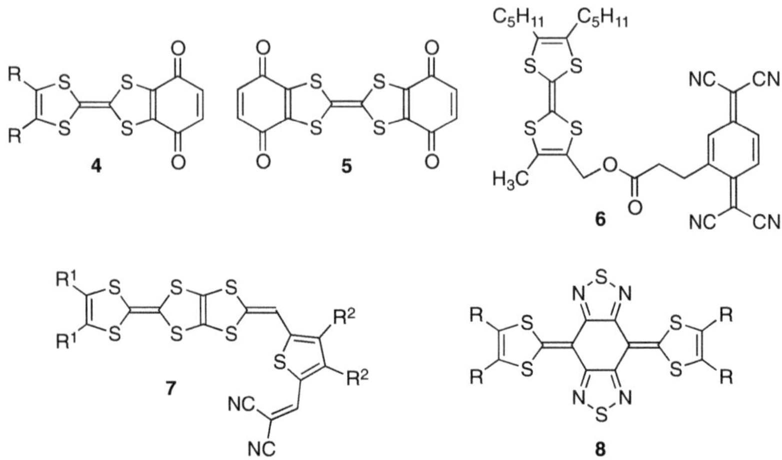

Figure 5.3 TTF and extended-TTF derivatives containing electron acceptor units.

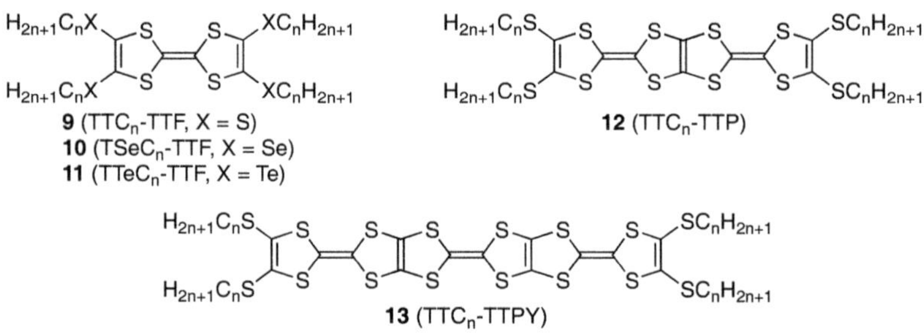

Figure 5.4 TTF, TTP, and TTPY derivatives, in which alkylthio groups are introduced.

To increase intermolecular interactions, two strategies for molecular modification were conducted; (i) the fastener effect of long alkyl chains was introduced to the TTF-type electron donors (Fig. 5.4) and (ii) multiple heavy atoms, such as selenium, tellurium, and iodine, were introduced in the organic π system (Fig. 5.5). Inokuchi et al. and Saito et al. synthesized a series of TTC$_n$-TTF derivatives (**9–11**) according to the former strategy (i), and the low energy gap (E_g) was estimated in TTC$_{10}$-TTF (0.26 eV) [22–32]. Long alkyl chains were also introduced in bis-fused TTF, 2,5-bis(1,3-dithiol-2-ylidene)-1,3,4,6-tetrathiapentalene (BDT-TTP, or simply TTP), and tris-fused TTF, 2,2′-bi(1″,3″-dithiol-2″-ylidene-1,3,4,6-tetrathiapentalene-5-

ylidene) (BDT-TTPY, or simply TTPY), by Mori et al. and Misaki et al. [33–35]. In systems **9–13**, organic semiconductors with relative high conductivity were yielded; however, organic metals and superconductors were not obtained.

According to the latter strategy (ii), molecules **14–17,** containing multiple heavy atoms, such as selenium, tellurium, and iodine atoms, were expected to produce strong intermolecular interactions between the heavy atoms (Fig. 5.5). In this system, single-component molecular metals were obtained under hydrostatic pressure [36–44]. Moreover, the iodine-containing molecules **16** (T_C = 2 K at 52 GPa) and **17** (T_C = 2.3 K at 58 GPa) exhibited superconductivity under high pressure [43, 44]. The structural and conducting properties of organic superconductors based on the closed-shell molecules and related materials are described in Section 5.2.

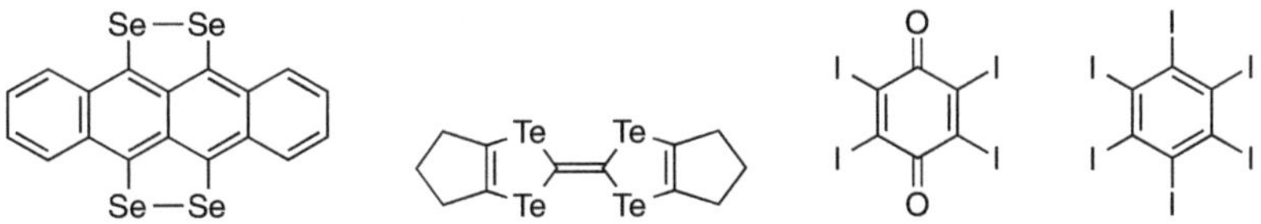

14 **15** **16** **17**

Figure 5.5 Molecules containing heavy atoms, such as selenium, tellurium, and iodine.

5.1.2 Molecular Conductors Based on Open-Shell and Partially Oxidized (Reduced) Molecules

Another idea to produce organic conducting materials is to remove electrons from the valence band or to add electrons to the conduction band (Fig. 5.6). The partially filled energy band of a neutral radical solid is composed of a singly occupied molecular orbital (SOMO). When a band width (W) is larger than an on-site coulomb repulsive energy (U), a half-filled band is formed. Neutral radicals are expected to show metallic conductivity in this case. However, most neutral radicals show semiconducting or insulating behavior. The U of common neutral π radicals is larger than their W; therefore, the band is separated by a Mott–Hubbard gap [45]. Figure 5.7 shows representative organic neutral π radicals (**18–35**) with a relative high conductivity and unique magnetic properties [46–81].

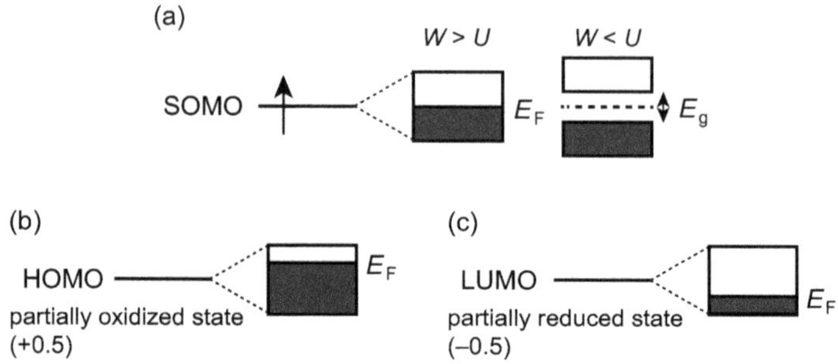

Figure 5.6 Schematic illustration of the band structure. (a) In the case of $W > U$, open-shell organic molecules form half-filled band structures. In the case of $W < U$, an energy gap is generated due to Mott–Hubbard instability. (b, c) A partially oxidized molecule and a partially reduced molecule (+0.5 and −0.5) form a metallic three-quarters-filled and a metallic quarter-filled band structure, respectively.

On the other hand, a solid composed of partially oxidized (reduced) organic molecules affords molecular metals as well as superconductors. In this system, two kinds of organic molecules, "electron donors" and "electron acceptors," collaborate to play a significant role. The typical electron donors and acceptors are listed in Fig. 5.8. The first molecular metal, which is a charge transfer (CT) complex of TTF (**1**) [82] with 7,7,8,8-tetracyanoquinodimethane (TCNQ, **43**) [83], was developed by Ferraris et al. [84]. The first molecular superconductor $(TMTSF)_2PF_6$ was reported by Bechgaard et al., where TMTSF is tetramethyltetraselenafulvalene (**39**) [85]. Since then, a large number of metallic CT complexes and molecular superconductors have been synthesized. However, only four compounds composed of purely organic components have shown superconductivity [86–90]. Two simple systems to develop purely organic superconductors are (i) radical salts and (ii) D-A-type CT complexes, comprised exclusively of organic species in either case. As to the former system (i), the first purely organic superconductor under ambient pressure, β''-$(BEDT\text{-}TTF)_2SF_5CH_2CF_2SO_3$, was synthesized by Geiser et al. and Schlueter et al. (T_C = 5.2 K), where BEDT-TTF is bis(ethylenedithio)tetrathiafulvalene (**40**) [86, 87]. This compound is a cation radical salt composed of a partially oxidized organic electron donor $(BEDT\text{-}TTF)^{0.5+}$ and organic

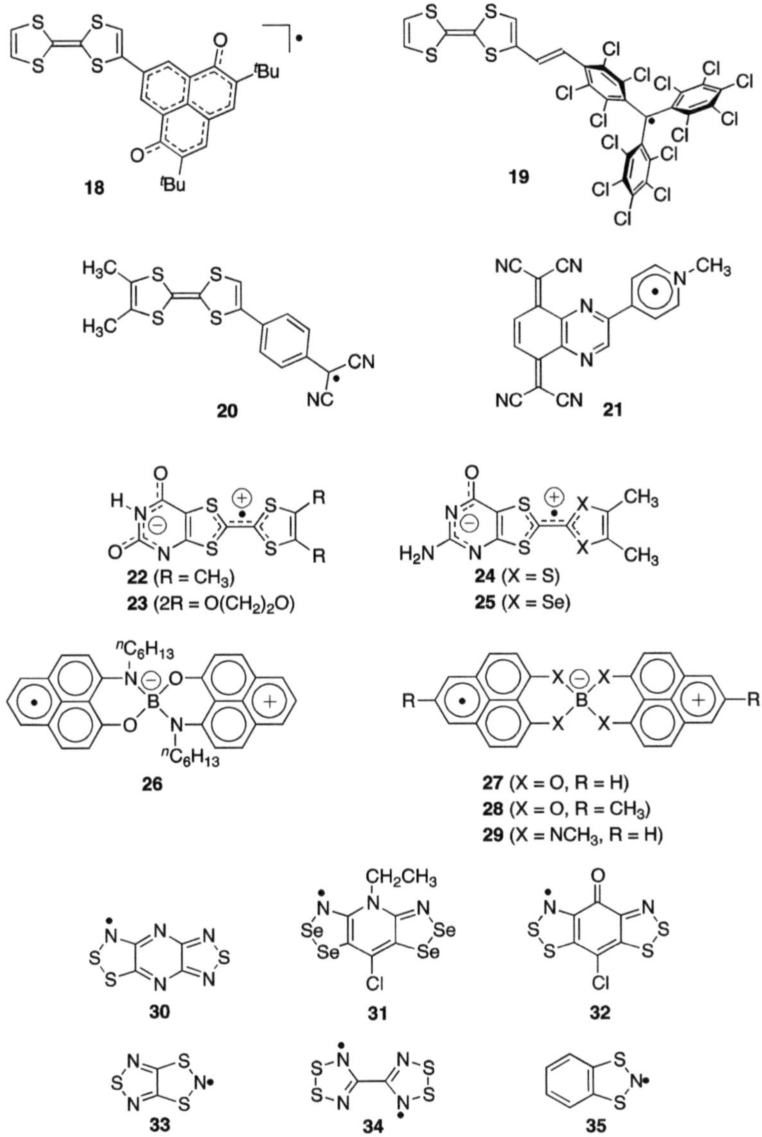

Figure 5.7 Organic π radical conductors.

monoanion $SF_5CH_2CF_2SO_3^-$. Recently, Ito et al. reported pressure-induced superconductivity in κ-(BEDT-TTF)$_2$CF$_3$SO$_3$, which is composed of the organic electron donor BEDT-TTF and the organic monoanion $CF_3SO_3^-$ [88]. As for the latter system (ii), Kondo et al. reported the superconducting transition of $(BETS)_2(Cl_2TCNQ)$ under hydrostatic pressure (T_C = 1.3 K at 0.35 GPa) [89], where BETS is bis(ethylenedithio)tetraselenafulvalene (**41**) and Cl_2TCNQ is 2,5-dichloro-7,7,8,8-tetracyanoquinodimethane (**44**). Recently, the present author and coworkers successfully synthesized a new, purely organic superconductor (EtDTET)(TCNQ) under ambient pressure (T_C = 5.5 K) [90], where EtDTET is 2-(3-pentylidene)-1,3-dithiolo[4,5-*d*]-4,5-ethylenedithiotetrathiafulvalene (**42**). The structural and conducting properties of organic superconductors based on the partially oxidized molecules and related materials are described in Section 5.3.

Electron donors

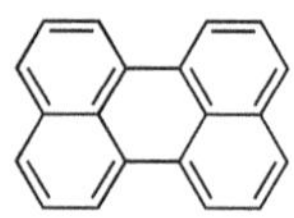

36 (perylene)

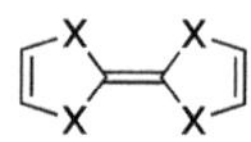

1 (TTF, X = S)
37 (TSF, X = Se)

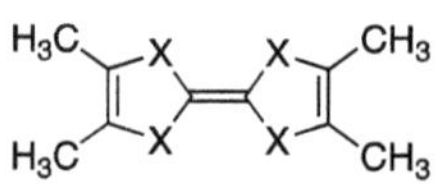

38 (TMTTF, X = S)
39 (TMTSF, X = Se)

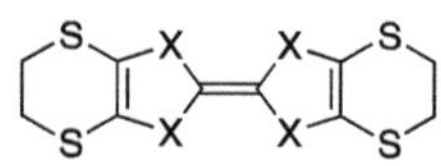

40 (BEDT-TTF, X = S)
41 (BETS, X = Se)

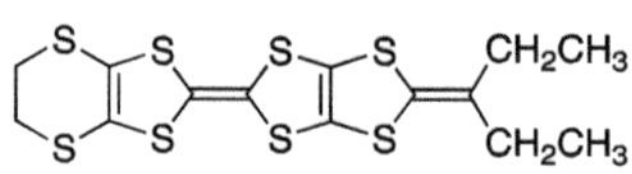

42 (EtDTET)

Electron acceptors

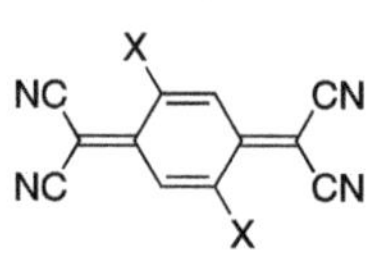

43 (TCNQ, X = H)
44 (Cl_2TCNQ, X = Cl)

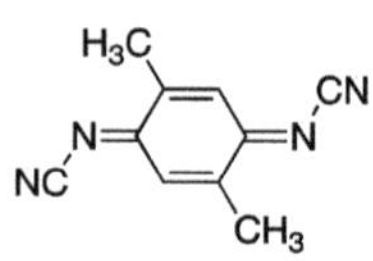

45 (DMe-DCNQI)

Figure 5.8 Typical component molecules for organic metals and superconductors. Compounds **40, 41, 42, 43,** and **44** afforded purely organic superconductors.

5.2 Molecular Superconductors Based on Closed-Shell Organic Molecules Only

5.2.1 Donor-Acceptor-Type Organic Molecules

As mentioned in Section 5.1, small HOMO-LUMO gaps (ΔEs) and the enhancement of the intermolecular overlap of MOs are indispensable for producing organic conducting materials. A large number of TTF derivatives containing electron acceptor units were designed to reduce ΔE [16–21, 91–94]. Table 5.1 summarizes properties of D-A-combined-type organic molecules.

Table 5.1 Properties of the representative D-A-combined-type molecules

Molecule	σ_{rt}/S cm^{-1}	ΔE/eV	Refs.
4 (R = $S(CH_2)_4CH_3$)	10^{-9} (single crystal)	1.37	[16, 17]
5	9×10^{-9} (single crystal)	1.46	[16]
6	9×10^{-9}–10^{-7} (LB film)	0.29[a] 0.70[b]	[18]
7 (R^1 = SCH_3, R^2 = H)	8.5×10^{-6} (pellet)	2.72	[19]
8a (R = H)	3.7×10^{-6} (single crystal)	2.97[c]	[20]
8b (R = CH_3)	2×10^{-9} (single crystal)	2.95[c]	[21]

[a]Extended conformation.
[b]Intramolecularly complexed conformation (see Fig. 5.10).
[c]ΔE was not given in Refs. [20] and [21]. The author estimated it by DFT calculation at the B3LYP/6-31G(d,p) level of theory.

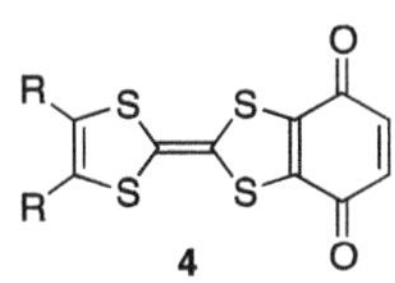

4

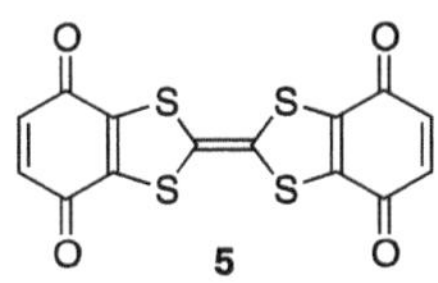

5

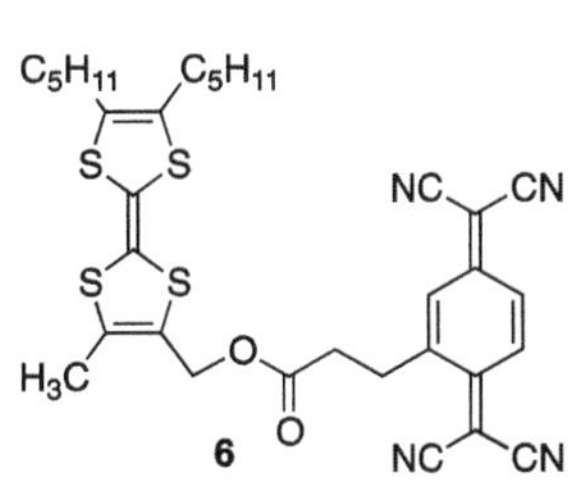

6

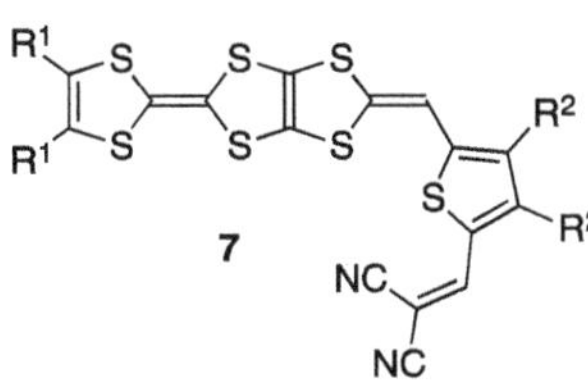

7

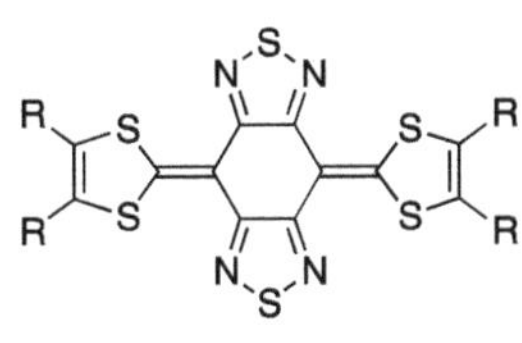

8

Hudhomme et al. and Liu et al. reported the synthesis of a fused D-A diad and triad composed of TTF and *p*-benzoquinone **4** and **5** [16, 17]. The redox behavior of the diad **4** and the triad **5** was investigated by cyclic voltammetry (CV) in a CH_2Cl_2:acetonitrile = 9:1 solution for **4** and in an *o*-dichlorobenzene:acetonitrile = 19:1 solution containing 0.1 M $^nBu_4N{\cdot}PF_6$ as a supporting electrolyte. The diad **4** (R = $S(CH_2)_4CH_3$) showed two pairs of one-electron reversible oxidation waves at +0.74 and +1.11 V versus Ag/AgCl, corresponding to the oxidation of the TTF unit. Two reduction processes of the *p*-benzoquinone moiety were observed, at −0.21 and −1.20 V, versus Ag/AgCl. Although the first reduction wave showed a reversible process, an irreversible reduction process was observed at the second reduction process. The triad **5** showed two pairs of reversible one-electron oxidation processes, at +0.97 and +1.34 V, and two pairs of reversible one-electron reduction processes, at −0.33 and −0.40 V, and an irreversible one- or two-electron reduction process at −1.16 V. The frontier MOs of **4** and **5** were calculated by the density functional theory (DFT) method. The HOMO-LUMO gap energies ΔEs of these compounds were calculated to be 1.37 eV for **4** (R = $S(CH_2)_4CH_3$, HOMO = −5.04, and LUMO = −3.67 eV) and 1.46 eV for **5** (HOMO = −5.35 and LUMO = −3.89 eV). The CT band was observed in a single-crystal infrared (IR) spectrum of **5,** and an electronic broad absorption centered at ca. 13,000 cm^{-1} (1.6 eV) was consistent with ΔE estimated by the DFT calculation. The crystal structure of the A-D-A triad **5** is shown in Fig. 5.9. Molecules **5** are stacked in a ring-over-bond manner, and they form a columnar structure with an interplanar distance of 3.60 Å along the crystallographic *a* axis. The donor unit was sandwiched by acceptor units. Consequently, an A···D···A-type "triplet" was encountered in the molecular stack. The mixed stack of the donor and acceptor units caused intermolecular CT, and the intermolecular CT band could be also observed at around 13,000 cm^{-1}. Therefore, the observed electronic band in the IR spectrum could be attributed to intra- and intermolecular CT. The electrical conductivities were 10^{-9} for the D-A diad **4** and 9×10^{-9} S cm^{-1} for the A-D-A triad **5** measured on compressed pellets using a four-probe technique. The low conductivity of the triad **5** originated from the disadvantageous molecular packing.

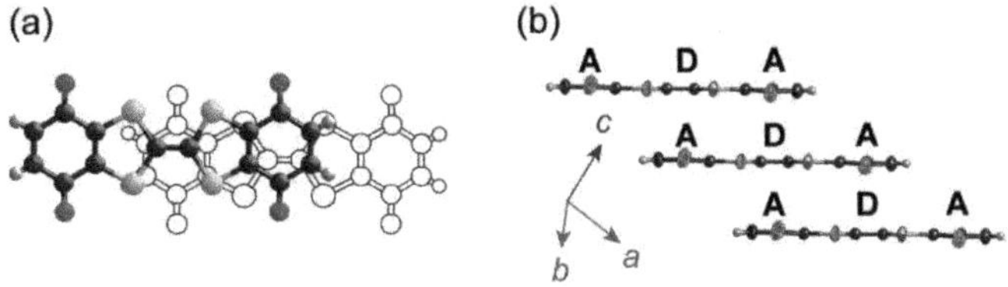

Figure 5.9 (a) Overlapping mode of the A-D-A triad **5** in the solid state. (b) Columnar structure of **5**. An A-D-A-type alternate stack is formed.

Several TTF-TCNQ diads were reported [91–94]; however, few of them included reliable experimental investigation. Bryce et al. reported the synthesis of a well-characterized TTF-σ-TCNQ diad **6** [18]. The diad **6** showed two pairs of reversible single-electron oxidation processes (–0.03 and +0.37 V versus ferrocene/ferrocenium [Fc/Fc^+]) and two pairs of reversible single-electron reduction processes (–0.26 and –0.77 V versus Fc/Fc^+). The first oxidation and reduction potentials were slightly shifted compared to related TTF and TCNQ molecules. As a result, the mutual interaction between TTF and TCNQ moieties was small. The difference between the first oxidation and reduction potentials (0.23 V) is expected to be low. The DFT calculations of model compound **6′**, where alkyl groups of **6** were replaced by hydrogen, demonstrated that the ΔE was estimated to be 0.29 eV in an extended conformation and to be 0.70 eV in a head-to-tail intramolecularly complexed conformation (Fig. 5.10). The electrical conductivity was measured by the Langmuir–Blodgett (LB) films of **6**, and the dc conductivity was estimated to be 10^{-7} (in-plane direction) and 2×10^{-11} S cm^{-1} (through-plane direction). The poor conductivities were associated with the molecular orientations of the TTF and TCNQ moieties, in which neither TTF nor TCNQ moieties could form face-to-face stacking.

Although a low ΔE was accomplished in compounds **4–6**, their electrical conductivities were not as high as we had expected, due to the nonideal molecular arrangement. Misaki et al. reported the synthesis of D-A diads **7**, which have a 2-methylidene-1,3-dithiolo[4,5-*d*]-TTF (DT-TTF, **46**) moiety (Fig. 5.11) [19]. A cyano (C≡N) stretching vibration (ν_{CN}) in the IR spectra of diads **7** was observed at 2210–2211 cm^{-1}. A low wavenumber of ν_{CN} compared with the neutral state of TCNQ (2225 cm^{-1}) indicated the existence of an intramolecular charge transfer (ICT) state and the contribution of a polarized structure (Fig. 5.12). The electrical conductivity of **7** (R^1 = SCH_3 and R^2 = H) showed a relatively high conductivity (8.6×10^{-6} S cm^{-1}) despite the fact that the measurement was performed on a compressed pellet using a two-probe method. The electrical conductivity of **7** is higher by 3 orders of magnitude than the electrical conductivities of **4–6**. DT-TTF molecules tend to form two-dimensional molecular arrangements in their cation radical salts and to afford highly conductive organic materials [95–102]. These observations indicated that the molecules of **7** align in appropriate molecular arrangements; however, the poor crystal quality of **7** prevented them from clarifying the molecular packing patterns.

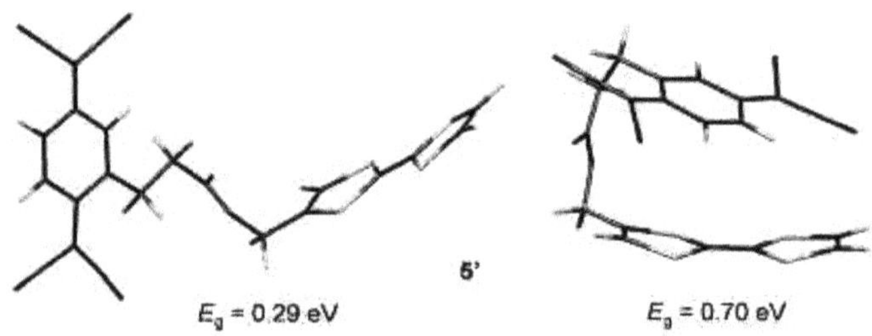

Figure 5.10 Geometries and HOMO-LUMO gaps (E_g means ΔE in this chapter) of an extended (left) and a head-to-tail intramolecularly complexed (right) conformation of a model TTF-TCNQ diad **5′** (written as **6′** in this chapter) and calculated at the B3LYP/6-31G(d) level. Reproduced from Ref. [18] with permission from John Wiley & Sons.

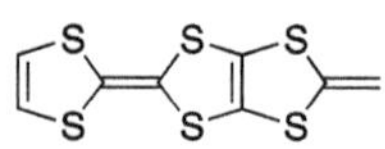

46 (DT-TTF)

Figure 5.11 Structure of DT-TTF (**46**).

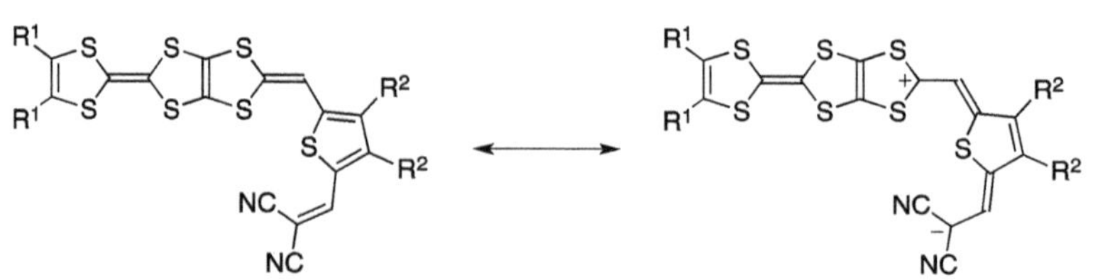

Figure 5.12 Resonance structures of D-A diads **7**.

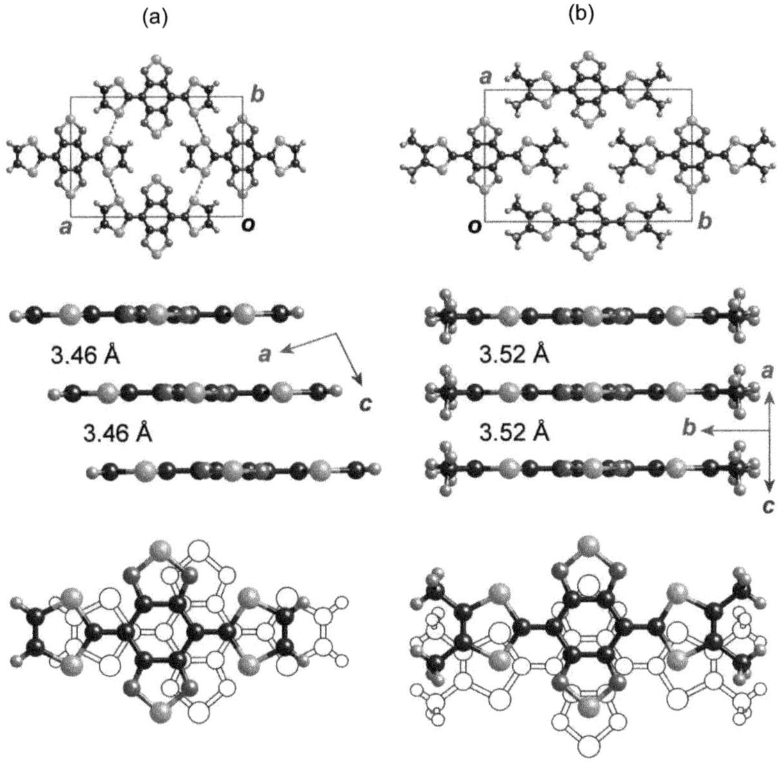

Figure 5.13 Crystal structures of (a) the parent **8a** (R = H) and (b) the tetramethyl-substituted **8b** (R = CH_3). Molecular packing viewed along the crystallographic *c* axis (upper), the columnar structure viewed along the molecular short axis (middle), and overlapping modes (bottom).

In contrast, a clear relationship between electrical conductivity and molecular packing was demonstrated in the study of compounds **8**. The molecules of the parent **8a** (R = H) and the tetramethyl derivative **8b** (R = CH_3) were stacked along the crystallographic *c* axis, and the resulting molecular columns were arranged on the *ab* plane; in other words, their crystal structures resemble each other (Fig. 5.13) [21]. However, there are two distinct differences. The first difference is the presence or absence of intermolecular contacts. In the crystal of **8a**, short S···S contacts (3.26 Å), less than the sum of the van der Waals radii [103] of sulfur atoms, occurred between molecular columns along the *b* axis. On the other hand, no remarkable intermolecular atomic contact was observed in the crystal of **8b**.

The second difference is overlapping modes of the molecules. The interplanar distance of **8a** (3.46 Å) was slightly shorter than that of **8b** (3.52 Å). The molecules of **8a** slipped along the direction of the extended-TTF skeleton, the so-called ring-over-bond-type overlapping mode. In contrast, the molecules of **8b** slid along the direction of the bis[1,2,5]-thiadiazolo-*p*-quinoid spacer unit, the so-called ring-over-atom-type overlapping mode. The difference in the molecular orientations resulted in a disparity of transfer integrals between **8a** and **8b** [104] and the higher conductivity for **8a** (3.7×10^{-6} S cm^{-1}) compared with that of **8b** (2×10^{-9} S cm^{-1}).

5.2.2 Utilization of van der Waals Interaction (Fastener Effect)

As mentioned in Section 5.2.1, molecular arrangement plays a critical role in realizing high conductivity. TTC$_n$-TTF (**9**), TSeC$_n$-TTF (**10**), TTeC$_n$-TTF (**11**), TTC$_n$-TTP (**12**), and TTC$_n$-TTPY (**13**) derivatives bearing long alkyl chains were synthesized to apply van der Waals interaction, which is the so-called fastener effect, by Inokuchi et al., Saito et al., Misaki et al., and Mori et al. [22–35]. Table 5.2 summarizes the electrical conductivities of **9–13**. Relatively high conductivities were measured: 10^{-4}–10^{-6} S cm^{-1} for **9** (n = 10 and 11), **10** (n = 1, 8, and 9), **11** (n = 1), **12** (n = 3–5), and **13** (n = 3–5). A typical crystal structure of TTC$_n$-TTP (**12**, n = 4) is shown in Fig. 5.14 for a series of **9–13**. The π-skeleton and alkyl groups were aligned alternately along the crystallographic *b* axis. The molecules were stacked along the crystallographic *c* axis, and the π–π stacking of the TTP skeleton of **12** (n = 4) was enhanced by the fastener effect of the butylthio groups. In contrast, the pristine BDT-TTP molecule adopts a nonplanar chair-type conformation in the absence of remarkable intermolecular interactions (Fig. 5.14c) [105]. The planar TTP skeleton of **12** (n = 4) is evidence of the existence of the fastener effect. The fastener effect of the long alkyl chain plays an important role in the high conductivity of this system, except for TSeC$_1$-TTF (**10**, n = 1) and TTeC$_1$-TTF (**11**, n = 1). In the case of **10** (n = 1) and **11** (n = 1), the intermolecular interactions between heavy selenium and tellurium atoms result in the high conductivity [27, 28]. In this system, the enhancement of the intermolecular interactions as well as the transfer integrals of MO were accomplished; however, metallic conductivity and superconductivity have not been observed.

Table 5.2 Room-temperature electrical conductivities (σ_{rt}/S cm^{-1}) of TTC$_n$-TTF (**9**), TSeC$_n$-TTF (**10**), TTeC$_n$-TTF (**11**), TTC$_n$-TTP (**12**), and TTC$_n$-TTPY (**13**)

n	TTC$_n$-TTF (9)	TSeC$_n$-TTF (10)	TTeC$_n$-TTF (11)	TTC$_n$-TTP (12)	TTC$_n$-TTPY (13)
1	3.4×10^{-11a}	1.0×10^{-6a}	1.4×10^{-5a}	—	—
2	8.3×10^{-11a}	5.0×10^{-10a}	4.3×10^{-10a}	1.7×10^{-8a}	—
3	1.0×10^{-10a}	1.5×10^{-9a}	—	2.0×10^{-6a}	2.5×10^{-3a}
4	1.6×10^{-7a}	1.0×10^{-11a}	—	1.7×10^{-4a}	1.0×10^{-4a}
5	1.6×10^{-8a}	2.8×10^{-10a}	—	2.5×10^{-6a}	1.0×10^{-4a}
6	3.3×10^{-8a}	7.7×10^{-7a}	—	—	—
7	2.6×10^{-8a}	1.1×10^{-7a}	—	—	—
8	1.4×10^{-8a}	1.3×10^{-6a}	5.9×10^{-7a}	—	—
9	2.0×10^{-8a}	1.2×10^{-6b}	5.0×10^{-7b}	—	—
10	2.7×10^{-6a}	1.0×10^{-7b}	—	—	—
11	1.8×10^{-6a}	1.2×10^{-7b}	—	—	—
12	1.9×10^{-8b}	2.6×10^{-7b}	—	—	—
13	4.3×10^{-8b}	1.8×10^{-7b}	9.1×10^{-9b}	—	—
14	1.3×10^{-8b}	1.6×10^{-7b}	3.8×10^{-9b}	—	—
15	3.4×10^{-7b}	3.0×10^{-7b}	5.9×10^{-9b}	—	—
16	1.3×10^{-7b}	3.7×10^{-7b}	8.3×10^{-9b}	—	—
17	6.3×10^{-8b}	1.6×10^{-7b}	7.1×10^{-9b}	—	—
18	1.5×10^{-10b}	3.3×10^{-8b}	2.9×10^{-9b}	—	—

[a]Measured on a single crystal
[b]Measured on a compressed pellet

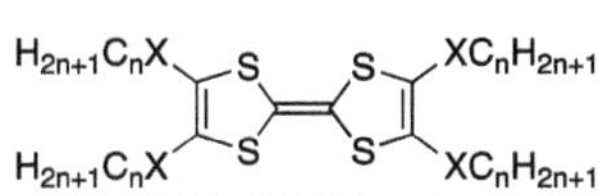

9 (TTC$_n$-TTF, X = S)
10 (TSeC$_n$-TTF, X = Se)
11 (TTeC$_n$-TTF, X = Te)

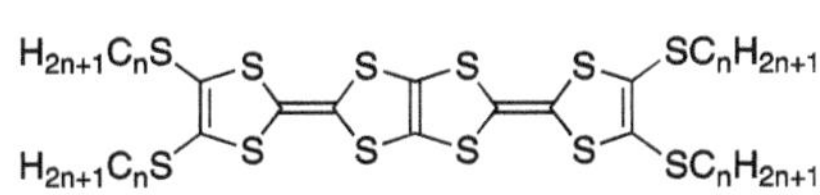

12 (TTC$_n$-TTP)

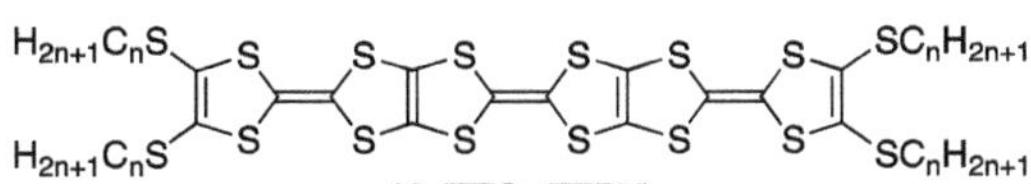

13 (TTC$_n$-TTPY)

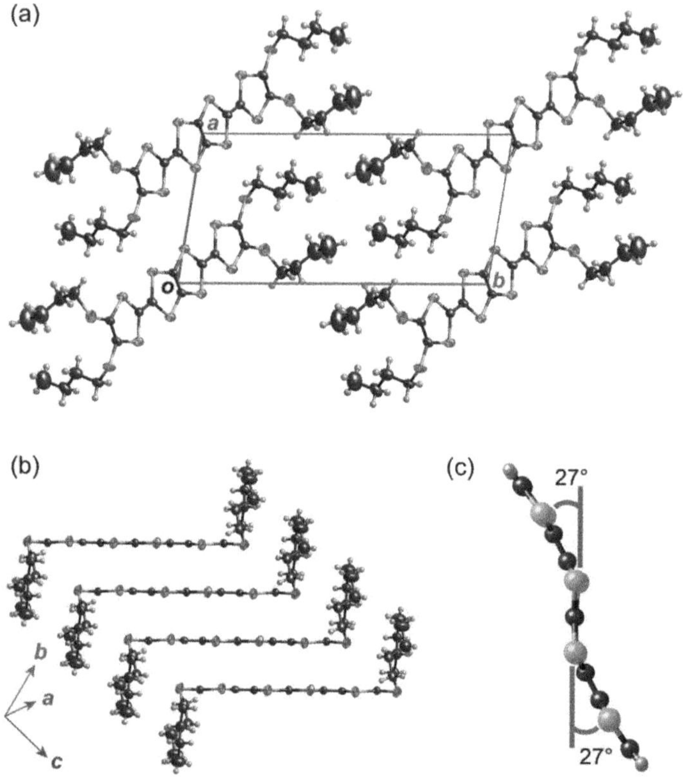

Figure 5.14 (a) Crystal structures of TTC$_n$-TTP (**12**, n = 4) viewed along the crystallographic c axis. (b) Columnar structure of TTC$_n$-TTP (**12**, n = 4) viewed along the molecular short axis. (c) Molecular structure of pristine BDT-TTP.

In Sections 5.2.1 and 5.2.2, several molecules as candidates for superconductors composed of organic molecules only were described in terms of a chemical approach. These results indicate that metallic conductivity and superconductivity are hardly realized under ambient conditions. The next section deals with research in terms of a physical approach.

5.2.3 Applying Physical Pressure to Organic Molecules Containing Heavy Atoms

In the 1960s, Drickamer et al. reported the electrical conductivity of fused-ring aromatic hydrocarbons on applying physical pressure

[106, 107]. The metallic conductivity of the organic molecule was first observed in pentacene under 21.3 GPa down to around 200 K [106]. Then, transport properties of organic compounds containing multiple heavy atoms, such as selenium, tellurium, and iodine atoms (**14–17**), were investigated. [36–44]. The transport properties of **14–17** and pentacene are summarized in Table 5.3. Compound **14** contains four selenium atoms on the tetracene unit. This compound showed a relative high conductivity of 8×10^{-7} S cm^{-1} at ambient pressure [36, 37]. On applying hydrostatic pressure of 20 GPa, the conductivity of **14** increased up to 1×10^{1} S cm^{-1}. Tellurium atoms were introduced to the TTF skeleton as a heavy chalcogen atom. Compound **15** showed high conductivity (10^{-3} S cm^{-1}) at around 13 GPa [38]. There are intermolecular short Te···Te contacts in the crystal [39]. The molecules formed face-to-face dimers with the intradimer Te(4)···Te(4) contact (3.743 Å), and the dimers were linked orthogonally by the interdimer Te(3)···Te(3) contact (3.583 Å). The formation of dimers gives rise to a midgap in the band structure. The HOMO and LUMO bands of the dimer system hardly overlap with each other compared with the uniform system, even if high pressure is applied. Therefore, metallic conductivity has not been observed in this compound under pressure.

Table 5.3 Properties of selected organic molecules containing heavy selenium, tellurium, and iodine atoms and related compound pentacene

Molecule	σ_{rt}/S cm^{-1}	Pressure/GPa	Behavior
14	8×10^{-7}	Ambient pressure	Semiconducting
	1×10^{1}	20	Semiconducting
15	10^{-5}	Ambient pressure	Semiconducting
	10^{-3}	ca. 13	Semiconducting
16	9×10^{-12}	Ambient pressure	Insulating
	2×10^{1}	25	Semiconducting
		35	Metallic
		52	Superconducting (T_C = 2 K)
17	10^{-14}	Ambient pressure	Insulating
		35	Metallic
		58	Superconducting (T_C = 2.3 K)
Pentacene	3×10^{-14}	Ambient pressure	Semiconducting
		21.3	Metallic

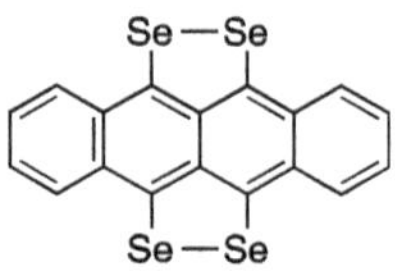

14

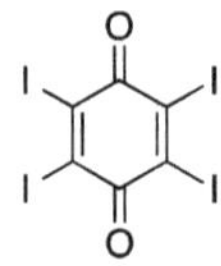

16

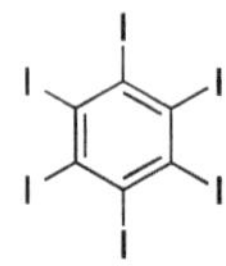

17

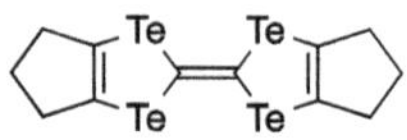

15

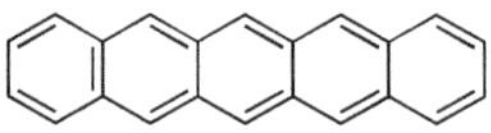

pentacene

On the other hand, in the 1970s, Shirotani et al. reported the electrical conductivity of iodine-containing organic molecules such as iodanil (**16**) and hexaiodobenzene (**17**) on applying physical pressure [108]. The electrical conductivities of **16** and **17** showed an insulating nature (10^{-12}–10^{-14} S cm^{-1}) under ambient conditions. Nevertheless, these compounds showed an increase in conductivity (2×10^1 S cm^{-1} for **16**) on applying a pressure of 25 GPa and exhibited metallic behavior under 35 GPa for **16** and **17**. The drop of resistance of **16** was observed at 2 K under a pressure of 52–78 GPa [43]. The magnetic field dependence of the resistance of **16** was measured, and the temperature showed that the resistance drop decreased as an increasing magnetic field was applied. These transport properties clearly indicated superconducting transition of **16**, and this compound is the first single-component molecular superconductor. Moreover, a similar temperature dependence of the resistivity of **17** was observed at 1–2.3 K under a pressure of 33–58 GPa (Fig. 5.15). The resistance drop was suppressed on applying a magnetic field (Fig. 5.15, inset), which evidences the superconductivity of **17** under pressure.

Figure 5.16 shows the crystal structure of iodanil (**16**) at ambient pressure [109]. The almost planar molecules were stacked uniformly with an interplanar distance of 3.694 Å along the crystallographic *b* axis. This packing motif is quite different from that of the molecules of compound **15**, which form face-to-face dimers. The molecules were linked by the short I···I and O···I contacts *d1*, *d2*, and *d3* and formed a three-dimensional heteroatom network (*d1* = 3.790(3), *d2* = 3.892(4), and *d3* = 3.23(2) Å). On the contrary, no π–π interaction between benzoquinoid moieties was observed (Fig. 5.16b). The crystal structure of hexaiodobenzene (**17**) is similar to **16** (Fig. 5.17) [110]. The almost-planar molecules **17** were stacked along the crystallographic *b* axis and formed molecular columns. These columns were linked by the I···I contacts (*d1* = 3.7045(7) and *d2* = 3.7125(6) Å). There is a difference in the overlap modes of **17** compared with those of **16**, where molecules **17** formed an ideal π–π stack with an interplanar distance of 3.804 Å and a tilted angle of 51.6°.

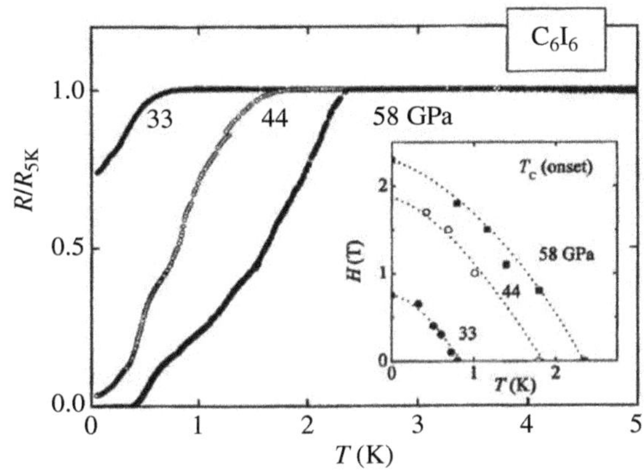

Figure 5.15 Superconducting transition of hexaiodobenzene (**17**) on applying a pressure of 33, 44, and 58 GPa. Onset temperature of superconducting transition increases with pressure (1, 1.9, and 2.3 K, respectively). The inset shows the critical magnetic field dependence as a function of temperature at each value of pressure. Reproduced from Ref. [44], Copyright (2001), with permission from Elsevier.

Observed uniform molecular arrangements and three-dimensional heteroatom networks are advantageous to high conductivity under pressure. However, it was in question whether the crystal and the molecules remained intact under such extremely high pressure, because crystalline iodine shows metallic and superconducting behavior [111–113]. The crystal structure of iodanil (**16**) under a pressure of around 10 GPa was investigated, and moreover, IR study of **16** was carried out to exclude the possibility of structural phase transitions and decomposition of the molecules [114, 115]. The results of these investigations revealed that the decomposition of molecule **16** was avoided by CT interactions through I···O contacts, which also played a key role in the metallic nature. In contrast, although no structural phase transition was observed in hexaiodobenzene (**17**) below 9.7 GPa [116], the details of the crystal and molecular structures under extremely high pressure remain unknown.

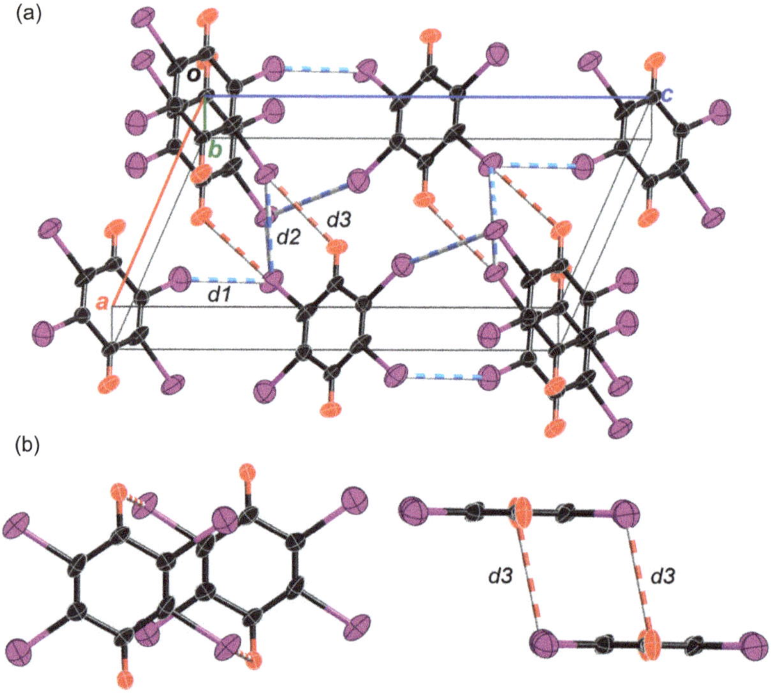

Figure 5.16 (a) Crystal structure of iodanil (**16**). The broken blue and red lines indicate intermolecular I···I contacts *d1* = 3.790(3) Å and *d2* = 3.892(4) Å and O···I contact *d3* = 3.23(2) Å, respectively, shorter than the sum of the van der Waals radii. (b) Top (left) and side views (right) of the overlapping mode.

5.3 Purely Organic Molecular Superconductors Based on Partially Oxidized Organic Molecules

5.3.1 Neutral π Radicals

Partially filled valence and conduction bands allow the production of metallic conductivity and superconductivity. As mentioned in Section 5.1, neutral π radicals are among the candidates for organic metals and superconductors based on organic molecules only. The neutral π radicals form a half-filled band because the valence band is composed of SOMO. Table 5.4 summarizes conducting properties of neutral π radicals.

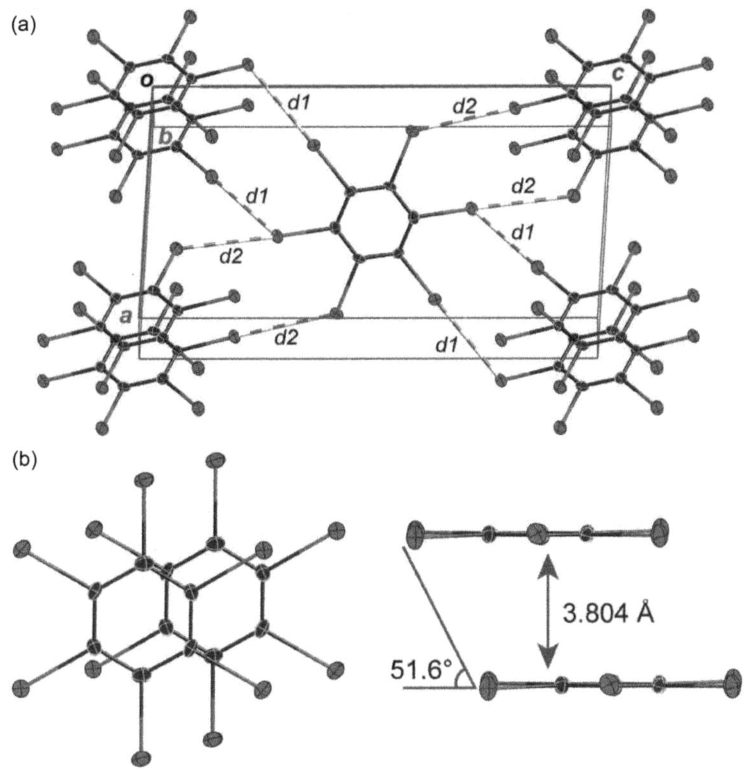

Figure 5.17 (a) Crystal structure of hexaiodobenzene (**17**). The broken lines indicate intermolecular I···I contacts *d1* = 3.7045(7) Å and *d2* = 3.7125(6) Å, shorter than the sum of the van der Waals radii. (b) Top (left) and side views (right) of the overlapping mode.

Table 5.4 Conducting properties of neutral π radicals

Molecule	σ_{rt}/S cm^{-1}	E_g/eV	Refs.
20	5.7 × 10^{-6a}	—	[50]
21	3.2 × 10^{-5a}	—	[51, 52]
22 (R = CH_3)	1.2 × 10^{-3a}	0.44	[53, 54]
23 (2R = $O(CH_2)_2O$)	5 × 10^{-4a}	0.51	[55]
24 (X = S)	1.4 × 10^{-1a}	0.23–0.46	[56]
25 (X = Se)	5 × 10^{-2a}	0.18–0.23	[56]
26	5 × 10^{-2b}	0.26	[57]
27 (X = O, R = H)	1 × 10^{-1b}	0.48	[62, 63]
28 (X = O, R = CH_3)	3 × 10^{-1b}	0.24	[62, 63]

(*Continued*)

Table 5.4 *(Continued)*

Molecule	σ_{rt}/S cm^{-1}	E_g/eV	Refs.
29 (X = NCH_3, R = H)	4×10^{-2b}	0.30	[60]
30	1×10^{-4b}	0.44	[64]
31	3.0×10^{-4b}	0.38	[65]
32	4.0×10^{-3b}	0.32	[66]
32·CH_3CN	3.0×10^{-3b}	0.22	[66]
33	10^{-8b}	0.66	[76]
34	$<10^{-6a}$	—	[70]
	10^{-9c}	—	[79]

[a]Measured on a compressed pellet
[b]Measured on a single crystal
[c]Measured on a thin film

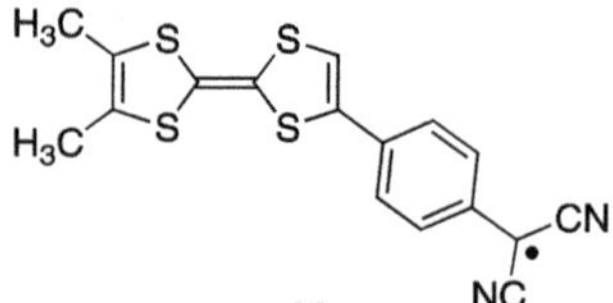

20

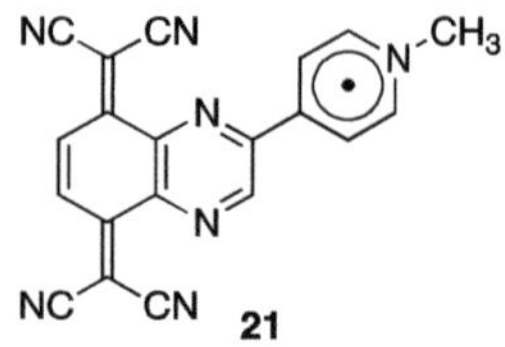

21

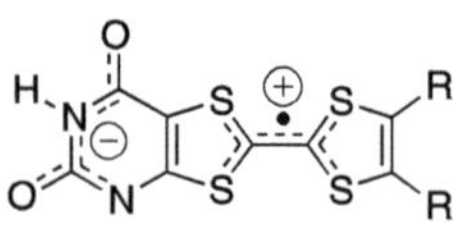

22 (R = CH_3)
23 (2R = $O(CH_2)_2O$)

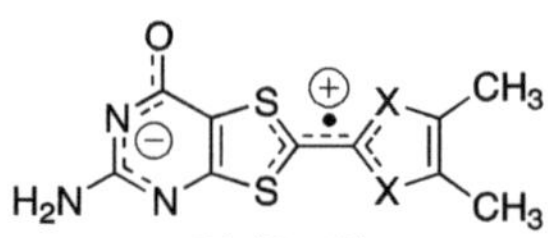

24 (X = S)
25 (X = Se)

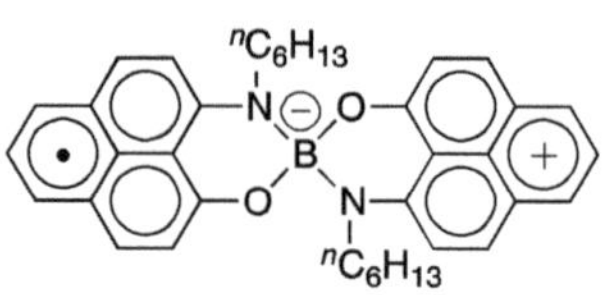

26

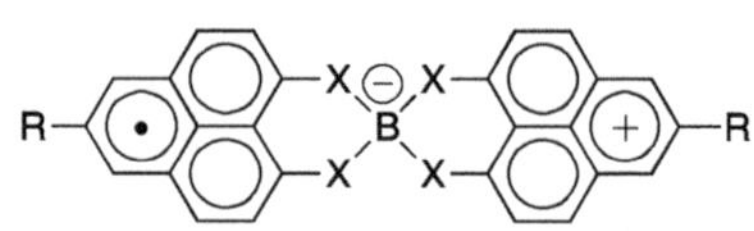

27 (X = O, R = H)
28 (X = O, R = CH_3)
29 (X = NCH_3, R = H)

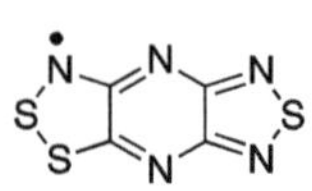

30

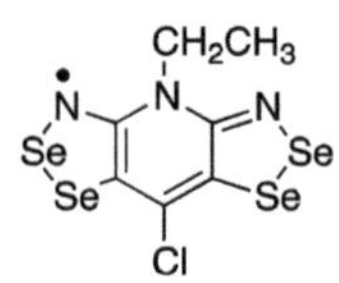

31

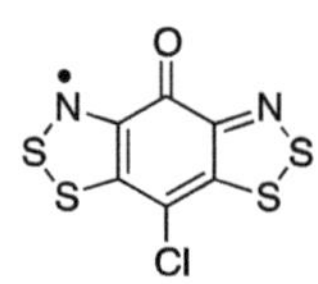

32

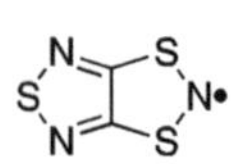

33

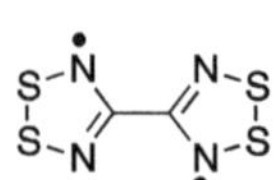

34

The TTF derivatives bearing π radical moieties showed a polarized contribution due to intramolecular electron transfer (IET) and intramolecular charge transfer (ICT). The IET and ICT of this kind of molecules were clearly described by Morita et al. [46] and Veciana et al. [47–49, 117]. The TTF connected to a 6-oxophenaloxyl (6OP) radical (**18**) and polychlorotriphenylmethyl (PTM) radical **19** possessed two electronic states—the neutral radical state (**18A** and **19A**) and the zwitterionic radical state (**18B** and **19B**)—which were obtained by IET and ICT processes (Fig. 5.18). Compound **19** showed an insulating behavior because the PTM radical is too large to enhance the transfer integrals. Recently, the semiconducting behavior of the molecular analog of **19** was reported under high pressure [118]. Miyashi et al., Yamashita et al., and Suzuki et al. synthesized simple TTF and TCNQ derivatives **20** and **21** with π radical moieties [50–52]. In these systems, the polarized structures were also observed, and Fig. 5.19 shows a neutral radical state **21A** and a zwitterionic state **21B** as a representative. The relatively high conductivities 5.7 × 10^{-6} and 3.2 × 10^{-5} S cm^{-1} were measure on a compressed pellet of compounds **20** and **21**, respectively, because there is no bulky group in these molecules and the polarized contribution plays an important role in reducing the on-site coulomb repulsive energy *U* [38].

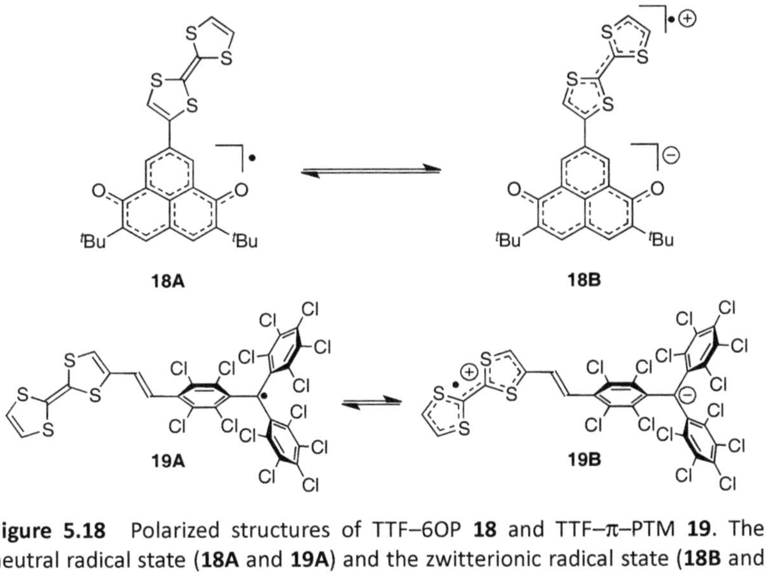

Figure 5.18 Polarized structures of TTF–6OP **18** and TTF–π–PTM **19**. The neutral radical state (**18A** and **19A**) and the zwitterionic radical state (**18B** and **19B**) were generated by the IET and ICT processes.

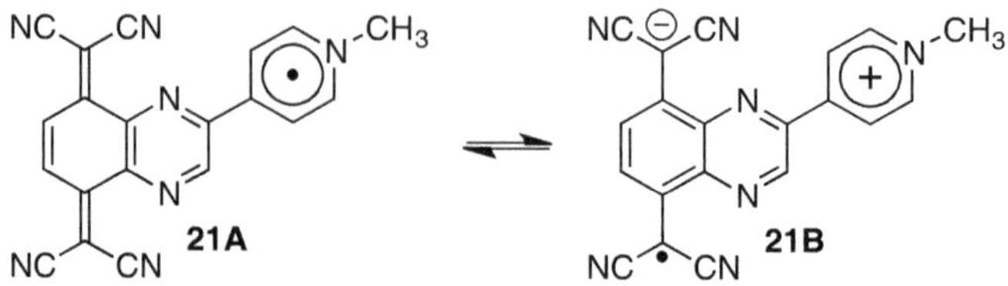

Figure 5.19 Polarized structures of TCNQ with the pyridinium radical **21**. The neutral radical state **21A** and the zwitterionic radical state **21B** were generated by the IET and ICT processes.

Taking advantage of the polarized structure for higher electrical conductivity, twitterionic (betainic) organic radicals were developed. Saito et al. and Neilands et al. reported betainic radicals of pyrimido-fused TTF [53–56]. This system showed high electrical conductivities (1.4×10^{-1}–5×10^{-4} S cm^{-1}) despite the measurement being carried out on a compressed pellet. The ultraviolet-visual-near-infrared (UV-vis-NIR) spectra of **23–25** showed absorption maxima at around 5×10^{3} cm^{-1}, which were lower than the 1:1 cation radical salts of TTF derivatives (ca. 10^{4} cm^{-1}) [38, 56]. It is known that this absorption band is associated with *U*. This result indicated that the *U* of the betainic radical system was reduced compared with the neutral radical and cation (anion) radical system. Therefore, the conductivities of the betainic radical TTF derivatives were much higher despite a purely organic single-component system.

Similar results were obtained by the phenaleny-based zwitterionic radicals, which were developed by Haddon et al. [57–63]. These zwitterionic radicals adopt a nonplanar structure due to the spiro ring and molecule **26** has bulky hexyl groups, as shown in Fig. 5.20. Nevertheless, compounds **26–29** showed high electrical conductivities (3×10^{-1}–5×10^{-2} S cm^{-1}) and semiconducting behavior with the activation energy of 0.13 (**26**), 0.24 (**27**), 0.12 (**28**), and 0.15 eV (**29**). The twitterionic radicals **27** and **28** formed a one-dimensional chain, as shown in Fig. 5.21. The unpaired electrons were distributed across the phenalenyl moieties in the molecule, and resonance of the electrons took place (Fig. 5.21a). In this case, the effective *U* was reduced and it is consistent with the high conductivity of the twitterionic radicals **27** and **28** [62, 63]. The semiconducting behavior of these twitterionic radicals can be described by the resonating valence bond (RVB) state [61, 119, 120].

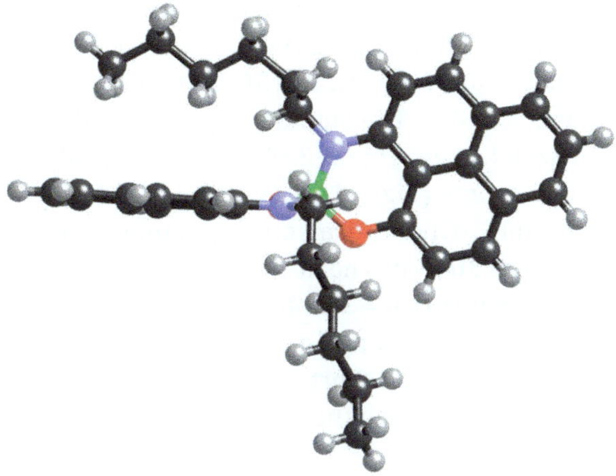

Figure 5.20 Molecular structure of twitterionic radical **26**. Two phenalenyl moieties are orthogonally arranged.

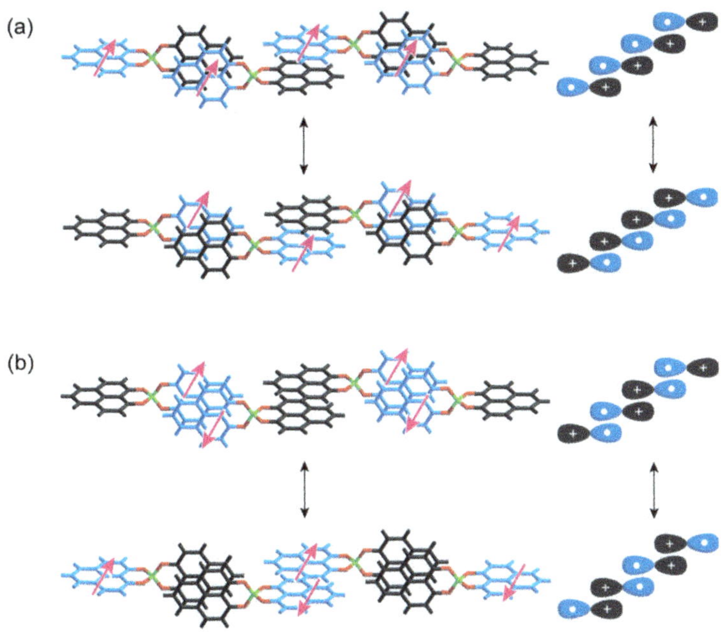

Figure 5.21 (a) Radical resonance structure and (b) resonating valence bond (RVB) resonance structure for **27**. The phenalenyl moieties with radical distribution are represented by blue atoms and bonds.

Introduction of heavy atoms to the neutral π radicals is also effective in reducing *U*. Oakley et al., Haddon et al., and Wolmershäuser et al. developed neutral π-radicals (**30–35**) containing sulfur and selenium atoms [64–72]. The spin density of 1,2,5-thiadiazolo[3,4-*b*]-1,2,3-dithiazolo[3,4-*b*]pyrazin-2-yl (**30**) was mainly distributed on the nitrogen atoms (Fig 5.22). As a result, the polymerization associated with the C–C bond formation was suppressed. As shown in Fig. 5.23, molecules **30** form a head-to-tail dimer, which is connected by a short S···S contact ($d1$ = 3.233(2) Å) in the crystal. The dimer formed a molecular column along the crystallographic *a* axis, and the columns were linked by short S···S contacts ($d2$ = 3.361(2) and $d3$ = 3.582(2) Å), shorter than the sum of the van der Waals radii.

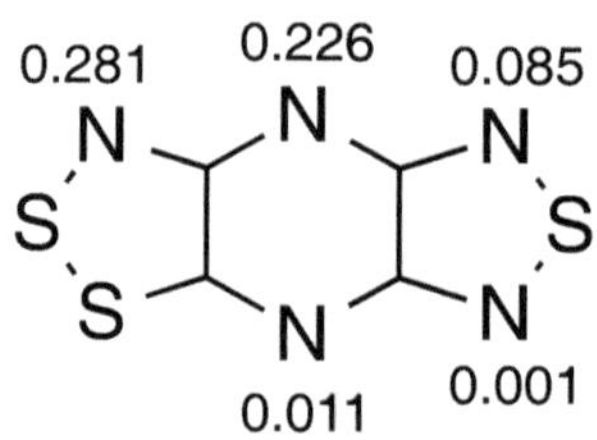

Figure 5.22 Spin density of the neutral π radicals **30** calculated by the semiempirical MNDO method.

The electrical conductivity of the neutral π radical **30** was measured to be 1 × 10^{-4} S cm^{-1} and decreased with a decreasing temperature. Radical **30** is a band insulator due to strong dimerization; however, this compound shows a semiconducting behavior. A one-dimensional hopping transport was predicted because the logarithm of conductivity (log(σ)) of **30** had a proportional relation with the square root of inverse temperature ($T^{-1/2}$). The neutral π radicals **31** and **32** showed a semiconducting behavior with high conductivities of 3.0 × 10^{-4} for **31**, 4.0 × 10^{-3} for **32**, and 3.0 × 10^{-3} S cm^{-1} for **32**·CH_3CN and low activation energies of 0.19 for **31**, 0.16 for **32**, and 0.11 eV for **32**·CH_3CN. Radical **31** showed a weak metallic behavior at around 7 GPa [71]. Organic superconductors based on neutral π radicals are yet to be obtained; however, Awaga et al. and Fujita et al. disclosed an interesting magnetic property of thiazyl radicals **33–35** in addition to potential application in optoelectronic devices [73–81]. 1,3,5-Trithia-2,4,6-triazapentalenyl (**33**) exhibited a first-order magnetic phase transition between high-temperature paramagnetic and low-temperature diamagnetic phases, which occurs accompanied by a wide thermal hysteresis in the temperature range of 230–305 K. Therefore, **33** possesses magnetic bistability at room temperature (290 K) [73, 76]. 4,4-Bis(1,2,3,5-dithiadiazoyl) (**34**) forms a two-dimensional honeycomb monolayer film on the surface of Cu(111). The presence of a Dirac point in the band structure of the monolayer was suggested by first principles calculation. Moreover, the van Hove singularities are located near the Fermi level. These results are expected to produce high-T_C superconductors [80].

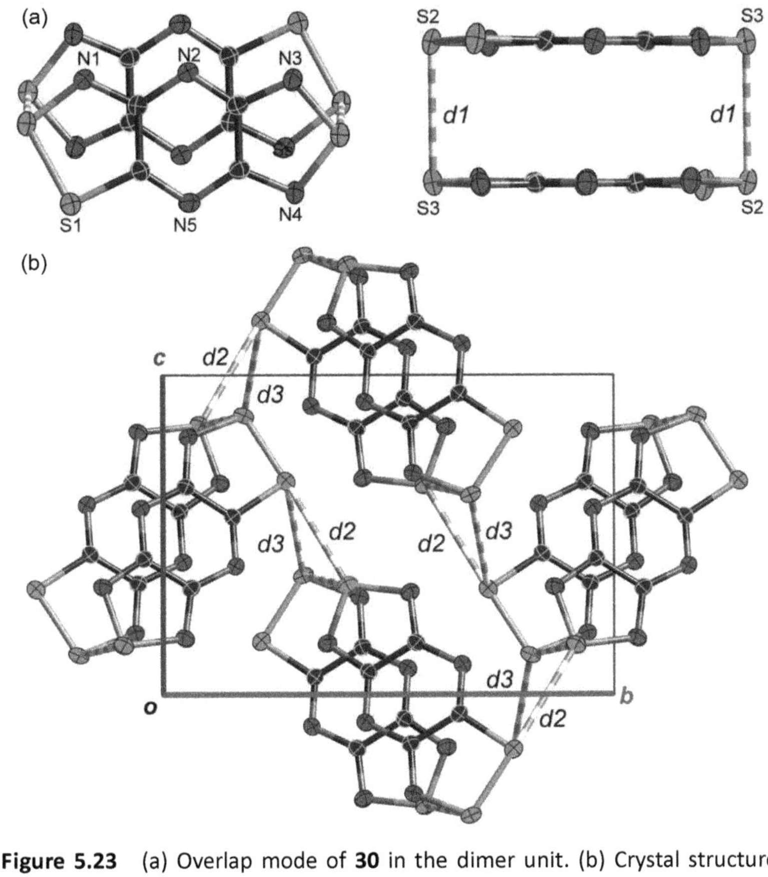

Figure 5.23 (a) Overlap mode of **30** in the dimer unit. (b) Crystal structure of **30** viewed along the crystallographic *a* axis. The broken lines represent S···S contacts, shorter than the sum of the van der Waals radii.

5.3.2 CT Type of Organic Solids

5.3.2.1 Historical background of conducting CT complexes

The most simple and successful method for producing partially filled valence and/or conduction bands is to remove an electron from the valence band or to add an electron to the conduction band. A large number of highly conducting CT complexes and cation (anion) radical salts have been developed by this strategy. Among them, more than 130 superconductors have been produced [38, 121–136]. In this section, the author mentions the historical background of the development of organic superconductors based on the CT complexes and cation (anion) radical salts.

Akamatu, Inokuchi, and Matsunaga reported a highly conductive CT complex of perylene (**36**) with bromine as the first organic semiconductor [137]. This discovery opened up a field of conducting organic materials based on the CT-type materials. Two kinds of organic molecules—electron donor and electron acceptor—play an important role in producing a partially filled metallic band. Typical electron donors and acceptors are shown in Fig. 5.24. The most representative electron acceptor, 7,7,8,8-tetracyanoquinodimethane (TCNQ, **43**), was synthesized by Acker and Hertler in 1962 [83], and the conducting properties of CT complexes of TCNQ were investigated [138]. On the other hand, Wudl et al. and Coffen et al. simultaneously reported the synthesis of the most popular electron donor, TTF (**1**), in 1969 [82, 139]. TTF has an isoelectronic structure with heptafulvalene and is easily oxidized to cation radical and dication states, in which the five-membered ring adopts the aromatic 1,3-dithiolyllium ring (Fig. 5.25).

The development of TTF led to the discovery of the first organic metal (TTF)(TCNQ) in 1973 [84], followed by active research for organic functional materials [91–94, 140]. The TTF and TCNQ molecules formed face-to-face segregated stacks in the solid of the CT complex (TTF)(TCNQ) (Fig. 5.26). There are strong interactions within the TTF and the TCNQ columns, respectively, yet no significant intermolecular interactions between TTF and TCNQ columns; in consequence a one-dimensional metallic band was constructed. X-ray scattering measurement revealed that the degree of CT was 0.59 per molecule, indicating a partially filled band structure [141, 142]. Consequently, (TTF)(TCNQ) exhibited high conductivity and a metallic nature. However, the conductivity of (TTF)(TCNQ) dramatically decreased at 54 K, because (TTF)(TCNQ) underwent a metal-insulator (MI) transition. The MI transition was found to originate from the Peierls instability, a typical property for one-dimensional metallic compounds [143, 144].

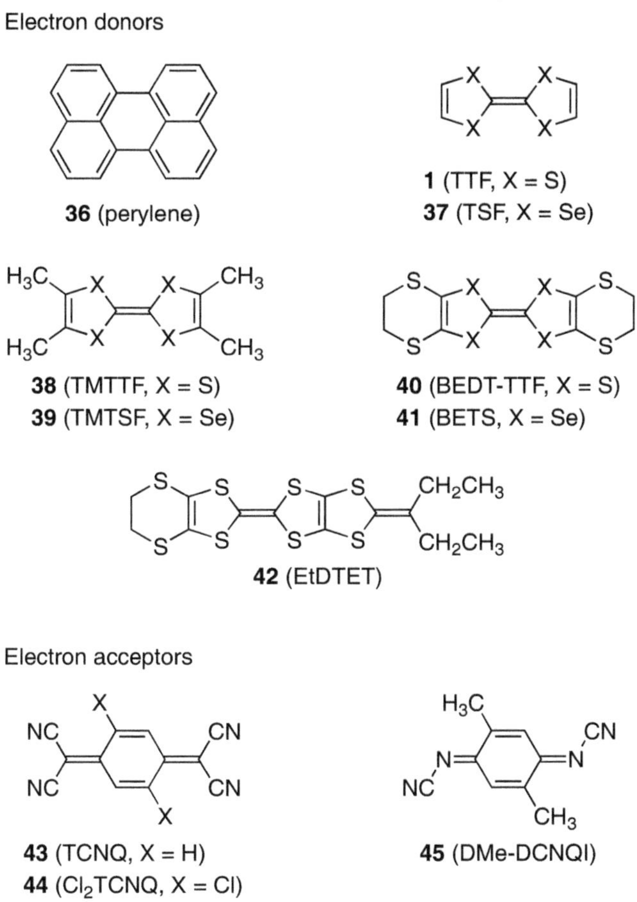

Figure 5.24 Typical electron donors and acceptors for organic conducting materials.

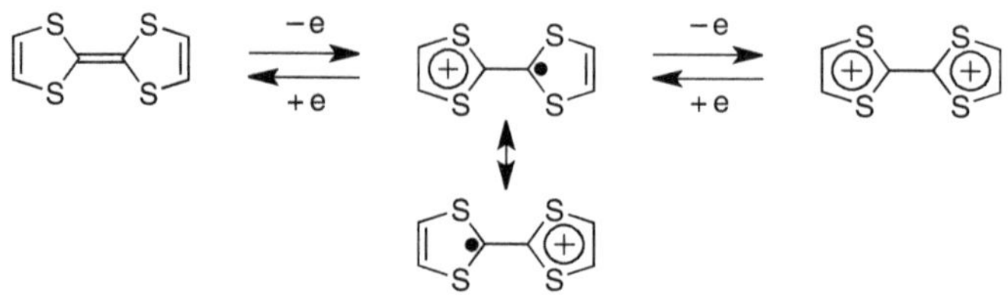

Figure 5.25 Redox scheme of TTF (**1**).

(a)

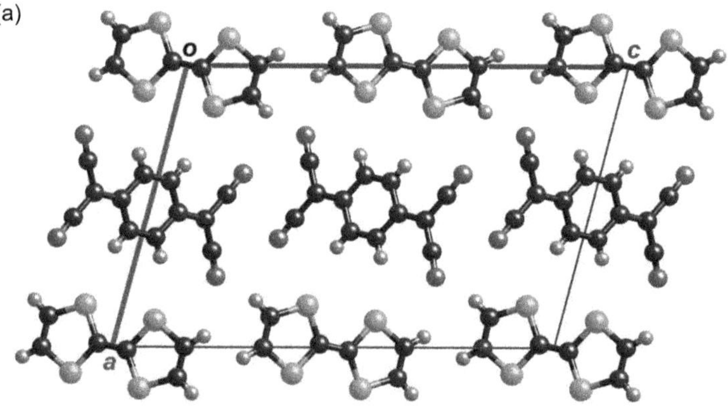

(b)

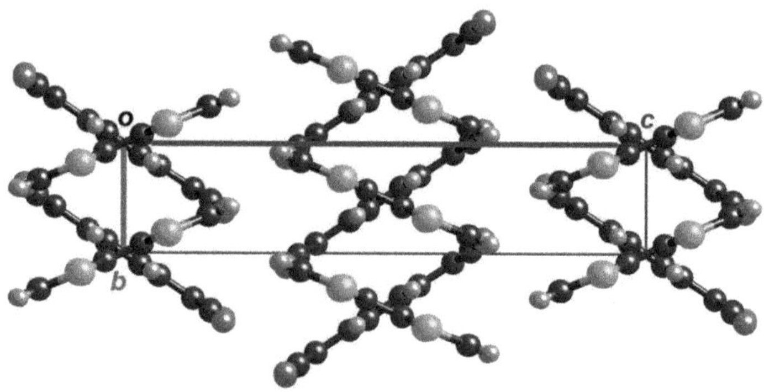

Figure 5.26 Crystal structure of (TTF)(TCNQ) viewed along the crystallographic (a) *b* axis and (b) *a* axis.

At an early stage in the development of molecular conductors, including superconductors, derivatives of TTF, such as tetraselenafulvalene (TSF) (**37**) [145], tetramethyltetrathiafulvalene

(TMTTF) (**38**) [146], tetramethyltetraselenafulvalene (TMTSF) (**39**) [147], BEDT-TTF (**40**) [148], and bis(ethylenedithio) tetraselenafulvalene (BETS) (**41**) [149, 150], and also TCNQ derivatives, such as Cl_2TCNQ (**44**) [151] and 2,5-dimethyl-*N,N′*-dicyanoquinodiimine (DMe-DCNQI, **45**) [152, 153], were synthesized (Fig. 5.24). TSF (**37**) and TMTSF (**39**) were designed to enhance the intermolecular interactions by using Se···Se contacts. (TSF)(TCNQ) exhibited a higher conductivity, of 800 S cm^{-1}, than that of (TTF)(TCNQ) (652 S cm^{-1}), and the MI transition temperature (40 K) is lower by 14 K compared with (TTF)(TCNQ) (54 K) [145]. The replacement of sulfur atoms by selenium atoms significantly enhanced the conductivity; however, the MI transition cannot be suppressed due to a one-dimensional electronic structure. To increase the dimensionality of the electronic structure, cation radical and anion radical salts were prepared with small inorganic anions/cations, respectively. A series of anion radical salts of DCNQI with copper, such as Cu(DMe-DCNQI)$_2$, exhibited a three-dimensional electronic structures [154]. This system creates most successful materials in terms of increasing the dimensionality and realizing the coupling of conduction π electrons and localized d spins. However, organic superconductors were not obtained in the anion radical salts of DCNQI derivatives. Bechgaard et al. reported the first organic superconductor (TMTSF)$_2$PF$_6$ (T_C = 0.9 K, at 0.12 GPa) as well as the ambient-pressure superconductor (TMTSF)$_2$ClO$_4$ (T_C = 1.4 K) [85, 155]. TMTSF molecules formed one-dimensional molecular stacks, which were linked by many Se···Se contacts, as shown in Fig. 5.27. Consequently, the Fermi surfaces of the (TMTSF)$_2$X system showed a quasi-one-dimensional nature along the molecular stacking direction [156].

BEDT-TTF (**40**) possessing peripheral sulfur atoms was developed in order to increase the electronic dimensionality and was synthesized by Mizuno et al. in 1978 [148]. BEDT-TTF is expected to form intermolecular interactions not only in the stacking direction but also in the transverse direction, the so-called side-by-side interactions, as shown in Fig. 5.28. Actually, two-dimensional conducting sheet structures were accomplished in the β-, β'-, and β''-type molecular arrangement in the BEDT-TTF salts (Fig. 5.29a–c) [157]. Two-dimensional molecular sheets were also observed in the θ-type and κ-type arrangements (Fig. 5.29d,e). The inclined molecules are arranged in herringbone-type molecular packing in the θ-type molecular sheet, and the face-to-face dimers are arranged orthogonally in a κ-type arrangement [158]. BEDT-TTF produced many organic superconductors, and the highest T_C for the BEDT-TTF-based superconductors was reported on κ-(BEDT-TTF)$_2$Cu[N(CN)$_2$]Br (T_C = 11.6 K) [159] at ambient pressure. Deuterium substitution of eight hydrogen atoms of BEDT-TTF led to an increase in the T_C of BEDT-TTF-based organic superconductors, and the T_C was enhanced up to 12.3 K in κ-(BEDT-TTF-d$_8$)$_2$Cu(CN)[N(CN)$_2$] [160]. Furthermore, the T_C of β'-(BEDT-TTF)$_2$ICl$_2$ reached 14.2 K under a hydrostatic pressure of 8.2 GPa [161].

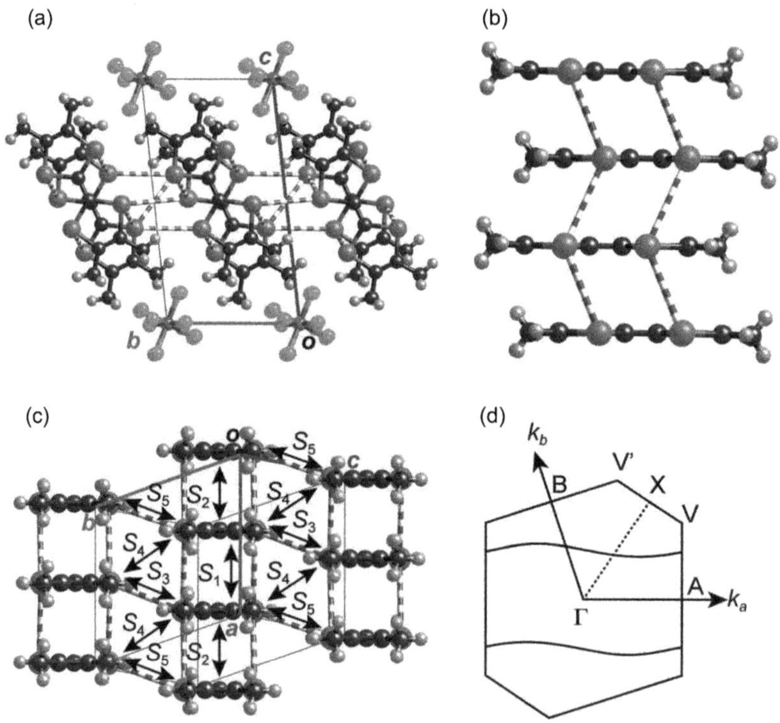

Figure 5.27 (a) Crystal structure of (TMTSF)$_2$PF$_6$ viewed along the crystallographic *a* axis. (b) Stacking structure of (TMTSF)$_2$PF$_6$ viewed along the molecular short axis. (c) Molecular arrangement of (TMTSF)$_2$PF$_6$ viewed along the molecular long axis. The broken lines represent Se···Se contacts, shorter than the sum of the van der Waals radii. (d) Calculated Fermi surfaces of (TMTSF)$_2$PF$_6$. Overlap integrals given in Ref. [156] were employed in the calculation (S_1 = 34.766, S_2 = 38.455, S_3 = −3.556, S_4 = −3.959, and S_5 = −1.607 × 10^{-3}).

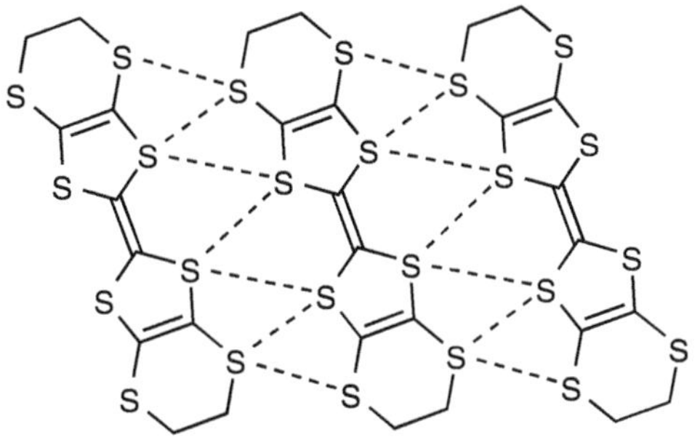

Figure 5.28 Schematic drawing of the side-by-side interactions of BEDT-TTF.

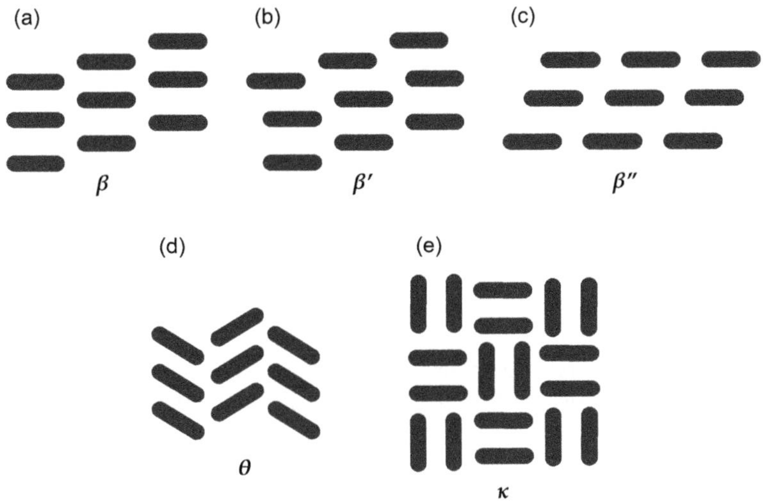

Figure 5.29 Schematic illustration of the two-dimensional (a) β-, (b) β′-, (c) β″-, (d) θ-, and (e) κ-type molecular arrangements in BEDT-TTF salts.

Schumaker et al. reported the synthesis of BETS (**41**) in 1983 [149], and then an improved synthetic procedure was established by Kato et al. [150]. BETS is a TSF analog of BEDT-TTF, namely, four sulfur atoms of the central TTF skeleton in BEDT-TTF were replaced with selenium atoms. A series of cation radical salts of BETS with magnetic FeX_4 anions (X = Cl and Br) afforded peculiar systems associated with the interaction between an itinerant π electron and a localized *d*-spin [127, 130, 162–168]. For example, λ-$(BETS)_2FeCl_4$ underwent a sharp MI transition at 8.5 K, where the insulating ground state was a π-d coupled antiferromagnetic phase [165]. The close Se···Cl contacts (3.528 Å) between the BETS molecule and the $FeCl_4$ anion in the crystal structure play an important role for the π-d coupling in λ-$(BETS)_2FeCl_4$ (Fig. 5.30). This compound showed superconductivity on applying hydrostatic pressure (T_C = ca. 2 K at 0.35 GPa) [166] and on applying a magnetic field above 17 T [167]. Moreover, coexistence of antiferromagnetic ordering and superconductivity was demonstrated in κ-$(BETS)_2FeX_4$ (X = Cl and Br) [130, 168].

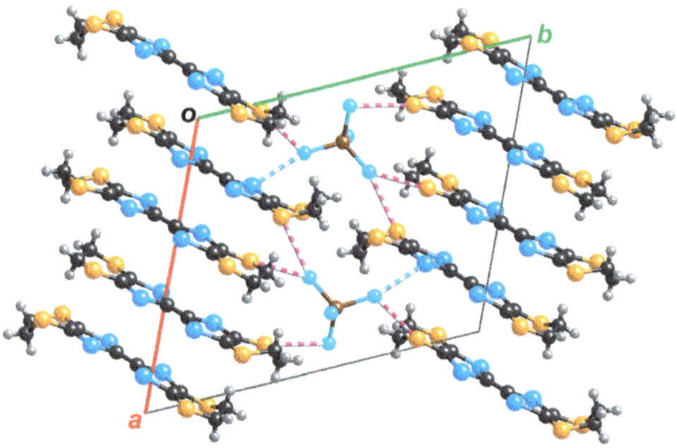

Figure 5.30 Crystal structure of λ-$(BETS)_2FeCl_4$ viewed along the crystallographic *c* axis. The broken pink and blue lines represent short S···Cl and Se···Cl contacts, respectively, shorter than the sum of the van der Waals radii.

The study on the superconductors based on TMTSF, BEDT-TTF, and BETS proposed effective molecular strategies for producing organic metals and superconductors. A bis-fused TTF donor 2,5-bis(1′,3′-dithiol-2′-ylidene)-1,3,4,6-tetrathiapentalene (BDT-TTP, **47**) has ladder-like eight sulfur atoms, as in BEDT-TTF (Fig. 5.31). Schumaker et al. reported the synthesis of BDT-TTP derivatives in 1980. However, their synthetic method was limited to the derivatives bearing the electron-withdrawing groups [169]. Misaki et al. reported improved synthesis that is applicable to the synthesis of various substituents, including pristine BDT-TTP [105, 170]. BDT-TTP and its derivatives tend to form a two-dimensional β-type molecular arrangement in their cation radical salts regardless of the size and shape of the counteranions (Fig. 5.32). Consequently, they showed a metallic conducting behavior down to a low temperature [135, 171–183]. The derivatives of DT-TTF (**46**) also tend to exhibit a metallic behavior down to a low temperature [95–102]. In these fused-donor systems, an ambient-pressure superconductor $(DTEDT)_3[Au(CN)_2]$ was developed, where DTEDT is 2-(1,3-dithiol-2-ylidene)-5-(2-ethanediylidene-1,3-dithiole)-1,3,4,6-tetrathiapentalene (**48**) [135, 184].

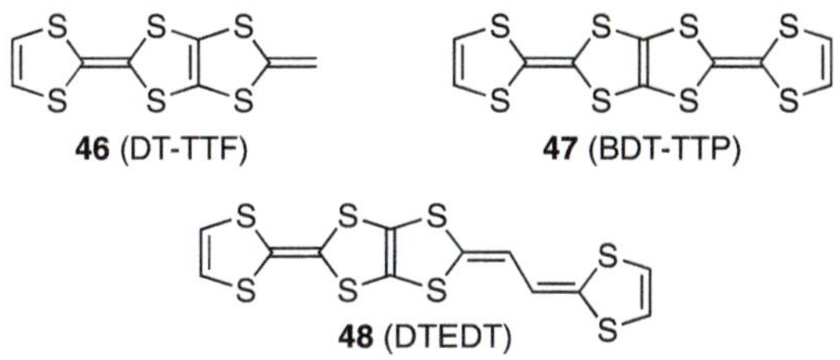

Figure 5.31 Fused TTF donors DT-TTF (**46**), BDT-TTP (**47**), and DTEDT (**48**).

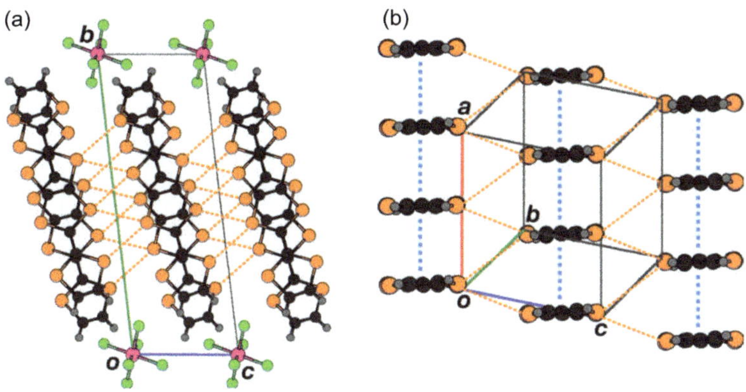

Figure 5.32 (a) The crystal structure of $(BDT\text{-}TTP)_2SbF_6$ and (b) the donor sheet structure viewed along the molecular long axis. The dotted orange lines represent S···S contacts, shorter than the sum of the van der Waals radii.

Most of the superconductors, as mentioned above, include organic electron donors and inorganic anions, that is, they are not purely organic materials. There are two straightforward strategies to yield organic superconductors containing organic molecules only; the first strategy is to prepare cation radical salts of organic electron donors with organic anions, and the second strategy is to prepare CT complexes composed of organic donors and organic acceptors. Four organic superconductors have been developed so far on the basis of these strategies (Table 5.5), and the following sections introduce these purely organic superconductors.

Table 5.5 Charge transfer (CT)-type organic superconductors containing organic molecules only

	T_C/K	P_C/GPa	Refs.
(BEDT-TTF)$_2$SF$_5$CH$_2$CF$_2$SO$_3$	5.1	—[a]	[86, 87]
(BEDT-TTF)$_2$CF$_3$SO$_3$	4.8	1.3	[88]
(BETS)$_2$(Cl$_2$TCNQ)	1.3	0.35	[89]
(EtDTET)(TCNQ)	5.5	—[a]	[90]

[a]Ambient pressure

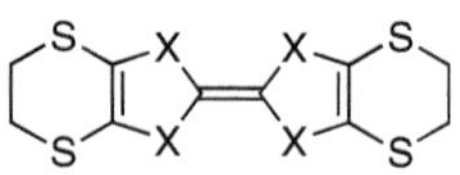

40 (BEDT-TTF, X = S)
41 (BETS, X = Se)

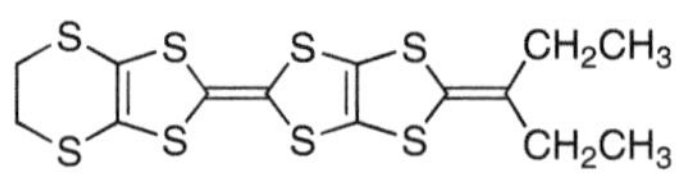

42 (EtDTET)

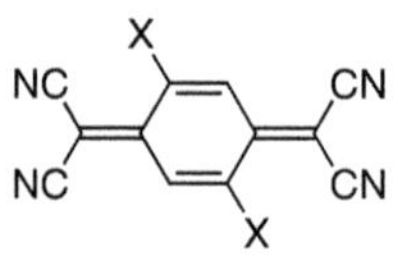

43 (TCNQ, X = H)
44 (Cl$_2$TCNQ, X = Cl)

$SF_5CH_2CF_2SO_3^{\ominus}$
$CF_3SO_3^{\ominus}$

5.3.2.2 Organic superconductors composed of BEDT-TTF and organic anions, β''-(BEDT-TTF)$_2$SF$_5$CH$_2$SF$_2$SO$_3$ and κ-(BEDT-TTF)$_2$CF$_3$SO$_3$

In general, cation radical salts are prepared by electrocrystallization, where electron donors are oxidized electrochemically in the solution containing an appropriate supporting electrolyte. The charge-compensating anion is provided by supporting electrolytes and, therefore, electrochemical stability of anions is indispensable for obtaining cation radical salts with organic anions. Although organic anions containing carboxylate groups are popular in chemistry, very few cation radical salts with carboxylate anions have been obtained, due to the instability of the carboxylate group under the condition of electrocrystallization [185]. On the other hand, many cation radical salts were reported with electrochemical stable anions, such as sulfonate and polycyano anions (Fig. 5.33) [87, 186, 187].

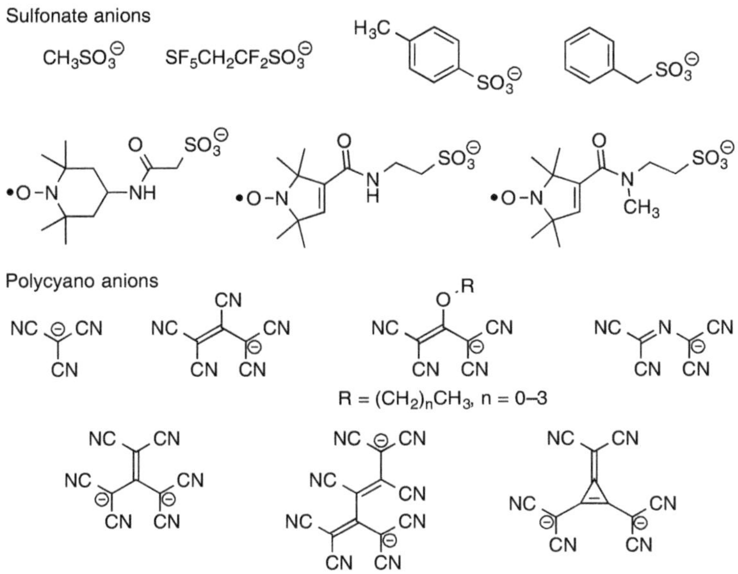

Figure 5.33 Organic anions.

Chasseau et al. reported BEDT-TTF salts with trifluoromethylsulfonate (triflate) and *p*-tolulenesulfonate (tosylate) in 1988 [188]. (BEDT-TTF)$_2$(*p*-$CH_3C_6H_4SO_3$) exhibited a semiconducting behavior with a room-temperature conductivity of 10^{-3} S cm^{-1} and activation energy of 0.26 eV. A complex conducting behavior was observed in (BEDT-TTF)$_2$(CF_3SO_3), which underwent a series of phase transitions at 306 and 220 K for type A and at 277 and 220 K for type B [189]. The crystal structure of (BEDT-TTF)$_2$(CF_3SO_3) possesses a δ-type molecular arrangement, in which four BEDT-TTF molecules are stacked with two overlap modes—"twisted" and "ring-over-atom"—alternatively [190]. Another phase of the BEDT-TTF salt of CF_3SO_3 anion was prepared by Fettouhi et al. in 1999 [191]. The molecular arrangement of the BEDT-TTF molecules is the so-called κ-type. The dimer of BEDT-TTF molecules, in which a spin (S = 1/2) was located, formed a quasi-one-dimensional spin lattice (t'/t = 1.79), where t and t' are interdimer transfer integrals in a triangular spin lattice [192]. κ-type molecular conductors show a variety of electronic and spin phases, such as superconducting, antiferromagnetic insulating, and quantum spin liquid (QSL) phases. Particularly, the discovery of the fascinating QSL phase in κ-(BEDT-TTF)$_2$Cu$_2$(CN)$_3$ led to the active investigation of the physical properties of the triangular spin lattices. Very recently, the electronic and magnetic properties of κ-(BEDT-TTF)$_2$(CF$_3$SO$_3$) were also investigated by Ito et al. and superconducting transition was observed by applying a pressure of 1.3 GPa (T_C = 4.8 K), as shown in Fig. 5.34 [88].

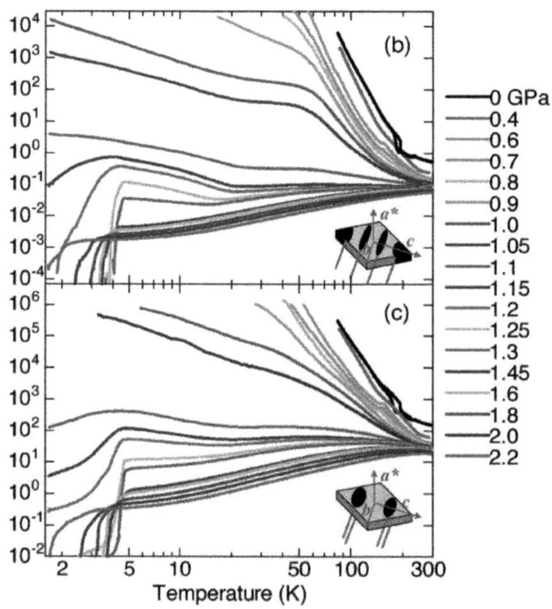

Figure 5.34 Temperature dependence of *c* axis (top panel) and *a** axis (bottom panel) resistivities under hydrostatic pressures of up to 2.2 GPa. Reprinted (figure) with permission from Ref. [88]. Copyright (2016) by the American Physical Society.

Geiser et al. and Schlueter et al. reported, in 1996, a superconducting transition of β″-(BEDT-TTF)$_2$SF$_5$CH$_2$CF$_2$SO$_3$ probed by the electrical resistivity and ac susceptibility measurement [86, 193]. This compound is the first ambient-pressure organic superconductor containing only organic components. The resistivity increased with a decreasing temperature and showed a broad maximum around 100 K. The semiconducting behavior was derived from the charge ordered (CO) state [194]. On further cooling, the resistivity showed a metallic behavior down to 6 K and then, a sharp drop in resistivity was observed at 5.1 K. The resistivity drop was suppressed on applying a magnetic field, indicating the superconducting transition [195]. As for the ac magnetic susceptibility, a drop of real part (χ') and a maximum of imaginary part (χ''), which were characteristic changes for the superconducting transition, were observed. The crystal structure is shown in Fig. 5.35. Two BEDT-TTF molecules (A and B) were crystallographically independent. Each BEDT-TTF molecule was stacked along the *a* axis in the A–A··· and B–B··· manner. The β''-type molecular arrangement was formed in the *ab* plane [157]. The conducting donor layer and the insulating organic anion layer were repeated alternatingly along the crystallographic *c* axis. The calculated Fermi surfaces consisted of hole-like pockets and quasi-one-dimensional pairs [196]. The Fermi surface of β''-(BEDT-TTF)$_2$SF$_5$CH$_2$CF$_2$SO$_3$ was also investigated by Shubnikov-de Haas (SdH) and angle-dependent magnetoresistance oscillations (AMROs) [197–199]. Recently, Cho et al. and Beyer et al. found a signature of the phase transition from the superconducting vortex state to the Fulde–Ferrell–Larkin–Ovchinnikov (FFLO) state, probed by upper critical field measurements and magnetic-field and angular-dependent high-resolution specific-heat measurements [200, 201].

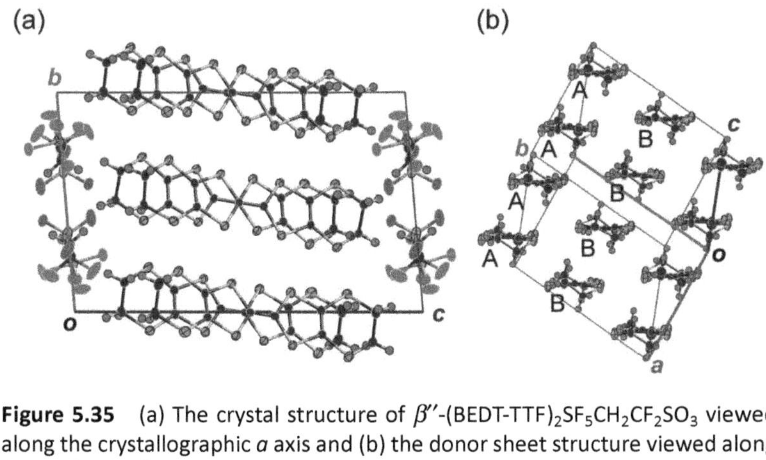

Figure 5.35 (a) The crystal structure of β''-(BEDT-TTF)$_2$SF$_5$CH$_2$CF$_2$SO$_3$ viewed along the crystallographic *a* axis and (b) the donor sheet structure viewed along the molecular long axis.

5.3.2.3 Organic superconductor composed of BETS and Cl_2TCNQ, $(BETS)_2(Cl_2TCNQ)$

The CT complex composed of organic electron donors and organic electron acceptors is a candidate for the purely organic superconductor. As mentioned above, (TTF)(TCNQ) underwent a MI transition due to the Peierls instability [141, 142]. Chemical modification of the donor and acceptor molecules was conducted in order to suppress the MI transition. Transition metal dithiolene complexes were successfully used as the electron acceptor to obtain organic superconductors, for example, $M(dmit)_2$ (**50**: M = Ni and **51**: N = Pd) and $Au(mnt)_2$ (**52**) complexes such as $[TTF][Ni(dmit)_2]_2$ [202], α-$[TTF][Pd(dmit)_2]_2$ [203], α'-$[TTF][Pd(dmit)_2]_2$ [204], $[EDT\text{-}TTF][Ni(dmit)_2]$ [205], and $(perylene)_2[Au(mnt)_2]$ [206], showing superconducting transition, where dmit = 1,3-dithiole-2-thione-4,5-dithiolate, mnt = maleonitrile dithiolate, and EDT-TTF (**49**) = ethylenedithio-tetrathiafulvalene (Fig. 5.36). These compounds were not purely organic materials because of the existence of a transition metal in the acceptor units.

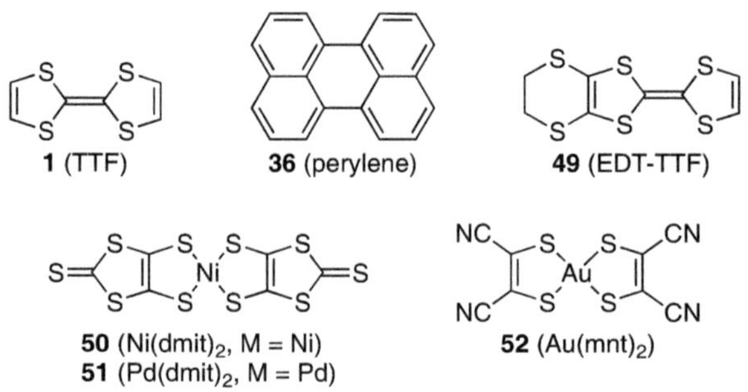

Figure 5.36 Component molecules for CT-complex-type organic superconductors.

Although BEDT-TTF provided many organic superconductors with inorganic anions, its CT complexes with organic acceptors did not show superconductivity. For example, there are three polymorphisms of (BEDT-TTF)(TCNQ) [148, 207–210]. The molecular packing motif of the monoclinic phase is a mixed-stack structure [209]. As a result, this phase exhibited a semiconducting behavior. On the other hand, there were two triclinic phases, each consisting of BEDT-TTF and TCNQ layers, in which the BEDT-TTF

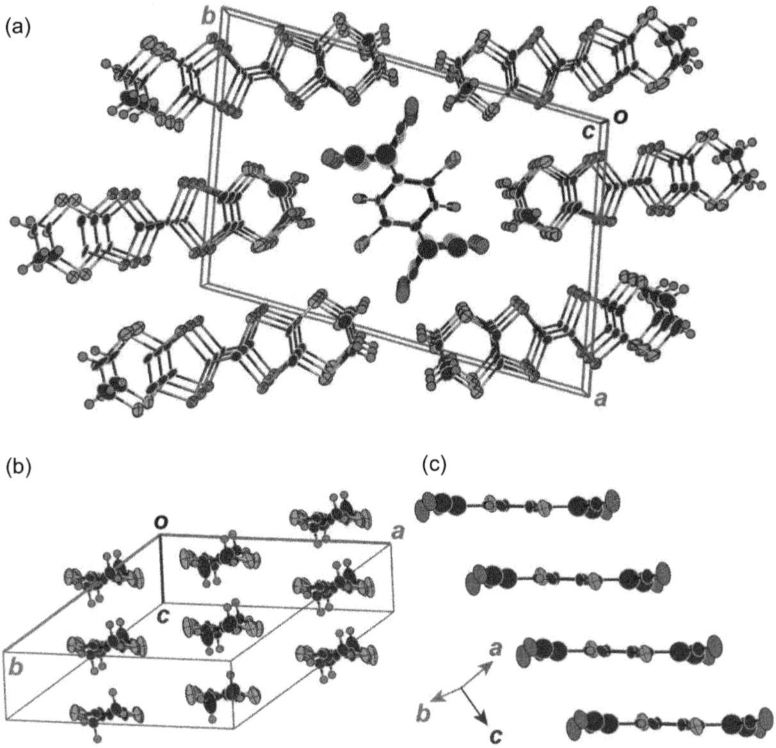

Figure 5.37 (a) The crystal structure of $(BETS)_2(Cl_2TCNQ)$ viewed along the crystallographic *c* axis, (b) the donor sheet structure viewed along the molecular long axis, and (c) the molecular packing of the Cl_2TCNQ molecules viewed along the molecular short axis.

molecules formed two-dimensional β'- and β''-type molecular arrangements [208, 210]. The β'-phase exhibited a MI transition at around 330 K [211]. In contrast, the β''-phase showed a stable metallic state down to 0.5 K [212, 213]. In 1999, Kondo et al. discovered the first organic superconductor $(BETS)_2(Cl_2TCNQ)$, composed of an organic electron donor and an organic electron acceptor [89]. $(BETS)_2(Cl_2TCNQ)$ exhibited superconducting transition at around 1 K on applying a pressure of 0.10–0.85 GPa. The maximum T_C = 1.3 K was recorded under 0.35 GPa [214]. The superconducting phase of this compound was contiguous to an insulating phase, as the insulating behavior was observed under a higher pressure in the resistivity measurement. The insulating behavior of this compound induced by pressure was associated with a spin density wave. The molecular packing motif of $(BETS)_2(Cl_2TCNQ)$ is shown in Fig. 5.37. One BETS molecule and a half of the Cl_2TCNQ molecule were crystallographically unique, namely, they were located on a general position and at the center of inversion, respectively. A unit cell contained two BETS molecules and one Cl_2TCNQ molecule, affording a BETS-to-Cl_2TCNQ ratio of 2:1. The BETS molecules were stacked uniformly along the *c* axis. The BETS columns aligned along the *a* axis with two modes, that is, a dimerization of the donor column resulted. The Cl_2TCNQ molecules were also stacked along the *c* axis, and the one-dimensional Cl_2TCNQ column was surrounded by the donor columns. The valence of the Cl_2TCNQ molecule was estimated to be almost −1 from the C≡N stretching vibration (ν_{CN} = 2180 cm^{-1}), where the ν_{CN} of neutral Cl_2TCNQ and $K(Cl_2TCNQ)$ was observed at 2225 and 2200 cm^{-1}, respectively [215]. As a result, the formal charge of the BETS molecule was estimated to be +0.5 per molecule. These structural and spectroscopic features imply that $(BETS)_2(Cl_2TCNQ)$ might be understood as a cation radical salt of BETS with the organic Cl_2TCNQ monoanion.

5.3.2.4 Organic superconductors composed of EtDTET and TCNQ, (EtDTET)(TCNQ)

The study of $(BETS)_2(Cl_2TCNQ)$ stimulated our discovery of superconductivity in (EtDTET)(TCNQ) (T_C = 5.5 K) [90]. This CT complex is the first purely organic ambient-pressure superconductor. For exploring purely organic superconductors, we focus on the fused-TTF donors of BDT-TTP and DT-TTF derivatives, because they tend to form two-dimensional molecular metals regardless of the size and shape of the counteranions in their cation radical salts. We anticipate that these donors afford two-dimensional molecular conductors even with the organic electron acceptor molecules. However, the metallic state of the molecular conductors based on BDT-TTP and DT-TTF derivatives was too stable to undergo a superconducting transition and, therefore, it is essential to develop chemically modified analogs of BDT-TTP and DT-TTF molecules. According to the phase diagrams of BEDT-TTF-based and TMTSF-based molecular conductors, the superconducting phase is adjacent to the metallic and insulating phases and the electronic states are scaled by U/W [216–219]. A promising approach of being accessible to the superconducting phase is to realize an increase in *U* and/or a decrease in the band width *W* for the molecular conductors of BDT-TTP and DT-TTF. To decrease the band width, a new DT-TTF derivative bearing ethyl groups, EtDTET (**42**), was developed by Fujiwara et al. [100].

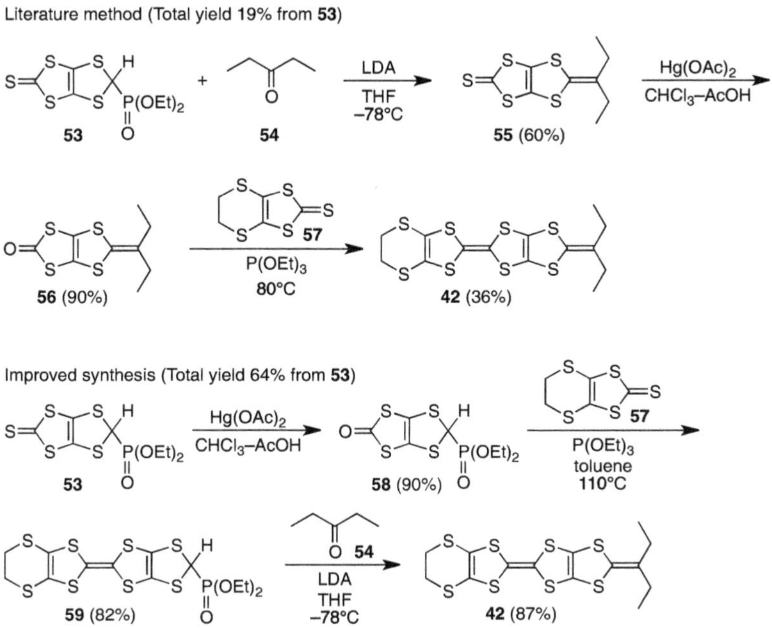

Figure 5.38 Synthesis of EtDTET (**42**).

The synthesis of EtDTET (**42**) is shown in Fig. 5.38. According to the literature method [100], EtDTET (**42**) was synthesized with an overall yield of 19% from **53**. The triethyl phosphite–mediated cross-coupling reaction between **56** and **57** at 80°C afforded an EtDTET (**42**) yield of 36%. In this reaction, homocoupling products of **56** and **57** were also generated and it was difficult to separate EtDTET (**42**) from the reaction mixture because they have similar polarities. To solve the low overall yield and the difficulty in purification, an improved synthetic method was applied [220, 221]. The triethyl phosphite–mediated cross-coupling reaction was carried out between 1,3-dithiol-2-one with the phosphonate moieties **58** and **57** in toluene at 110°C and afforded a good yield of the TTF derivative (**59**) (82%). A few homocoupling products of **58** and **57** were generated in this reaction. However, the target compound **59** was easy to separate from the reaction mixture due to the difference in polarities of **59** and the generated by-products. The Horner-Wadsworth-Emmons reaction of **59** with 3-pentanone (**54**) in the presence of lithium diisopropylamide (LDA) in tetrahydrofuran (THF) at −78°C gave an EtDTET (**42**) yield of 87%. The overall yield of the new synthetic method is successfully improved by the yield of 64% from **53**.

Cation radical salts of EtDTET (**42**) with AsF_6, SbF_6, and TaF_6 anions were prepared by electrocrystallization. The room-temperature conductivities of $(EtDTET)_2X$ (X = AsF_6, SbF_6, and TaF_6) are 2.4×10^0, 1.3×10^1, and 1.2×10^0 S cm^{-1}, respectively. Figure 5.39 shows temperature dependence of the resistivity. The EtDTET salts exhibited a semiconducting behavior with the activation energy (E_a) of 0.12 eV, 0.11 eV, and 0.14 eV for the AsF_6, SbF_6, and TaF_6 salts, respectively.

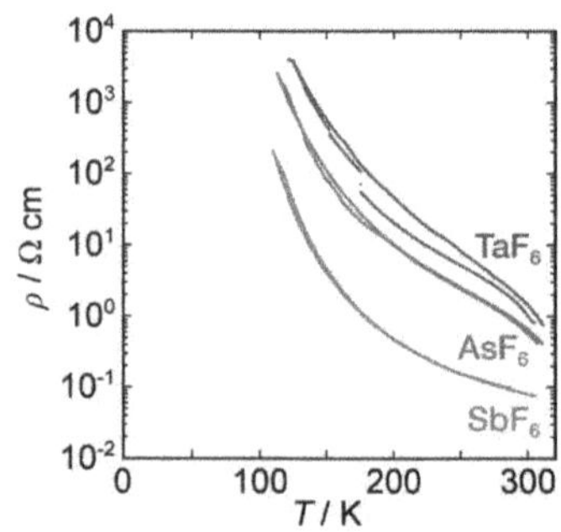

Figure 5.39 Temperature dependence of the resistivity of $(EtDTET)_2X$ (X = AsF_6, SbF_6, and TaF_6).

Relatively low conductivity and semiconducting behavior of the EtDTET salts implied a moderate decrease in intermolecular interactions. Judging from the conducting behavior of the EtDTET salts, we decided to develop the CT complex of EtDTET with organic electron acceptors and successfully obtained single crystals of (EtDTET)(TCNQ). This CT complex was prepared by mixing each hot solution of EtDTET and TCNQ in chlorobenzene, and then the mixture was slowly cooled to room temperature so that the crystal quality was enhanced. The resistivity decreased from room temperature with a lowering temperature but increased at around 13 K (Fig. 5.40a). An abrupt resistivity drop was observed at 5.5 K (midpoint, onset temperature at 6.8 K) (Fig. 5.40b). As for magnetic behavior, the static magnetic susceptibility exhibited Curie-like behavior in an applied field of 10 kOe (Fig. 5.40c). A diamagnetic transition was detected at around 5 K (Fig. 5.40d). The volume fraction of superconductivity at 2.0 K was about 45% of the perfect diamagnetism.

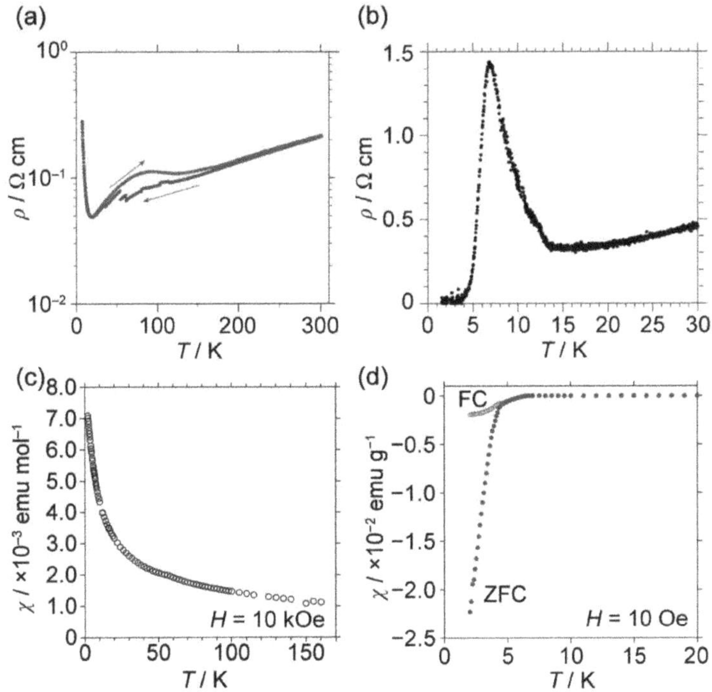

Figure 5.40 Temperature dependence of the resistivity of (EtDTET)(TCNQ) (a) in the temperature region of 7–300 K and (b) in the region of 1.6–30 K. (c) Temperature dependence of the static magnetic susceptibility of (EtDTET)(TCNQ) in an applied field of 10 kOe. (d) Temperature dependence of zero-field cooling (ZFC, closed circle) and field cooling (FC, open circle) DC magnetization of (EtDTET)(TCNQ) in an applied field of 10 Oe.

An X-ray crystal structure analysis revealed that the EtDTET molecules were stacked along the crystallographic *b* axis in a head-to-tail manner and formed a two-dimensional molecular array on the *ab* plane (Fig. 5.41). On the other hand, the TCNQ molecules formed face-to-face dimers and the quasi-stacking direction of the dimers was (*a*–*b*) axis. As a result, the EtDTET column and the TCNQ column were solid-crossing with a dihedral angle of 75°. The bond lengths of the TTF and TCNQ skeletons are sensitive to the molecular valence [222–225]. The degree of CT was estimated to be 0.5 by the bond length of the EtDTET and TCNQ molecules.

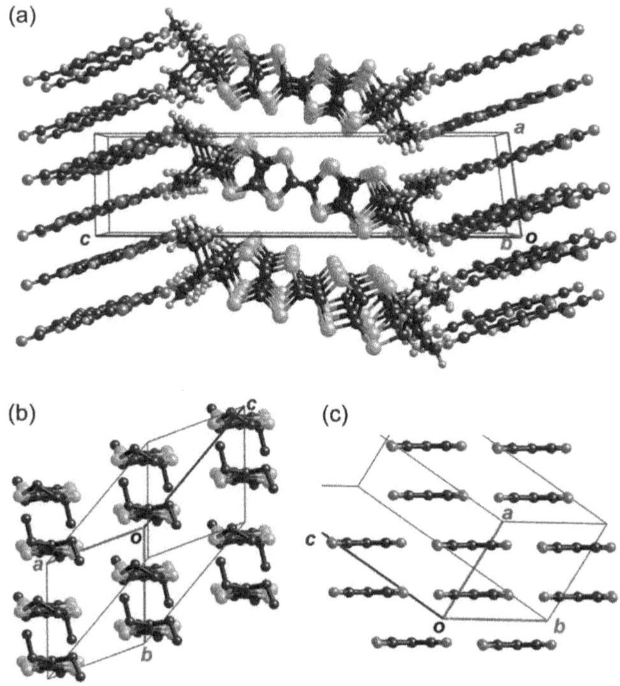

Figure 5.41 (a) Crystal structure of (EtDTET)(TCNQ) viewed along the crystallographic *b* axis. (b) Molecular arrangement of the EtDTET molecules viewed along the molecular long axis. (c) Molecular arrangement of the TCNQ molecules viewed along the molecular long axis.

The dimerization of the TCNQ molecules and the molecular valence of −0.5 per one TCNQ molecule result in a half-filled band structure derived from the LUMO band of TCNQ. Therefore, the TCNQ layer is to be in a Mott insulating state. On the other hand, the EtDTET molecules form a β-type two-dimensional molecular arrangement similar to the arrangements of the two-dimensional metals based on the DT-TTF derivative. We have concluded that the EtDTET layer can contribute electrical conductivity, including superconductivity, and the TCNQ layer can bring about magnetism. (EtDTET)(TCNQ) has a possibility as a magnetic conductor [133, 226, 227]. Further investigation into a mutual relation between an itinerant π-electron in the EtDTET layer and localized π-spin in the TCNQ layer is currently underway in our group.

5.4 Summary and Outlook

From this review, it is clear that superconducting materials can be developed using organic components only. Superconductivity of the neutral molecule was observed in the iodine-containing molecules iodanil (**16**) and hexaiodobenzene (**17**) under high pressure. However, metallic conductivity and superconductivity hardly occur under ambient pressure in neutral organic molecules, whose strategy is to reduce the HOMO-LUMO gap (ΔE) or to enhance the intermolecular interactions. If purely organic molecules satisfy both strategies, like transition metal complexes with TTF-dithiolato ligands (**2** and **3**), superconductivity of neutral organic molecule under ambient pressure will be achieved in the future. High T_C superconductivity at 203 K under an extremely high pressure of 155 GPa has been reported for hydrogen sulfide H_3S, which formed from H_2S by decomposition due to high pressure [228]. A high pressure could provide new superconductors based on neutral organic molecules; however, molecular decomposition can be a problem.

CT-type purely organic superconductors, β''-(BEDT-TTF)$_2$$SF_5CH_2CF_2SO_3$, κ-(BEDT-TTF)$_2$$CF_3SO_3$, (BETS)$_2$($Cl_2$TCNQ), and (EtDTET)(TCNQ), were successfully developed. The strategy of these organic superconductors is simple, namely, they consist of a combination of an organic electron donor and an organic monoanion or an electron acceptor. Closely related materials of these superconductors were also yielded, such as β''-(BEDT-TTF)$_2$$SF_5CHFCF_2SO_3$, β''-(BEDT-TTF)$_2$$SF_5CH_2SO_3$, β''-(BEDT-TTF)$_2$$SF_5CHFSO_3$, and (BETS)$_2$($Br_2$TCNQ) [214, 229–232]. However, the signature of superconductivity has not been detected in the related materials thus far. These results indicate the difficulty in producing purely organic superconductors. Recently, brand-new molecular conductors have been reported. Single-component organic units connected by hydrogen bonding exhibited high electrical conductivity and metallic behavior under pressure [233]. Metallic behavior down to 4 K under ambient conditions was probed in ammonium salts of the TTP donor bearing a carboxylate group using solid-state ^{1}H nuclear magnetic resonance (NMR) measurement [234]. These brand-new conductors will become candidates for purely organic superconductors.

Multicomponent organic superconductors, such as $(BEDT\text{-}TTF)_2M(CF_3)_4$(solvent) (M = Cu, Ag, and Au; solvent = 1,1,2-trihaloethane), $(BEDT\text{-}TTF)_n[AM(C_2O_4)_3]$·Solvent (A = H_3O^+ and K^+; M = Cr and Fe; solvent = benzonitrile, nitrobenzene, and bromobenzene), and $(DMEDO\text{-}TTF)_2[Au(CN)_4]$(solvent) (solvent = cyclic ether) have been developed, and their T_C are tunable by varying the solvent of crystallization [87, 133, 235–237]. Their related materials composed of all-organic molecules will also become candidates for new, purely organic superconductors. Especially, the role for electrical conduction, the role for magnetic properties, and the role for the modulation of transfer integrals could be shared by the electron donors, the electron acceptors, and the solvent molecules, respectively, in multicomponent CT complexes. This review concludes that the diversity of the component molecules, including their combinations, is advantageous to developing superconductors and that purely organic superconductors should provide multifunctional materials.

Acknowledgments

Several parts of our works were conducted with the help of the Grant-in-Aid for Research Promotion and the Collaborative Research Grant at the Graduate School of Science and Engineering and the Research Unit for Power Generation and Storage Materials (PGeS) at Ehime University. The works were also supported by JSPS (KAKENHI) Grant Numbers JP23550155, JP26410095, and JP15H00948 and by MEXT (KAKENHI) Grant Number JP15H03798. We thank the Integrated Center for Sciences (INCS), Ehime University, for technical assistance. We also appreciate the contribution from Prof. T. Mori and Dr. T. Kawamoto in the Tokyo Institute of Technology and from Dr. R. Kato and Dr. H.-B. Cui in RIKEN.

References

1. Tanaka, H., Okano, Y., Kobayashi, H., Suzuki, W. and Kobayashi, A. (2001). A three-dimensional synthetic metallic crystal composed of single-component molecules, *Science*, **291**, pp. 285–287.
2. Kobayashi, A., Fujiwara, E. and Kobayashi, H. (2004). Single-component molecular metals with extended-TTF dithiolate ligands, *Chem. Rev.*, **104**, pp. 5243–5264.
3. Nakano, M., Kuroda, A., Maikawa, T. and Matsubayashi, G. (1996). Synthesis and electrical conductivities of some metal complexes with the extended dithiolato ligand having a C_8S_8 skeleton, *Mol. Cryst. Liq. Cryst.*, **284**, pp. 301–305.
4. Ueda, K., Goto, M., Iwamatsu, M., Sugimoto, T., Endo, S., Toyota, N., Yamamoto, K. and Fujita, H. (1998). Highly-conducting neutral copper complexes substituted with two tetrathiafulvalenyldithiolato groups, *J. Mater. Chem.*, **8**, pp. 2195–2198.
5. Ueda, K., Kamata, Y., Iwamatsu, M., Sugimoto, T. and Fujita, H. (1999). A neutral bis(diethylthiotetrathiafulvalenyldiselenolato)nickel complex: a superior organic conductor to the corresponding dithiolato derivative, *J. Mater. Chem.*, **9**, pp. 2979–2983.
6. Misaki, Y., Tani, Y., Taniguchi, M., Maitani, T., Tanaka, K. and Bechgaard, K. (2000). Preparation and properties of gold complexes with TTF dithiolato ligands, *Mol. Cryst. Liq. Cryst.*, **343**, pp. 59–64.
7. Sasa, M., Fujiwara, E., Kobayashi, A., Ishibashi, S., Terakura, K., Okano, Y., Fujiwara, H. and Kobayashi, H. (2005). Crystal structures and physical properties of single-component molecular conductors consisting of nickel and gold complexes with bis(trifluoromethyl) tetrathiafulvalenedithiolate ligands, *J. Mater. Chem.*, **15**, pp. 155–163.
8. Llusar, R., Triguero, S., Polo, V., Vicent, C., Gómez-García, C. J., Jeannin, O. and Fourmigué, M. (2008). Trinuclear Mo_3S_7 clusters coordinated to dithiolate or diselenolate ligands and their use in the preparation of magnetic single component molecular conductors, *Inorg. Chem.*, **47**, pp. 9400–9409.
9. Fujiwara, E., Zhou, B. Kobayashi, A., Kobayashi, H., Fujishiro, Y., Nishibori, E., Sakata, M., Ishibashi, S. and Terakura, K. (2009). Structures and physical properties of highly conducting single-component molecular conductors containing Se atoms, *Eur. J. Inorg. Chem.*, **2009**, pp. 1585–1591.
10. Tenn, N., Bellec, N., Jeannin, O., Piekara-Sady, L., Auban-Senzier, P., Íñiguez, J., Canadell, E. and Lorcy, D. (2009). A single-component

molecular metal based on a thiazole dithiolate gold complex, *J. Am. Chem. Soc.*, **131**, pp. 16961–16967.

11. Zhou, B., Yajima, H., Kobayashi, A., Okano, Y., Tanaka, H., Kumashiro, T., Nishibori, E., Sawa, H. and Kobayashi, H. (2010). Single-component molecular conductor [Cu(tmdt)$_2$] containing an antiferromagnetic Heisenberg chain, *Inorg. Chem.*, **49**, pp. 6740–6747.

12. Yzambart, G., Bellec, N., Nasser, G., Jeannin, O., Roisnel, T. Fourmigué, M., Auban-Senzier, Íñiguez, J., Canadell, E. and Lorcy, D. (2012). Anisotropic chemical pressure effects in single-component molecular metals based on radical dithiolene and diselenolene gold complexes, *J. Am. Chem. Soc.*, **134**, pp. 17138–17148.

13. Gushchin, A. L., Llusar, R., Vicent, C., Abramov, P. A. and Gómez-Garcia, C. J. (2013). Mo_3Q_7 (Q = S, Se) clusters containing dithiolate/diselenolate ligands: synthesis, structures, and their useas precursors of magnetic single-component molecular conductors, *Eur. J. Inorg. Chem.*, **2013**, pp. 2615–2622.

14. Tanaka, H., Tokumoto, M., Ishibashi, S. Graf, D., Choi, E. S., Brooks, J. S., Yasuzuka, S., Okano, Y., Kobayashi, H. and Kobayashi, A. (2004). Observation of three-dimensional Fermi surfaces in a single-component molecular metal, [Ni(tmdt)$_2$], *J. Am. Chem. Soc.*, **126**, pp. 10518–10519.

15. Cui, H.-B., Kobayashi, H., Ishibashi, S., Sasa, M., Iwase, F., Kato, R. and Kobayashi, A. (2014). A single-component molecular superconductor, *J. Am. Chem. Soc.*, **136**, pp. 7619–7622.

16. Dumur, F., Gautier, N., Gallego-Planas, N., Şahin, Y., Levillain, E., Mercier, N. and Hudhomme, P. (2004). Novel fused D-A dyad and A-D-A triad incorporating tetrathiafulvalene and *p*-benzoquinone, *J. Org. Chem.*, **69**, pp. 2164–2177.

17. Dumur, F., Guégano, X., Gautier, N., Liu, S.-X., Neels, A., Decurtins, S. and Hudhomme, P. (2009). Approaches to fused tetrathiafulvalene/tetracyanoquinodimethane systems, *Eur. J. Org. Chem.*, **2009**, pp. 6341–6354.

18. Perepichka, D. F., Bryce, M. R., Pearson, C., Petty, M. C., McInnes, E. J. L. and Zhao, J. P. (2003). A covalent tetrathiafulvalene–tetracyanoquinodimethane diad: extremely low HOMO–LUMO gap, thermoexcited electron transfer, and high-quality Langmuir–Blodgett films, *Angew. Chem. Int. Ed.*, **42**, pp. 4636–4639.

19. Nakamura, K.-i., Shirahata, T., Miyamoto, H. and Misaki, Y. (2011). Synthesis and properties of a new donor-acceptor diad composed of

DT-TTF and dicyanomethylidene group, *Heterocycles*, **83**, pp. 2115–2126.

20. Yamashita, Y., Tanaka, S., Imaeda K. and Inokuchi, H. (1991). Tetrathio-derivatives of *p*-quinodimethanes fused with 1,2,5-thiadiazoles. A novel type of organic semiconductors, *Chem. Lett.*, **20**, pp. 1213–1216.
21. Yamashita, Y., Tanaka, S., Imaeda, K., Inokuchi, H. and Sano, M. (1992). Preparation and properties of bis[1,2,5]thiadiazolo-*p*-quinobis(1,3-dithiole) (BTQBT) and its derivatives. Novel organic semiconductors, *J. Org. Chem.*, **57**, pp. 5517–5522.
22. Katayama, C., Honda, M., Kumagai, H., Tanaka, J., Saito, G. and Inokuchi, H. (1985). Crystal structures of complexes between hexacyanobutadiene and tetramethyltetrathiafulvalene and tetramethylthiotetrathiafulvalene, *Bull. Chem. Soc. Jpn.*, **58**, pp. 2272–2278.
23. Wu, P., Saito, G., Imaeda, K., Shi, Z., Mori, T., Enoki, T. and Inokuchi, H. (1986). Uncapped alkylthio substituted tetrathiafulvalens (TTC_n-TTF) and their charge transfer complexes, *Chem. Lett.*, **15**, pp. 441–444.
24. Inokuchi, H., Saito, G., Wu, P., Seki, K., Tang, T. B., Mori, T., Imaeda, K., Enoki, T., Higuchi, Y., Inaka, K. and Yasuoka, N. (1986). A novel type of organic semiconductors. Molecular fastener, *Chem. Lett.*, **15**, pp. 1263–1266.
25. Seki, K., Tang, T. B., Mori, T., Ji, W. P., Saito, G. and Inokuchi, H. (1986). Valence electronic structures of tetrakis(alkylthio)tetrathiafulvalenes, *J. Chem. Soc., Faraday Trans. 2*, **82**, pp. 1067–1075.
26. Okada, N., Yamochi, H., Shinozaki, F., Oshima, K. and Saito, G. (1986). Tetrakis(n-alkyltelluro)tetrathiafulvalene ($TTeC_n$-TTF), *Chem. Lett.*, **15**, pp. 1861–1864.
27. Imaeda, K., Enoki, T., Shi, Z., Wu, P., Okada, N., Yamochi, H., Saito, G. and Inokuchi, H. (1987). Electrical conductivities of tetrakis(alkylthio) tetrathiafulvalene (TTC_n–TTF) and tetrakis(alkyltelluro) tetrathiafulvalene ($TTeC_n$–TTF), *Bull. Chem. Soc. Jpn.*, **60**, pp. 3163–3167.
28. Yamochi, H., Iwasawa, N., Urayama, H. and Saito, G. (1987). Synthesis and properties of tetrakis(alkylseleno)tetrathiafulvalene, *Chem. Lett.*, **16**, pp. 2265–2268.
29. Shi, Z., Enoki, T., Imaeda, K., Seki, K., Wu, P., Inokuchi, H. and Saito, G. (1988). Thermal properties of tetrakis(alkylthio)tetrathiafulvalenes, *J. Phys. Chem.*, **92**, pp. 5044–5048.

30. Wang, P., Enoki, T., Imaeda, K., Iwasawa, N., Yamochi, H., Urayama, H., Saito, G. and Inokuchi, H. (1989). Thermal and electrical properties of tetrakis(alkylseleno)tetrathiafulvalenes ($TSeC_n$-TTF, n = 1-18), *J. Phys. Chem.*, **93**, pp. 5947–5952.

31. Li, Y. F., Nakano, C., Imaeda, K., Inokuchi, H., Maruyama, Y., Iwasawa, N. and Saito, G. (1990). Charge-carrier drift mobilities and phase transition in tetrakis(octylthio)tetrathiafulvalene, TTC_8–TTF, crystal, *Bull. Chem. Soc. Jpn.*, **63**, pp. 1857–1859.

32. Otsuka, A., Saito, G., Ohfuchi, K. and Konno, M. (1992). C_1TET-TTF and its related compounds as single component molecular conductors, *Phosphorus, Sulfur Silicon Relat. Elem.*, **67**, pp. 333–338.

33. Misaki, Y., Kawakami, K., Higuchi, N., Nishikawa, H., Miura, T. and Yamabe, T. (1996). Synthesis and properties of higher homologues of bis-fused TTF, *Mol. Cryst. Liq. Cryst.*, **284**, pp. 337–344.

34. Kimura, S., Kurai, H. and Mori, T. (2002). 2,5-bis(1,3-dithiol-2-ylidene)-1,3,4,6-tetrathiapentalene (TTP) derivatives having four long alkylthio chains, *Tetrahedron*, **58**, pp. 1119–1124.

35. Ashizawa, M., Kimura, S., Mori, T., Misaki, Y. and Tanaka, K. (2004). Tris-fused tetrathiafulvalenes (TTF): highly conducting single-component organics and metallic charge-transfer salt, *Synth. Met.*, **141**, pp. 307–313.

36. Shirotani, I., Kamura, Y., Inokuchi, H. and Hirooka, T. (1976). Electrical conductivity, photoemission and absorption spectrum of tetraselenonaphthacene, $C_{18}H_8Se_4$, *Chem. Phys. Lett.*, **40**, pp. 257–260.

37. Onodera, A., Shirotani, I., Inokuchi, H. and Kawai, N. (1974). Conductive tetraselenonaphthacene and iodanil under very high pressure, *Chem. Phys. Lett.*, **25**, pp. 296–298.

38. Saito, G. and Yoshida, Y. (2007). Development of conductive organic molecular assemblies: organic metals, superconductors, and exotic functional materials, *Bull. Chem. Soc. Jpn.*, **80**, pp. 1–137.

39. Carroll, P. J., Lakshmikantham, M. V., Cava, M. P., Wudl, F., Aharon-Shalom, E. C. and Sherman, D. (1982). The crystal and molecular structure of bi(1,2,4,5,6-pentahydro-1,3-dutekkyrapentalen-2-ylidene)(hexamethylenetetratellurafulvalene): the shortest observed intermolecular Te–Te contact, *J. Chem. Soc., Chem. Commun.*, pp. 1316–1318.

40. Matsuzaki, S., Okumura, H., Takenouchi, H., Kyouda, T. and Sano, M. (1991). Electrical and optical properties of hexamethylenetetratellurafulvalene crystals and their pressure dependence, *Synth. Met.*, **39**, pp. 385–395.

41. Yokota, T., Takeshita, N., Shimizu, K., Amaya, K., Onodera, A., Shirotani, I. and Endo, S. (1996). Pressure-induced superconductivity of iodanil, *Czech. J. Phys.*, **46**(suppl S2), pp. 817–818.

42. Amaya, K., Shimizu, K., Eremets, M. I., Kobayashi, T. C. and Endo, S. (1998). Observation of pressure-induced superconductivity in the megabar region, *J. Phys.: Condens. Matter*, **10**, pp. 11179–11190.

43. Shirotani, I., Hayashi, J., Yakushi, K., Takeda, K., Yokota, T., Shimizu, K., Amaya, K., Nakayama, A. and Aoki, K. (2001). Pressure-induced insulator-to-metal transition and superconductivity in iodanil, $C_6I_4O_2$, *Physica B*, **304**, pp. 6–11.

44. Iwasaki, E., Shimizu, K., Amaya, K., Nakayama, A., Aoki, K. and Carlón, R. P. (2001). Metallization and superconductivity in hexaiodobenzene under high pressure, *Synth. Met.*, **120**, pp. 1003–1004.

45. Mott, N. F. (1990). *Metal-Insulator Transitions*, 2nd Ed., Taylor and Francis, London.

46. Nishida, S., Morita, Y., Fukui, K., Sato., K., Shiomi, D., Takui, T. and Nakasuji, K. (2005). Spin transfer and solvato-/thermochromism induced by intramolecular electron transfer in a purely organic open-shell system, *Angew. Chem. Int. Ed.*, **44**, pp. 7277–7280.

47. Guasch, J., Grisanti, L., Lloveras, V., Vidal-Gancedo, J., Souto, M., Morales, D. C., Vilaseca, M., Sissa, C., Painelli, A., Ratera, I., Rovira, C. and Veciana, J. (2012). Induced self-assembly of a tetrathiafulvalene-based open-shell dyad through intramolecular electron transfer, *Angew. Chem. Int. Ed.*, **51**, pp. 11024–11028.

48. Souto, M., Guasch, J., Lloveras, V., Mayorga, P., Navarrete, J. T. L., Casado, J., Ratera, I., Rovira, C., Painelli, A. and Veciana, J. (2013). Thermomagnetic molecular system based on TTF-PTM radical: switching the spin and charge delocalization, *J. Phys. Chem. Lett.*, **4**, pp. 2721–2726.

49. Guasch, J., Grisanti, L., Souto, M., Lloveras, V., Vidal-Gancedo, J., Ratera, I., Painelli, A., Rovira, C. and Veciana, J. (2013). Intra- and intermolecular charge transfer in aggregates of tetrathiafulvalene-triphenylmethyl radical derivatives in solution, *J. Am. Chem. Soc.*, **135**, pp. 6958–6967.

50. Suzuki, T., Yamada, M., Ohkita, M. and Tsuji, T. (2001). Generation and amphoteric redox properties of novel neutral radicals with the TTF-TCNQ hybrid structure, *Heterocycles*, **54**, pp. 387–394.

51. Tsubata, Y., Suzuki, T., Miyashi, T. and Yamashita, Y. (1992). Single-component organic conductors based on neutral radicals containing the pyrazino-TCNQ skeleton, *J. Org. Chem.*, **57**, pp. 6749–6755.

52. Suzuki, T., Miyanari, S., Kawai, H., Fujiwara, K., Fukushima, T., Miyashi, T. and Yamashita, Y. (2004). Pyrazino-tetracyanonaphthoquinodimethanes: sterically deformed electron acceptors affording zwitterionic radicals, *Tetrahedron*, **60**, pp. 1997–2003.

53. Neilands, O., Tilika, V., Sudmale, I., Grigorjeva, I., Edzina, A., Fonavs, E. and Muzikante, I. (1997). Dioxo- and aminooxopyrimidotetrathiafulvalenes: π-electron donors for design of conducting materials containing intramolecular hydrogen bonds of nucleic acid base pair type, *Adv. Mater. Opt. Electron.*, **7**, pp. 39–43.

54. Neilands, O. (2001). Dioxo-and aminooxopyrimido-fused tetrathiafulvalenes–base compounds for novel organic semiconductors and for design of sensors for recognition of nucleic acid components, *Mol. Cryst. Liq. Cryst.*, **355**, pp. 331–349.

55. Balodis, K., Khasanov, S., Chong, C., Maesato, M., Yamochi, H., Saito, G. and Neilands, O. (2003). Single component betainic conductor: pyrimido-fused TTF derivatives having ethylenedioxy group, *Synth. Met.*, **133–134**, pp. 353–355.

56. Saito, G., Balodis, K., Yoshida, Y., Maesato, M., Yamochi, H., Khasanov, S. and Murata, T. (2007). Mesomeric fused betainic radicals as organic conductors, in *Multifunctional Conducting Molecular Materials*, Saito, G., Wudl, F., Haddon, R. C., Tanigaki, K., Enoki, T., Katz, H. E. and Maesato, M. (eds.), RSC Publishing, Cambridge, pp. 3–14.

57. Chi, X., Itkis, M. E., Patrick, B. O., Barclay, T. M., Reed, R. W., Oakley, R. T., Cordes, A. W. and Haddon, R. C. (1999). The first phenalenyl-based neutral radical molecular conductor, *J. Am. Chem. Soc.*, **121**, pp. 10395–10402.

58. Chi, X., Itkis, M. E., Kirschbaum, K., Pinkerton, A. A., Oakley, R. T., Cordes, A. W. and Haddon, R. C. (2001). Dimeric phenalenyl-based neutral radical molecular conductors, *J. Am. Chem. Soc.*, **123**, pp. 4041–4048.

59. Itkis, M. E., Chi, X., Cordes, A. W. and Haddon, R. C. (2002). Magneto-opto-electronic bistability in a phenalenyl-based neutral radical, *Science*, **296**, pp. 1443–1445.

60. Mandal, S. K., Itkis, M. E., Chi, X., Samanta, S., Lidsky, D., Reed, R. W., Oakley, R. T., Tham, F. S. and Haddon, R. C. (2005). New family of aminophenalenyl-based neutral radical molecular conductors: synthesis, structure, and solid state properties, *J. Am. Chem. Soc.*, **127**, pp. 8185–8196.

61. Pal, S. K., Itkis, M. E., Tham, F. S., Reed, R. W., Oakley, R. T. and Haddon, R. C. (2005). Resonating valence-bond ground state in a phenalenyl-based neutral radical conductor, *Science*, **309**, pp. 281–284.

62. Mandal, S. K., Samanta, S., Itkis, M. E., Jensen, D. W., Reed, R. W., Oakley, R. T., Tham, F. S., Donnadieu, B. and Haddon, R. C. (2006). Resonating valence bond ground state in oxygen-functionalized phenalenyl-based neutral radical molecular conductors, *J. Am. Chem. Soc.*, **128**, pp. 1982–1994.

63. Pal, S. K., Itkis, M. E., Tham, F. S., Reed, R. W., Oakley, R. T. and Haddon, R. C. (2008). Trisphenalenyl-based neutral radical molecular conductor, *J. Am. Chem. Soc.*, **130**, pp. 3942–3951.

64. Barclay, T. M., Cordes, A. W., Haddon, R. C., Itkis, M. E., Oakley, R. T., Reed, R. W. and Zhang, H. (1999). Preparation and characterization of a neutral π-radical molecular conductor, *J. Am. Chem. Soc.*, **121**, pp. 969–976.

65. Robertson, C. M., Leitch, A. A., Cvrkalj, K., Reed, R. W., Myles, D. J. T., Dube, P. A. and Oakley, R. T. (2008). Enhanced conductivity and magnetic ordering in isostructural heavy atom radicals, *J. Am. Chem. Soc.*, **130**, pp. 8414–8425.

66. Yu, X., Mailman, A., Lekin, K., Assoud, A., Robertson, C. M., Noll, B. C., Campana, C. F., Howard, J. A. K., Dube, P. A. and Oakley, R. T. (2012). Semiquinone-bridged bisdithiazolyl radicals as neutral radical conductors, *J. Am. Chem. Soc.*, **134** pp. 2264–2275.

67. Wolmershäuser, G. and Johann, R. (1989). 1,3,5-trithia-2,4,6-triazapentalenyl—a stable sulfur-nitrogen radical, *Angew. Chem. Int. Ed. Engl.*, **28**, pp. 920–921.

68. Bryan, C. D., Cordes, A. W., Haddon, R. C., Hicks, R. G., Oakley, R. T., Palstra, T. T. M. and Perel, A. J. (1994). Preparation and solid state characterization of 4,4′-bis (1,2,3,5-dithiadiazolyl), *J. Chem. Soc., Chem. Commun.*, pp. 1447–1448.

69. Wolmershäuser, G., Schnauber, M. and Wilhelm, T. (1984). Highly conducting charge transfer complexes of benzo-1,3,2-dithiazol-2yl and its derivatives with teracyanoquinodimethane, *J. Chem. Soc., Chem. Commun.*, pp. 573–574.

70. Bryan, C. D., Cordes, A. W., Goddard, J. D., Haddon, R. C., Hicks, R. G., MacKinnon, C. D., Mawhinney, R. C., Oakley, R. T., Palstra, T. T. M. and Perel, A. S. (1996). Preparation and characterization of the disjoint diradical 4,4′-bis(1,2,3,5-dithiadiazolyl)[S_2N_2C-CN_2S_2] and its iodine charge transfer salt [S_2N_2C-CN_2S_2][I], *J. Am. Chem. Soc.*, **118**, pp. 330–338.

71. Leitch, A. A., Lekin, K., Winter, S. M., Downie, L. E., Tsuruda, H., Tse, J. S., Mito, M., Desgreniers, S., Dube, P. A., Zhang, S., Liu, Q., Jin, C., Ohishi, Y. and Oakley, R. T. (2011). From magnets to metals: the response of

tetragonal bisdiselenazolyl radicals to pressure, *J. Am. Chem. Soc.*, **133**, pp. 6051–6060.

72. Awere, E. G., Burford, N., Haddon, R. C., Parsons, S., Passmore, J., Waszczak, J. V. and White, P. S. (1990). X-ray crystal structures of the 1, 3, 2-benzodithiazolyl dimer and 1, 3, 2-benzodithiazolium chloride sulfur dioxide solvate: comparison of the molecular and electronic structures of the 10-π-electron $C_6H_4S_2N^+$ cation and the $C_6H_4S_2N^{\bullet}$ radical and dimer and a study of the variable-temperature magnetic behavior of the radical, *Inorg. Chem.*, **29**, pp. 4821–4830.

73. Fujita, W. and Awaga, K. (1999). Room-temperature magnetic bistability in organic radical crystals, *Science*, **286**, pp. 261–262.

74. Fujita, W., Awaga, K., Matsuzaki, H. and Okamoto, H. (2002). Room-temperature magnetic bistability in organic radical crystals: paramagnetic-diamagnetic phase transition in 1,3,5-trithia-2,4,6-triazapentalenyl, *Phys. Rev. B*, **65**, pp. 064434 (9 pages).

75. Matsuzaki, H., Fujita, W., Awaga, K. and Okamoto, H. (2003). Photoinduced phase transition in an organic radical crystal with room-temperature optical and magnetic bistability, *Phys. Rev. Lett.*, **91**, pp. 017403 (4 pages).

76. Awaga, K., Tanaka, T., Shirai, T., Fujimori, M., Suzuki, Y., Yoshikawa, H. and Fujita, W. (2006). Multi-dimensional crystal structures and unique solid-state properties of heterocyclic thiazyl radicals and related materials, *Bull. Chem. Soc. Jpn.*, **79**, pp. 25–34.

77. Vela, S., Mota, F., Deumal, M., Suizu, R., Shuku, Y., Mizuno, A., Awaga, K., Shiga, M., Novoa J. J. and Ribas-Arino, J. (2014). The key role of vibrational entropy in the phase transitions of dithiazolyl-based bistable magnetic materials, *Nat. Commun.*, **5**, pp. 4411 (9 pages).

78. Suizu, R., Iwasaki, A., Shuku, Y. and Awaga, K. (2015). Spatially inhomogeneous, stepwise phase transitions in a thiazyl diradical: a structural mismatch induced by lattice transformation, *J. Mater. Chem. C*, **3**, pp. 7968–7977.

79. Iwasaki, A., Hu, L., Suizu, R., Nomura, K., Yoshikawa, H., Awaga, K., Noda, Y., Kanai, K., Ouchi, Y., Seki, K. and Ito, H. (2009). Interactive radical dimers in photoconductive organic thin films, *Angew. Chem. Int. Ed.*, **48**, pp. 4022–4024.

80. Yamamoto, M., Suizu, R., Dutta, S., Mishra, P., Nakayama, T., Sakamoto, K., Wakabayashi, K., Uchihashi, T. and Awaga, K. (2015). Self-assembled honeycomb lattice in the monolayer of cyclic thiazyl diradical BDTDA (= 4,4′-bis(1,2,3,5-dithiadiazolyl)) on Cu (111) with a zero-bias tunneling spectra anomaly, *Sci. Rep.*, **5**, pp. 18359 (7 pages).

81. Fujita, W., Awaga, K., Nakazawa, Y., Saito, K. and Sorai, M. (2002). Complex phase transitions in stable thiazyl radicals: spin-gap, antiferromagnetic ordering and double melting, *Chem. Phys. Lett.*, **352**, pp. 348–352.

82. Wudl, F., Smith, G. M. and Hufnagel, E. J. (1970). Bis-1,3-dithiolium chloride: an unusually stable organic radical cation, *J. Chem. Soc. D, Chem. Commun.*, pp. 1453–1454.

83. Acker, D. S. and Hertler, W. R. (1962). Substituted quinodimethans. I. Preparation and chemistry of 7,7,8,8-tetracyanoquinodimethan, *J. Am. Chem. Soc.*, **84**, pp. 3370–3374.

84. Ferraris, J. P., Cowan, D. O., Walatka, V. and Perstein, J. H. (1973). Electron transfer in a new highly conducting donor-acceptor complex, *J. Am. Chem. Soc.*, **95**, pp. 948–949.

85. Jérome, D., Mazaud, A., Ribault, M. and Bechgaard, K. (1980). Superconductivity in a synthetic organic conductor $(TMTSF)_2PF_6$, *J. Phys. Lett.*, **41**, pp. L95–L98.

86. Geiser, U., Schlueter, J. A., Wang, H. H., Kini, A. M., Williams, J. M., Sche, P. P., Zakowicz, H. I., VanZile, M. L., Dudek, J. D., Nixon, P. G., Winter, R. W., Gard, G. L., Ren, J. and Whangbo, M.-H. (1996). Superconductivity at 5.2 K in an electron donor radical salt of bis(ethylenedithio) tetrathiafulvalene (BEDT-TTF) with the novel polyfluorinated organic anion $SF_5CH_2CF_2SO_3^-$, *J. Am. Chem. Soc.*, **118**, pp. 9996–9997.

87. Geiser, U. and Schlueter, J. A. (2004). Conducting organic radical cation salts with organic and organometallic anions, *Chem. Rev.*, **104**, pp. 5203–5241.

88. Ito, H., Asai, T., Shimizu, Y., Hayama, H., Yoshida, Y. and Saito, G. (2016). Pressure-induced superconductivity in the antiferromagnet κ-$(ET)_2CF_3SO_3$ with quasi-one-dimensional triangular spin lattice, *Phys. Rev. B*, **94**, pp. 020503 (5 pages).

89. Kondo, R., Hasegawa, T., Mochida, T., Kagoshima, S. and Iwasa, Y. (1999). Donor-acceptor type superconductor, $(BETS)_2(Cl_2TCNQ)$, *Chem. Lett.*, **28**, pp. 333–334.

90. Shirahata, T., Furuta, K., Hibara, Y., Misaki, Y., Kawamoto, T. and Mori, T. (14–19 July 2013). *Synthesis of New Organic Superconductor Based on Purely Organic Donor-Acceptor System*, paper presented at the 10th International Symposium on Crystalline Organic Metals Superconductors and Ferromagnets (ISCOM2013), Montréal.

91. Segura, J. L. and Martín, N. (2001). New concepts in tetrathiafulvalene chemistry, *Angew. Chem. Int. Ed.*, **40**, pp. 1372–1409.

92. Bryce, M. R. (1999). Tetrathiafulvalenes as π-electron donors for intramolecular charge-transfer materials, *Adv. Mater.*, **11**, pp. 11–23.

93. Bryce, M. R. (2000). Functionalised tetrathiafulvalenes: new applications as versatile π-electron systems in materials chemistry, *J. Mater. Chem.*, **10**, pp. 589–598.

94. Yamada, J. and Sugimoto, T. (2004). *TTF Chemistry: Fundamentals and Applications of Tetrathiafulvalene*, 1st Ed., Kodansha Springer, Tokyo.

95. Aonuma, S., Okano, Y., Sawa, H., Kato, R. and Kobayashi, H. (1992). Stable molecular metals based on a novel unsymmetrical diselenadithiafulvalene, *J. Chem. Soc., Chem. Commun.*, pp. 1193–1195.

96. Misaki, Y., Nishikawa, H., Fujiwara, H., Kawakami, K., Yamabe, T., Yamochi, H. and Saito, G. (1992). (2-Methylidene-1,3-dithiolo[4,5-*d*]) tetrathiafulvalene (DT-TTF): new unsymmetrical TTFs condensed with 1,3-dithiol-2-ylidene moieties, *J. Chem. Soc., Chem. Commun.*, pp. 1408–1409.

97. Misaki, Y., Kawakami, K., Nishikawa, H., Fujiwara, H., Yamabe, T. and Shiro, M. (1993). Preparation, crystal structure and electrical properties of 2-cyclopentanylidene-1,3-dithiolo[4,5-*d*]tetrathiafulvalene derivatives, *Chem. Lett.*, **22**, pp. 445–448.

98. Misaki, Y., Nishikawa, H., Yamabe, T., Mori, T., Inokuchi, H., Mori, H. and Tanaka, S. (1993). Structure and electrical properties of MeDTET salts, *Chem. Lett.*, **22**, pp. 1341–1344.

99. Nishikawa, H., Kawakami, K., Fujiwara, H., Uehara, T., Misaki, Y., Yamabe, T., Mori, T. and Shiro, M. (1993). Synthesis and properties of new tetrathiafulvalenes condensed with 1,3-dithiol-2-ylidenes, *Synth. Met.*, **56**, pp. 1983–1988.

100. Fujiwara, H., Miura, T., Misaki, Y., Yamabe, T., Mori, T., Mori H. and Tanaka, S. (1996). Electrical properties of DT-TTF salts, *Mol. Cryst. Liq. Cryst.*, **284**, pp. 329–336.

101. Misaki, Y., Taniguchi, M., Miura, T., Fujiwara, H., Yamabe, T., Kawamoto, T. and Mori, T. (1997). Structures and properties of MeDTDM salts, *Adv. Mater.*, **9**, pp. 633–635.

102. Fujiwara, H., Misaki, Y., Taniguchi, M., Yamabe, T., Kawamoto, T., Mori, T., Mori H. and Tanaka, S. (1998). Preparation, structures and physical properties of κ-type two-dimensional conductors based on unsymmetrical extended tetrathiafulvalene: 2-cyclopentanylidene-1,3-dithiolo[4,5-*d*]-4,5-ethylenedithiotetrathiafulvalene (CPDTET), *J. Mater. Chem.*, **8**, pp. 1711–1717.

103. Bondi, A. (1964). van der Waals volumes and radii, *J. Phys. Chem.*, **68**, pp. 441–451.

104. Hasegawa, S., Tanaka, S., Yamashita, Y., Inokuchi, H., Fujimoto, H., Kamiya, K., Seki, K. and Ueno, N. (1993). Molecular orientation in thin films of bis(1,2,5-thiadiazolo)-*p*-quinobis(1,3-dithiole) on graphite studied by angle-resolved photoelectron spectroscopy, *Phys. Rev. B*, **48**, pp. 2596–2600.

105. Misaki, Y., Matsui, T., Kawakami, K., Nishikawa, H., Yamabe, T. and Shiro, M. (1993). 2,5-bis(1′,3′-dithiol-2′-ylidene)-1,3,4,6-tetrathiapentalene and its related unsymmetrical donors, *Chem. Lett.*, **22**, pp. 1337–1340.

106. Aust, R. B., Bentley W. H. and Drickamer, H. G. (1964). Behavior of fused–ring aromatic hydrocarbons at very high pressure, *J. Chem. Phys.*, **41**, pp. 1856–1871.

107. Samara, G. A. and Drickamer, H. G. (1962). Effect of pressure on the resistance of fused–ring aromatic compounds, *J. Chem. Phys.*, **37**, pp. 474–478.

108. Shirotani, I., Onodera, A., Kamura, Y., Inokuchi, H. and Kawai, N. (1976). Electrical conductivity of hexaiodobenzene and tetrahalo-*p*-benzoquinone under very high pressures, *J. Solid State Chem.*, **18**, pp. 235–239.

109. Kobayashi, H., Danno, T. and Shirotani, I. (1974). The crystal structure of tetraiodo-*p*-benzoquinone, *Bull. Chem. Soc. Jpn.*, **47**, pp. 2333–2334.

110. Ghosh, S., Reddy C. M. and Desiraju, G. R. (2007). Hexaiodobenzene: a redetermination at 100 K, *Acta Cryst. E*, **63**, pp. o910–o911.

111. Riggleman, B. M. and Drickamer, H. G. (1963). Approach to the metallic state as obtained from optical and electrical measurements, *J. Chem. Phys.*, **38**, pp. 2721–2724.

112. Takemura, K., Minomura, S., Shimomura, O., Fujii, Y. and Axe, J. D. (1982). Structural aspects of solid iodine associated with metallization and molecular dissociation under high pressure, *Phys. Rev. B*, **26**, pp. 998–1004.

113. Shimizu, K., Yamauchi, T., Tamitani, N., Takeshita, N., Ishizuka, M., Amaya, K. and Endo, S. (1994). The pressure-induced superconductivity of iodine, *J. Supercond.*, **7**, pp. 921–924.

114. Nakayama, A., Fujihisa, H., Aoki, K. and Shirotani, I. (1999). Structural change of iodanil under high pressure, *Synth. Met.*, **103**, pp. 1901–1902.

115. Nakayama, A., Aoki, K., Matsushita, Y. and Shirotani, I. (1999). Infrared study of iodanil under very high pressure, *Solid State Commun.*, **110**, pp. 627–632.

116. Nakayama, A., Fujihisa, H., Aoki, K. and Carlón, R. P. (2000). Structural study of hexaiodobenzene up to 9.7 GPa, *Phys. Rev. B*, **62**, pp. 8759–8765.

117. Souto, M., Solano, M. V., Jensen, M., Bendixen, D., Delchiaro, F., Girlando, A., Painelli, A., Jeppesen, J. O., Rovira, C., Ratera, I. and Veciana, J. (2015). Self-assembled architectures with segregated donor and acceptor units of a dyad based on a monopyrrolo-annulated TTF–PTM radical, *Chem. Eur. J.*, **21**, pp. 8816–8825.

118. Souto, M., Cui, H.B., Peña-Álvarez, M., Baonza, V. G., Jeschke, H. O., Tomic, M., Valentí, R., Blasi, D., Ratera, I., Rovira, C. and Veciana, J. (2016). Pressure-induced conductivity in a neutral nonplanar spin-localized radical, *J. Am. Chem. Soc.*, **138**, pp. 11517–11525.

119. Pauling, L. (1948). The metallic state, *Nature*, **161**, pp. 1019–1020.

120. Anderson, P.W. (1973). Resonating valence bonds: a new kind of insulator?, *Mater. Res. Bull.*, **8**, pp. 153–160.

121. Ishiguro, T., Yamaji, K. and Saito, G. (1998). *Organic Superconductors*, 2nd Ed., Springer-Verlag, Heidelberg.

122. Williams, J. M., Ferraro, J. R., Thorn, R. J., Carlson, K. D., Geiser, U., Wang, H. H., Kini, A. M. and Whangbo, M.-H. (1992). *Organic Superconductors (Including Fullerenes) Synthesis, Structure, Properties, and Theory*, 1st Ed., Prentice-Hall, New Jersey.

123. Mori, H. (2006). Materials viewpoint of organic superconductors, *J. Phys. Soc. Jpn.*, **75**, pp. 051003 (15 pages).

124. Jérome, D. (2008). The development of organic conductors: organic superconductors, *Solid State Sci.*, **10**, pp. 1692–1700.

125. Saito, G. and Yoshida, Y. (2011). Organic superconductors, *Chem. Rec.*, **11**, pp. 124–145.

126. Ardavan, A., Brown, S., Kagoshima, S., Kanoda, K., Kuroki, K., Mori, H., Ogata, M., Uji, S. and Wosnitza, J. (2012). Recent topics of organic superconductors, *J. Phys. Soc. Jpn.*, **81**, pp. 011004 (27 pages).

127. Kobayashi, H., Kobayashi, A. and Cassoux, P. (2000). BETS as a source of molecular magnetic superconductors (BETS = bis(ethylenedithio) tetraselenafulvalene), *Chem. Soc. Rev.*, **29**, pp. 325–333.

128. Mori, T. (2004). Organic conductors with unusual band fillings, *Chem. Rev.*, **104**, pp. 4947–4970.

129. Yamada, J.-I., Akutsu, H., Nishikawa, H. and Kikuchi, K., (2004) New trends in the synthesis of π-electron donors for molecular conductors and superconductors, *Chem. Rev.*, **104**, pp. 5057–5084.

130. Kobayashi, H., Cui, H. B. and Kobayashi, A. (2004). Organic metals and superconductors based on BETS (BETS = bis(ethylenedithio) tetraselenafulvalene), *Chem. Rev.*, **104**, pp. 5265–5288.

131. Kato, R. (2004). Conducting metal dithiolene complexes: structural and electronic properties, *Chem. Rev.*, **104**, pp. 5319–5346.

132. Shibaeva, R. P. and Yagubskii, E. B. (2004). Molecular conductors and superconductors based on trihalides of BEDT-TTF and some of its analogues, *Chem. Rev.*, **104**, pp. 5347–5378.

133. Coronado, E. and Day, P. (2004). Magnetic molecular conductors, *Chem. Rev.*, **104**, pp. 5419–5448.

134. Jérome, D. (2004). Organic conductors: from charge density wave TTF–TCNQ to superconducting $(TMTSF)_2PF_6$, *Chem. Rev.*, **104**, pp. 5565–5592.

135. Misaki, Y. (2009). Tetrathiapentalene-based organic conductors, *Sci. Technol. Adv. Mater.*, **10**, pp. 024301 (22 pages).

136. Kawamoto, T. and Takimiya, K. (2009). Organic superconductors with an incommensurate anion structure, *Sci. Technol. Adv. Mater.*, **10**, pp. 024303 (18 pages).

137. Akamatu, H., Inokuchi, H. and Matsunaga, Y. (1954). Electrical conductivity of the perylene–bromine complex, *Nature*, **173**, pp. 168–169.

138. Hertler, W. R., Mahler, W., Melby, L. R., Miller, J. S., Putscher, R. E. and Webster, O. W. (1989). Cyanocarbons—their history from conducting to magnetic organic charge transfer salts, *Mol. Cryst. Liq. Cryst.*, **171**, pp. 205–216.

139. Coffen, D. L. and Garrett, P. E. (1969). Oxidation of thioacetals with DDQ, *Tetrahedron Lett.*, **10**, pp. 2043–2046.

140. Hasegawa, M. and Iyoda, M. (2016). Tetrathiafulvalene: a redox unit for functional materials and a building block for supramolecular self-assembly, in *Organic Redox Systems: Synthesis, Properties, and Applications*, Nishinaga, T. (ed.), Wiley, New Jersey, pp. 89–129.

141. Kagoshima, S., Anzai, H., Kajimura, K. and Ishiguro, T. (1975). Observation of the Kohn anomaly and the Peierls transition in TTF-TCNQ by X-ray scattering, *J. Phys. Soc. Jpn.*, **39**, pp. 1143–1144.

142. Kagoshima, S., Ishiguro, T. and Anzai, H. (1976). X-ray scattering study of phonon anomalies and superstructures in TTF-TCNQ, *J. Phys. Soc. Jpn.*, **41**, pp. 2061–2071.

143. Peierls, R. E. (2001). *Quantum Theory of Solids*, Rev. Ed., Oxford University Press, USA.

144. Toyota, N., Lang, M. and Müller, J. (2007). *Low-Dimensional Molecular Metals*, 1st Ed., Springer-Verlag, Berlin.

145. Engler, E. M. and Patel, V. V. (1974). Structure control in organic metals. Synthesis of tetraselenofulvalene and its charge transfer salt with tetracyano-*p*-quinodimethane, *J. Am. Chem. Soc.*, **96**, pp. 7376–7378.

146. Ferraris, J. P., Poehler, T. O., Bloch, A. N. and Cowan, D. O. (1973). Synthesis of the highly conducting organic salt tetramethyltetrathiofulvalenium tetracyano-*p*-quinodimethanide, *Tetrahedron Lett.*, **14**, pp. 2553–2556.

147. Bechgaard, K., Cowan, D. O. and Bloch, A. N. (1974). Synthesis of the organic conductor tetramethyltetraselenofulvalenium 7,7,8,8-tetracyano-*p*-quinodimethanide (TMTSF–TCNQ) [4,4′,5,5′-tetramethyl-$\Delta^{2,2'}$-bis-1,3-diselenolium 3,6-bis-(dicyanomethylene)cyclohexadienide], *J. Chem. Soc., Chem. Commun.*, pp. 937–938.

148. Mizuno, M., Garito, A. F. and Cava, M. P. (1978). 'Organic metals': alkylthio substitution effects in tetrathiafulvalene–tetracyanoquinodimethane charge-transfer complexes, *J. Chem. Soc., Chem. Commun.*, pp. 18–19.

149. Schumaker, R. R., Lee, V. Y. and Engler, E. M. (1983). New synthetic approaches to tetrathiafulvalene derivatives: systematic modifications of BEDT-TTF and TMTTF donors, *J. Phys. Colloques*, **44**, pp. C3-1139–C3-1145.

150. Kato, R., Kobayashi, H. and Kobayashi, A. (1991). Synthesis and properties of bis(ethylenedithio)tetraselenafulvalene (BEDT-TSeF) compounds, *Synth. Met.*, **42**, pp. 2093–2096.

151. Wheland, R. C. and Martin, E. L. (1975). Synthesis of substituted 7,7,8,8-tetracyanoquinodimethanes, *J. Org. Chem.*, **40**, pp. 3101–3109.

152. Aumüller, A. and Hünig, S. (1984). One-step entry to *N*-cyanimines and to *N,N′*-dicyanoquinonediimines, a novel class of electron-acceptors, *Angew. Chem., Int. Ed. Engl.*, **23**, pp. 447–448.

153. Aumüller, A., Erk, P., Klebe, G., Hünig, S., von Schütz, J. U. and Werner, H.-P. (1986). A radical anion salt of 2,5-dimethyl-*N,N′*-dicyanoquinonediimine with extremely high electrical conductivity, *Angew. Chem. Int. Ed. Engl.*, **25**, pp. 740–741.

154. Hünig, S. and Herberth, E. (2004). *N,N′*-dicyanoquinone diimines (DCNQIs): versatile acceptors for organic conductors, *Chem. Rev.*, **104**, pp. 5535–5563.

155. Bechgaard, K., Carneiro, K., Rasmussen, F. B., Olsen, M., Rindorf, G., Jacobsen, C. S., Pedersen, H. J. and Scott, J. C. (1981). Superconductivity in an organic solid. Synthesis, structure, and conductivity of bis(tetram ethyltetraselenafulvalenium) perchlorate, $(TMTSF)_2ClO_4$, *J. Am. Chem. Soc.*, **103**, pp. 2440–2442.

156. Mori, T., Kobayashi, A., Sasaki, Y. and Kobayashi, H. (1982). Band structure of the organic superconductor: $(TMTSF)_2X$, *Chem. Lett.*, **11**, pp. 1923–1926.

157. Mori, T. (1998). Structural genealogy of BEDT-TTF-based organic conductors I. Parallel molecules: β and β'' phases, *Bull. Chem. Soc. Jpn.*, **71**, pp. 2509–2526.

158. Mori, T., Mori, H. and Tanaka, S. (1999). Structural genealogy of BEDT-TTF-based organic conductors II. Inclined molecules: θ, α, and κ phases, *Bull. Chem. Soc. Jpn.*, **72**, pp. 179–197.

159. Kini, A. M., Geiser, U., Wang, H. H., Carlson, K. D., Williams, J. M., Kwok, W. K., Vandervoort, K. G., Thompson, J. E. and Stupka, D. L. (1990). A new ambient-pressure organic superconductor, κ-$(ET)_2Cu[N(CN)_2]Br$, with the highest transition temperature yet observed (inductive onset T_c = 11.6 K, resistive onset = 12.5 K), *Inorg. Chem.*, **29**, pp. 2555–2557.

160. Saito, G., Yamochi, H., Nakamura, T., Komatsu, T., Matsukawa, N., Inoue, T., Ito, H., Ishiguro, T., Kusunoki, M., Sakaguchi, K. and Mori, T. (1993). Structural and physical properties of two new ambient pressure κ-type BEDT-TTF superconductors and their related salts, *Synth. Met.*, **56**, pp. 2883–2890.

161. Taniguchi, H., Miyashita, M., Uchiyama, K., Satoh, K., Môri, N., Okamoto, H., Miyagawa, K., Kanoda, K., Hedo, M. and Uwatoko, Y. (2003). Superconductivity at 14.2 K in layered organics under extreme pressure, *J. Phys. Soc. Jpn.*, **72**, pp. 468–471.

162. Kobayashi, A., Udagawa, T., Tomita, H., Naito, T. and Kobayashi, H. (1993). New organic metals based on BETS compounds with MX_4^- anions (BETS = bis(ethylenedithio)tetraselenafulvalene; M=Ga, Fe, In; X=Cl, Br), *Chem. Lett.*, **22**, pp. 2179–2182.

163. Kobayashi, H., Tomita, H., Naito, T., Kobayashi, A., Sakai, F., Watanabe, T. and Cassoux, P. (1996). New BETS conductors with magnetic anions (BETS = bis(ethylenedithio)tetraselenafulvalene), *J. Am. Chem. Soc.*, **118**, pp. 368–377.

164. Otsuka, T., Kobayashi, A., Miyamoto, Y., Kiuchi, J., Wada, N., Ojima, E., Fujiwara, H. and Kobayashi, H. (2000). Successive antiferromagnetic and superconducting transitions in an organic metal, κ-$(BETS)_2FeCl_4$, *Chem. Lett.*, **29**, pp. 732–733.

165. Akutsu, H., Arai, E., Kobayashi, H., Tanaka, H., Kobayashi, A. and Cassoux, P. (1997). Highly correlated organic conductor with magnetic anions exhibiting a π–d coupled metal–insulator transition, λ-$(BETS)_2FeBr_xCl_{4-x}$ (BETS = bis(ethylenedithio)tetraselenafulvalene), *J. Am. Chem. Soc.*, **119**, pp. 12681–12682.

166. Tanaka, H., Adachi, T., Ojima, E., Fujiwara, H., Kato, K., Kobayashi, H., Kobayashi, A. and Cassoux, P. (1999). Pressure-induced superconducting transition of λ-$(BETS)_2FeCl_4$ with π–d coupled antiferromagnetic insulating ground state at ambient pressure [BETS = bis(ethylenedithio)tetraselenafulvalene], *J. Am. Chem. Soc.*, **121**, pp. 11243–11244.

167. Uji, S., Shinagawa, H., Terashima, T., Yakabe, T., Terai, Y., Tokumoto, M., Kobayashi, A., Tanaka H. and Kobayashi, H. (2001). Magnetic-field-induced superconductivity in a two-dimensional organic conductor, *Nature*, **410**, pp. 908–910.

168. Fujiwara, H., Fujiwara, E., Nakazawa, Y., Narymbetov, B. Z., Kato, K., Kobayashi, H., Kobayashi, A., Tokumoto, M. and Cassoux, P. (2001). A novel antiferromagnetic organic superconductor κ-$(BETS)_2FeBr_4$ [where BETS = bis(ethylenedithio)tetraselenafulvalene], *J. Am. Chem. Soc.*, **123**, pp. 306–314.

169. Schumaker, R. R. and Engler, E. M. (1980). Thiapen chemistry: synthesis of higher homologs of tetrathiafulvalene, *J. Am. Chem. Soc.*, **102**, pp. 6651–6652.

170. Misaki, Y., Nishikawa, H., Kawakami, K., Koyanagi, S., Yamabe, T. and Shiro, M. (1992). Methylthio substituted 2,5-bis(1′,3′-dithiol-2(-ylidene)-1,3,4,6-tetrathiapentalene. A bis-fused tetrathiafulvalene, *Chem. Lett.*, **21**, pp. 2321–2324.

171. Misaki, Y., Kochi, T., Yamabe, T. and Mori, T. (1998). Organic metals based on a selenium analogue of bis-fused TTF, *Adv. Mater.*, **10**, pp. 588–590.

172. Misaki, Y., Kochi, T., Taniguchi, M., Yamabe, T., Tanaka, K., Takimiya, K., Morikami, A., Otsubo, T. and Mori, T. (1999). Structural feature of radical cation salts based on TTP and its selenium analogues, *Synth. Met.*, **102**, pp. 1675.

173. Mori, T., Misaki, Y., Fujiwara, H. and Yamabe, T. (1994). Crystal structure and physical properties of $(BDT\text{-}TTP)_2ClO_4$, *Bull. Chem. Soc. Jpn.*, **67**, pp. 2685–2689.

174. Misaki, Y., Fujiwara, H., Yamabe, T., Mori, T., Mori, H. and Tanaka, S. (1994). Structure and conducting properties of BDT-TTP salts, *Chem. Lett.*, **23**, pp. 1653–1656.

175. Ito, T., Nakamura, K.-I., Shirahata, T., Kawamoto, T., Mori, T. and Misaki, Y. (2011). New molecular metals based on a tetrathiapentalene donor with peripheral methoxy groups, *Chem. Lett.*, **40**, pp. 81–83.

176. Nishiwaki, M., Nakamura, K.-I., Shirahata, T. and Misaki, Y. (2011). Novel molecular conductors based on cyclopenteno-TTP showing strong self-aggregation, *Chem. Lett.*, **40**, pp. 452–454.

177. Ashizawa, M., Akutsu, A., Noda, B., Nii, H., Kawamoto, T., Mori, T., Nakayashiki, T., Misaki, Y., Tanaka, K., Takimiya, K. and Otsubo, T. (2004). Synthesis and structures of highly conducting charge-transfer salts of selenium containing TTM-TTP derivatives, *Bull. Chem. Soc. Jpn.*, **77**, pp. 1449–1458.

178. Deluzet, A., Batail, P., Misaki, Y., Auban-Senzier, P. and Canadell, E. (2000). Donor slab robustness and band filling variations in BDT–TTP-based molecular conductors: β-$(BDT\text{-}TTP)_6[Re_6S_6Cl_8]\cdot(CH_2Cl\text{–}CHCl_2)_2$ and β-$(BDT\text{-}TTP)_6[Mo_6Cl_{14}]\cdot(CH_2Cl\text{–}CHCl_2)_2$, *Adv. Mater.*, **12**, pp. 436–439.

179. Perruchas, S., Boubekeur, K., Canadell, E., Misaki, Y., Auban-Senzier, P., Pasquier, C. and Batail, P. (2008). Modulating the framework negative charge density in the system $[BDT\text{–}TTP^{\bullet+}]/[Re_6S_5Cl_9{}^{1-}]/[Re_6(S/Se)_6Cl_8{}^{2-}]/[Re_6S_7Cl_7{}^{3-}]$: templating by isosteric cluster anions of identical symmetry and shape, variations of incommensurate band filling, and electronic structure in 2D metals, *J. Am. Chem. Soc.*, **130**, pp. 3335–3348.

180. Cui, H., Otsuka, T., Kobayashi, A., Takeda, N., Ishikawa, M., Misaki, Y. and Kobayashi, H. (2003). Structural, electrical, and magnetic properties of a series of molecular conductors based on BDT-TTP and lanthanoid nitrate complex anions (BDT-TTP = 2,5-Bis(1,3-dithiol-2-ylidene)-1,3,4,6-tetrathiapentalene), *Inorg. Chem.*, **42**, pp. 6114–6122.

181. Cui, H. B., Otsuka, T., Kobayashi, A., Misaki, Y. and Kobayashi, H. (2003). Structural and electrical properties of novel molecular conductors based on extended-TTF donors BDT-TTP and I^- anions, *Bull. Chem. Soc. Jpn.*, **76**, pp. 97–102.

182. Kusamoto, T., Fujiwara, E., Kobayashi, A., Cui, H., Otsuka, T., Okano, Y., Fujiwara, H. and Kobayashi, H. (2006). BDT-TTP-based π conductors

containing divalent magnetic and non-magnetic inorganic anions, $[M^{II}Cl_4]^{2-}$ (M = Co, Mn, Zn), *Bull. Chem. Soc. Jpn.*, **79**, pp. 527–536.

183. Otsuka, T., Cui, H. B., Kobayashi, A., Misaki, Y. and Kobayashi, H. (2002). Magnetism of metallic molecular crystals with rare-earth complex anions, *J. Solid State Chem.*, **168**, pp. 444–449.

184. Misaki, Y., Higuchi, N., Fujiwara, H., Yamabe, T., Mori, T., Mori, H. and Tanaka, S. (1995). (DTEDT)$[Au(CN)_2]_{0.4}$: an organic superconductor based on the novel π-electron framework of vinylogous bis-fused tetrathiafulvalene, *Angew. Chem. Int. Ed. Engl.*, **34**, pp. 1222–1225.

185. Giffard, M., Riou, A., Mabon, G., Mercier, N., Molinié, P. and Nguyen, T. P. J. (1999). Structural and electrical properties of bis(ethylenedithio) tetrathiafulvalenium hydrogen fumarate (2/1): a BEDT-TTF mixed-valency salt with a hydrogen-bonding organic anion, *J. Mater. Chem.*, **9**, pp. 851–853.

186. Akutsu, H., Yamashita, S., Yamada, J.-I., Nakatsuji, S., Hosokoshi, Y. and Turner, S. S. (2011). A purely organic paramagnetic metal, κ-β''-(BEDT-TTF)$_2$(PO-CONH$C_2H_4SO_3$), where PO = 2,2,5,5-tetramethyl-3-pyrrolin-1-oxyl free radical, *Chem. Mater.*, **23**, pp. 762–764.

187. Akutsu, H., Kawamura, A., Yamada, J.-I., Nakatsuji, S. and Turner, S. S. (2011). Anion polarity-induced dual oxidation states in a dual-layered purely organic paramagnetic charge-transfer salt, (TTF)$_3$(PO-CON(CH_3)$C_2H_4SO_3$)$_2$, where PO = 2,2,5,5-tetramethyl-3-pyrrolin-1-oxyl free radical, *CrystEngComm*, **13**, pp. 5281–5284.

188. Chasseau, D., Watkin, D., Rosseinsky, M. J., Kurmoo, M., Talham, D. R. and Day, P. (1988). Syntheses, crystal structures and physical properties of conducting salts (BEDT-TTF)$_2$X (X = CF_3SO_3 or *p*-$CH_3C_6H_4SO_3$), *Synth. Met.*, **24**, pp. 117–125.

189. Marsden, I. R., Obertelli, S. D., Allan, M. L., Friend, R. H., Kurmoo, M., Rosseinsky, M. J. and Day, P. (1991). Electronic and magnetic properties of a triflate salt of BEDT-TTF, (BEDT-TTF)$_2CF_3SO_3$, *Synth. Met.*, **42**, pp. 2151–2154.

190. Mori, T. (1999). Structural genealogy of BEDT-TTF-based organic conductors III. Twisted molecules: δ and α' phase, *Bull. Chem. Soc. Jpn.*, **72**, pp. 2011–2027.

191. Fettouhi, M., Ouahab, L., Gómez-García, C., Ducasse, L. and Delhaès, P. (1995). Structural and physical properties of κ-(BEDT-TTF)$_2$(CF_3SO_3), *Synth. Met.*, **70**, pp. 1131–1132.

192. Kanoda, K. and Kato, R. (2011). Mott physics in organic conductors with triangular lattices, *Annu. Rev. Condens. Matter Phys.*, **2**, pp. 167–188.

193. Schlueter, J. A., Geiser, U., Williams, J. M., Dudek, J. D., Kelly, M. E., Flynn, J. P., Wilson, R. R., Zakowicz, H. I., Sche, P. P., Naumann, D., Roy, T., Nixon, P. G., Winter, R. W. and Gard, G. L. (1997). Rational design of organic superconductors through the use of the large, discrete molecular anions $M(CF_3)_4^-$ (M= Cu, Ag, Au) and $SO_3CF_2CH_2SF_5^-$, *Synth. Met.*, **85**, pp. 1453–1456.

194. Kaiser, S., Dressel, M., Sun, Y., Greco, A., Schlueter, J. A., Gard, G. L. and Drichko, N. (2010). Bandwidth tuning triggers interplay of charge order and superconductivity in two-dimensional organic materials, *Phys. Rev. Lett.*, **105**, pp. 206402 (4 pages).

195. Su, X., Zuo, F., Schlueter, J. A., Williams, J. M., Nixon, P. G., Winter, R. W. and Gard, G. L. (1999). Anisotropic magnetoresistance in the organic superconductor β''-(BEDT-TTF)$_2$SF$_5$CH$_2$CF$_2$SO$_3$, *Phys. Rev. B*, **59**, pp. 4376–4381.

196. Koo, H.-J., Whangbo, M.-H., Dong, J., Olejniczak, I., Musfeldt, J. L., Schlueter, J. A. and Geiser, U. (1999). Analysis of polarized optical excitations by electronic band structure calculations: organic superconductor β''-(BEDT-TTF)$_2$SF$_5$CH$_2$CF$_2$SO$_3$, *Solid State Commun.*, **112**, pp. 403–407.

197. Beckmann, D., Wanka, S., Wosnitza, J., Schlueter, J. A., Williams, J. M., Nixon, P. G., Winter, R. W., Gard, G. L., Ren, J. and Whangbo, M.-H. (1998). Characterization of the Fermi surface of the organic superconductor - by measurements of Shubnikov-de Haas and angle-dependent magnetoresistance oscillations and by electronic band-structure calculations, *Eur. Phys. J. B*, **1**, pp. 295–300.

198. Wosnitza, J., Wanka, S., Hagel, J., Häussler, R., von Löhneysen, H., Schlueter, J. A., Geiser, U., Nixon, P. G., Winter, R. W. and Gard, G. L. (2000). Shubnikov-de Haas effect in the superconducting state of an organic superconductor, *Phys. Rev. B*, **62**, pp. R11973–R11976.

199. Wosnitza, J., Wanka, S., Qualls, J. S., Brooks, J. S., Mielke, C. H., Harrison, N., Schlueter, J. A., Williams, J. M., Nixon, P. G., Winter, R. W. and Gard, G. L. (1999). Fermiology of the organic superconductor β''-(BEDT-TTF)$_2$SF$_5$CH$_2$CF$_2$SO$_3$, *Synth. Met.*, **103**, pp. 2000–2001.

200. Cho, K., Smith, B. E., Coniglio, W. A., Winter, L. E., Agosta, C. C. and Schlueter, J. A. (2009). Upper critical field in the organic superconductor β''-(BEDT-TTF)$_2$SF$_5$CH$_2$CF$_2$SO$_3$: possibility of Fulde-Ferrell-Larkin-Ovchinnikov state, *Phys. Rev. B*, **79**, pp. 220507 (4 pages).

201. Beyer, R., Bergk, B., Yasin, S., Schlueter, J. A. and Wosnitza, J. (2012). Angle-dependent evolution of the Fulde-Ferrell-Larkin-Ovchinnikov

state in an organic superconductor, *Phys. Rev. Lett.*, **109**, pp. 027003 (5 pages).

202. Brossard, L., Ribault, M., Bousseau, M., Valade, L. and Cassoux, P. (1986). A new type of molecular superconductor: TTF[Ni(dmit)$_2$]$_2$, *C. R. Acad. Sci., Ser. II*, **302**, pp. 205–210.

203. Brossard, L., Ribault, M., Valade, L. and Cassoux, P. (1989). Pressure induced superconductivity in a molecular 4,5-dimercapto-1,3-dithiole-2-thione palladium complex (TTF(Pd(dmit)$_2$)$_2$), *J. Phys. (Paris)*, **50**, pp. 1521–1534.

204. Brossard, L., Hurdequint, H., Ribault, M., Valade, L., Legros, J. P. and Cassoux, P. (1988). Pressure-temperature phase diagram of α'-TTF [Pd(Dmit)$_2$]$_2$, *Synth. Met.*, **27**, pp. 157–162.

205. Tajima, H., Inokuchi, M., Kobayashi, A., Ohta, T., Kato, R., Kobayashi, H. and Kuroda, H. (1993). First ambient-pressure superconductor based on Ni(dmit)$_2$, α-EDT-TTF[Ni(dmit)$_2$], *Chem. Lett.*, **22**, pp. 1235–1238.

206. Graf, D., Brooks, J. S., Almeida, M., Dias, J. C., Uji, S., Terashima, T. and Kimata, M. (2009). Evolution of superconductivity from a charge-density-wave ground state in pressurized (Per)$_2$[Au(mnt)$_2$], *Europhys. Lett.*, **85**, pp. 27009 (5 pages).

207. Saito, G., Hayashi, H., Enoki, T. and Inokuchi, H. (1985). The study of charge transfer complexes of BEDT-TTF derivatives, *Mol. Cryst. Liq. Cryst.*, **120**, pp. 341–344.

208. Mori, T. and Inokuchi, H. (1986). Structural and electrical properties of (BEDT-TTF)(TCNQ), *Solid State Commun.*, **59**, pp. 355–359.

209. Mori, T. and Inokuchi, H. (1987). Crystal structure of the mixed-stacked salt of bis(ethylenedithio)tetrathiafulvalene (BEDT-TTF) and tetracyanoquinodimethane (TCNQ), *Bull. Chem. Soc. Jpn.*, **60**, pp. 402–404.

210. Yamamoto, H. M., Hagiwara, M. and Kato, R. (2003). New phase of (BEDT-TTF)(TCNQ), *Synth. Met.*, **133–134**, pp. 449–451.

211. Iwasa, Y., Mizuhashi, K., Koda, T., Tokura, Y. and Saito, G. (1994). Metal-insulator transition and antiferromagnetic order in bis(ethylenedithio) tetrathiafulvalene tetracyanoquinodimethane (BEDT-TTF)(TCNQ), *Phys. Rev. B*, **49**, pp. 3580–3583.

212. Yamamoto, H. M., Tajima, N., Hagiwara, M., Kato, R. and Yamaura, J.-I. (2003). Strange electric/magnetic behaviour of new (BEDT-TTF) (TCNQ), *Synth. Met.*, **135–136**, pp. 623–624.

213. Yasuzuka, S., Uji, S., Konoike, T., Terashima, T., Graf, D., Choi, E. S., Brooks, J. S., Yamamoto, H. M. and Kato, R. (2016). Shubnikov-de Haas

effect and angular-dependent magnetoresistance in layered organic conductor β''-(ET)(TCNQ), *J. Phys. Soc. Jpn.*, **85**, pp. 084701 (7 pages).

214. Kondo, R., Hasegawa, T., Mochida, T., Kagoshima, S., Iwasa, Y., Hiraki, K.-I. and Takahashi, T. (2001). Superconductivity and metal–insulator transition in twin-columnar organic superconductor, $(BETS)_2(X_2TCNQ)$ [X=Cl, Br], *J. Phys. Soc. Jpn.*, **70**, pp. 3023–3030.

215. Kondo, R., Hasegwa, T., Mochida, T., Kagoshima, S. and Iwasa, Y. (2001). Metal-insulator transition in donor-acceptor type superconductor, $(BETS)_2(X_2TCNQ)$, *Synth. Met.*, **120**, pp. 995–996.

216. Kanoda, K. (1997). Recent progress in NMR studies on organic conductors, *Hyperfine Interact.*, **104**, pp. 235–249.

217. Miyagawa, K., Kanoda, K. and Kawamoto, A. (2004). NMR studies on two-dimensional molecular conductors and superconductors: Mott transition in κ-$(BEDT\text{-}TTF)_2X$, *Chem. Rev.*, **104**, pp. 5635–5654.

218. Jérome, D. (1991). The physics of organic superconductors, *Science*, **252**, pp. 1509–1514.

219. Jérome, D. (2012). Organic superconductors: when correlations and magnetism walk in, *J. Supercond. Nov. Magn.*, **25**, pp. 633–655.

220. Ogi, D., Yoshimoto, S., Shirahata, T. and Misaki, Y. (2015). Synthesis, structures, and electrochemical properties of new extensively conjugated TTF, *Chem. Lett.*, **44**, pp. 554–556.

221. Takahashi, K., Tanioka, H., Fueno, H., Misaki, Y. and Tanaka, K. (2002). Preparation and characterization of novel aromatic-inserted tris-fused tetrathiafulvalenes, *Chem. Lett.*, **31**, pp. 1002–1003.

222. Umland, T. C., Allie, S., Kuhlmann, T. and Coppens, P. (1988). Relation between geometry and charge transfer in low-dimensional organic salts, *J. Phys. Chem.*, **92**, pp. 6456–6460.

223. Guionneau, P., Kepert, C. J., Bravic, G., Chasseau, D., Truter, M. R., Kurmoo, M. and Day, P. (1997). Determining the charge distribution in BEDT-TTF salts, *Synth. Met.*, **86**, pp. 1973–1974.

224. Guionneau, P., Chasseau, D., Howard, J. A. K. and Day, P. (2000). Neutral bis(ethylenedithio)tetrathiafulvalene at 100 K, *Acta Cryst. Sect. C*, **56**, pp. 453–454.

225. Furuta, K., Kohno, S., Shirahata, T., Yamasaki, K., Hino, S. and Misaki, Y. (2012). Synthesis and properties of 2-alkylidene-1,3-dithiolo[4,5-*d*]-4,5-ethylenediselenotetrathiafulvalene derivatives and crystal structures of their cation radical salts, *Crystals*, **2**, pp. 393–412.

226. Enoki, T. and Miyazaki, A. (2004). Magnetic TTF-based charge-transfer complexes, *Chem. Rev.*, **104**, pp. 5449–5478.

227. Fourmigué, M. and Ouahab, L. (2009). *Conducting and Magnetic Organometallic Molecular Materials*, 1st Ed., Springer-Verlag, Berlin.

228. Drozdov, A. P., Eremets, M. I., Troyan, I. A., Ksenofontov, V. and Shylin, S. I. (2015). Conventional superconductivity at 203 kelvin at high pressures in the sulfur hydride system, *Nature*, **525**, pp. 73–76.

229. Ward, B. H., Schlueter, J. A., Geiser, U., Wang, H.-H., Morales, E., Parakka, J. P., Thomas, S. Y., Williams, J. M., Nixon, P. G., Winter, R. W., Gard, G. L., Koo, H.-J. and Whangbo, M.-H. (2000). Comparison of the crystal and electronic structures of three 2:1 salts of the organic donor molecule BEDT-TTF with pentafluorothiomethylsulfonate anions $SF_5CH_2SO_3^-$, $SF_5CHFSO_3^-$, and $SF_5CF_2SO_3^-$, *Chem. Mater.*, **12**, pp. 343–351.

230. Wang, H. H., Geiser, U., Schlueter, J. A., Ward, B. H., Parakka, J. P., Kini, A. M., O'Malley, J. L., Thomas, S. Y., Morales, E., Dudek, J. D., Williams, J. M. and Gard, G. L. (1999). Novel sulfonate containing ET based synmetals, *Synth. Met.*, **102**, pp. 1666–1669.

231. Schlueter, J. A., Ward, B. H., Geiser, U., Wang, H. H., Kini, A. M., Parakka, J. P., Morales, E., Koo, H.-J., Whangbo, M.-H., Winter, R. W., Mohtasham, J. and Gard, G. L. (2001). Crystal structure, physical properties and electronic structure of a new organic conductor β''-(BEDT-TTF)$_2SF_5CHFCF_2SO_3$, *J. Mater. Chem.*, **11**, pp. 2008–2013.

232. Hiraki, K., Furuta, S., Takahashi, T., Kondo, R., Kagoshima, S., Hasegawa, T., Mochida, T. and Iwasa, Y. (2002). Metal-insulator transition of donor-acceptor-type organic charge-transfer complex (BETS)$_2$(Br_2TCNQ): site-selective NMR measurements, *Phys. Rev. B*, **66**, pp. 035104 (6 pages).

233. Isono, T., Kamo, H., Ueda, A., Takahashi, K., Nakao, A., Kumai, R., Nakao, H., Kobayashi, K., Murakami, Y. and Mori, H. (2013). Hydrogen bond-promoted metallic state in a purely organic single-component conductor under pressure, *Nat. Commun.*, **4**, pp. 1344 (6 pages).

234. Terauchi, T., Sumi, S., Kobayashi, Y., Nakamura, T., Furukawa, K. and Misaki, Y. (2014). A stable metallic state of (TTPCOO)$_2NH_4$ with a mobile dopant, *Chem. Commun.*, **50**, pp. 7111–7113.

235. Prokhorova, T. G., Zorina, L. V., Simonov, S. V., Zverev, V. N., Canadell, E., Shibaeva, R. P. and Yagubskii, E. B. (2013). The first molecular superconductor based on BEDT-TTF radical cation salt with paramagnetic tris(oxalato)ruthenate anion, *CrystEngComm*, **15**, pp. 7048–7055.

236. Shirahata, T., Kibune, M. and Imakubo, T. (2006.). New ambient pressure organic superconductors κ_H- and κ_L-(DMEDO-TSeF)$_2$[Au(CN)$_4$](THF), *Chem. Commun.*, pp. 1592–1594.

237. Shirahata, T., Kibune, M., Yoshino, H. and Imakubo, T. (2007). Ambient-pressure organic superconductors κ-(DMEDO-TSeF)$_2$[Au(CN)$_4$](solv.): T_c tuning by modification of the solvent of crystallization, *Chem. Eur. J.*, **13**, pp. 7619–7630.

Chapter 6

Multiple-Decker Metal Porphyrins

Shigeki Mori
Division of Material Science, Advanced Research Support Center, Ehime University, 2-5 Bunkyo-cho, Matsuyama, Ehime 790-8577, Japan
mori.shigeki.mu@ehime-u.ac.jp

6.1 Introduction

Porphyrins (pors) are macrocyclic compounds consisting of four pyrrole units and characterized by strong absorption bands in the visible region, redox behavior, and unique metal coordination ability due to their large π-conjugated system and N atom–inward structure. In particular, the metal coordination chemistry of pors and related compounds is one of the most fundamental and important issues [1]. As regards modification of ligands, intensive and extensive research has been carried out, in which many metal complexes contain the nitrogen atoms of the central moiety. Metals with small ionic radii can fit in the cavity pors with or without axial ligands (Fig. 6.1).

On the other hand, it is well known that complexation of pors with large metals such as with rare earth elements gives double-

Functional Materials: Advances and Applications in Energy Storage and Conversion
Edited by Toshio Naito

ISBN 978-981-4800-09-9 (Hardcover), 978-0-429-46813-1 (eBook)
www.panstanford.com

and triple-decker type compounds. In this chapter, we mainly focus on the synthesis and structural characterization of multiple-decker tetrapyrrolic compounds.

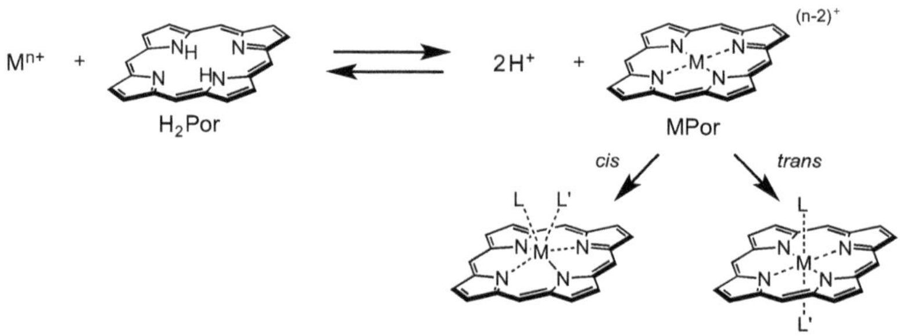

Figure 6.1 Metal-coordination manner of porphyrins: L and L′ are axial ligands (anionic or neutral).

The first synthesis of a sandwich-type tetrapyrrole metal complex, in 1936 by Linstead and coworkers, produced bis(phthalocyanine)-Sn(VI) instead of por (Fig. 6.2) [2]. Phthalocyanines are also macrocyclic molecules and have four inward nitrogen atoms. The double-decker structure of $Sn(Pc)_2$, where Pc is phthalocyanine, was proven by X-ray crystallographic analysis in 1973 (Fig. 6.3). [3]. The metal center is surrounded by the eight nitrogen atoms in a square antiprism geometry. The macrocyclic systems are rotated 42° with respect to one another, and the two square planes composed of coordinated four nitrogens are 2.70 Å apart.

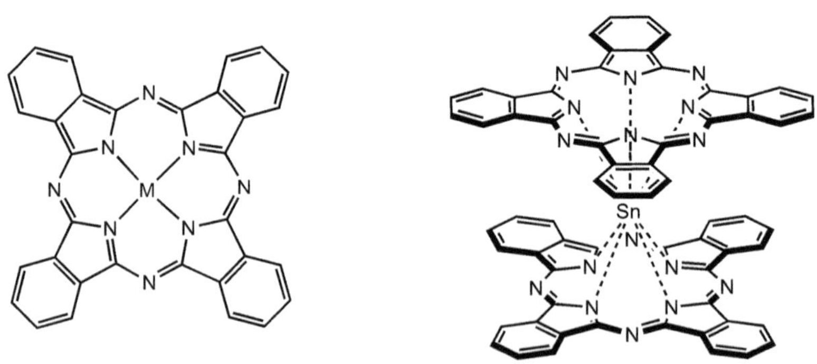

Figure 6.2 Phthalocyanine framework and its Sn double-decker complex.

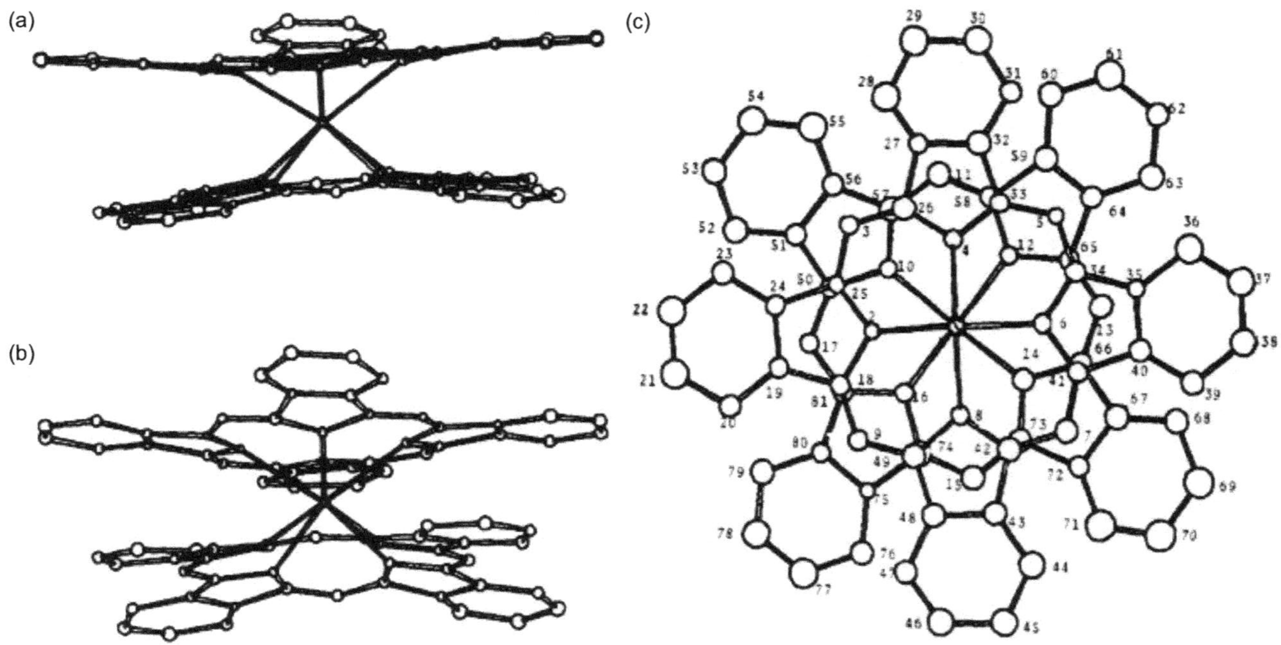

Figure 6.3 Molecular structure of $Sn(Pc)_2$. (a) Side view, (b) 10° oblique view, and (c) top view. Reprinted with permission from Ref. [3]. Copyright (1973) American Chemical Society.

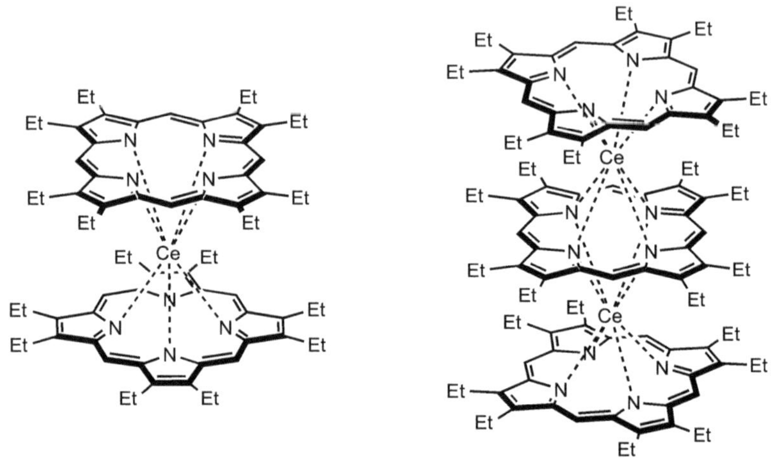

$Ce(OEP)_2$ $Ce_2(OEP)_3$

Figure 6.4 Octaethylporphyrin Ce double-decker and triple-decker complexes.

In the por derivatives, the first sandwich-type compound was synthesized in 1977 as a bis(porphyrin)-Ce complex, in which a *meso*-aryl type por was used [4]. The single-crystal X-ray structural determination of por double- and triple-decker complexes was performed in 1986 by Buchler and coworkers (Figs. 6.4 and 6.5) [5]. In the crystalline state of $Ce(OEP)_2$ (OEP stands for 2,3,7,8,12,13,17,18-octaethylporphyrin), the Ce^{VI} ion is coordinated by the eight nitrogen atoms in a nearly antiprism geometry. The por squares–defined four nitrogen atoms (4N plane) are also rotated by an angle of 41.8° with respect to their eclipsed position, like $Sn(Pc)_2$, and the distance between the 4N planes of two pors is 2.75 Å. The crystal structure of triple-decker $Ce_2(OEP)_3$ includes two 1,2,4-trichlorobenzene (1,2,4-TCB), but the cocrystalized solvent molecules are not involved in any intermolecular interaction. Each $Ce_2(OEP)_3$ lies on a crystallographic inversion center located in the middle of two Ce ions. The coordination atmosphere of Ce^{III} is distorted from the ideal square antiprism with a rotational angle of 24.5° between outer and inner por macrocycles. The separation between the outer 4N plane and the inner one amounts to 3.27 Å. The bond lengths of Ce–N are divided into the following two types: Ce-outer por nitrogens and Ce-inner ones. The mean lengths of $Ce–N_{outer}$ is 2.501(2) Å and $Ce–N_{inner}$ is 2.758(2) Å, respectively. Therefore, the Ce–Ce distance is 3.75 Å. In the study above, the sandwich complexes were prepared from the reaction of the corresponding por ligand

and acetylacetonate (acac) lanthanide salt in 1,2,4-TCB under reflux condition. The synthetic method is applicable in a wide range of por metalation.

The formation of a phthalocyanine triple-decker complex was reported in 1985–1986 [6, 7], and the first crystal structure of the triple-decker with unsubstituted phthalocyanines was determined as $In_2(Pc)_3$ [8]. The molecular planes of the two outer macrocycles are eclipsed by each other, while the inner ligand is rotated about 37.5° in relation to the peripheral rings. The intramolecular distance between phthalocyanines is 2.95 Å. The intramolecular distance between two In ions is 4.04 Å. Various double-decker and triple-decker complexes have been synthesized and explored so far [9–12].

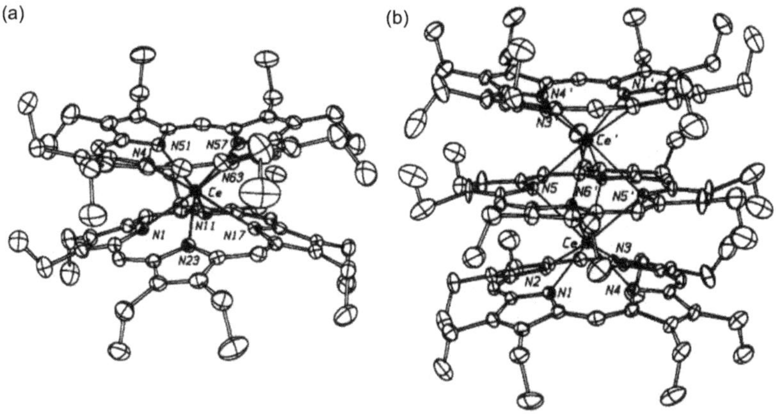

Figure 6.5 Ortep plots of (a) $Ce(OEP)_2$ and (b) $Ce_2(OEP)_3$. Ellipsoids are scaled to enclose 50% of the electronic density. Hydrogen atoms are omitted. Reprinted with permission from Ref. [8]. Copyright (1986) American Chemical Society.

In 2003, Ishikawa and coworkers discovered that bis(phthalocyanine) lanthanide (Ln = Tb or Dy) double-decker complexes exhibited properties of a single-molecule magnet (SMM) in a higher temperature range than any other SMMs (Figs. 6.6 and 6.7) [13]. SMMs are categorized as paramagnetic molecules, and their typical character is quantum tunneling of magnetization and slow spin relaxation [14]. The metal center is trivalent in this double-decker complex and an unpaired electron is delocalized over the two phthalocyanine ligands. After this seminal study, studies for a deep understanding and improvement of SMM materials using sandwich-type rare earth complexes were widely explored. Recent progress of the tetrapyrrole multidecker as a SMM was reviewed by Gao et al. [15].

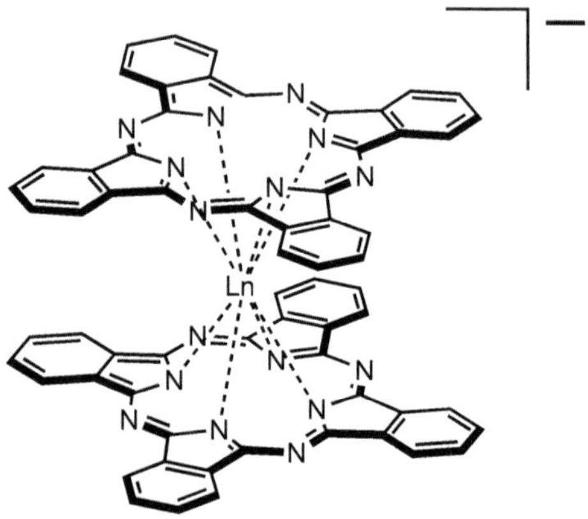

$Ln(Pc)_2^-$

Figure 6.6 Bis(phthalocyanine) lanthanide double-decker complexes (Ln = Tb, Dy, Ho, Er, Tm, or Yb).

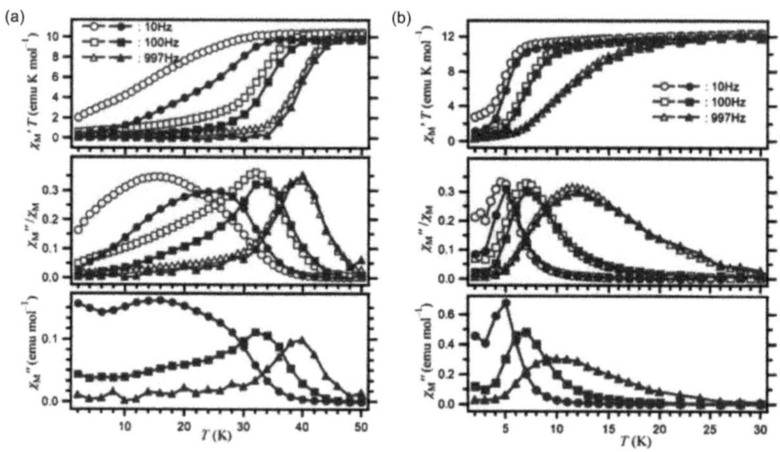

Figure 6.7 Plots of (top) $\chi_M'T$ and (bottom) χ_M''/χ_M against temperature T, where χ_M', χ_M'', and χ_M are in-phase AC, out-of-phase AC, and DC molar magnetic susceptibilities, respectively, for a powder sample of (a) $[Tb(Pc)_2]^- \cdot TBA^+$ and (b) $[Dy(Pc)_2]^- \cdot TBA^+$ (open marks) and that diluted in $[Y(Pc)_2]^- \cdot TBA^+$ (filled marks) measured in a 3.5 G AC magnetic field oscillating at indicated frequencies. TBA is for tetra-*n*-butylammonium. Reprinted with permission from Ref. [13]. Copyright (2003) American Chemical Society.

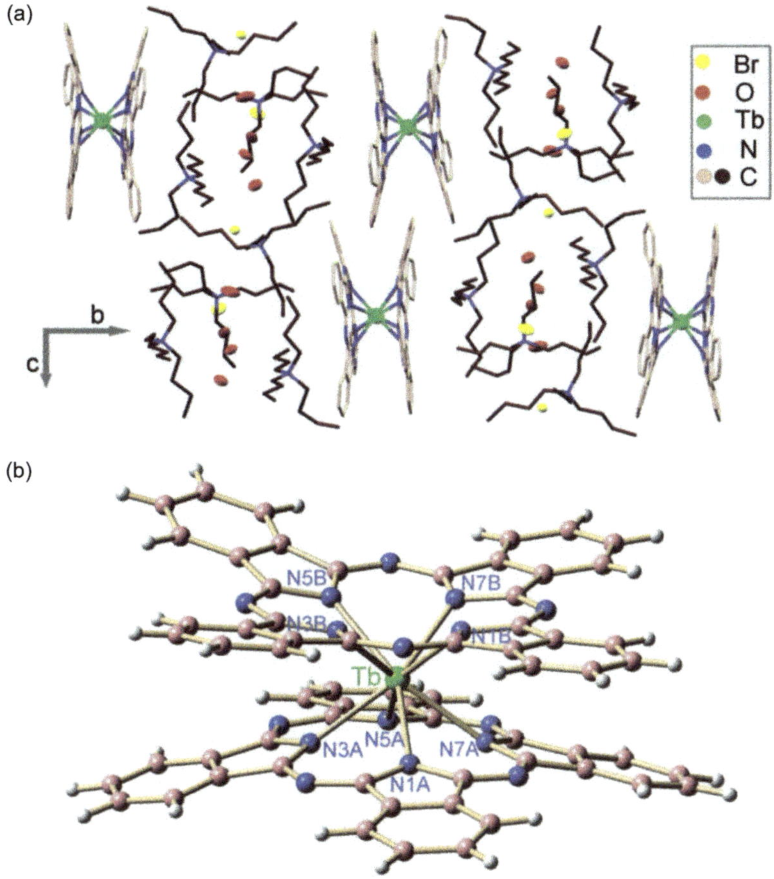

Figure 6.8 (a) Packing motif of $[Pc_2Tb]^-[TBA]^+ \times 3$ [TBA]Br $\times$ 3 H_2O, showing the allocation of the $[Pc_2Tb]^-$ complex anions in the unit cell. (b) Molecular structure of the $[Pc_2Tb]^-$ complex anion showing the Tb^{III} metal ion (green) coordinated by the phthalocyanine ligands (carbons: pink; hydrogen: gray; nitrogens: blue). Relevant Tb-N (intramolecular) distances: Tb(1)-N(1A), 2.548(6) Å; Tb(1)-N(3A), 2.340(5) Å; Tb(1)-N(5A), 2.509(6) Å; Tb(1)-N(7A), 2.357(5) Å; Tb(1)-N(1B), 2.535(6) Å; Tb(1)-N(3B), 2.313(5) Å; Tb(1)-N(5B), 2.573(6) Å; Tb(1)-N(7B), 2.338(6) Å. Reprinted with permission from Ref. [16]. Copyright (2009) American Chemical Society.

The crystal structures of negatively charged $Tb(Pc)_2$ (Fig. 6.8) [16] and other $Ln(Pc)_2$ (Ln = Nd, Gd, Ho, or Lu) [17] were analyzed by X-ray crystallography. The distance between the two phthalocyanines has a linear relationship to the ionic radii of central metals. This structural information suggests a significant correlation between the strength of the π-π interaction of the two phthalocyanines and the twist angle of the 4N plane with respect to one another.

In 2012, Ishikawa et al. found that the single-molecule magnetic properties of meso-tetraphenylporphyrin-based Tb^{III} double-decker complex was switchable by protonation and deprotonation (Fig. 6.9) [18]. Both heptacoordinated ($Tb(TPP)_2$-hepta), where TPP stands for 5,10,15,20-tetraphenylporphyrin, and octacoordinated ($Tb(TPP)_2$-octa) Tb^{III} double-decker complexes were structurally determined by X-ray diffraction analysis. In contrast to no SMM character of $Tb(TPP)_2$-hepta, $Tb(TPP)_2$-octa exhibits a clear magnetic property as a SMM. Other rare earth metal complexes of the protonated form were also elucidated by X-ray diffraction analysis [19].

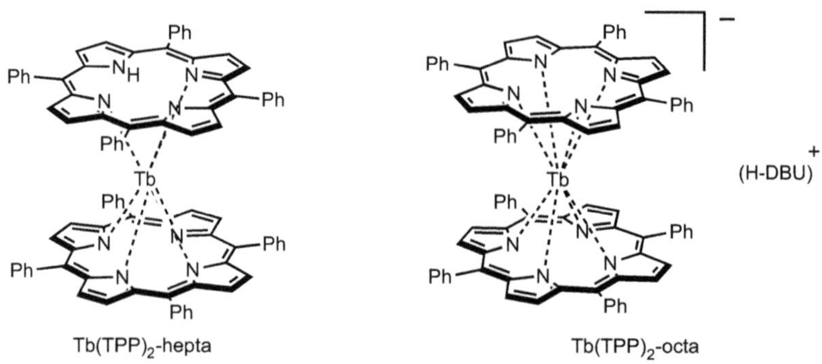

Figure 6.9 Heptacoordinated and octacoordinated bis(meso-phenylporphyrin) Tb^{III} double-decker complexes.

Liu et al. succeeded in the synthesis and characterization of novel binuclear phthalocyanine sandwich-type rare earth complexes based on benzene-linked bis(phthalocyanine) and two phthlocyanines (Figs. 6.10 and 6.11) [20]. The complex exhibits intramolecular delocalized biradical-based antiferromagnetic interactions, which probably lead to the observation of magnetic hysteresis up to 5 K, the highest ever measured for the sandwich-type tetrapyrrole-Dy SMMs.

The magnetization behavior of mixed porphyrin-phthalocyanine double-decker was also elucidated [21, 22].

Multidecker complexes containing four or more tetrapyrrole ligands were also reported in the 2010s. Fukuda et al., in 2010, synthesized a discrete phthalocyanine quadruple-decker complex from the reaction of bis(phthalocyanine) luthetium anionic species and cadmium acetate (Fig. 6.12) [23].

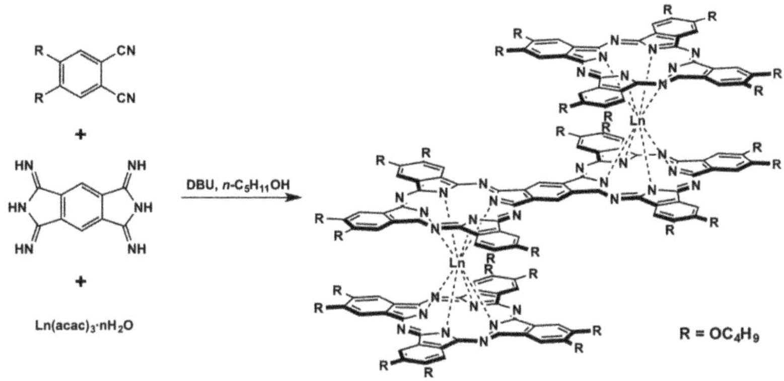

Figure 6.10 Synthesis of binuclear phthalocyanine-based sandwich-type rare earth complexes. (Ln = Dy and Y)

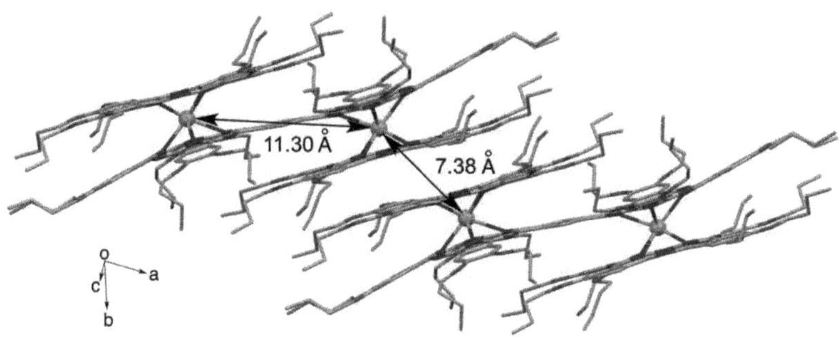

Figure 6.11 Crystal packingof bis-Dy bisphthalocyanine complex along the *a* axis. All hydrogen atoms are omitted for clarity. The values in the figure are the intramolecular and intermolecular $Dy^{III}–Dy^{III}$ distances. The figure was produced with the deposited data in Ref. [20], downloaded from the Cambridge Crystallographic Data Centre.

This strategy using a preorganized bis(phthalocyanine) metal complex is versatile in the synthesis of higher multidecker analogs. Jiang and colleagues prepared a quintuple-decker complex in 2012 [24] and a sextuple-decker complex in 2013 with a similar procedure (Fig. 6.13) [25]. Katoh and coworkers revealed the crystal structure of Tb^{III}-Cd^{II}-Tb^{III} and Dy^{III}-Cd^{II}-Dy^{III} tetrakis(phthalocyanine) quadruple-decker complexes (Fig. 6.14) [26] and the Tb^{III}-Cd^{II}-

Cd^{II}-Tb^{III} pentakis(phthalocyanine) quintuple-decker complex by synchrotron radiation (Fig. 6.15) [27].

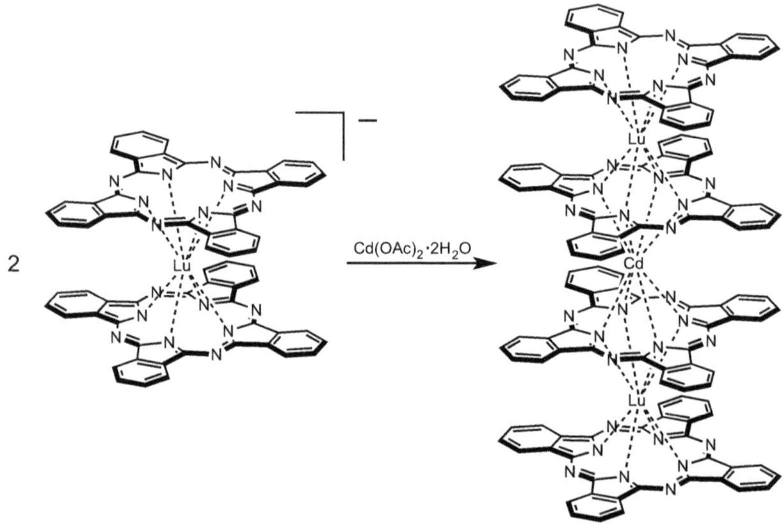

Figure 6.12 Synthesis of phthalocyanine quadruple-decker complex.

Tetrapyrrolic compounds, especially phthalocyanines, have been investigated as molecular materials for potential applications in, for example, organic field effect transistors (OFETs), nonlinear optics, and solar cells [28–30]. These investigations have been also conducted on multidecker metal complexes. Among them, studies for OFET application have been extensively performed. In 1988, *p*-type OFETs with a NiPc/Sc(Pc)$_2$ mixed active layer were fabricated by Clarisse et al. They show hole mobility of about 1×10^{-3} cm^2·V^{-1}·s^{-1} and a threshold of 1 V [31]. The same group reported the fabrication of OFETs with monomeric phthalocyanine (MPc; M = H_2, Ni, or Cu) and bis(phthalocyanine) rare earth metal complexes (Ln(Pc)$_2$; Ln = Y or Sc) [32]. The hole mobility is in the range of 1 to 10×10^{-4} cm^2·V^{-1}·s^{-1}, depending on the combination of metal ions coordinated by monomeric phthalocyanines and double-decker compounds. Jiang and colleagues fabricated OFETs with bis(phthalocyanine) rare earth metal complexes consisting of two different phthalocyanines via the Langmuir–Blodgett (LB) technique [33]. The LB method overcame the difficulty of ordering of molecules in film states. The high performance (0.24 to 0.60 cm^2·V^{-1}·s^{-1}) of LB film-based OFETs with triple-decker phthalocyanine rare earth metal complexes ($Ln_2(Pc)_3$; Ln = Eu, Ho, or Lu) was elucidated in 2005 [34]. Chen and coworkers achieved the fabrication of solvent vapor annealing over quasi-Langmuir–Shäfer film-based OFET devices from two different phthalocyanine Eu triple-deckers, which exhibited a high and balanced ambipolar performance with mobilities of 1.71 (holes) and 1.25 (electrons) $cm^2 \cdot V^{-1} \cdot s^{-1}$ under ambient conditions [35]. The multidecker metal complexes exhibit several interesting properties arising from their large and stacked π-conjugation systems, stable multiredox states, and adjacent (adj) metal ions, but their other directions remain to be explored.

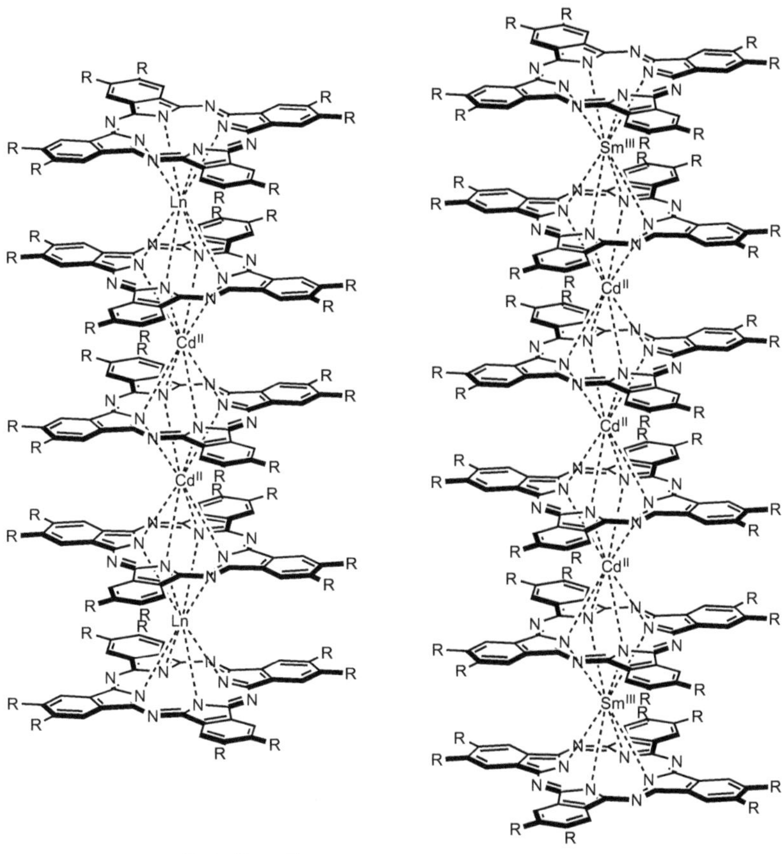

Ln = Nd^{III}, Sm^{III}, Eu^{III}
R = OC_4H_9

R = OC_4H_9

Figure 6.13 Phthalocyanine quintuple-decker and sextuple-decker complexes.

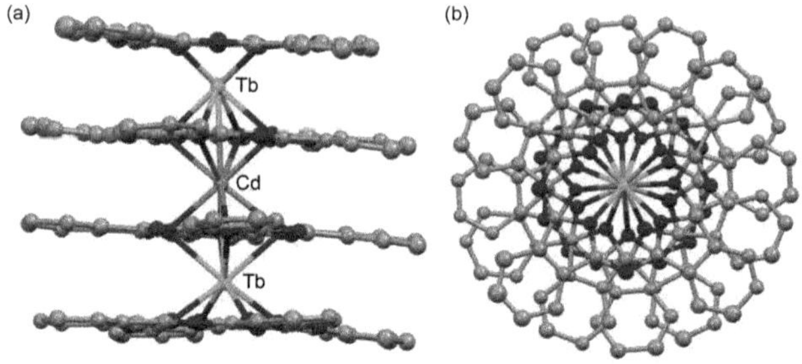

Figure 6.14 Molecular structure of Tb-Cd-Tb tetrakis(phthalocyanine) quadruple-decker: (a) side view and (b) top view with all hydrogen atoms and *n*-butyloxy chains omitted for clarity. The figure was produced with the deposited data in Ref. [26], downloaded from the Cambridge Crystallographic Data Centre. Reprinted from Ref. [26] with permission of The Royal Society of Chemistry.

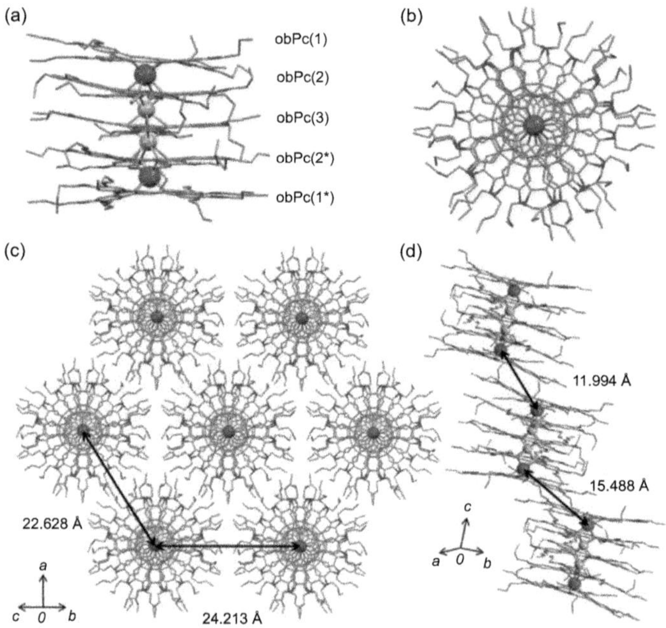

Figure 6.15 (a) Side view and (b) top view of the crystal structure of Tb-Cd-Cd-Tb pentakis(phthalocyanine) quintuple-decker. Butoxy chains are highly disordered, and only the chains with the highest occupancy are shown for clarity. The crystal packing is shown as (c) a two-dimensional array extended in the *ab* plane and (d) a slipped column along the *c* axis. The values in the figure are the intermolecular Tb^{III}–Tb^{III} distances. Reprinted with permission from Ref. [27]. Copyright (2015) American Chemical Society.

6.2 Benzoporphyrin Triple-Decker Complex

6.2.1 π-Extended Porphyrin

π-Extended pors have attracted much attention in the fields of organic chemistry, metal coordination chemistry, and material science because their strong absorption in the visible and near-infrared (NIR) region, redox activity, and other physical properties can be controlled by peripheral substituents. To enhance the property of π-conjugated molecules, extension of the π-system is a straightforward strategy. From this point of view, por with a fused benzene ring in the pyrrole exterior region (benzoporphyrin) is one of the simplest extended derivatives. Indeed, the chemistry of benzoporphyrin was launched in 1937 [36]. And then, several methods for benzoporphyrin synthesis were reported, which required a severe condition (about 350°C; Fig. 6.16) [37–40]. Despite their large, extended π-system, intriguing optical and electrochemical properties, and thermal and chemical stability, benzoporphyrins have been less explored than phthalocyanines. The main reasons are probably the severe reaction conditions, their poor solubility, and the instability of isoindole as a precursor [41].

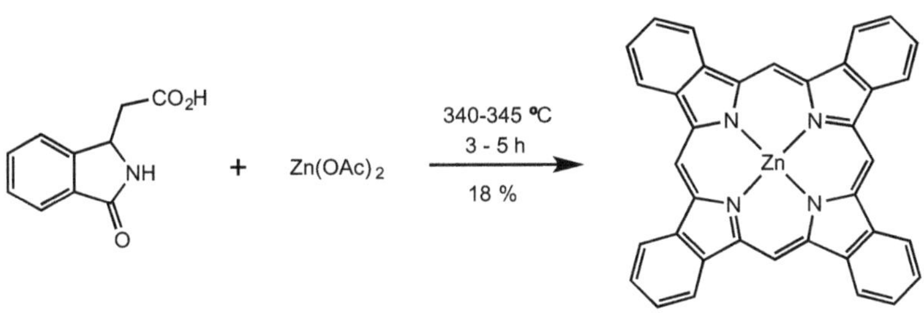

Figure 6.16 Optimal synthetic protocol of tetrabenzoporphyrin by Linstead and Weiss [37].

Ono and coworkers synthesized pyrroles and pors fused with bicyclo[2.2.2]octadiene (BCOD) and other polycyclic skeletons [42]. And soon, a new method of synthesis of highly pure benzoporphyrins based on the final conversion of their precursors by the thermal retro Diels–Alder reaction of the BCOD moieties was discovered by the same authors (Fig. 6.17) [43]. Pors with a bulky BCOD group are soluble in common organic solvents and easy to purify with conventional solution processes.

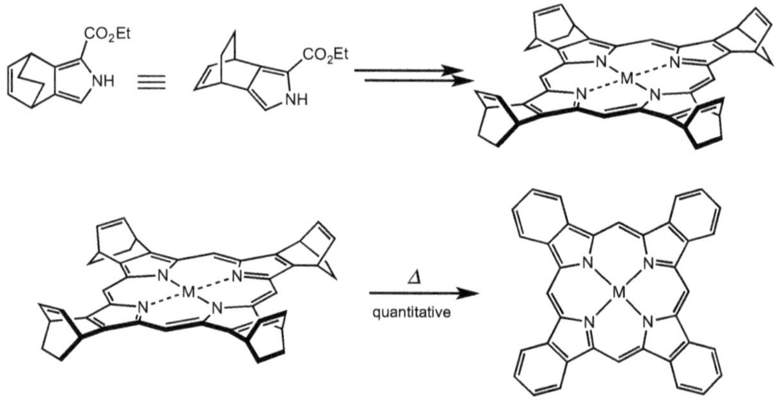

Figure 6.17 Synthesis of tetrabenzoporphyrin from BCOD porphyrin by thermal conversion.

The above-mentioned method of synthesis of benzoporphyrins from BCOD-fused pors is virtually the most important and simplest, and BCOD-fused pors was prepared via a different synthetic route by Lim and coworkers [44].

The retro Diels–Alder reaction of the BCOD moiety is a promising method for constructing a novel π-conjugation network. This methodology is widely applicable to π-conjugated compounds such as larger extended pors [45–47], oligoporphyrins [48–52], core-modified pors [53], boron dipyrromethenes (BODIPYs) [54–56], and so on [57, 58]. Properties of benzoporphyrins as OFETs have also been examined. The details are mentioned in Chapter 8. The direct meso-halogenation of tetra-BCOD-fused por—$(BCOD)_4$-P—and subsequent thermal conversion by the retro Diels–Alder reaction was achieved by Ito's group [59, 60]. On the other hand, the rigid framework of BCOD unit gave characteristic three-dimensional structures [61–64].

6.2.2 Benzoporphyrins

Benzoporphyrin is a genuine counterpart of phthalocyanine from the molecular framework point of view. Distinct differences between pors and phthalocyanines are their electronic structures and the

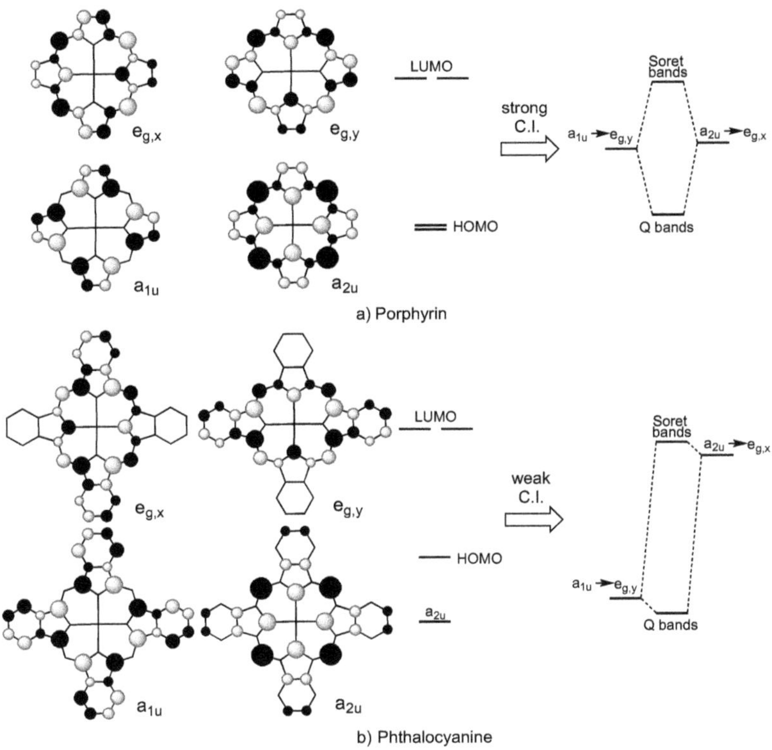

Figure 6.18 Frontier molecular orbital diagrams and configuration interactions (CI) of (a) porphyrin and (b) phthalocyanine.

size of their cavities. In por's case, the highest occupied molecular orbital (HOMO) (a_{2u}) and next-HOMO (a_{1u}) are almost degenerated; thus a strong configuration interaction occurs in the transition from these orbitals to the lowest unoccupied molecular orbital (LUMO) (eg). In contrast, the a_{2u} orbital is stabilized by nitrogen atoms in phthalocyanine. It is explained in terms of electronegativity. As a result, the configuration interaction becomes weaker and it leads to intense Q band absorption (Fig. 6.18) [65, 66]. The coordination cavity of pors is larger than that of phthalocyanines, so the metal ion is located closer to the macrocyclic plane. Therefore, stronger interaction between metal ion and macrocycle can be expected. When the distance between two π-conjugated planes is shorter than 3.4 Å, it is considered that π-π interaction can occur due to an orbital overlap. In general, the distance between macrocycles in a multidecker por complex is shorter than 3.4 Å. To discuss the electronic and magnetic properties in neutral and oxidized states or in photoexcited states, we have planned to synthesize multidecker metal benzoporphyrins. Because the methodology for the modification of por is now well established, the numbers and positions of fused benzene rings in benzoporphyrins are selectively accessible. The electronic and magnetic properties can be influenced by the difference of π-π interaction arising from the degree of π extension and symmetry difference of pors. These electronic and structural differences can create different delocalization behaviors of electrons and spins over pors. Oligoporphyrin as a ligand has the advantage of construction of well-defined multinuclear metal complexes in alignment with metal ions. In this paper, we present an efficient synthesis method of lanthanum and terbium triple-decker benzoporphyrin complexes in a one-pot reaction and their X-ray crystal structures. The lanthanum trivant ion is diamagnetic; so its structures in solution can be discuss by ^{1}H nuclear magnetic resonance (NMR) measurement. Attempts at synthesis of other lanthanide triple-decker complexes and triple-decker complexes with disporphyrins are also mentioned.

6.2.3 Synthesis of Metal-Free BCOD Porphyrins

All types of benzoporphyrins were prepared in a manner similar to the literature [67–69]. We adopted alkylated pyrrole units for improvement in solubility. Mono-, two isomers of di-, and tri-BCOD-fused por [BCOD-P, *adj*-$(BCOD)_2$-P, *opp*-$(BCOD)_2$-P [where opp is for opposite], and $(BCOD)_3$-P] were synthesized from the stepwise reaction of the corresponding precursors (Fig. 6.19). BCOD-P was prepared from the reaction of BCOD pyrrole and diformylated β-alkylated tripyrrane under acidic conditions. For easy purification, zinc metalation was carried out. Separation over a silica gel column chromatography provided a BCOD-P zinc complex. The metal-free por was obtained upon the treatment of trifluoroacetic acid (TFA). *Adj*-$(BCOD)_2$-P was prepared from the reaction of BCOD-fused dipyrromethane and diformylated beta-ethylated dipyrromethane under acidic conditions. After zinc metalation, silica-gel separation gave an *adj*-$(BCOD)_2$-P zinc complex and demetalation with TFA was smoothly proceeded. For preparation of *opp*-$(BCOD)_2$-P, tripyrrane consisting of two BCOD pyrroles and diethyl pyrrole

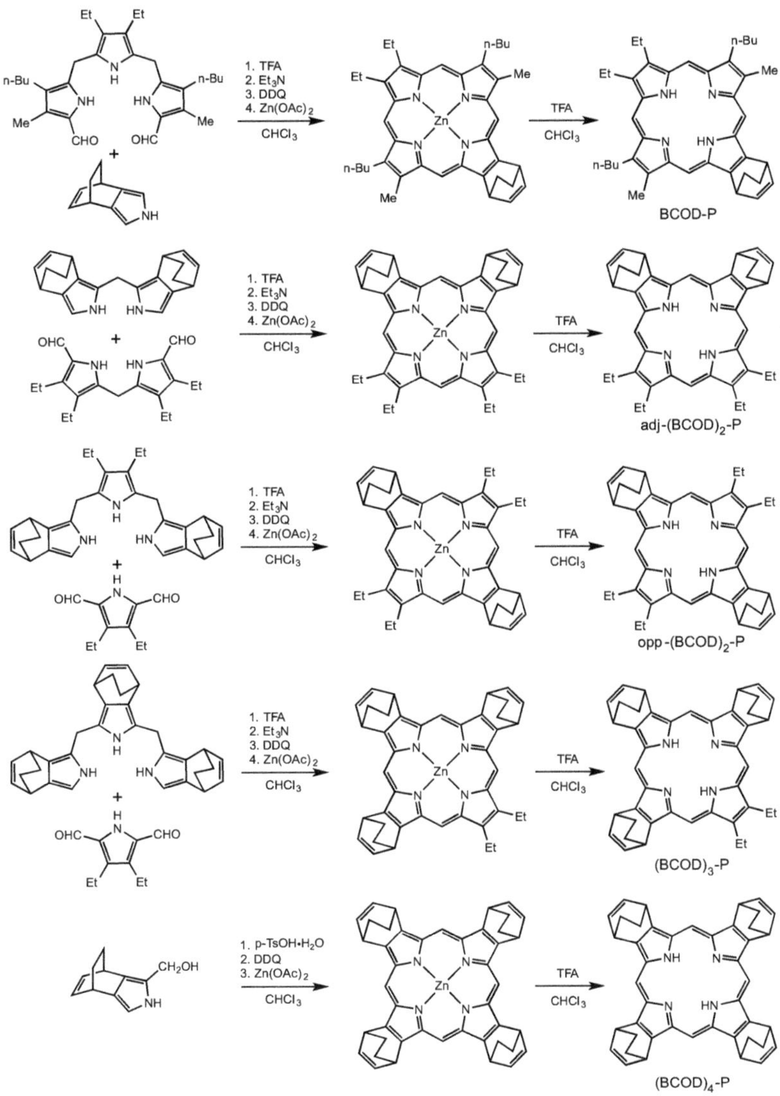

Figure 6.19 Synthesis of mono-, di-, tri-, and tetra-BCOD-fused porphyrins.

was synthesized in advance. *Opp*-$(BCOD)_2$-P was prepared from the reaction of tripyrrane and diformylated diethylpyrrole under acidic conditions. A similar way of zinc-metalation, silica-gel separation, and demetalation generated the corresponding metal-free derivative. $(BCOD)_3$-P was prepared from the reaction of BCOD-fused tripyrrane and diformylated diethylpyrrole under acidic conditions. After purification of a zinc complex, metal-free por was isolated upon the treatment of TFA. $(BCOD)_4$-P was prepared from the acid-catalyzed cyclotetramerization of BCOD pyrrole bearing a hydroxymethyl group. After zinc metalation, silica-gel separation gave a tetra-BCOD por zinc complex and metal-free por was obtained upon the treatment of TFA. The isolation and purity of all BCOD-fused pors were checked by spectroscopic measurement, including mass spectrometry (MS), ^{1}H NMR, ultraviolet-visible (UV-Vis) absorption, and fluorescence.

6.2.4 Synthesis and Characterization of Benzoporphyrin La^{III} Triple-Decker Complexes

The reaction of monobenzoporphyrin (MBP) with lanthanum acac under various conditions has been attempted, which, however, gave a merely incomplete metalation mixture, including metal-free MBP, a lanthanum double-decker complex, and a bis-lanthanum triple-decker complex, revealed by thin-layer chromatography (TLC) and matrix-assisted laser desorption ionization–time of flight–mass spectrometry (MALDI-TOF-MS) analysis. Next, we tried the retro Diels–Alder reaction of the BCOD moiety and metal coordination in a one-pot reaction. The treatment of BCOD-P with lanthanum salt in 1,2,4-TCB under reflux condition for three days, followed by the removal of the solvent in vacuo, gave a triple-decker complex, which was isolated by alumina column chromatography, giving a yield of 72% (Fig. 6.20).

By MALDI-TOF-MS, the molecular weight of $La_2(MBP)_3$ was determined to be m/z = 1941.1 ([M]$^+$); (for $C_{114}H_{126}La_2N_{12}$, it was calculated to be 1940.8). The ^{1}H NMR spectrum of the diamagnetic complex $La_2(MBP)_3$ exhibited four singlet peaks, at 8.46, 8.45, 8.14, and 7.91 parts per million (ppm), assigned as *meso* proton; four resonances due to fused benzene proton, at 9.46, 8.78, 8.72, and 7.94 ppm; and proton signals of peripheral alkyl groups in the range of 3.81–0.64 ppm (Fig. 6.21). The upper and bottom pors were equivalent, and the three pors of $La_2(MBP)_3$ showed the same symmetry as metal-free MBP [70].

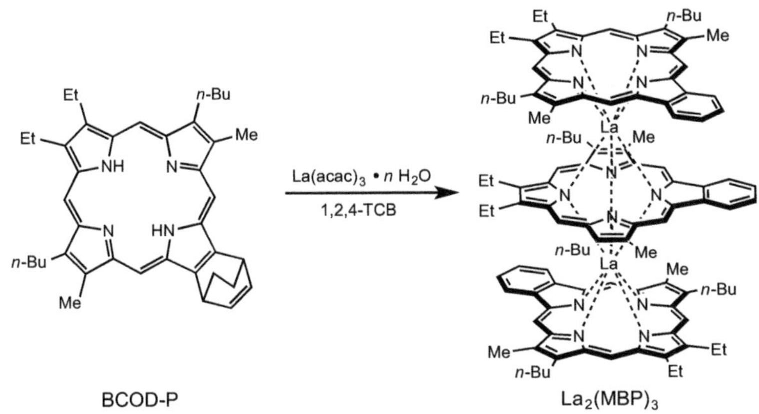

Figure 6.20 Synthesis of a monobenzoporphyrin La^{III} triple-decker complex.

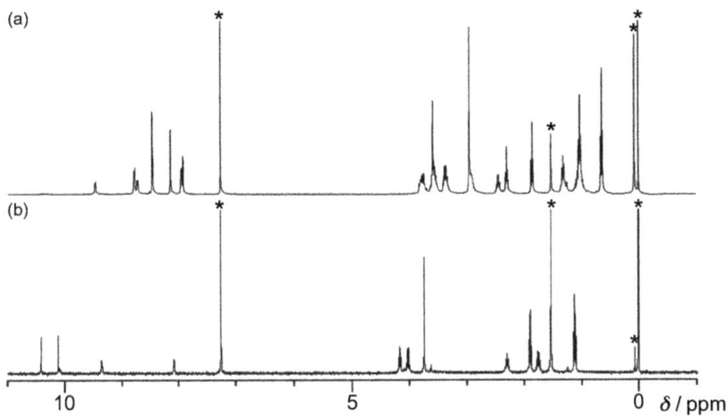

Figure 6.21 ^{1}H NMR spectra of (a) $La_2(MBP)_3$ and (b) metal-free MBP in $CDCl_3$. An inner N-H proton signal appeared at −3.59 ppm (not shown here). The asterisks indicate solvents and impurities.

Diffraction-grade crystals of $La_2(MBP)_3$ were obtained by vapor diffusion of EtOH into a solution of $La_2(MBP)_3$ in ethyl acetate. The crystallographic data are as follows: $C_{114}H_{126}La_2N_{12}$, a = 6.0150(3) Å, b = 25.3670(5) Å, c = 24.0394(4) Å, β = 104.112(2)°, monoclinic, $P2_1/n$, and Z = 4 (Figs. 6.22 and 6.23) [70]. The two La^{III} atoms are coordinated in a twisted antiprism geometry, and the twisted angles of upper and bottom pors against the central por are 29.6° and 34.4°. The La–N bond lengths of upper and bottom pors are in the range of 2.491(2)–2.517(2) Å and the La–N bond lengths of central pors are in the range of 2.751(2)–2.807(2) Å. The two La ions are 3.75 Å apart, and the distances between the outer and central 4N planes are 3.25 Å and 3.27 Å.

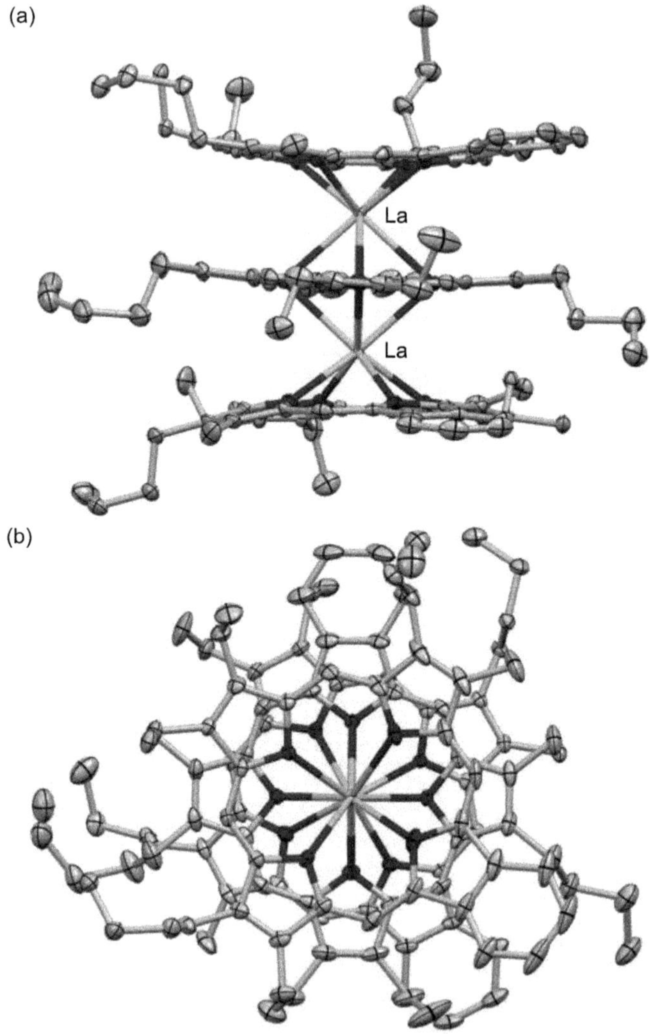

Figure 6.22 X-ray crystal structure of $La_2(MBP)_3$ obtained from EtOAc/EtOH condition: (a) side view and (b) top view. The thermal ellipsoids are scaled to the 50% probability level. Hydrogen atoms are omitted for clarity. Several peripheral alkyl groups are disordered. Only one conformation of disordered structures is shown.

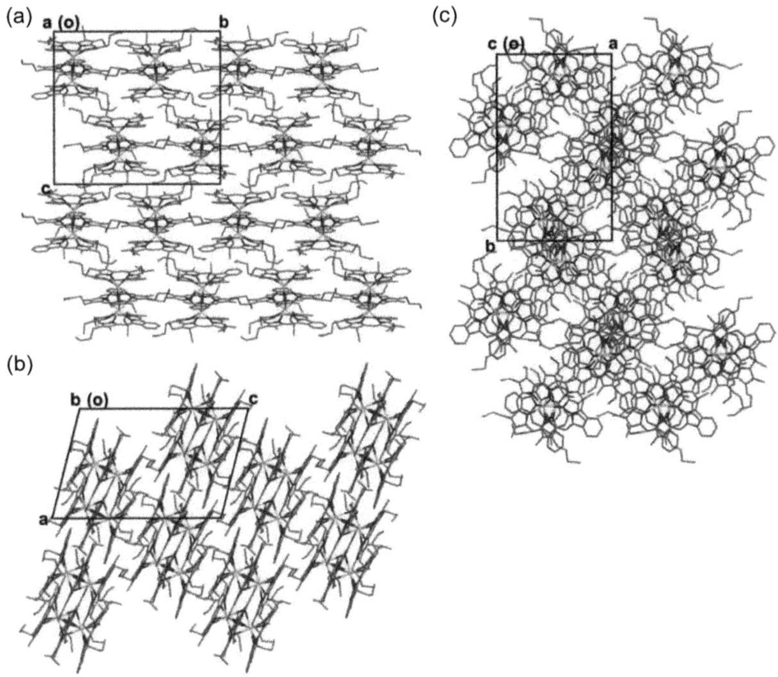

Figure 6.23 Packing structures of $La_2(MBP)_3$ obtained from EtOAc/EtOH condition: views along the (a) *a* axis, (b) *b* axis, and (c) *c* axis.

For possible influence of solvent on the crystal quality, various crystallization conditions were examined. X-ray crystallographic data of crystals were obtained from the $CHCl_3$/MeOH condition; the cell data are as follows: $C_{114}H_{126}La_2N_{12}$, 2 ($CHCl_3$), a = 16.0663(3) Å, b = 24.3422(4) Å, c = 27.4031(5) Å, orthorhombic, *Pbca*, and Z = 4 (Figs. 6.24 and 6.25). The complex is centrosymmetric, with an inversion center in the middle point of two La atoms. In this crystal, both La atoms are coordinated in an antiprism geometry and the $CHCl_3$ molecules are cocrystallized above and below the triple-decker complex in a disordered fashion. The La–La distance is 3.71 Å.

The reaction condition above is applicable to *adj*-dibenzoporphyrin (*adj*-DBP), *opp*-DBP, tribenzoporphyrin (TriBP), and tetrabenzoporphyrin (TBP) (Fig. 6.26). When *adj*-$(BCOD)_2$-P was used instead of $(BCOD)_2$-P, the triple-decker complex La_2(*app*-DBP$)_3$ was isolated (an 87% yield). The triple-decker complex of

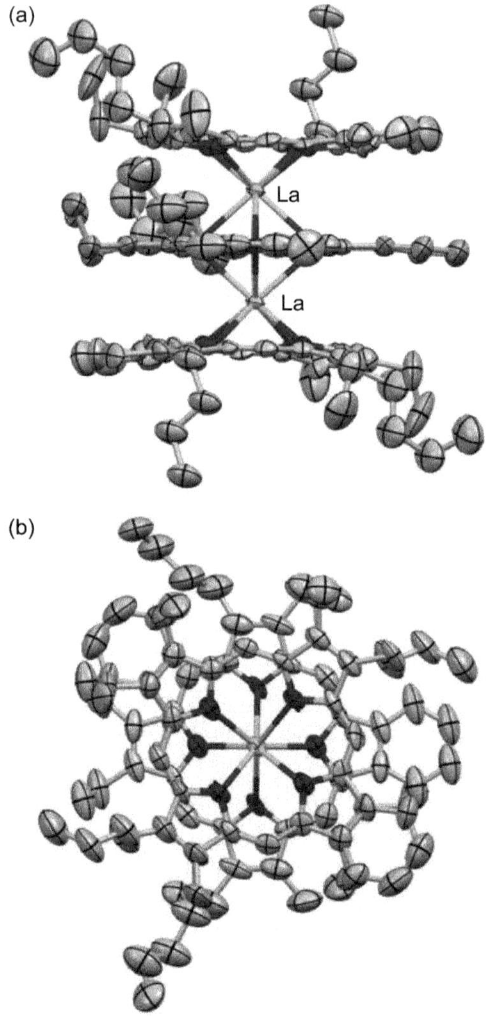

Figure 6.24 X-ray crystal structure of $La_2(MBP)_3$ obtained from $CHCl_3$/MeOH condition: (a) side view and (b) top view. The thermal ellipsoids are scaled to the 50% probability level. Solvent molecules and hydrogen atoms are omitted for clarity. One fused benzene moiety and several peripheral alkyl groups are disordered. Only one conformation of disordered structures is shown.

La_2(*adj*-DBP)$_3$ exhibits the parent ion peak at m/z = 1838.4 ($[M]^+$); (for $C_{108}H_{96}La_2N_{12}$, it was calculated to be 1838.6). Single crystals suitable for X-ray diffraction analysis were grown from vapor diffusion of iPrOH into a tetrahydrofuran (THF) solution. The cell parameters are similar to those of $La_2(MBP)_3$ obtained from a $CHCl_3$/MeOH solution and as follows: $C_{108}H_{96}La_2N_{12}$, 2 (THF), a = 16.5100(3) Å, b = 22.5738(3) Å, c = 25.1412(4) Å, orthorhombic, *Pbca*, and Z = 4 (Fig. 6.27). Like $La_2(MBP)_3$, this crystal structure also has a center of symmetry, so a half of the complex and one THF molecule consist of the asymmetric unit of the crystal. The two equivalent La^{III} atoms are coordinated in a twisted antiprism geometry, and the twisted angle of an outer por against the central por is 43.5°. The La–N bond lengths of outer pors are in the range of 2.491(5)–2.519(5) Å and La–N bond lengths of central pors are in the range of 2.736(5)–2.779(4) Å. The two La ions are 3.73 Å apart, and the distance between the outer and central 4N planes is 3.25 Å.

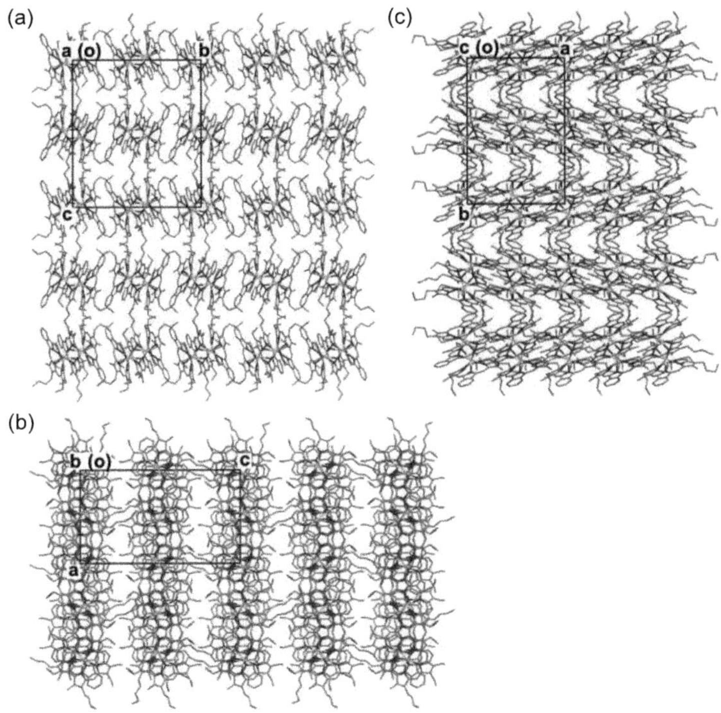

Figure 6.25 Packing structures of $La_2(MBP)_3$ obtained from $CHCl_3$/MeOH condition: views along the (a) a axis, (b) b axis, and (c) c axis.

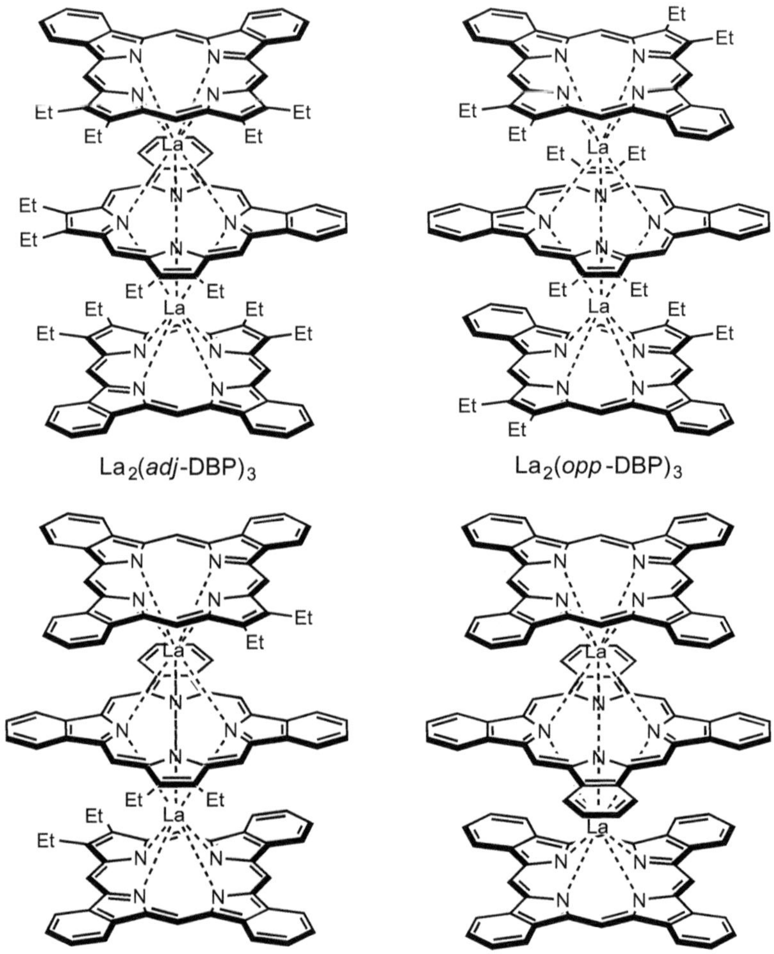

Figure 6.26 Schematic structures of La_2(*adj*-DBP)$_3$, La_2(*opp*-DBP)$_3$, La_2(TriBP)$_3$, and La_2(TBP)$_3$.

La^{III} metalation was conducted with *opp*-(BCOD)$_2$-P; a bis-La complex was isolated (a 95% yield). The complex exhibits the parent ion peak at $m/z = 1838.5$ ([M]$^+$); (for $C_{108}H_{96}La_2N_{12}$, it was calculated to be 1838.6). Single crystals from a $CHCl_3$/MeOH solution were analyzed by an X-ray diffraction technique. The cell parameters are as follows: $C_{108}H_{96}La_2N_{12}$, 2 ($CHCl_3$), $a = 19.1079(5)$ Å, $b = 21.0595(6)$ Å, $c = 23.4083(5)$ Å, orthorhombic, *Pbca*, and $Z = 4$ (Fig. 6.28). As might be expected, the crystal structure of La_2(*opp*-DBP)$_3$ is similar to that of La_2(*adj*-DBP)$_3$. The two equivalent La^{III} atoms are coordinated in a twisted antiprism geometry, and the twisted angle of an outer por against the central por is 43.7°. The La–N bond lengths of outer pors are in the range of 2.498(5)–2.509(5) Å and La–N bond lengths of central pors are in the range of 2.736(5)–2.770(4) Å. The two La ions are 3.69 Å apart, and the distance between the outer and central 4N planes is 3.25 Å.

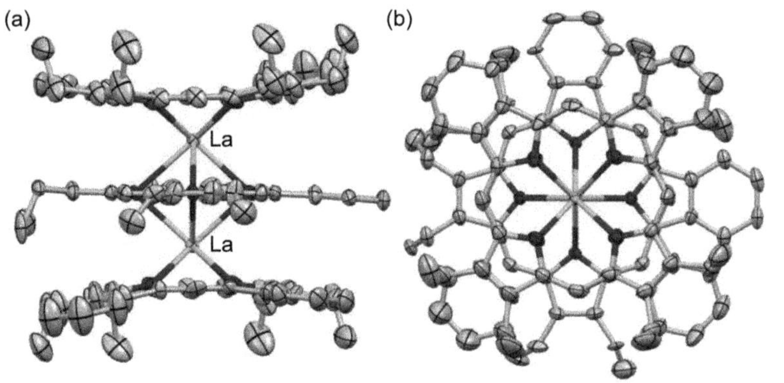

Figure 6.27 X-ray crystal structure of La_2(*adj*-DBP)$_3$: (a) side view and (b) top view. The thermal ellipsoids are scaled to the 50% probability level. Solvent molecules and hydrogen atoms are omitted for clarity. One fused benzene moiety and several peripheral alkyl groups are disordered. Only one conformation of disordered structures is shown.

Next, La^{III} metalation of (BCOD)$_3$-P was investigated in the conditions above; the corresponding bis-La complex was obtained at a 92% yield. The complex displayed the parent ion peak at m/z = 1820.5 ([M]$^+$); (for $C_{108}H_{78}La_2N_{12}$, it was calculated to be 1820.5). Single crystals suitable for X-ray diffraction analysis were also obtained from the $CHCl_3$/MeOH solution. The cell parameters are as follows: $C_{108}H_{78}La_2N_{12}$, 2 ($CHCl_3$), a = 18.6771(3) Å, b = 21.2957(3) Å, c = 22.7434(7) Å, orthorhombic, *Pbca*, and Z = 4 (Fig. 6.29). The crystal structure of La_2(TriBP)$_3$ is also similar to the above-mentioned structures. The two equivalent La^{III} atoms are coordinated in a twisted antiprism geometry, and the twisted angle of an outer por against the central por is 43.7°. The La–N bond lengths of outer pors are in the range of 2.503(3)–2.512(3) Å and La–N bond lengths of central pors are in the range of 2.750(2)–2.767(3) Å. The two La ions are 3.69 Å apart, and the distance between the outer and central 4N planes is 3.24 Å. The outer pors of La_2(MBP)$_3$, La_2(*adj*-DBP)$_3$, La_2(*opp*-DBP)$_3$, and La_2(TriBP)$_3$ are more bent than those of metal-free pors due to minimization of repulsion of peripheral substituents.

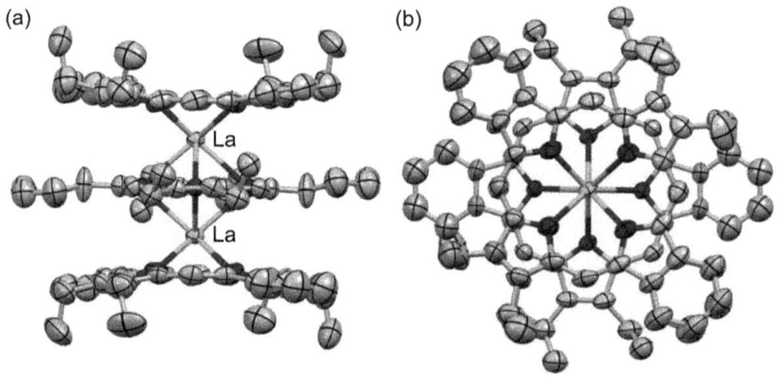

Figure 6.28 X-ray crystal structure of La_2(*opp*-DBP)$_3$: (a) side view and (b) top view. The thermal ellipsoids are scaled to the 50% probability level. Solvent molecules and hydrogen atoms are omitted for clarity. One fused benzene moiety and several peripheral alkyl groups are disordered. Only one conformation of disordered structures is shown.

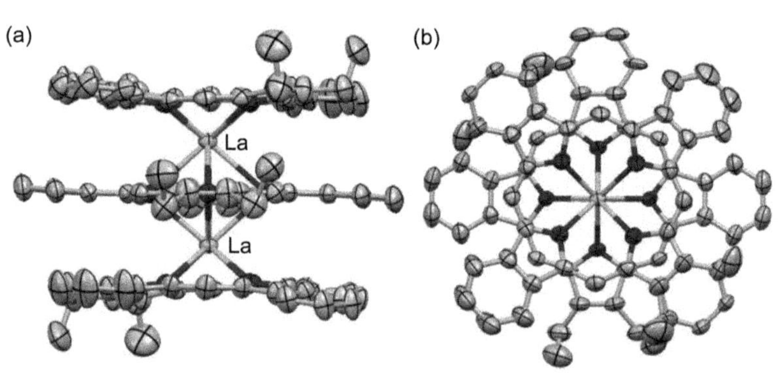

Figure 6.29 X-ray crystal structure of La_2(TriBP)$_3$: (a) side view and (b) top view. The thermal ellipsoids are scaled to the 50% probability level. Solvent molecules and hydrogen atoms are omitted for clarity. One fused benzene moiety and several peripheral alkyl groups are disordered. Only one conformation of disordered structures is shown.

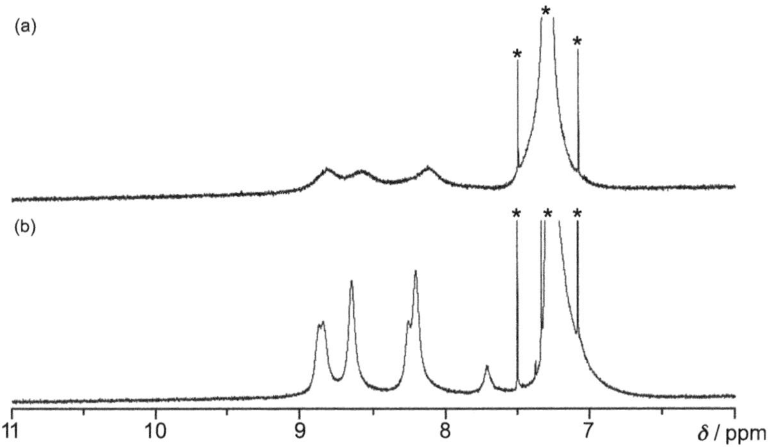

Figure 6.30 ^{1}H NMR spectra of $La_2(TBP)_3$ in $CDCl_3$ at (a) 25°C and (b) −60°C. The asterisks indicate solvents and impurities.

Finally, complexation of $(BCOD)_4$-P with La salt was examined in the same condition, which proceeded successfully, with an 84% yield. MALDI-TOF-MS measurement revealed the parent ion peak of $La_2(TBP)_3$ at m/z = 1802.2 ($[M]^+$) (for $C_{108}H_{60}La_2N_{12}$, it was calculated to be 1802.3). The ^{1}H NMR spectrum of $La_2(TBP)_3$ at 25°C displayed several broad peaks in the range of 9–8 ppm. The observed peaks sharpened at -60°C, but assignment has not been successful thus far (Fig. 6.30). The single crystals suitable for X-ray diffraction analysis were grown by slow evaporation in CS_2/pyridine/MeOH. The cell parameters are as follows: $C_{108}H_{60}La_2N_{12}$, $a = b$ = 14.1310(3) Å, c = 19.8722(10) Å, tetragonal, $I4/m$, and Z = 2 (Figs. 6.31 and 6.32) [70]. The molecular four-fold axis is along the c axis. The two identical La^{III} ions are coordinated in a twisted antiprism geometry, and the twisted angle of an outer por against the central por is 37.5°. The distance between the outer and central 4N planes is 3.20 Å. The La–N bond length of an outer por is 2.490(4) Å and that of the central por is 2.769(4) Å. The distance of intramolecular La ions is 3.70 Å, and the nearest intermolecular La ions is 11.8 Å. The conformation of exterior pors for $La_2(TBP)_3$ is more planar than that of other benzoporphyrins.

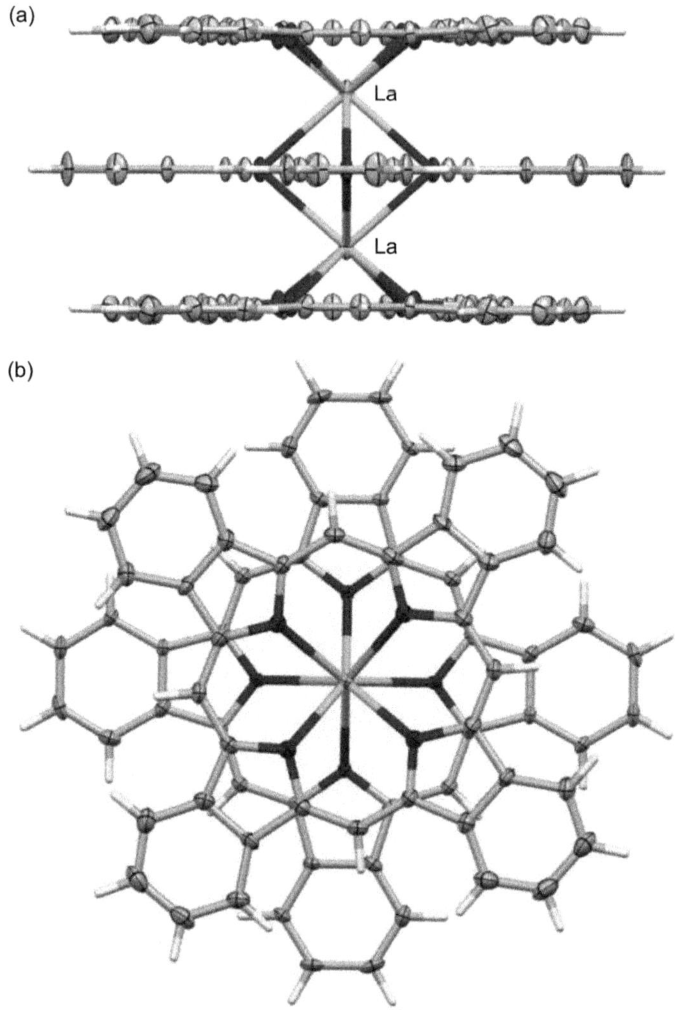

Figure 6.31 X-ray crystal structure of $La_2(TBP)_3$: (a) side view and (b) top view. The thermal ellipsoids are scaled to the 50% probability level.

The absorption spectra of bis-La triple-decker complexes in CH_2Cl_2 showed a Soret band around 400 nm (ε = 310,000–420,000) and Q bands in the range of 470–670 nm (Fig. 6.33). Both absorption maxima were red-shifted as the fused benzene rings increased in number. These results indicate no interaction between pors or por–La in the neutral state.

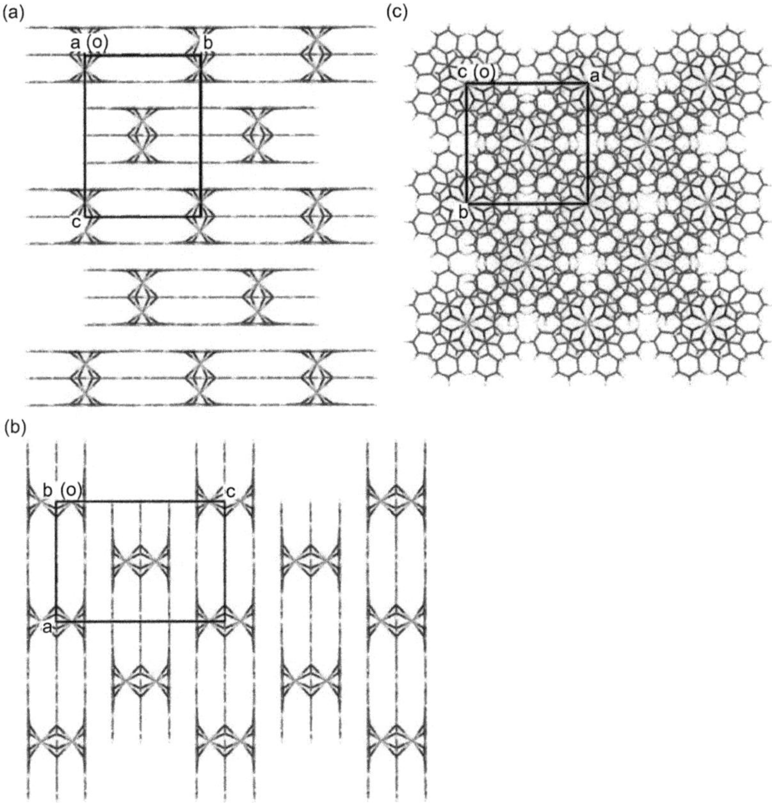

Figure 6.32 Packing structures of $La_2(TBP)_3$: views along the (a) *a* axis, (b) *b* axis, and (c) *c* axis.

Table 6.1 Distances between metals, porphyrins (4N planes), and the twisted angle between porphyrins of La_2Por_3

Ligand	MBP[a]	MBP[b]	*adj*-DBP	*opp*-DBP	TriBP	TBP
Distance of La–La (Å)	3.75	3.71	3.73	3.69	3.69	3.70
Distance of por–por (Å)	3.25, 3.27	3.26	3.25	3.25	3.24	3.20
Twisted angle (°)	29.6, 34.4	36.8	43.5	43.7	43.7	37.5

[a]Crystallization from ethyl acetate and ethanol

[b]Crystallization from $CHCl_3$ and methanol

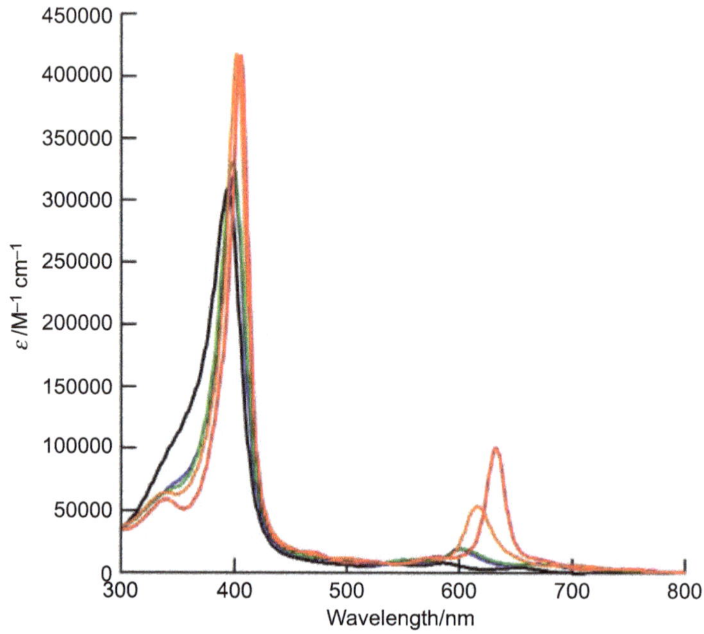

Figure 6.33 The absorption spectra of $La_2(MBP)_3$ (black), La_2(*adj*-DBP)$_3$ (blue), La_2(*opp*-DBP)$_3$ (green), $La_2(TriBP)_3$ (orange), and $La_2(TBP)_3$ (red) in CH_2Cl_2.

The redox behavior of bis-La triple-decker complexes was investigated by cyclic voltammetry measurement. Only the data of $La_2(MBP)_3$, La_2(*adj*-DBP)$_3$, and La_2(*opp*-DBP)$_3$ were collected due to poor solubility of other complexes. Four reversible oxidation waves were observed, which were assigned to be macrocyclic oxidation on the basis of a previous study for $La_2(OEP)_3$ [71], and no reduction was observed in the potential range of measurement (Fig. 6.34). The more conjugated system extended, the more negative the oxidation potential shifted.

6.2.5 Synthesis and Characterization of Benzoporphyrin Tb^{III} Triple-Decker Complexes

The established complexation method in the section above was also valid for Tb^{III} metalation of benzoporphyrins, except for the reaction time. In Tb^{III} metalation, it took four days to convert the metal-free pors into triple-decker complexes (Fig. 6.35).

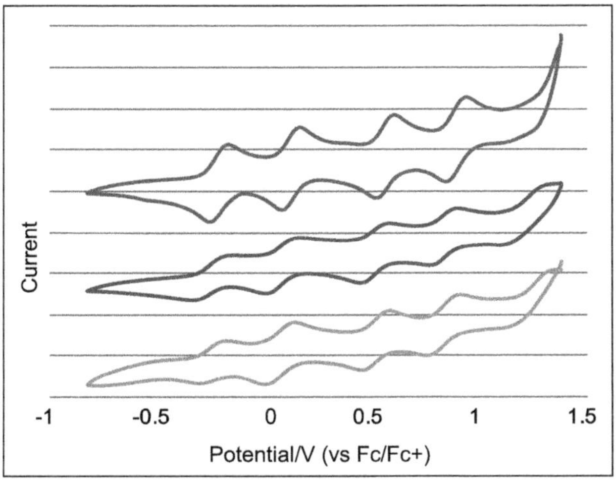

Figure 6.34 Cyclic voltammograms of $La_2(MBP)_3$ (top), $La_2(\textit{adj}\text{-DBP})_3$ (middle), and $La_2(\textit{opp}\text{-DBP})_3$ (bottom) in anhydrous CH_2Cl_2 containing 0.1 M $TBAPF_6$ on a Pt working electrode (scan rate 0.1 V/s).

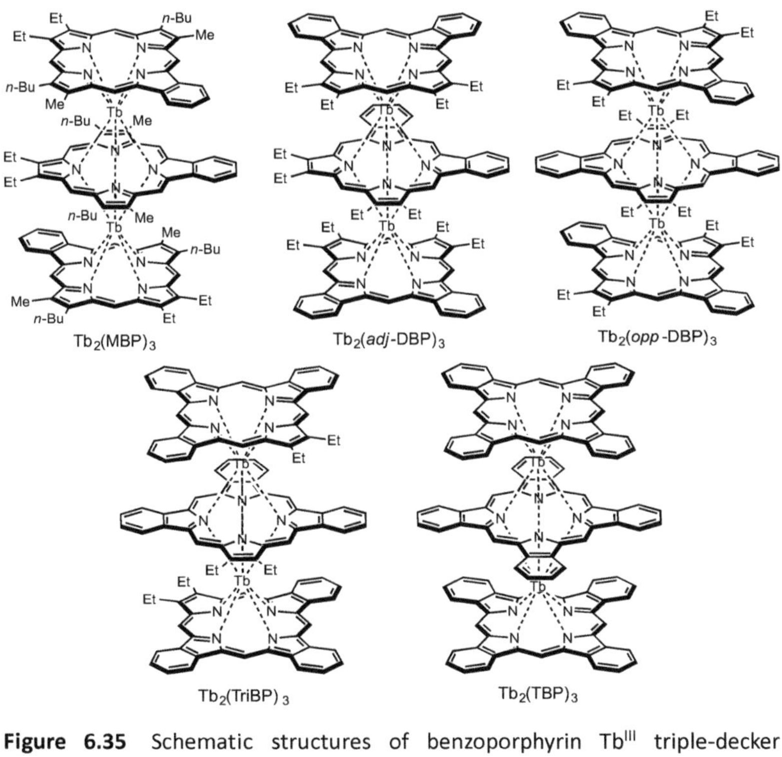

Figure 6.35 Schematic structures of benzoporphyrin Tb^{III} triple-decker complexes.

The isolated yields of $Tb_2(MBP)_3$, Tb_2(*adj*-DBP)$_3$, Tb_2(*opp*-DBP)$_3$, $Tb_2(TriBP)_3$, and $Tb_2(TBP)_3$ were 73%, 87%, 85%, 77%, and 65%, respectively. The isolation and identification of all five triple-decker complexes were elucidated by MALDI-TOF-MS and elemental analysis.

Diffraction-grade crystals of $Tb_2(MBP)_3$ were obtained by vapor diffusion of MeOH into a solution of $Tb_2(MBP)_3$ in $CHCl_3$. The crystallographic data are as follows: $C_{114}H_{126}N_{12}Tb_2$, 2 ($CHCl_3$), a = 15.9247(2) Å, b = 23.9196(3) Å, c = 27.4661(4) Å, orthorhombic, *Pbca*, and Z = 4 (Fig. 6.36) [70]. The two equivalent Tb^{III} atoms are coordinated in a twisted antiprism geometry, and the twisted angle of an outer por against the central por is 40.4°. The Tb–N bond lengths of outer pors are in the range of 2.401(3)–2.415(4) Å, and the Tb–N bond lengths of central pors are in the range of 2.658(3)–2.689(3) Å. The two Tb ions are 3.55 Å apart, and the distance between the outer and central 4N planes is 3.04 Å.

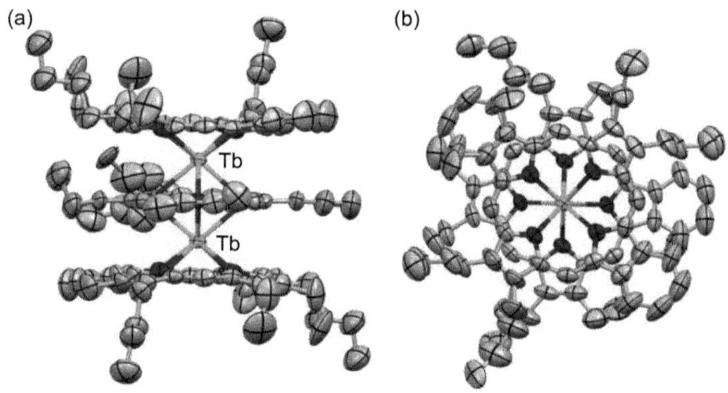

Figure 6.36 X-ray crystal structure of $Tb_2(MBP)_3$: (a) side view and (b) top view. The thermal ellipsoids are scaled to the 50% probability level. Solvent molecules and hydrogen atoms are omitted for clarity. Several peripheral alkyl groups are disordered. Only one conformation of disordered structures is shown.

Single crystals of Tb_2(*adj*-DBP)$_3$ were also obtained from a THF/iPrOH solution. The crystallographic data are as follows: $C_{108}H_{96}N_{12}Tb_2$, 2 (THF), a = 16.4568(3) Å, b = 22.2224(5) Å, c = 25.0879(5) Å, orthorhombic, *Pbca*, and Z = 4 (Fig. 6.37). The two equivalent Tb^{III} atoms are coordinated in a twisted antiprism geometry, and the twisted angle of an outer por against the central por is 43.2°. The Tb–N bond lengths of outer pors are in the range of 2.393(4)–2.424(3) Å, and the Tb–N bond lengths of central pors are in the range of 2.650(3)–2.715(3) Å. The two Tb ions are 3.57 Å apart, and the distance between the outer and central 4N planes is 3.03 Å.

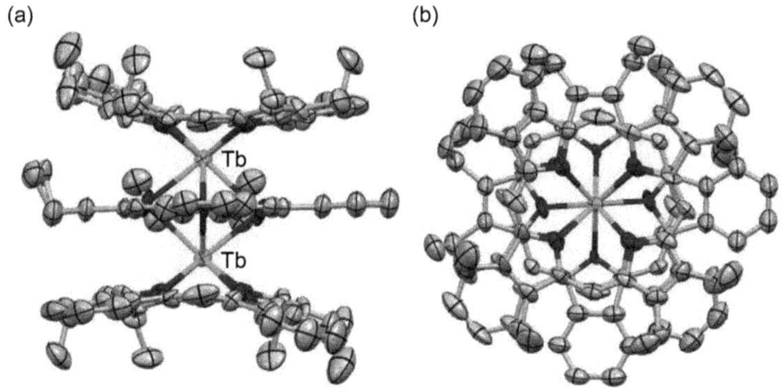

Figure 6.37 X-ray crystal structure of Tb_2(*adj*-DBP)$_3$: (a) side view and (b) top view. The thermal ellipsoids are scaled to the 50% probability level. Solvent molecules and hydrogen atoms are omitted for clarity. Only one conformation of disordered structures is shown.

Tb_2(*opp*-DBP)$_3$ was crystalized by vapor diffusion of iPrOH into a solution of Tb_2(*opp*-DBP)$_3$ in THF. The crystallographic data are as follows: $C_{108}H_{96}N_{12}Tb_2$, 2 (THF), a = 18.8704(4) Å, b = 21.1421(5) Å, c = 23.4588(4) Å, orthorhombic, *Pbca*, and Z = 4 (Fig. 6.38). The two equivalent Tb^{III} atoms are coordinated in a twisted antiprism geometry, and the twisted angle of an outer por against the central por is 44.0°. The Tb–N bond lengths of outer pors are in the range of 2.405(7)–2.416(6) Å, and the Tb–N bond lengths of central pors are in the range of 2.664(6)–2.718(5) Å. The two Tb ions are 3.56 Å apart, and the distance between the outer and central 4N planes is 3.02 Å.

Diffraction-grade crystals of $Tb_2(TriBP)_3$ were obtained by vapor diffusion of MeOH into a solution of $Tb_2(TriBP)_3$ in $CHCl_3$. The crystallographic data are as follows: $C_{108}H_{78}N_{12}Tb_2$, 2 ($CHCl_3$), a = 18.4364(4) Å, b = 21.0472(5) Å, c = 22.7734(5) Å, orthorhombic, *Pbca*, and Z = 4 (Fig. 6.39). The two equivalent Tb^{III} atoms are coordinated in a twisted antiprism geometry, and the twisted angle of an outer por against the central por is 44.2°. The Tb–N bond lengths of outer pors are in the range of 2.409(4)–2.416(4) Å, and the Tb–N bond lengths of central pors are in the range of 2.665(4)–2.696(4) Å. The two Tb ions are 3.54 Å apart, and the distance between the outer and central 4N planes is 3.02 Å.

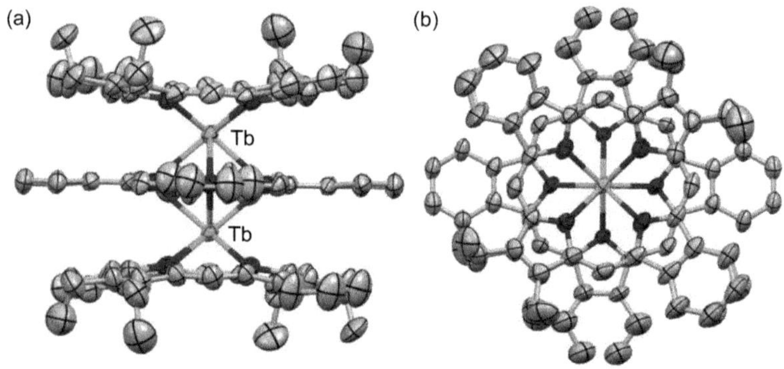

Figure 6.38 X-ray crystal structure of Tb_2(*opp*-DBP)$_3$: (a) side view and (b) top view. The thermal ellipsoids are scaled to the 50% probability level. Solvent molecules and hydrogen atoms are omitted for clarity. One fused benzene moiety and several peripheral alkyl groups are disordered. Only one conformation of disordered structures is shown.

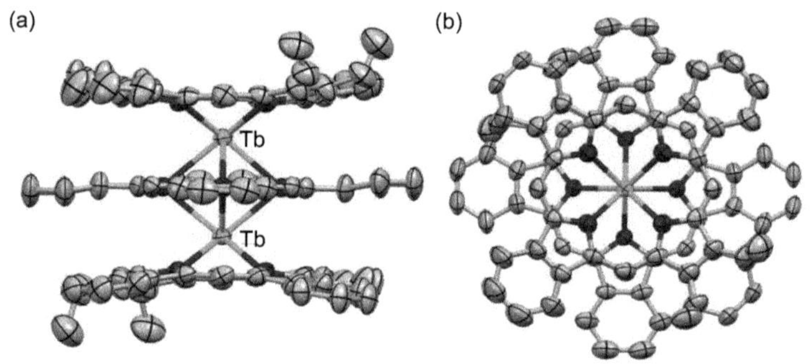

Figure 6.39 X-ray crystal structure of Tb_2(TriBP)$_3$: (a) side view and (b) top view. The thermal ellipsoids are scaled to the 50% probability level. Solvent molecules and hydrogen atoms are omitted for clarity. One fused benzene moiety and several peripheral alkyl groups are disordered. Only one conformation of disordered structures is shown.

These layered structures of Tb complexes are similar to those of phthalocyanines complexes that exhibit SMM properties. The relationship between structure and magnetic property is under study.

Table 6.2 Distances between metals, porphyrins (4N planes), and the twisted angle between pophyrins of Tb_2Por_3

Ligand	MBP	*adj*-DBP	*opp*-DBP	TriBP
Tb–Tb distance (Å)	3.55	3.57	3.56	3.54
Por–por distance (Å)	3.04	3.03	3.02	3.02
Twisted angle (°)	40.4	43.2	44.0	44.2

The absorption spectra of bis-Tb triple-decker complexes in CH2Cl2 showed a Soret band around 400 nm (e = 230,000–340,000) and Q bands in the range of 470–670 nm (Fig. 6.40). Both absorption maxima were red-shifted as the fused benzene rings increased in number. These results indicated no interaction between pors or por–Tb in the neutral state.

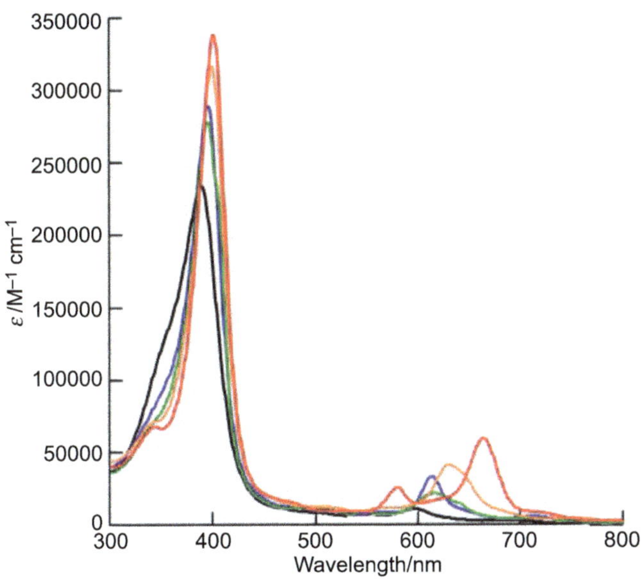

Figure 6.40 The absorption spectra of $Tb_2(MBP)_3$ (black), Tb_2(*adj*-DBP)$_3$ (blue), Tb_2(*opp*-DBP)$_3$ (green), $Tb_2(TriBP)_3$ (orange), and $Tb_2(TBP)_3$ (red) in CH_2Cl_2.

The redox behavior of bis-Tb triple-decker complexes was investigated by cyclic voltammetry measurement. Only the data of

$Tb_2(MBP)_3$, Tb_2(*adj*-DBP)$_3$, and Tb_2(*opp*-DBP)$_3$ were collected due to poor solubility of other complexes. Four reversible oxidation waves were observed, which were assigned to be macrocyclic oxidation on the basis of the previous study for $La_2(OEP)_3$, and no reduction was observed in the potential range of measurement (Fig. 6.41). The more conjugated system extended, the more negative the oxidation potential shifted in common with the La complexes.

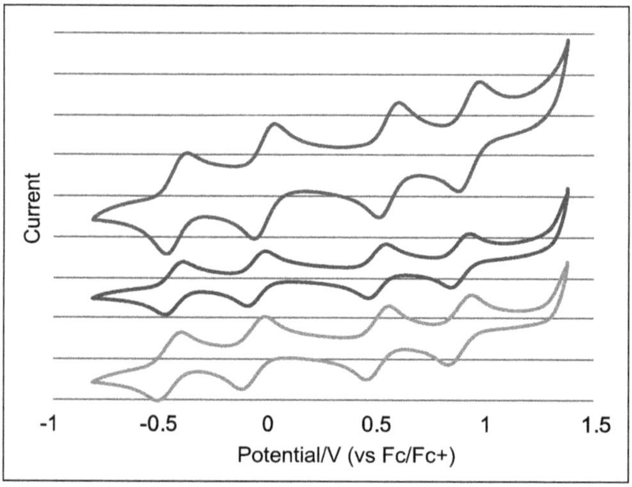

Figure 6.41 Cyclic voltammograms of $Tb_2(MBP)_3$ (top), Tb_2(*adj*-DBP)$_3$ (middle), and Tb_2(*opp*-DBP)$_3$ (bottom) in anhydrous CH_2Cl_2 containing 0.1 M $TBAPF_6$ on a Pt working electrode (scan rate 0.1 V/s).

6.2.6 Monobenzoporphyrin and Other Lanthanide Triple-Decker Complexes

The established complexation method in the section above was applied to Eu^{III}, Gd^{III}, and Dy^{III} metalation of benzoporphyrins (Fig. 6.42).

At present, synthesis and structural characterization of these three triple-decker complexes were accomplished. The isolated yields of $Eu_2(MBP)_3$, $Gd_2(MBP)_3$, and $Dy_2(MBP)_3$ were 60%, 62%, and 58%, respectively. All three complexes showed similar properties, and single crystals suitable for X-ray diffraction analysis were obtained by MeOH diffusion into the corresponding $CHCl_3$ solution.

The three triple-decker complexes crystallized in the orthorhombic system in the space group *Pbca*. The cell parameters are as follows:

- $Eu_2(MBP)_3$: $C_{114}H_{126}Eu_2N_{12}$, 2 ($CHCl_3$), a = 16.001(4) Å, b = 23.906(7) Å, c = 27.380(8) Å, and Z = 4 (Fig. 6.43) [70].
- $Gd_2(MBP)_3$: $C_{114}H_{126}Gd_2N_{12}$, 2 ($CHCl_3$), a = 15.9359(2) Å, b = 23.9162(3) Å, c = 27.4153(3) Å, and Z = 4 (Fig. 6.44) [68].
- $Dy_2(MBP)_3$: $C_{114}H_{126}Dy_2N_{12}$, 2 ($CHCl_3$), a = 15.9132(2) Å, b = 23.8357(3) Å, c = 27.4522(4) Å, and Z = 4 (Fig. 6.45).

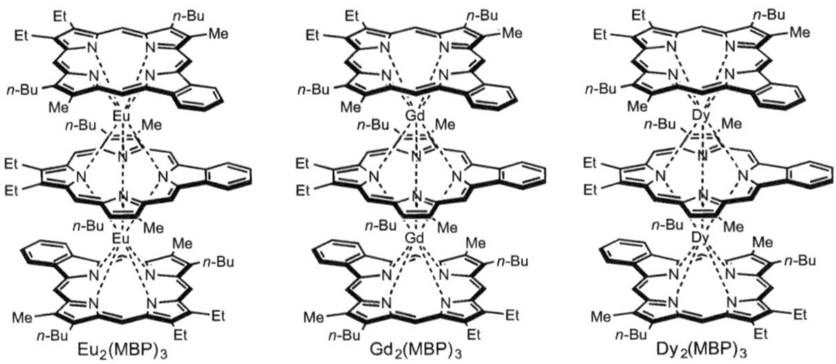

Figure 6.42 Schematic structures of monobenzoporphyrin Eu^{III}, Gd^{III}, and Dy^{III} triple-decker complexes.

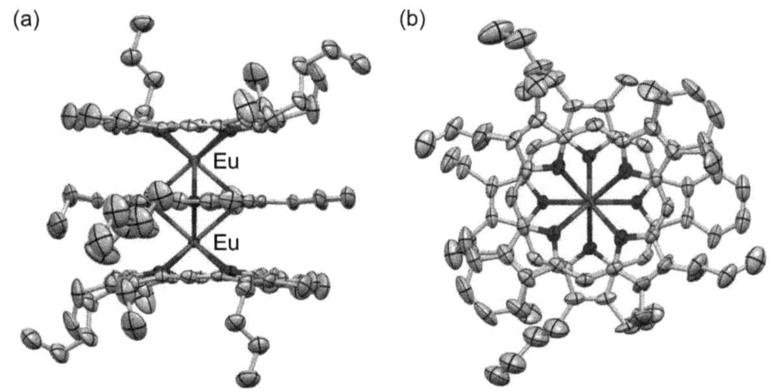

Figure 6.43 X-ray crystal structure of $Eu_2(MBP)_3$: (a) side view and (b) top view. The thermal ellipsoids are scaled to the 50% probability level. Solvent molecules and hydrogen atoms are omitted for clarity. Several peripheral alkyl groups are disordered. Only one conformation of disordered structures is shown.

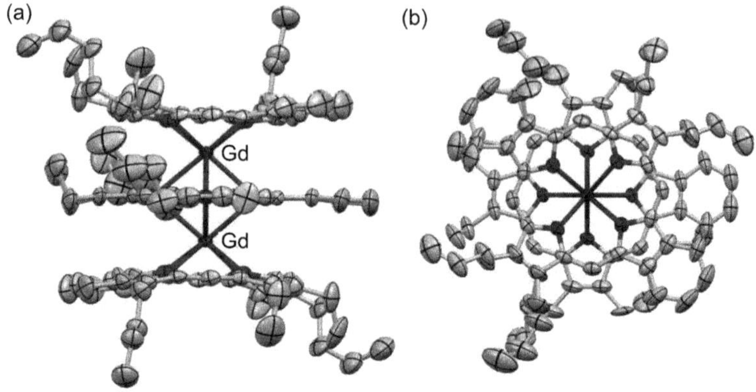

Figure 6.44 X-ray crystal structure of $Gd_2(MBP)_3$: (a) side view and (b) top view. The thermal ellipsoids are scaled to the 50% probability level. Solvent molecules and hydrogen atoms are omitted for clarity. Several peripheral alkyl groups are disordered. Only one conformation of disordered structures is shown.

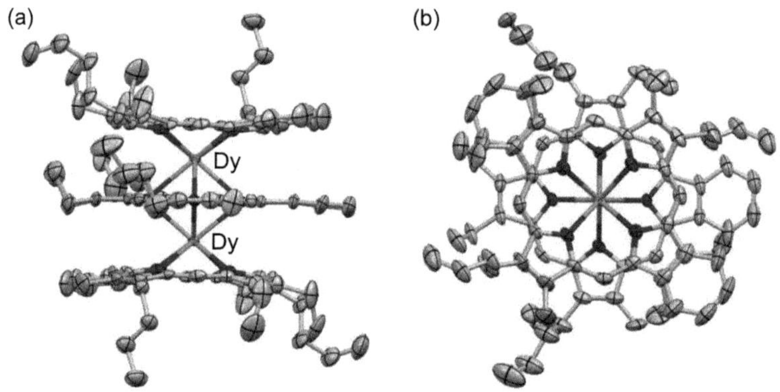

Figure 6.45 X-ray crystal structure of $Dy_2(MBP)_3$: (a) side view and (b) top view. The thermal ellipsoids are scaled to the 50% probability level. Solvent molecules and hydrogen atoms are omitted for clarity. Several peripheral alkyl groups are disordered. Only one conformation of disordered structures is shown.

The two equivalent Eu^{III} atoms are coordinated in a twisted antiprism geometry, and the twisted angle of an outer por against the central por is 41.0°. The Eu–N bond lengths of outer pors are in the range of 2.427(4)–2.429(4) Å, and the Eu–N bond lengths of central pors are in the range of 2.679(4)–2.714(4) Å. The two Eu ions are 3.57 Å apart, and the distance between the outer and central 4N planes is 3.07 Å.

The two Gd^{III} atoms are coordinated in a twisted antiprism geometry, and the twisted angle of an outer por against the central por is 40.9°. The Gd–N bond lengths of outer pors are in the range of 2.413(3)–2.426(3) Å, and the Gd–N bond lengths of central pors are in the range of 2.666(3)–2.695(3) Å. The two Gd ions are 3.56 Å apart, and the distance between the outer and central 4N planes is 3.05 Å.

Table 6.3 Distances between metals, porphyrins (4N planes), and the twisted angle between pophyrins of Ln_2Por_3

Ligand	**$Eu_2(MBP)_3$**	**$Gd_2(MBP)_3$**	**$Dy_2(MBP)_3$**
Ln–Ln distance (Å)	3.57	3.56	3.54
Por–por distance (Å)	3.07	3.05	3.01
Twisted angle (°)	41.0	40.9	41.3

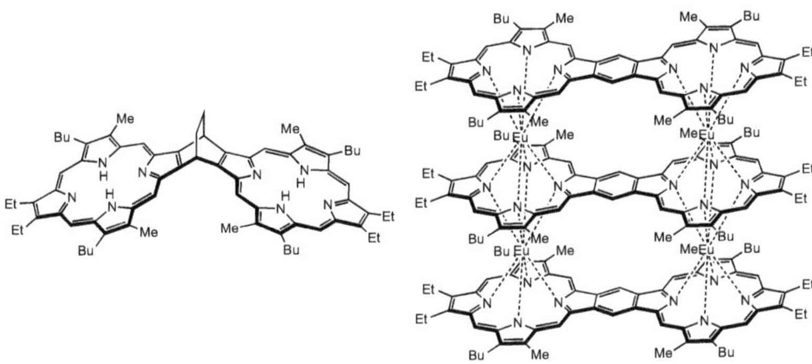

Figure 6.46 Schematic structures of BCOD-bridged bisporphyrin and bisporphyrin Eu^{III} triple-decker complexes.

The two Dy^{III} atoms are coordinated in a twisted antiprism geometry, and the twisted angle of an outer por against the central por is 41.3°. The Dy–N bond lengths of outer pors are in the range of 2.393(4)–2.411(4) Å, and the Dy–N bond lengths of central pors are in the range of 2.652(3)–2.690(4) Å. The two Dy ions are 3.54 Å apart, and the distance between the outer and central 4N planes is 3.01 Å.

6.2.7 Attempt for a Triple-Decker Complex with Bisporphyrin

The construction of a multidecker complex with benzene-bridged bisporphyrin was attempted using BCOD-bridged bisporphyrin [48] as a precursor (Fig. 6.46). The reaction of BCOD-bridged bisporphyrin with Eu salt was conducted in a similar way to the complexation of benzoporphyrins.

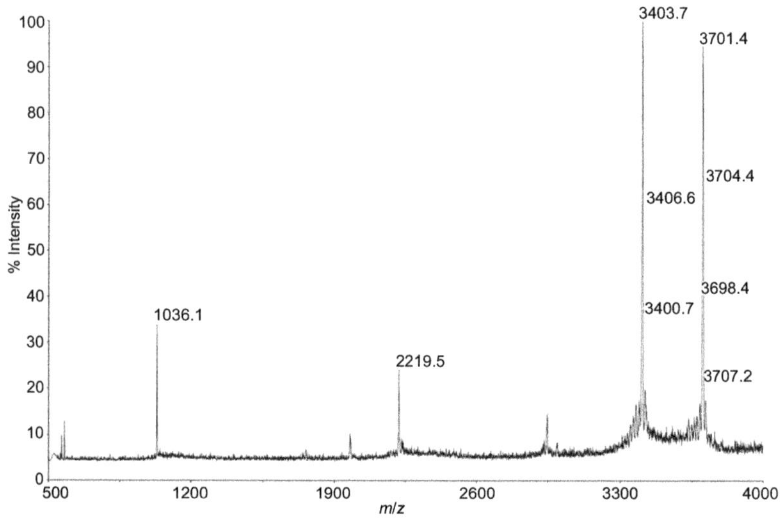

Figure 6.47 MALDI-TOF-MS chart of bisporphyrin Eu^{III} triple-decker complexes.

The MALDI-TOF-MS chart of the reaction mixture showed two intense signals for the triple-decker tetranuclear complex ion, *m/z* = 3701.4 $[M]^+$ (for $C_{210}H_{234}Eu_4N_{24}$, it was calculated to be 3701.6), and the triple-decker dinuclear complex ion, *m/z* = 3403.7 $[M]^+$ (for $C_{210}H_{240}Eu_2N_{24}$, it was calculated to be 3403.8). (Fig. 6.47). The minor peaks assigned as the double-decker mononuclear complex ion, *m/z* = 2219.5 $[M]^+$ (for $C_{140}H_{160}EuN_{16}$, it was calculated to be 2219.2), and metal-free bisporphyrin, *m/z* = 1036.1 $[M+H]^+$ (for $C_{70}H_{82}N_8$, it was calculate to be 1035.7), indicated the instability of the triple-decker tetranuclear complex. Actually, the crude mixture including the tetranuclear complex converted into the dinuclear complex as per MALDI-TOF-MS analysis. One of the reasons for the instability is considered to be the steric hindrance arising from the coordination environment of metal ions. In this case, outer and central ligands in the triple-decker tetranuclear complex are the same molecule (benzene-bridged bisporphyrin), which results in a complete prism geometry. The survey of synthesis of a triple-decker with two different bisporphyrins is now underway.

6.3 Summary and Outlook

The complexation behavior of pors fused with one to four benzene rings with rare earth metals has been reviewed. The reaction of BCOD-fused pors with lanthanide salts smoothly proceeded, and desired triple-decker complexes with π-extended pors were selectively isolated at high yield. A diamagnetic and lower-symmetry complex of $La_2(MBP)_3$ was characterized by ^1H NMR. Their triple-decker structures were unambiguously revealed by X-ray diffraction analysis. Triple-decker complexes with bisporphyrin were also accessible via BCOD-bridged bisporphyrin. By a one-pot reaction with BCOD-fused precursors, various multiple-decker complexes with π-extended pors and bisporphyrins can be obtained. For fundamental research, elucidation on interaction among lanthanide(s) and delocalized radical(s) (f-π, f-f, and π-π) is important and challenging. From this point of view, these triple-decker complexes serve as a good platform for the construction of f- and π-electron collaborative magnetic materials. Multidecker metal pors exhibit several metastable oxidation states and characteristic absorption in the visible region. Investigation of their magnetic and conducting behavior in a partially oxidized and photoexcitation state can lead to the development of new materials.

Acknowledgments

The authors appreciate Prof. T. Naito (Ehime University) for the useful discussion. The analyses including MS, NMR, IR absorption, UV-Vis absorption, and X-ray were conducted at the Integrated Center for Sciences (INCS) and the Advanced Research Support Center (ADRES), Ehime University. The series of work here was supported in part by JSPS KAKENHI Grant Numbers 24750042 and 15K17825 and Grant-in-Aid for Research Promotion, Ehime University.

References

1. Kadish, K. M., Smith, K. M. and Guilard, R. (2000). *The Porphyrin Handbook, Vol. 3: Inorganic, Organometallic and Coordination Chemistry*, Academic Press, San Diego, CA.
2. Barrett, P. A., Dent, C. E. and Linstead, R. P. (1936). Phthalocyanines. Part VII. Phthalocyanine as a co-ordinating group. A general investigation of the metallic derivatives, *J. Chem. Soc.*, pp. 1719–1736.
3. Bennett, W. E., Broberg, D. E. and Baenziger, N. C. (1973). Crystal structure of stannic phthalocyanine, an eight-coordinated tin complex, *Inorg. Chem.*, **12**, pp. 930–936.
4. Buchler, J. W., Kapellmann, H.-G., Knoff, M., Lay, K.-L. and Pfeifer, S. (1983). Metal complexes with tetrapyrrole ligands, XXXI [1] neutral and anionoid bisporphinates of cerium and praeseodymium, *Z. Naturforsch.*, **38B**, pp. 1339–1345.
5. Buchler, J. W., Cian, A. D., Fischer, J., Kihn-Botulinski, M., Paulus, H. and Weiss, R. (1986). Cerium(IV) bis(octaethylporphyrinate) and dicerium(III) tris(octaethy1porphyrinate): parents of a new family of lanthanoid double-decker and triple-decker molecules, *J. Am. Chem. Soc.*, **108**, pp. 3652–3659.
6. M'sadak, M., Roncali, J. and Garnier, F. (1985). Rare-earth substitution effect on the electrochemical properties of lanthanide diphthalocyanines, *J. Electroanal. Chem. Interfacial Electrochem.*, **189**, pp. 99–111.
7. Kasuga, K., Ando, M., Morimoto, H. and Isa, K. (1986). Preparation of new phthalocyanine complexes of yttrium(III) and some lanthanoid(III) ions, *Chem. Lett.*, pp. 1095–1098.
8. Janczak, J. and Kubiak, R. (1993). Crystal and molecular structure of diindiurn triphthalocyaninate at 300 K, *J. Chem. Soc., Dalton Trans.*, **12**, pp. 3809–3812.
9. Buchler, J. W. and Ng, D. K. P. (2000). Metal tetrapyrrole double- and triple-deckers with special emphasis on porphyrin systems, in *The Porphyrin Handbook*, Kadish, K. M., Smith, K. M. and Guilard, R. (eds.), Vol. 3, J Academic Press, San Diego, pp. 245–294.
10. Jiang, J. and Ng, D. K. P. (2009). A decade journey in the chemistry of sandwich-type tetrapyrrolato-rare earth complexes, *Acc. Chem. Res.*, **42**, pp. 79–88.

11. Bian, Y., Zhang, Y., Ou, Z. and Jiang, J. (2010). Chemistry of sandwich tetrapyrrole rare earth complexes, in *Handbook of Porphyrin Science*, Kadish, K. M., Smith, K. M. and Guilard, R. (eds.), Vol. 14, World Scientific, Singapore, pp. 249–460.

12. Pushkarev, V., Tomilova, L. G. and Nemykin, V. N. (2016). Historic overview and new developments in synthetic methods for preparation of the rare-earth tetrapyrrolic complexes, *Coord. Chem. Rev.*, **319**, pp. 110–179.

13. Ishikawa, N., Sugita, M., Ishikawa, T., Koshihara, S. and Kaizu, Y. (2003). Lanthanide double-decker complexes functioning as magnets at the single-molecular level, *J. Am. Chem. Soc.*, **125**, pp. 8694–8695.

14. Lee, S. and Ogawa, T. (2017). Molecular design for single-molecule magnetism of lanthanide complexes, *Chem. Lett.*, **46**, pp. 10–18.

15. Wang, H., Wang, B.-W., Bian, Y., Gao, S. and Jiang, J. (2016). Single-molecule magnetism of tetrapyrrole lanthanide compounds with sandwich multiple-decker structures, *Coord. Chem. Rev.*, **306**, pp. 195–216.

16. Branzoli, F., Carretta, P. Filibian, M., Zoppellaro, G., Graf, M. J., Galan-Mascaros, J. R., Fuhr, O., Brink, S. and Ruben, M. (2009). Spin dynamics in the negatively charged terbium (III) bis-phthalocyaninato complex, *J. Am. Chem. Soc.*, **131**, pp. 4387–4396.

17. Koike, N., Uekusa, H. Ohashi, U., Harnoode, C., Kitamura, F., Ohsaka, T. and Tokuda, K. (1996). Relationship between the skew angle and interplanar distance in four bis(phthalocyaninato)lanthanide(III) tetrabutylammonium salts ($[NBu^n_4][Ln^{III}Pc_2]$; Ln = Nd, Gd, Ho, Lu), *Inorg. Chem.*, **35**, pp. 5798–5804.

18. Tanaka, D., Inose, T. Tanaka, H., Lee, S., Ishikawa, N. and Ogawa, T. (2012). Proton-induced switching of the single molecule magnetic properties of a porphyrin based Tb^{III} double-decker complex, *Chem. Commun.*, **48**, pp. 7796–7798.

19. Yamashita, K., Sakata, N. and Ogawa, T. (2016). Systematic structural elucidation for the protonated form of rare earth bis(porphyrinato) double-decker complexes: direct structural evidence of the location of the attached proton, *Inorg. Chem.*, **55**, pp. 8935–8942.

20. Wang, K., Qi, D., Wang, H., Cao, W., Li, W., Liu, T., Duan, C. and Jiang, J. (2013). Binuclear phthalocyanine-based sandwich-type rare earth complexes: unprecedented two π-bridged biradical-metal integrated SMMs, *Chem. Eur. J.*, **19**, pp. 11162–11166.

21. Wang, H., Wang, K., Tao, J. and Jiang, J. (2012). Twist angle perturbation on mixed (phthalocyaninato)(porphyrinato) dysprosium(III) double-decker SMMs, *Chem. Commun.*, **48**, pp. 2973–2975.
22. Wang, H., Qian, K., Qi, D., Cao, W., Wang, K., Gao, S. and Jiang, J. (2014). Co-crystallized fullerene and mixed (phthalocyaninato)(porphyrinato) dysprosium double-decker SMM, *Chem. Sci.*, **5**, pp. 3214–3220.
23. Fukuda, T., Biyajima, T. and Kobayashi, N. (2010). A discrete quadruple-decker phthalocyanine, *J. Am. Chem. Soc.*, **132**, pp. 6278–6279.
24. Wang, H., Kobayashi, N. and Jiang, J. (2012). New sandwich-type phthalocyaninato–metal quintuple-decker complexes, *Chem. Eur. J.*, **18**, pp. 1047–1049.
25. Wang, H., Qi, D., Xie, Z., Cao, W., Wang, K., Shang, H. and Jiang, J. (2013). A sandwich-type phthalocyaninato metal sextuple-decker complex: synthesis and NLO properties, *Chem. Commun.*, **49**, pp. 889–891.
26. Katoh, K., Horii, Y., Yasuda, N., Wernsdorfer, W., Toriumi, K., Breedlove, B. K. and Yamashita, M. (2012). Multiple-decker phthalocyaninato dinuclear lanthanoid(III) single-molecule magnets with dual-magnetic relaxation processes, *Dalton. Trans.*, **41**, pp. 13582–13600.
27. Horii, Y., Katoh, K., Yasuda, N., Breedlove, B. K. and Yamashita, M. (2015). Effects of f–f interactions on the single-molecule magnet properties of terbium(III)–phthalocyaninato quintuple-decker complexes, *Inorg. Chem.*, **54**, pp. 3297–3305.
28. Sirringhaus, H. (2014). 25th anniversary article: organic field-effect transistors: the path beyond amorphous silicon, *Adv. Mater.*, **26**, pp. 1319–1335.
29. Senge, M. O., Fazekas, M., Notaras, E. G. A., Blau, W. J., Zawadzka, M., Locos, O. B. and Mhuircheartaigh, E. M. N. (2007). Nonlinear optical properties of porphyrins, *Adv. Mater.*, **19**, pp. 2737–2774.
30. Higashino, T. and Imahori, H. (2015). Porphyrins as excellent dyes for dye-sensitized solar cells: recent developments and insights, *Dalton. Trans.*, **44**, pp. 448–463.
31. Clarisse, C., Riou, M. T., Gauneau, M. and Contellec, M. L. (1988). Field-effect transistor with diphthalocyanine thin film, *Electron. Lett.*, **24**, pp. 674–675.
32. Clarisse, C. and Riou, M. (1991). The operation and characteristics of diphthalocyanine field effect transistors, *J. Appl. Phys.*, **69**, pp. 3324–3327.
33. Su, W., Jiang, J., Xiao, K., Chen, Y., Zhao, Q., Yu, G. and Liu, Y. (2005). Thin-film transistors based on Langmuir-Blodgett films of heteroleptic

bis(phthalocyaninato) rare earth complexes, *Langmuir*, **21**, pp. 6527–6531.

34. Chen, Y., Su, W., Bai, M., Jiang, J., Li, X., Liu, Y., Wang, L. and Wang, S. (2005). High performance organic field-effect transistors based on amphiphilic tris(phthalocyaninato) rare earth triple-decker complexes, *J. Am. Chem. Soc.*, **127**, pp. 15700–15701.
35. Kong, X., Zhang, X., Gao, D., Qi, D., Chen, Y. and Jiang, J. (2015). Air-stable ambipolar field-effect transistor based on a solution-processed octanaphthoxy-substituted tris(phthalocyaninato) europium semiconductor with high and balanced carrier mobilities, *Chem. Sci.*, **6**, pp. 1967–1972.
36. Helberger, J. H. (1937). Über die Einwirkung von Kuofer-1-cyanid auf o-Halobenacetophenone (in German), *Liebigs Ann. Chem.*, **529**, pp. 205–218.
37. Linstead, R. and Weiss, F. T. (1950). Phthulocyanines and related compounds. Part XX. Further investigations on tetrabenzporphin and allied substances, *J. Chem. Soc.*, pp. 2975–2981.
38. Edwards, L., Gouterman, M. and Rose, C. B. (1976). Synthesis and vapor spectrum of zinc tetrabenzporphine, *J. Am. Chem. Soc.*, **98**, pp. 7638–7641.
39. Bonnett, R. P. and McManus, K. A. (1996). Approaches to the stepwise syhthesis of benzoporphyrins and phthalocyanines. Part 1. Synthesis of *opp*-dibenzoporphyrins (dibenzo[*g,q*]porphyrins), *J. Chem. Soc., Perkin Trans. 1*, pp. 2461–2466.
40. Vicente, M. G. H., Tomé, A. C., Walter, A. and Cavaleiro, J. A. S. (1997). Approaches to the stepwise syhthesis of benzoporphyrins and phthalocyanines. Part 1. Synthesis of *opp*-dibenzoporphyrins (dibenzo[g,q]porphyrins), *Tetrahedron Lett.*, **38**, pp. 3639–3642.
41. Bonnett, R. P. and Brown, R. F. C. (1972). Isoindole, *J. Chem. Soc., Chem. Commun.*, pp. 393–395.
42. Ito, S., Murashima, T. and Ono, N. (1997). A new synthesis of pyrroles fused with polycyclic skeletons, *J. Chem. Soc., Perkin Trans. 1*, pp. 3161–3165.
43. Ito, S., Murashima, T., Uno, H. and Ono, N. (1998). A new synthesis of benzoporphyrins using 4,7-dihydro-4,7-ethano-2H-isoindole as a synthon of isoindole, *Chem. Commun.*, pp. 1661–1662.
44. Jeong, S.-D., Min, B., Cho, S. Y., Lee, C., Park, B. K., An, K.-S. and Lim, J. (2012). Method for synthesis of tetrabenzoporphyrin precursor for use in organic electronic devices, *J. Org. Chem.*, **77**, pp. 8329–8331.

45. Ito, S., Ochi, N., Uno, H., Murashima, T. and Ono, N. (2000). A new synthesis of [2,3]naphthoporphyrins, *Chem. Commun.*, pp. 893–894.

46. Ito, S., Uno, H., Murashima, T. and Ono, N. (2001). Synthesis of benzoporphyrins functionalized with octaester groups, *Tetrahedron Lett.*, **42**, pp. 45–47.

47. Yamada, H., Kuzuhara, D., Takahashi, T., Shimizu, Y., Uota, K., Okujima, T., Uno, H. and Ono, N. (2008). Synthesis and characterization of tetraanthroporphyrins, *Org. Lett.*, **10**, pp. 2947–2950.

48. Ito, S., Nakamoto, K., Uno, H., Murashima, T. and Ono, N. (2001). Synthesis of a gable bis-porphyrin linked with a bicyclo[2.2.2]octadiene ring and its conversion into a conjugated planar bis-porphyrin, *Chem. Commun.*, pp. 2696–2697.

49. Uno, H., Masumoto, A. and Ono, N. (2003). Hexagonal columnar porphyrin assembly by unique trimeric complexation of a porphyrin dimer with π-π stacking: remarkable thermal behavior in a solid, *J. Am. Chem. Soc.*, **125**, pp. 12082–12083.

50. Uno, H., Nakamoto, K., Kuroki, K., Fujimoto, A. and Ono, N. (2007). Synthesis of porphyrin dimers fused with a benzene unit, *Chem. Eur. J.*, **13**, pp. 5773–5784.

51. Uoyama, H., Kim, K. S., Kuroki, K., Shin, J.-Y., Nagata, T., Okujima, T., Yamada, H., Ono, N., Kim, D. and Uno, H. (2010). Highly pure synthesis, spectral assignments, and two-photon properties of cruciform porphyrin pentamers fused with benzene units, *Chem. Eur. J.*, **16**, pp. 4063–4074.

52. Uno, H., Tagawa, K., Kawamoto, N., Nakamura, J., Mori, S., Okujima, T. and Nakae, T. (2014). Synthesis and properties of diporphyrins fused with PAH, *ICPP-8: 8th International Conference on Porphyrins and Phthalocyanines* (July 22–27, 2014, Istanbul, Turkey), p. 394.

53. Uno, H., Shimizu, Y., Uoyama, H., Tanaka, Y., Okujima, T. and Ono, N. (2008). Thermal behavior of free-base and core-modified bicyclo[2.2.2] octadiene- fused porphyrins, *Eur. J. Org. Chem.*, pp. 87–98.

54. Wada, M., Ito, S., Uno, H., Murashima, T., Ono, N., Urano, T. and Urano, Y. (2001). Synthesis and optical properties of a new class of pyrromethene–BF_2 complexes fused with rigid bicyclo rings and benzo derivatives, *Tetrahedron Lett.*, **42**, pp. 6711–6713.

55. Nakamura, M., Tahara, H., Takahashi, K., Nagata, T., Uoyama, H., Kuzuhara, D., Mori, S., Okujima, T., Yamada, H. and Uno, H. (2012). π-Fused bis-BODIPY as a candidate for NIR dyes, *Org. Biomol. Chem.*, **10**, pp. 6840–6849.

56. Nakamura, M., Kitatsuka, M., Takahashi, K., Nagata, T., Mori, S., Kuzuhara, D., Okujima, T., Yamada, H., Nakae, T. and Uno, H. (2014). Yellow NIR dye: π-fused bisbenzoBODIPYs with electron-withdrawing groups, *Org. Biomol. Chem.*, **12**, pp. 1309–1317.

57. Inokuma, Y., Matsunari, T., Ono, N., Uno, H. and Osuka, A. (2005). A doubly N-fused benzohexaphyrin and its rearrangement to a fluorescent macrocycle upon DDQ oxidation, *Angew. Chem. Int. Ed.*, **44**, pp. 1856–1860.

58. Okujima, T., Guangnan, J., Matsumoto, N., Mack, J., Mori, S., Ohara, K., Kuzuhara, D., Ando, C., Ono, N., Yamada, H., Uno, H. and Kobayashi, N. (2011). Cyclo[8]isoindoles: ring-expanded and annelated porphyrinoids, *Angew. Chem. Int. Ed.*, **50**, pp. 5699–5703.

59. Ito, S., Phong, L. T., Komatsu, T., Igarashi, N., Otsubo, S., Sakai, Y., Ohno, A., Aramaki, S., Tanaka, Y., Uno, H., Oba, T. and Hiratani, K. (2009). Synthesis of *meso*-chlorinated tetrabenzoporphyrins, *Eur. J. Org. Chem.*, pp. 5373–5382.

60. Ito, S., Terada, N., Seino, K., Makihata, D., Sasaki, A. and Oba, T. (2013). Synthesis of *meso*-fluorotetrabenzoporphyrins, *Tetrahedron Lett.*, **54**, pp. 5916–5919.

61. Uno, H., Furukawa, M., Fujimoto, A., Uoyama, H., Watanabe, H., Okujima, T., Yamada, H., Mori, S., Kuramoto, M., Iwamoto, T., Hatae, N., Tani, F. and Komatsu, N. (2011). Porphyrin molecular tweezers for fullerenes, *J. Porphyrins Phthalocyanines*, **15**, pp. 951–963.

62. Tagawa, K., Mori, S., Takase, M., Okujima, T., Hisaki, I. and Uno, H. (2016). Synthesis of hexagonal shape-persistent cyclophane with D_{2h} symmetry, *Tetrahedron Lett.*, **57**, pp. 4079–4081.

63. Oki, K., Tagawa, K., Mori, S., Takase, M., Okujima, T. and Uno, H. (2017). One-pot synthesis of a rice ball-shaped cyclophane with *syn*-diethanoanthracene-fused dipyrrole and hexafluorobenzene, *Chem. Lett.*, **46**, pp. 243–244.

64. Kawamoto, N., Mori, S., Nakae, T., Okujima, T., Uno, H. (2014). Molecular recognition of fullerenes utilizing diporphyrin with chiral π-space, *ICPP-8: 8th International Conference on Porphyrins and Phthalocyanines* (July 22–27, 2014, Istanbul, Turkey), p. 577.

65. Gouterman, M. (1959). Study of the effects of substitution on the absorption spectra of porphin, *J. Chem. Phys.*, **30**, pp. 1139–1161.

66. Gouterman, M. (1961). Spectra of porphyrins, *J. Mol. Spectrosc.*, **6**, pp. 138–163.

67. Ito, S., Ochi, N., Murashima, T., Uno, H. and Ono, N. (2000). A new synthesis of benzoporphyrins using 4,7-dihydro-4,7-ethano-2*H*-isoindole as an isoindole equivalent, *Heterocycles*, **52**, pp. 399–411.

68. Okujima, T., Hashimoto, Y., Jin, G., Yamada, H. and Ono, N. (2009). Synthesis of mono-, di- and tribenzoporphyrins from their soluble precursors, *Heterocycles*, **77**, pp. 1235–1248.

69. Uoyama, H., Takiue, T., Tominaga, K., Ono, N. and Uno, H. (2009). Synthesis, structures, and properties of BCOD-fused porphyrins and benzoporphyrins, *J. Porphyrins Phthalocyanines*, **13**, pp. 122–135.

70. Mori, S., Watanabe, Y., Nakae, T., Okujima, T. and Uno, H. (2014). Synthesis and properties of lanthanide complexes of π-conjugation extended porphyrins, *ICPP-8: 8th International Conference on Porphyrins and Phthalocyanines* (July 22–27, 2014, Istanbul, Turkey), p. 621.

71. Duchowski, J. K. and Bocian, D. F. (1990). Spectroscopic characterization of triple-decker lanthanide porphyrin sandwich complexes. Effects of strong ππ interactions in extended assemblies, *J. Am. Chem. Soc.*, **112**, pp. 8807–8811.

Chapter 7

Solid Oxide Fuel Cells: Electrode Materials and Membrane Formations

Yoshiteru Itagaki and Hidenori Yahiro
Department of Materials Science and Biotechnology, Graduate School of Science and Engineering, Ehime University, 3 Bunkyo-cho, Matsuyama, Ehime 790-8577, Japan
yahiro.hidenori.me@ehime-u.ac.jp

Solid oxide fuel cells (SOFCs) are next-generation power generators because of their high-energy conversion efficiency and fuel flexibility because of their high operating temperatures (800–900°C). Ensuring their stable operation in a long period is important for introducing SOFCs into practical applications. For this reason, many efforts have been devoted to the material development and fabrication process for lowering the temperature. This article introduces a part of the work implemented in Ehime University about preparation and electrode membrane formation techniques aiming at lowering the operating temperature.

Functional Materials: Advances and Applications in Energy Storage and Conversion
Edited by Toshio Naito

ISBN 978-981-4800-09-9 (Hardcover), 978-0-429-46813-1 (eBook)
www.panstanford.com

7.1 Solid Oxide Fuel Cells

A fuel cell is an electric power generator that directly converts the chemical energy of hydrogen into electric energy. The energy conversion efficiency of fuel cells is 50–60%, which is considerably higher than that of internal combustion engines ($\eta \approx 30\%$). In 2009, Japan had started the commercial production of household fuel cells, the so-called ENE-FARM, which can afford to take care of ca. 50% of the annual electricity usage of a standard home. Fuel cells are categorized into several types in terms of their components, as summarized in Table 7.1: the polymer electrolyte membrane fuel cell (PEMFC), the phosphoric acid fuel cell (PAFC), the molten carbonate fuel cell (MCFC), and the solid oxide fuel cell (SOFC). Currently, the most common type of fuel cells are the PEMFCs, in which proton exchange membranes such as Nafion are used as the polymeric electrolyte. Nafion possesses a high protonic conductance at 70–110°C under sufficient humidity.

Table 7.1 Characteristics of conventional fuel cells

	PEMFC	PAFC	MCFC	SOFC
Electrolyte in general	Nafion	H_3PO_4	$(Na,K)_2CO_3$	$(Zr,Y)O_{2-\delta}$
Fuels	H_2, CH_3OH	H_2	H_2, CO	H_2, CO
Temperature/°C	70–110	150–250	500–700	700–1000
Mobile ion	$(H_2O)_nH^+$	H^+	CO_3^{2-}	O^{2-}
Efficiency/% (LHV)	40–50	40–45	45–60	50–60

The electrodes of PEMFCs usually consist of conductive carbon loaded with fine Pt particles. PEMFCs are now introduced in the household power generating systems and fuel cell vehicles (FCVs). These cells can operate at near room temperature because of the high conductivity of the polymer electrolyte and good catalytic activity of the Pt/carbon electrode. Pt catalysts in the PEMFCs are expensive materials and require high-grade hydrogen fuel; in the case of FCVs at least 99.99% purity is required. Recently, a next-gen-type fuel cell, the SOFC, has been put to practical application. SOFCs are entirely composed of solid oxides having high thermal durability. The most worthy point of this type of cell is its high energy conversion efficiency because of its high operation temperature, above 700°C. A combination of SOFC and micro gas turbine (MGT), called the SOFC-MGT-combined cycle system, where waste heat from SOFC is utilized for MGT, has a potential capacity of 70% efficiency. Moreover, SOFCs have high fuel flexibility, where CO gas, which causes catalytic poisoning in PEMFC, can be used as a fuel. This is a great advantage for the hydrogen production and purification process. SOFCs are generally based on oxide ion conductors. The typical electrolyte material is 8 mol% Y_2O_3–doped ZrO_2, which is generally called yttria-stabilized zirconia (YSZ), where oxide ions migrate through the fluorite crystal lattice via oxygen defects. The most attractive feature of YSZ is its great stability in a wide range of oxygen partial pressures even at a high temperature: the transportation number is nearly 1 in $P(O_2) = 1–10^{-20}$ atm at 1000°C, [1]. YSZ has a relatively high activation energy for the oxygen hopping migration, which is about 1 eV, [2]. Therefore, a YSZ-based SOFC needs a sufficiently high temperature (~900°C) for stable migration of the oxide ions. Other oxide ion conductors that have higher ionic conductivities have been developed for intermediate-temperature operation, 700–800°C, for example, scandia-doped zirconia (ScSZ), doped-cerias, and lanthanum gallates [2].

7.2 Problems of SOFCs

On the basis of the high operating temperature, SOFCs have the highest-energy conversion efficiency among fuel cells. However, there are several problems due to the high operating temperature. At a high temperature, chemical reaction rates related to the operation increase whereas subreactions causing cell deterioration may be also induced. Possible unstable factors are listed below.

- Coagulation of the Ni particle in the anode, which reduces the activity [2]
- Formation of an insulating layer at the interface solid electrolyte/electrode and delamination at the interface [3]
- Reoxidation of Ni metallic particles, which induces volume expansion of the anode [4, 5]
- Catalyst poisoning by sulfur from the fuel and/or chrome from metallic components [6, 7]

In addition, metallic components should be composed of thermal-resistant materials, such as Ni alloys, which are expensive.

All the above problems can cause instability of the SOFC operation. In addition, the high operating temperature forces us to use thermally durable materials that are costly, such as nickel-based alloys. NEDO (New Energy and Industrial Technology Development Organization) of Japan is aiming for 90,000 h for the lifetime of an SOFC system as a final target [8]. To solve these problems, efforts have been devoted to reducing the operating temperature without reducing its efficiency. At present, reduced operating temperatures of 750–800°C have been achieved. Key points of further reducing the operating temperature are how to enhance the electrode activity and reduce the electrolyte resistance at a lower temperature. In the following sections, the preparation of electrode materials having high activity and the morphological control of electrode films are introduced.

7.3 Materials for a Fuel Electrode

7.3.1 Ceria-Based Anodes

Up till now, many efforts have been devoted to developing functional anodes to obtain a better output at a lower temperature. Ni-YSZ (8 mol% Y_2O_3-ZrO_2) has prevailed as an anode material, whereas doped cerias instead of YSZ have been also examined because of their advantage over YSZ. In the Ni-YSZ cermet, the anode reactions occur at the interface between Ni, YSZ, and fuel gas, that is, the three-phase boundary (TPB) [9]. In the Ni-YSZ anodes, Ni particles are moderately agglomerated to form the electronic conduction* pathways, but on the other hand, such a structure tends to reduce the surface area of the Ni catalyst, reducing the activity during the cell operation. To maintain the Ni-dispersed anode structure, using mixed ionic electronic conductors (MIECs) as a support for Ni particles is one of the strategies. Doped cerias (CeO_2-R_2O_3; R = rare earth elements) possess higher oxide-ionic conductivity than YSZ, and furthermore, they will be also electronic conductors under a reducing atmosphere

*In this chapter, "electronic conduction" is defined as the electrical conduction by electrons as charge carriers. The term of "electrical conduction" is used for any kind of carriers.

[10]. When a Ce^{4+} ion in CeO_2 with a fluorite structure is substituted with divalent or trivalent cations (M^{3+}, M^{2+}), such as Sm^{3+}, Gd^{3+}, Y^{3+}, or Ca^{2+}, the ionic conductivity of CeO_2 is significantly enhanced because of the formation of an oxygen vacancy, as expressed by the following reaction:

$$2MO_{1.5} \xrightarrow{CeO_2} 2M''_{Ce} + O_o + V_O^{\cdot\cdot}, \tag{7.1}$$

where M''_{Ce} is the Ce^{4+} sites replaced by the metal cation, O_o is the occupied oxygen site, and $V_O^{\cdot\cdot}$ is an oxygen vacancy, and effective charges are –2, 0, and +2, respectively. Sm_2O_3-doped CeO_2 (samaria-doped ceria, SDC) was initially developed as an electrolyte of SOFC. Yahiro and coworkers [11] reported the oxide-ionic conductivities of CeO_2 doped with a series of rare earth metal ions, and they found that the samarium ions were the best additives to improve the ionic conductivity of CeO_2. The oxide-ionic conductivity of SDC was much higher than that of YSZ in air, suggesting that SDC was a potential electrolyte candidate for SOFCs [11]. However, in a reduced atmosphere, Ce^{4+} was easily reduced to Ce^{3+}, indicating that electronic conductivity exceeds oxide-ionic conductivity. The electronic conductivity is significantly dependent on the partial pressure of oxygen. Under high oxygen partial pressure ($>10^{-5}$ atm), the oxide-ionic conduction of SDC is preferable because of the constant electrical conductivity [11]. The electrical conductivity of SDC increased on decreasing the oxygen partial pressure. This increase in electrical conduction is attributed to the appearance of n-type conduction due to the reduction of CeO_2. For this reason, SDC-containing anodes, such as Ni/SDC [12–14], Fe-Co-Ni/SDC [15], and Ni-CaO/SDC [16], have been so far reported to modify the anodic performance. Meanwhile, in the Ni/SDC anode, agglomerated SDC particles can assume the function of electronic conduction paths. Therefore, the Ni-dispersed structure in Ni/SDC is worth considering.

7.3.2 Ni-Dispersed SDC Anodes

In ordinary Ni-cermet anodes, such as Ni-YSZ, 50–60 vol% Ni particles are contained to yield sufficient electronic conductivity. Therefore, Ni aggregation is essentially required. However, as mentioned above, a ceria-based anode can impose the electronic condition on the ceria phase and a reduced amount of Ni is to be loaded on the SDC support by some chemical route, such as the impregnation method, to achieve Ni dispersion. Figure 7.1 shows the chemical process for Ni-SDC preparation [17]. The SDC support was prepared by the ammonia reverse coprecipitation process, with an ethanol solution containing stoichiometric amounts of nitrates as the starting material. The Ni particles were loaded on the SDC support by the impregnation process in the aqueous solution of $Ni(NO_3)_2{\cdot}6H_2O$.

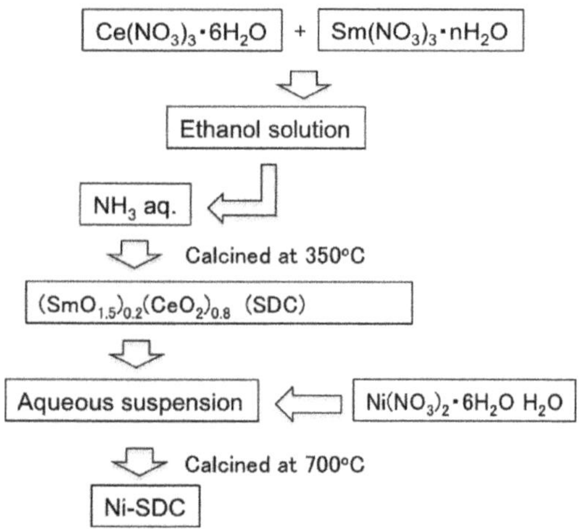

Figure 7.1 Preparation procedure of Ni-SDC.

Figure 7.2 shows the anodic properties of the Ni-SDC anodes with different Ni contents [17]. The anodic property is sensitively changed with Ni content, and 20 wt% of Ni is the maximum in the anodic property. In the 40 wt% Ni-SDC anode, Ni particles may start to aggregate to decrease the number of reaction sites, that is, the TBP. Lee et al. [9] reported the electrical conductivities of the Ni-YSZ cermets with different Ni contents. According to their result, the conductivity rapidly increased above 30 vol% of Ni content, and the conductivity change was explained by the percolation theory.

Therefore, 40 wt% (~33 vol%) Ni content would be above the percolation limit, where Ni particles are already aggregated in high order. By using SDC as a Ni support, the Ni-dispersed structure will be acceptable because the network of the SDC support plays the role of electronic conduction paths, instead of Ni. Since the anodic reactions occur at the TPBs, the highly dispersed Ni particles can drastically increase the number of reaction sites. As mentioned above, Ni impregnation of SDC particles confirmed that only 20 wt% of Ni content was the optimal composition for anodic performance [13]. It was deduced that impregnation of such low content of Ni gave the dispersed structure. In addition, Ni dispersion might be related to the surface area of the SDC support. To establish the Ni-impregnation method for anode preparation, the authors investigated how the surface area of the SDC support affects the particle size distribution of Ni and anodic properties. Figure 7.3 shows the scanning electron microscopy (SEM) and SEM-EDX (EDX is energy dispersive X-ray spectrometry) images of the Ni-SDC particles prepared by Ni impregnation [17]. In the SEM-EDX picture highly dispersed Ni particles (~48 nm) are clearly observed. Since the Ni particles directly impregnate the SDC support, Ni particle distribution may depend on the surface area of the SDC support. Table 7.2 shows the Ni particle sizes on the different SDC surface areas. The surface area of SDC was controlled by the calcination temperature for the precursor prepared by the reverse coprecipitation method. The surface area of SDC ranges between 12 and 93 m^2g^{-1} and decreases with an increase in the calcination temperature. After the impregnation process, the dried powders were again calcinated at 900°C. The final average Ni particle size ranged between 48 and 80 nm.

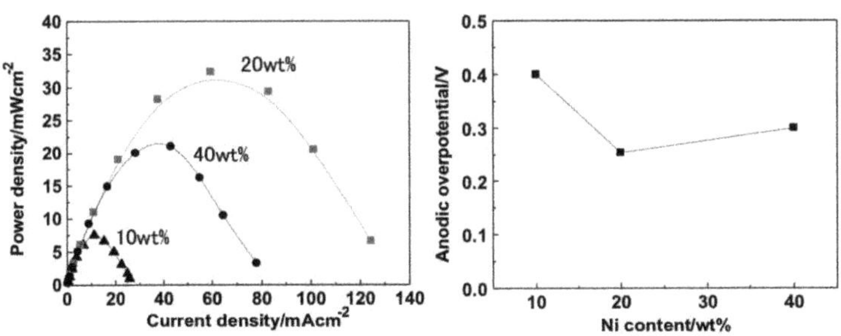

Figure 7.2 The *i-p* characteristics of the cells with Ni-SDC anodes with different Ni contents (right) and the anodic overpotentials of the Ni-SDC anodes (left) [16].

Figure 7.3 SEM and SEM-EDX images of the Ni-SDC particle: the bright spots in the SEM-EDX picture are the Ni particles loaded on the SDC support. The average particle size of Ni is 48 nm.

Table 7.2 Specific surface areas, Ni crystallite sizes, and average sizes of Ni particles for Ni/SDC-x samples

	Specific surface areas of SDC/m^2g^{-1}			
Samples	**As prepared by calcination at 700°C**	**After calcination at 900°C[a]**	**Ni crystallite size[b]/nm**	**Average size of Ni particle[c]/nm**
Ni/SDC-350	93	7	41	73
Ni/SDC-600	65	12	27	48
Ni/SDC-700	41	13	30	54
Ni/SDC-900	-	-	36	63
Ni/SDC-1200	16	-	43	74
Ni/SDC-1500	12	-	45	80

Source: Reprinted from Ref. [17], Copyright (2014), with permission from Elsevier.
[a]The calcination at 900°C was performed without Ni-loading for this experiment.
[b]Ni crystallite sizes were calculated from Scherrer's equation using an XRD peak of Ni(111).
[c]The average sizes of Ni particles were estimated from FE-SEM images.

The Ni particle size is the minimum when the surface area of the SDC support is 65 m^2g^{-1}. This surface area is second to the maximum value. The maximum value is 93 nm, when the calcination temperature is 350°C. However, the surface area significantly decreases after calcination at 900°C. Therefore, Ni particles may diffuse on the SDC particles during the grain-growth process of the SDC support. The relationship between Ni particle size and maximum power density (MPD) in H_2 and CH_4 is shown in Fig. 7.4. For both fuels, MPD decreased with an increase in the Ni particle size.

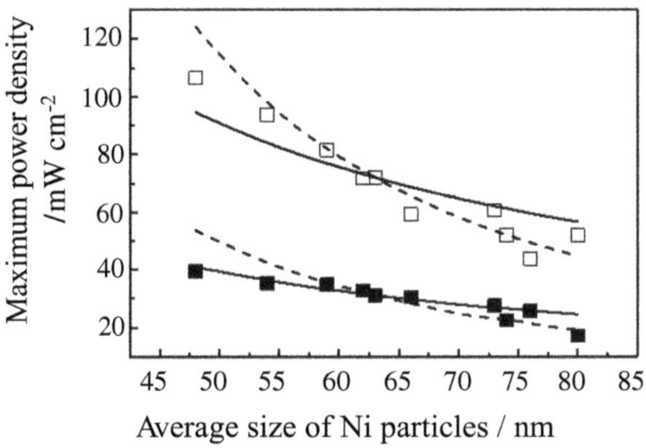

Figure 7.4 Correlations between the maximum power densities of the cell with the Ni/SDC-x anodes and the average size of Ni particles under H_2 (■) and CH_4 fuels (□). Calculated curves derived with the assumptions that the anodic reactions would be controlled by the surface of Ni particles (dotted curves) and by the TPB (solid curves) are also inserted in the figure. Reprinted from Ref. [17], Copyright (2014), with permission from Elsevier.

This correlation was theoretically analyzed with several assumptions: (i) all Ni particles are loaded on the SDC surfaces in a hemispherical shape *R* in diameter (Fig. 7.5), (ii) the effective electrode thicknesses where the oxidation reactions occur are identical, and (iii) the MPD depends only on the Ni particle sizes. We considered two cases: (case 1) that the MPD is dependent on the surface area of Ni and (case 2) that the MPD is dependent on the TPB length. By comparing with the MPD of Ni-SDC-900 (31.2 $mWcm^{-2}$ for H_2 and 72.0 $mWcm^{-2}$ for CH_4) where the mean Ni particle size is 64 nm, the MPD at each Ni particle size was calculated as follows:

$$\text{MPD}(R) = \text{MPD}(64) \times (64/R)^2 \text{ for case 1} \quad (7.2)$$

$$\text{MPD}(R) = \text{MPD}(64) \times (64/R) \text{ for case 2} \quad (7.3)$$

Here MPD(64) is the MPD of Ni-SDC-900 (R = 64 nm). For H_2 fuel, better fitting was obtained for case 1 (dotted curve), while case 2 (solid curve) is more suitable for the case of CH_4 fuel. The MPD for

H_2 fuel was controlled by the surface area of Ni. On the other hand, MPD for CH_4 fuel was controlled by the TPB length. The dissociative adsorption of H_2 onto a Ni particle might be the rate-determining step. Meanwhile the CH_4 molecule might easily migrate and directly reacts at the TPB. In any case, the fine-dispersed metallic Ni promotes the efficient oxidation of fuel, with a subsequent increase in the power density.

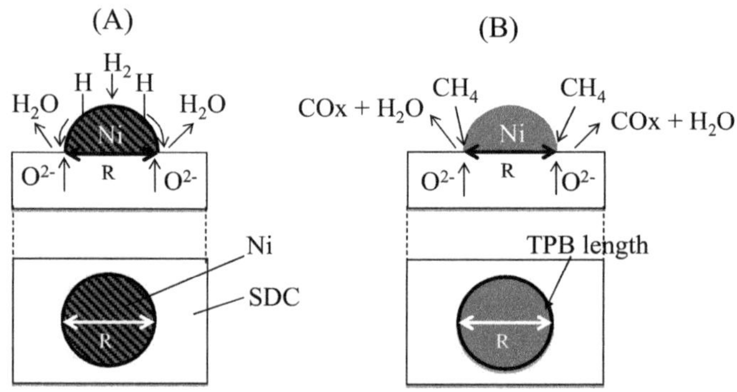

Figure 7.5 Correlations between the maximum power densities of the cell with the Ni/SDC-x anodes and the average size of Ni particles when using H_2 fuel (A) and CH_4 fuel (B). Calculated curves derived with the assumptions that the anodic reactions would be controlled by the surface of Ni particles (dotted curves) and by the TPB (solid curves) are also inserted in the figure. Reprinted from Ref. [17], Copyright (2014), with permission from Elsevier.

7.3.3 SDC-Based Anodes with Ni-Fe Alloys

Bimetallic anodes, such as Ni-M (M = Fe, Co, or Cu), are known to be effective for further enhancing the catalytic activity of hydrogen and hydrocarbon fuels [18–23]. Ni-Fe anodes with a high concentration Ni-Fe have been already reported [18–20]. Lu et al. [20] evaluated the 50 wt% $Ni_{1-x}Fe_x$/SDC anode in hydrogen fuel and reported its best performance at $x = 0.25$. They assigned the modified performance to the electronic interaction between Ni and Fe. An et al. [21] calculated the chemisorption of atomic species O, S, C, and H on the (111) surface of Ni-Fe/YSZ. They concluded that Fe addition increased the binding strength of O atom to a favorable level. The highly dispersed Ni-Fe particles are expected to further increase the length of the TPB, and

Itagaki el al. investigated the anodic properties of a series of 20 wt% $Ni_{1-x}Fe_x$/SDC [24]. Figure 7.6 shows X-ray diffraction (XRD) patterns of the series of the $Ni_{1-x}Fe_x$/SDC powder after hydrogen reduction. The XRD patterns were composed of the signals arising from face-centered cubic (fcc) Ni for $x = 0$ and body-centered cubic (bcc) Fe phases for $x = 1$, in addition to that of SDC. At $x = 0$, XRD signals were observed from the (100) planes of fcc. The Ni lattice was clearly observed at 44.5°. On increasing the Fe content, the signals shifted to the lower angles. No signals from pure Fe were observed, which means that Ni and Fe formed alloys. Since the atomic sizes of Ni and Fe were 1.49 and 1.56 Å, respectively, introducing Fe atom will expand a Ni crystal lattice and shift the XRD signal to lower angles. The signal shift was not observed from $x = 0.5$ to $x = 0.8$. There are two possibilities for this phenomenon. One is that Fe is not soluble in Ni beyond $x = 0.5$, and the other is that the crystal structure varied above $x = 0.5$. It has been reported that Ni and Fe form an alloy at any composition [24].

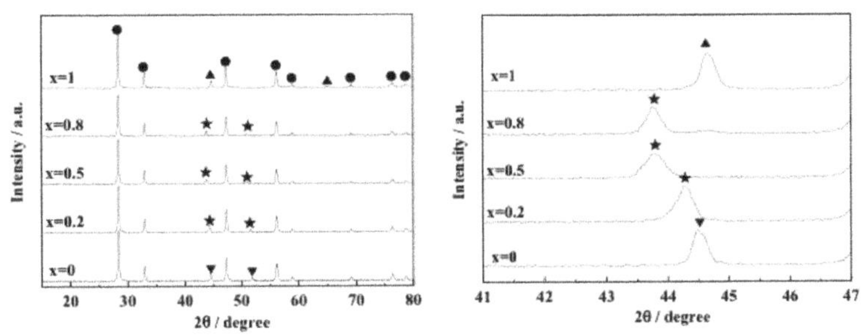

Figure 7.6 XRD patterns of the $Ni_{1-x}Fe_x$/SDC powders after hydrogen reduction (left) and the magnified images of the (110) signals (right): ● represents SDC, ▼ represents Ni, ▲ represents Fe, and □ represents Ni-Fe alloys. Reproduced from Ref. [24] with permission from the Electrochemical Society.

Actually, no signals from a deposited Fe metal were observed. Furthermore, the crystal structure of pure Fe belongs to bcc, where XRD peaks from the (110) planes of Fe ($x = 0$) lattice are observed at the higher angle of 44.6°. Therefore, it was concluded that $x = 0.8$ forms a solid solution based on the crystal structure close to bcc Fe. Figure 7.7 shows the Nyquist plots of the $Ni_{1-x}Fe_x$/SDC anodes, with $x = 0$ and 0.5. Both of them consist of at least three depressed semicircles, indicating that the electrode polarization is controlled by at least three rate-limiting processes. The Nyquist plots were then analyzed by the curve fitting using the equivalent circuit, including constant phase elements (CPEs). The ohmic resistances are almost equivalent, and they are due to the resistance of the electrolyte substrate. On the other hand, the polarization resistances at $x = 0.5$ is apparently smaller than that of $x = 0$. Therefore, the addition Fe into Ni is effective for reducing the polarization resistance.

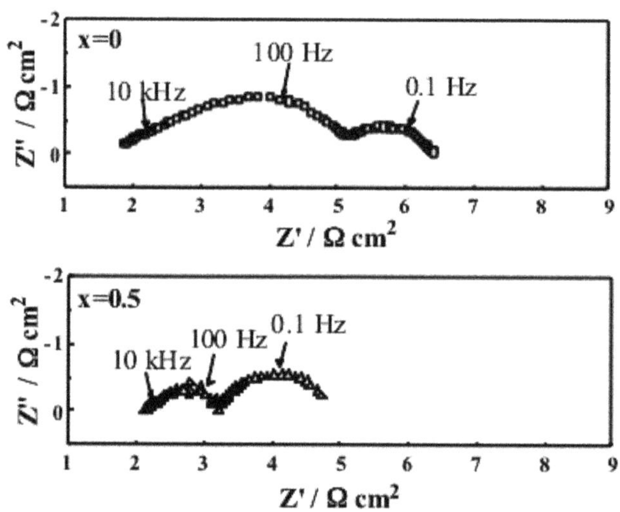

Figure 7.7 Nyquist plots of $Ni_{1-x}Fe_x$/SDC anodes, where $x = 0$ (upper) and 0.5 (lower) in H_2 under OCV conduction at 800°C. Reproduced from Ref. [24] with permission from the Electrochemical Society.

Figure 7.8 shows the dependence of MPDs and crystallite sizes on the Fe content, *x*. It seems that the MPD values are correlated to the crystallite size of $Ni_{1-x}Fe_x$ particles; the smaller the crystallite size, the higher the MPD value. Lu et al. [20] reported $Ni_{1-x}Fe_x$/SDC anodes containing about 50 wt% Ni-Fe. Their work resulted in a maximum anodic performance of around $x = 0.2$, whereas Fig. 7.8 represents the maximum performance at $x = 0.2$ or 0.5 for the $Ni_{1-x}Fe_x$/SDC anodes with 20 wt% Ni-Fe. The superiority of $x = 0.2$ was also supported by Fig. 7.8, but Ni-Fe particle size effect would emerge when the Ni-Fe amount is so small.

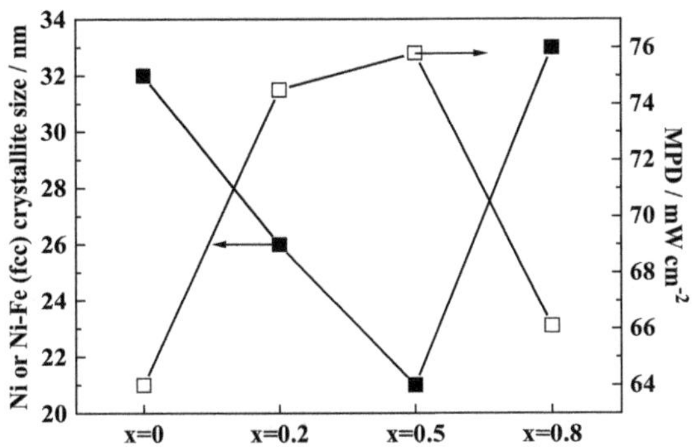

Figure 7.8 Relationship between the Fe content, *x*, and the crystallite sizes of Ni-Fe and between *x* and MPD at 800°C. Reproduced from Ref. [24] with permission from the Electrochemical Society.

7.3.4 Perovskite-Based Cathode Materials

Perovskite oxides (ABO3) are currently accepted as a SOFC cathode material. Among them, the perovskite oxides containing La at the A-site and transition metals such as Mn, Co, and Fe at the B-site have been commonly used because they possess high electrical conductivity, catalytic activity for oxygen dissociation, and oxygen transportability. (La,Sr)MnO_3 oxides (lanthanum strontium manganite, LSM) have been used for high-temperature cathodes. These oxides are stable in a wide range of oxygen partial pressures [25]. However, these oxides have a relatively low oxygen diffusion coefficient ($D = 3 \times 10^{-12}$ cm^2 s^{-1} at 900°C for $La_{0.5}Sr_{0.5}MnO_3$) [26], and cathodic overpotential tends to be large at a low temperature. Recently, (La,Sr)(Co,Fe)O_3 oxides (lanthanum strontium cobalt ferrite, LSCF) have been mostly adopted as a common cathode material for an intermediate-temperature solid oxide fuel cell (IT-SOFC). The LSCF oxides have a high oxygen transportation property ($D = 5 \times 10^{-7}$ cm^2 s^{-1} at 900°C for $La_{0.5}Sr_{0.5}Co_{0.8}Fe_{0.2}O_3$) [26]. However, these oxides are highly reactive, especially with YSZ. In fact, LSCF reacts with YSZ to form a resistive product, such as $SrZrO_3$ [27, 28]. To avoid a resistive phase such as this, a thin layer of gadolinia-doped ceria (GDC) is usually inserted as a diffusion barrier between LSCF and YSZ [29–31]. Thus the cathodic property strongly depends on the chemical composition of the perovskite structure. However, the electrode performance is also affected by the electrode powder property, such as particle size, which directly relates to the effective length of the TPB [32]. For this reason, several wet processes have been proposed for size control of the powders, such as the citrate process [33, 34], the sol-gel process [35, 36], and the coprecipitation process [37]. Figure 7.9 shows the particle size distributions of the LSM powders prepared by the sol-gel method using different calcination temperatures. The peak size tends to decrease with a reducing calcination temperature [38]. The cathodic films were prepared using LSM powders with different particle sizes. The SEM images of the films are shown in Fig. 7.10. After sintering of the green film at 1100°C, apparent particle growths were observed. The particle size distributions after the sintering at 1100°C were estimated to be 0.3–0.8 μm (LSM600), 0.25–0.5 μm (LSM800), and 0.5–0.6 μm (LSM1000) from SEM observations. The order of the particle size is LSM600 > LSM1000 > LSM800. This result is due to the fact that LSM600 with the smallest particle size exhibited the highest sinterability. Figure 7.11 shows the cathodic potential drop for each LSM cathode. The result is straightforward, in order of the particle size; LSM800 originally having the intermediate particle size resulted in the highest cathodic performance. The film thickness after the sintering process was proportional to the deposition time at a constant voltage of 15 V/cm.

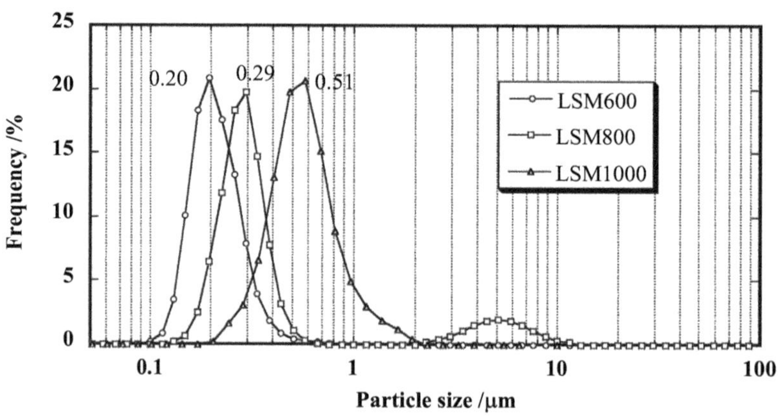

Figure 7.9 Particle size distributions of LSM powders prepared by the sol-gel route. The precursor was calcinated at 600°C (LSM600), at 800°C (LSM800), and at 1000°C (LSM1000) [38].

Figure 7.10 SEM micrographs of the LSM deposit surfaces formed by EPD at a voltage of 15 V/cm using (a, d) LSM600, (b, e) LSM800, and (c, f) LSM1000 powders. The images shown in (a, b, and c) and (d, e, and f) represent those before and after sintering at 1100°C, respectively [38].

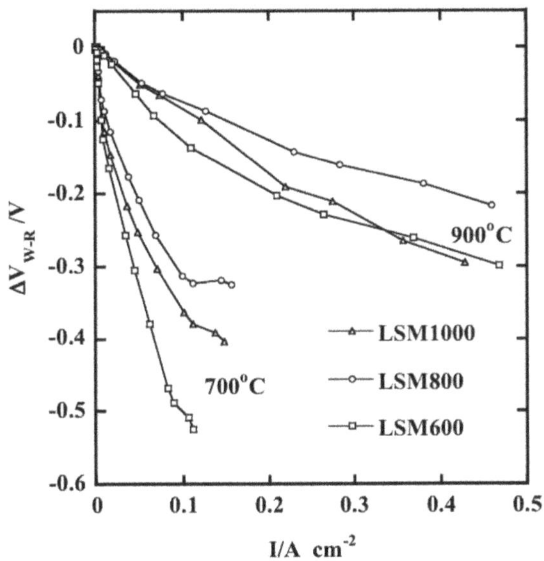

Figure 7.11 Cathodic voltage drops for the cells prepared with LSM600, LSM800, and LSM1000 films observed at 700 and 900°C [38].

7.4 Electrophoretic Deposition for SOFC Fabrication

7.4.1 Electrophoretic Deposition

SOFCs are fundamentally composed of a triple layer (anode/electrolyte/cathode) of different ceramics. Cell performance mostly depends on their morphologies, layer thicknesses, and interfaces. For the electrolyte layers, their ohmic resistances of ionic conduction within the bulks and grain boundaries should be suppressed. High conductivity is fundamental to a solid electrolyte material. Currently, the most general and reliable solid electrolyte for the component of SOFC is 8 mol% Y_2O_3-ZrO_2 (YSZ), which has a high oxide ionic conductivity (10^{-1} S·cm^{-1} at 1000°C), and the transport number of the oxide ion is close to 1 in a wide range of oxygen partial pressures ($P(O_2) = 1$–10^{-20} atm at 1000°C). Since the activation energy of oxide ion conduction is very high (~1 eV), ionic conductivity steeply drops at lower temperatures. Therefore, a SOFC using a thick YSZ membrane is usually limited to operating at a high temperature. Thin-film preparation on the anode support (i.e., an anode-supported SOFC) is an alternative to operating at reduced temperatures. Thus thin-film preparation is an important technique in SOFC fabrication. So far, slurry methods have been applied for preparing multilayered cathode films, such as screen printing [39]. Electrophoretic deposition (EPD) has been also developed for preparing thin electrolyte films in the micrometer range. EPD is a colloidal process, that can form microstructured ceramic films, and its kinetics are governed by several parameters. The kinetics of the EPD process have been systematically studied by Sarkar et al. [40]. The deposition rate is theoretically described by the following equation:

$$\frac{dy}{dx} = \mu c A \tag{7.4}$$

where dw/dt is the deposition rate, μ is the mobility of the particles, c is the concentration of the particles, A is the deposition area, and E is the applied electric field. μ is represented by the following equation:

$$\mu = \frac{\varepsilon_0 \varepsilon_r \zeta}{\eta}, \tag{7.5}$$

where ε_0 is the vacuum permittivity, ε_r is the relative dielectric constant of the suspension medium, ζ is the zeta potential of the particles, and η is the viscosity of the medium. EPD efficiency is largely dependent on the medium. The most important parameter for reproducibility is the zeta potential, which mostly affects the stability of the suspension. In an aqueous suspension, the zeta potential is easily controlled with pH. Lowering the pH value usually shifts the zeta potential to the positive direction. In this case, positively charged particles migrate and are deposited on the negatively charged substrate. However, at a high applied voltage (theoretically > 1.23 V), hydrogen gas bubbles will evolve on the substrate, which may interfere with homogeneous particle deposition. Therefore on a lab scale, organic solvents are widely used as dispersants.

7.4.2 EPD Techniques for SOFC Fabrication

EPD can be an effective technique for SOFC fabrication because it can form a highly dense and homogeneous film a few micrometers in thickness. EPD preparations of YSZ electrolyte thin films have been studied by a lot of researchers [41–46]. In general, the YSZ suspension is prepared using acetylacetone as a dispersant and I_2 as a charging agent. In the suspension the added I_2 reacts with acetylacetone to form HI and protons dissociated from the HI are adsorbed on the YSZ particles to make them positively charged.

$$CH_3COCH_2COCH_3 + I_2 \rightarrow CH_3COCH_2COCH_2I + HI \qquad (7.6)$$

Therefore, the amount of I_2 added sensitively affects the zeta potential of the suspension. Figure 7.12 shows the YSZ deposition dependent on I_2 concentration in the acetylacetone suspension. The maximum deposition amount was achieved at an I_2 concentration of 0.6 g/L. Ishihara et al. [41] optimized the I_2 concentration to obtain a positive zeta potential; +50 mV was attained at the I_2 concentration of >0.5 g/L. To fabricate a single cell, a YSZ thin film is formed on an anode substrate, which is called anode-supported SOFC. EPD techniques can be used for fabricating SOFCs of such a structure.

Using Ni-YSZ as the anode support, the YSZ film will be deposited on the prereduced NiO-YSZ substrate. However, both phases, NiO and YSZ, have insufficient electronic conductivities for an EPD substrate. Yamaji et al. [46] added graphite powder to the NiO-YSZ to make a substrate of sufficient conductivity. Figure 7.13 shows YSZ thin films formed on the NiO-YSZ substrate with different deposition times. Fairly dense YSZ thin layers were formed after the sintering process at 1300°C. The thickness of the films is proportional to the deposition time. Homogeneous films are formed even a few micrometers in thickness. Uchikoshi et al. [47] mentioned that densities of EPD films are nearly comparable to that of films after cold isostatic pressing (CIP) treatment. It is noteworthy that the anode substrates remain porous even after the sintering process. The graphite additive acts as a pore former as well as a conductive agent. Therefore, the performance of the single cells should be affected by the amount and particle size of the graphite additive.

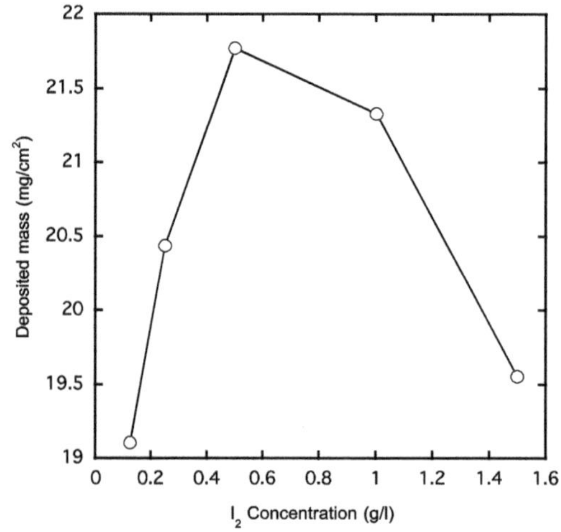

Figure 7.12 Deposition mass of the YSZ powder by EPD depending on the I_2 concentration in the acetylacetone suspension: YSZ concentration 8.0 g/L and applied voltage 15 V/cm.

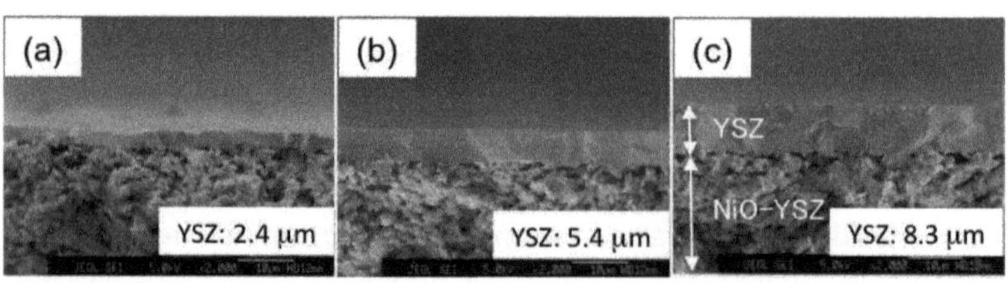

Figure 7.13 YSZ thin films formed on the NiO-YSZ substrate by EPD under an applied voltage of 10 V for (a) 1 min., (b) 2 min., and (c) 3 min. The films were sintered at 1300°C.

7.5 Electrode Film Fabrication by EPD

7.5.1 Preparation of Cathode Films by EPD

So far, EPD has been mainly used for preparing a thin film of a solid electrolyte such as YSZ on an electrode substrate to fabricate an electrode-supported SOFC. EPD gives homogeneous and dense films, and their thickness can be adjusted at the micrometer level. Thus EPD is a powerful technique for SOFC fabrication. Whereas there is a primary condition for EPD that film formation using EPD requires a conductive substrate on which colloidal particles get deposited during the electrochemical process, for an electrolyte-supported SOFC, electrode films are usually formed on a solid electrolyte surface. Although solid electrolytes commonly used for SOFCs are good ionic conductors, their electronic conductivities are negligible. Therefore, it is hard to prepare electrode films directly on a nontreated electrolyte surface, and the electrolyte surface should be treated to be conductive using some suitable method. Santillán et al. [48] successfully formed $La_{0.6}Sr_{0.4}Co_{0.8}Fe_{0.2}O_{3-\delta}$ (LSCF) films on Au- or Ag-sputtered $Ce_{0.9}Gd_{0.1}O_{1.95}$ and concluded that the microstructure of the LSCF film can be controlled by EPD parameters such as an applied voltage and deposition time. Yamaji et al. prepared the LSM cathodic films on the graphite carbon–coated YSZ surface by EPD [38]. Graphite coatings are implemented by carbon vapor deposition, or more simply, a graphite rod is directly applied onto the solid electrolyte surface. Anyway, the coated graphite layer can be eliminated by heating in air at around 500°C. Therefore, by the subsequent sintering process of the electrode film at around 1000°C, the graphite will be eliminated and the electrode films directly adhere to the electrolyte surface. Suzuki et al. [49] created a polypyrrole coating by chemical oxidation polymerization on the nonconductive NiO-YSZ substrate and formed a GDC/LSGM/GDC trilayer on the conductive polymer layer by EPD. They also pointed out that the EPD on the polypyrrole-coated surface gave better adhesion due to the lower removal temperature of the polymer than that of graphite.

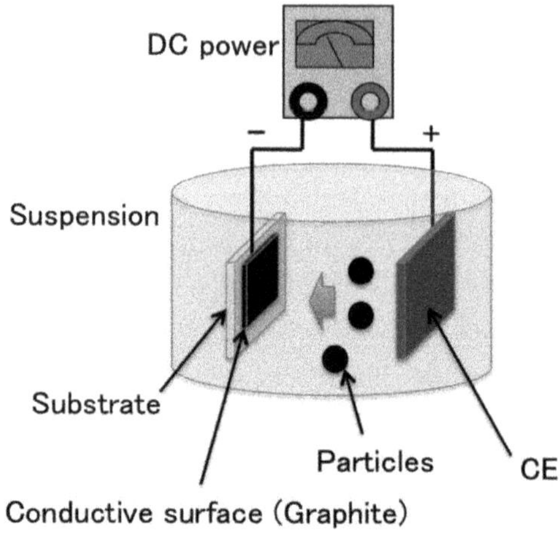

Figure 7.14 Illustration of the EPD setup for preparing an electrode film on a solid electrolyte substrate.

Figure 7.15 shows cathode films ($La_{0.8}Sr_{0.2}MnO_3$: LSM) formed by EPD. Similar to the case of electrolyte films, the film thickness is controlled well, in the range of micrometers, by varying the deposition time. In all the films fine particles are homogeneously sintered. Since the film thickness of the electrode layer is directly related to the electrode properties, it should be controlled. When the film is too thin, the electrode tends to lack the current collecting ability.

The amount of deposit is proportional not only to the deposition time but also to the applied electric field (V/cm) because the migration velocity increases with an increase in the applied electric field.

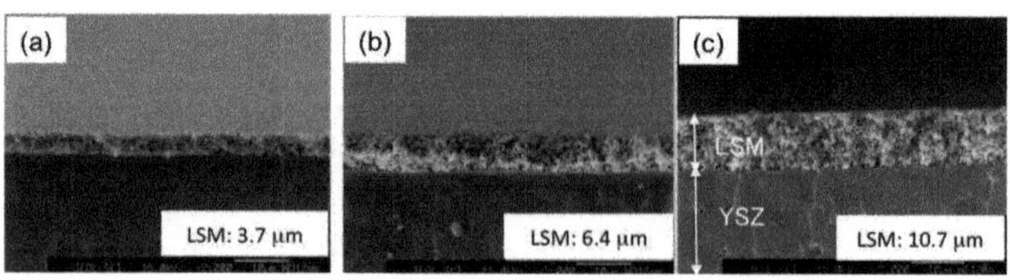

Figure 7.15 LSM thin films formed on the YSZ substrate by EPD under an applied voltage of 15 V for (a) 1 min., (b) 2 min., and (c) 3 min. The films were sintered at 1100°C.

Figure 7.16 shows cathodic films prepared by EPD with different applied electric fields. The difference in the applied electric fields is reflected in the difference in the morphologies. Apparently, the film prepared with the lower electric field has the denser structure. In the EPD process, the migrating particles will be attracted toward the vacancies on the deposition surface. This is because the electric field passing the deposition vacancies is stronger than that at the filled deposition surface. Figure 7.17 shows the cathodic potential drops of the LSM films prepared at different applied electric fields. The LSM cathode formed at 7.5 V/cm apparently exhibited a better performance than that formed at 15 V/cm, especially at a low temperature. The dense cathode film is superior in the interfacial contact area at the electrolyte/electrode interface and the current collecting ability. Thus the EPD method can control the electrode thickness and morphology.

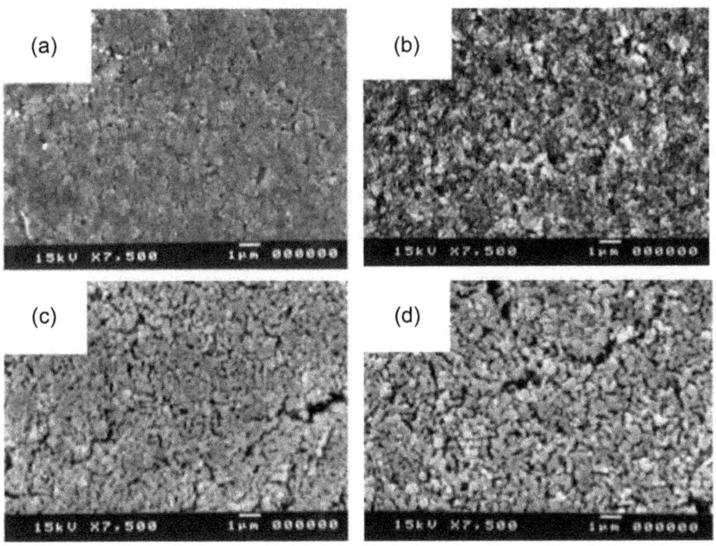

Figure 7.16 SEM images of LSM films prepared by EPD at the applied voltages of 7.5 (a and c) and 15 V/cm (b and d). The images (a and b) are taken before sintering, and the images (c and d) are taken after the sintering process at 1100°C.

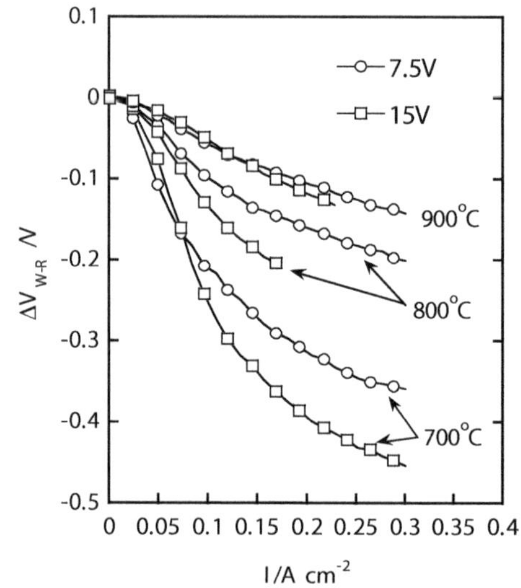

Figure 7.17 Cathodic potential drops of the half-cells Pt, LSM/YSZ/Pt with LSM cathode films formed by EPD at the applied electric fields of 7.5 and 15 V/cm.

Figure 7.18 shows the cathodic potential drop of the series of the monolayered cathodes with different thicknesses observed at 600 and 800°C [50]. In the figure, the monolayered cathodes of LSM and LSM-YSZ are denoted as LSM and LY, respectively, along with the values of thickness in micrometers (e.g., a LSM cathode with an 8 μm thickness is denoted as "LSM8"). It is widely accepted that the LSM-YSZ composite cathode, where an electrolyte component of YSZ is added as a secondary phase to the cathode material of LSM, can effectively reduce the cathodic polarization resistance [51–53]. In both cathode films, the cathodic potential drop tends to be suppressed with an increasing film thickness. In both LSM and LSM-YSZ cathodes, an increasing thickness tends to reduce ohmic and polarization resistances. The ohmic drop assigned to an electrode can be divided into in-plane and cross-plane drops. Here cylindrical pores are assumed to be present at the interface between electrode and porous current collector. The pores do not directly come in contact with the current collecting layer. When integration of the above-mentioned pores over the electrode surface can be considered as a single pore with a radius r, in-plane and cross-plane ohmic drops can be expressed by Eqs. 7.7 and 7.8 [54–56], where U = ohmic drop, I = current density, ρ = resistivity of electrode, th =

thickness of the electrode, and r = radius of the cylindrical pores. The ohmic drops are inversely proportional and directly proportional to the thickness of the electrode, respectively. Therefore, the decreases in the ohmic drop with an increasing electrode thickness, which can be commonly seen in LSM and LSM-YSZ cathodes, are probably because a decrease in the in-plane ohmic resistance is beyond an increase in the cross-plane ohmic resistance.

$$U_{\text{in-plane}} = (\rho_e i/4\text{th}) \times r^2 \tag{7.7}$$

$$U_{\text{cross-plane}} = \rho_e \times i \times \text{th} \tag{7.8}$$

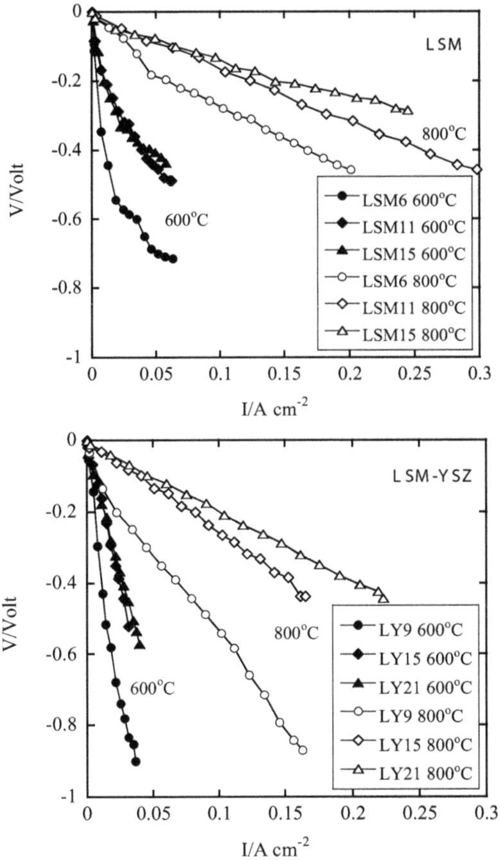

Figure 7.18 Cathodic potential drop versus applied current density measured for the LSM (upper graph) and LSM-YSZ (lower graph) cathodes with different thicknesses at 600°C and 800°C. Reprinted from Ref. [50], Copyright (2012), with permission from Elsevier.

Meanwhile, polarization drop may be related to the number of reaction sites. In the case of the LSM cathode, the reaction sites are limited to the area near the interface between electrode and electrolyte because ionic conductance of LSM is negligible. In this case, the LSM layer is exclusively utilized for an electron migration pathway. Therefore, electronic migration resistance in a LSM cathode seems to be an important factor for the polarization drop. The decrease in the polarization drop with an increasing electrode thickness is probably due to the enhancement of the in-plane electronic conduction. On the other hand, the number of reaction sites in the LSM-YSZ cathode is expected to increase to some extent with an increasing thickness. The effective thickness for a cathodic reaction would depend on electronic and ionic migration resistances in LSM-YSZ. The decrease in the cathodic potential of the LSM-YSZ cathodes appeared to be larger than that of the LSM cathodes. This difference became more significant at 800°C.

Since the electronic conductivity of YSZ was almost negligible at $p(O_2) = 1$ compared to that of LSM (200 Scm^{-1} at 800°C [26]), the high ohmic resistance of the LSM-YSZ cathode was probably due to the fact that the electronic conduction path through the electrode thickness was partially obstructed by the addition of YSZ [57]. The polarization resistance of LSM-YSZ was lower than that of LSM. Therefore, the addition of the YSZ phase has a certain degree of contribution in enhancing the ionic conductivity of LSM-YSZ. To verify the increase in ohmic resistance by adding YSZ, the electrical conductivities of LSM and LSM-YSZ were evaluated by the four-probe direct current (DC) method.

Figure 7.19 shows the temperature dependence of the dc conductivities of the materials sintered at 1100°C. Conductivity decreased remarkably with increasing YSZ content. The activation energies evaluated from the line slopes were 8.59×10^{-2} (LSM), 9.22×10^{-2} (LSM-YSZ 70:30), and 1.09×10^{-1} (LSM-YSZ 50:50) eV. The decrease in the conductivity is unambiguously due to the addition of the YSZ secondary phase, which has a much lower electrical conductivity than LSM. The addition of YSZ inhibits the grain growth of the LSM particles, which probably increases the grain-boundary resistance and its activation energy.

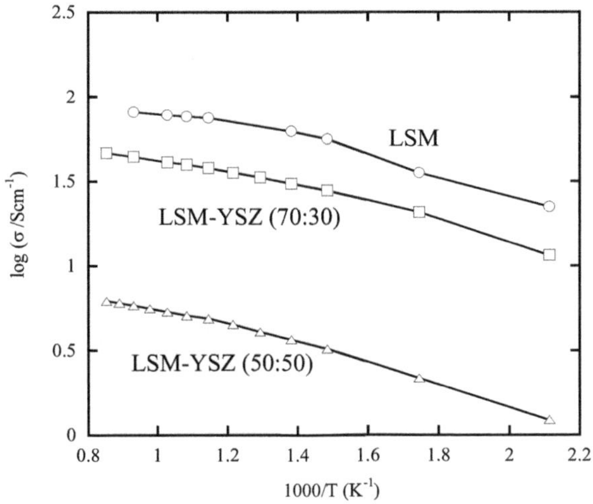

Figure 7.19 The dc conductivity measured for LSM and LSM-YSZ bulks sintered at 1100°C. Reprinted from Ref. [50], Copyright (2012), with permission from Elsevier.

7.5.2 Bilayered Cathode Films

Adding YSZ to LSM is believed to increase the number of interfaces between LSM and YSZ and extend the TPB from the interface between the electrolyte and the electrode to the inner parts of the electrode layer. On the other hand, as shown in Fig. 7.19, adding YSZ could cause an increase in ohmic resistance of the cathode because the electron conductivity of YSZ is much lower than that of LSM [56]. An increase in the electrode thickness will increase the total number of interfaces along the thickness direction of the electrode layer. However, the number of effective reaction sites of oxygen is not necessarily proportional to the electrode thickness; it will depend on the microstructure of the electrode [57]. If the thickness of the composite electrode is larger than the thickness of the effective reaction layer, the remaining part of the composite electrode will not function sufficiently; rather it will just act as an electron conducting layer with a high resistance. In such a case, it is reasonable that a current collective LSM layer substitutes for the poorly functioning part in the composite electrode. Multilayered or functionally graded cathodes have been studied by some researchers [58, 59]. Antunes et al. [60] previously prepared LSM/LSM-YSZ bilayered cathodes with layers of different thicknesses by screen-printing. They concluded that the optimal thickness of the LSM-YSZ layer was 6–12 μm. The EPD technique can be used for preparing such a bilayered cathode film with a controlled thickness [50].

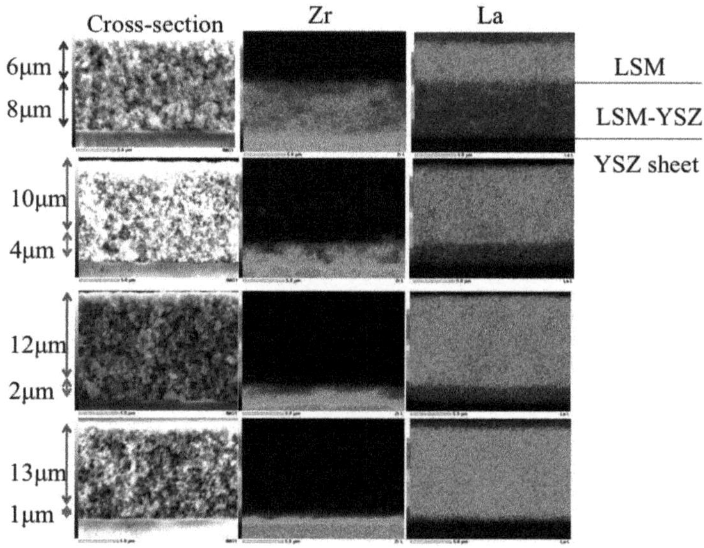

Figure 7.20 Zr- and La-elemental mappings of the fractured surfaces of the bilayered cathodes. Reprinted from Ref. [50], Copyright (2012), with permission from Elsevier.

Figure 7.20 shows the cross-sectional views of the bilayer cathodes consisting of LSM-YSZ as the bottom layer and LSM as the top layer. From the elemental mappings, the formation of the bilayer structure was apparently recognized. All the films have flat interfaces between LSM-YSZ and LSM layers, and the thickness is regulated at the micrometer level. Figure 7.21 shows that the cathodic potential drops at 600 and 800°C. The bilayer cathodes exhibited better performances compared to the monolayered cathodes shown in Fig. 7.18. The LSM10/LY4 and LSM18/LY8 cathodes have optimal structures at 600 and 800°C, respectively. It was suggested that the effective thickness of the oxygen reaction in the LY layer would be a decisive factor for cathode performance at 600°C. This result suggests that the effective reaction sites extend from the interface between the electrode and electrolyte up to 4 μm deep into the LY layer. The effective thickness is expected to increase at a high temperature, and 8 μm is the optimal thickness, at 800°C. The cathodic performance is enhanced with an increase in the thickness of LSM at both temperatures. It should be mentioned that the optimal thickness would also vary with morphology, particle size distribution, porosity, and grain-boundary structure.

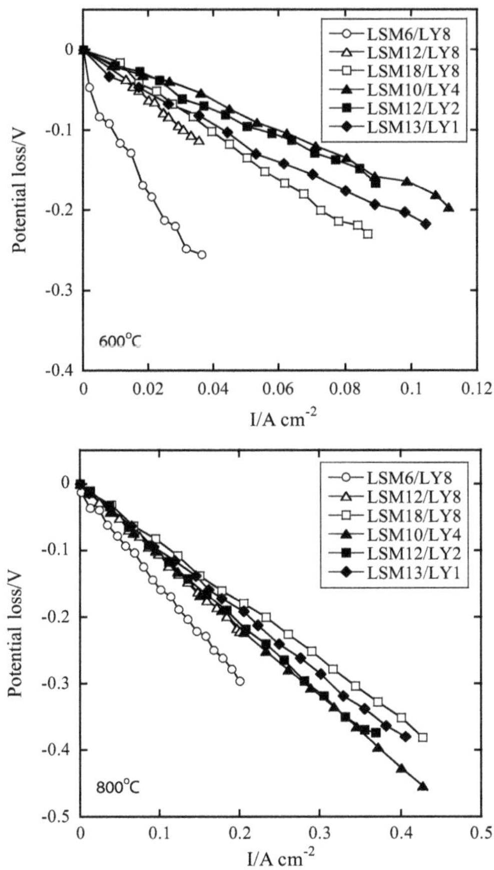

Figure 7.21 Cathodic potential drop versus applied current density measured for the series of bilayered cathodes at 600 and 800°C. Reprinted from Ref. [50], Copyright (2012), with permission from Elsevier.

For practical use, a SOFC should have chemical and mechanical stability over a long period. LSCF ($La_{0.6}Sr_{0.4}Co_{0.2}Fe_{0.8}O_3$) is one of the candidates for a cathode material of an IT-SOFC, which can be operated below 800°C, because it has high electronic and oxide-ionic conductivities at low temperatures [54].

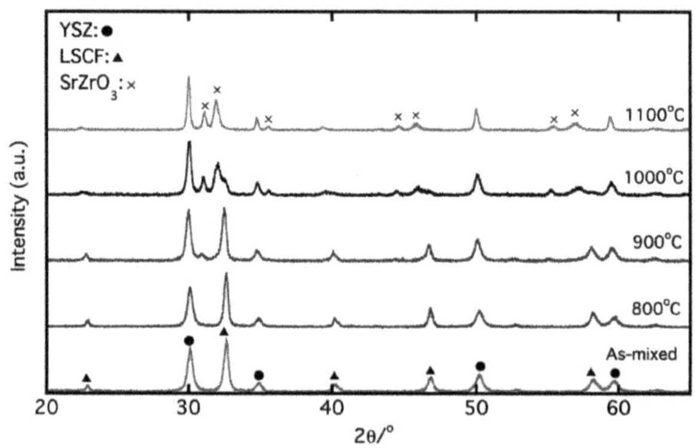

Figure 7.22 XRD patterns of the mixed powder of LSCF + YSZ and those heat-treated at 800 to 1100°C for 1 h.

However, direct contact of LSCF with YSZ causes the formation of insulating phases such as $SrZrO_3$ and $La_2Zr_2O_7$ at the interface [28, 29]. Figure 7.22 shows XRD patterns of the mixed powders of LSCF and YSZ heat-treated at different temperatures. Apparently a $SrZrO_3$ subphase was recognized by calcination at above 1000°C. To avoid such undesired chemical reactions, GDC is frequently inserted as a diffusion barrier between a LSCF cathode and a YSZ electrolyte [30, 31]. The EPD technique helped us to easily form a double-layer structure with an optimized layer thickness. Double-layer cathodes, each composed of a LSCF layer and a GDC interlayer, were formed on an YSZ sheet, where the thickness of the GDC layer was 5–10 μm, using EPD. Murray et al. [61] have clarified that the LSCF-GDC composite cathode effectively reduced the cathodic polarization resistance. The LSCF-GDC/GDC double-layer cathode prepared by the EPD method also gave both good initial and long-term properties. The single cells preheated at 1000°C for 100 h were evaluated at 700°C. The maximum power densities before and after the pretreatment are summarized in Table 7.3. By the pretreatment, a 70% reduction in the MPD was observed in the LSCF/YSZ cathode, whereas the reduction in the MPDs of the LSCF/GDC and LSCF-GDC/GDC cathodes was 25%. The large amount of power drop in LSCF/YSZ is probably due to an increase in the insulating phase, such as $SrZrO_3$, as predicted in Fig. 7.22.

Table 7.3 Power densities of cells with the series of cathode films preheated at 1000°C

Maximum power density at 700°C/mW cm^{-2}				
Heating time/h	**LSCF**	**LSCF/GDC**	**LSCF/YSZ**	**GDC-LSCF/GDC**
0	4.5	12.4	33.5	34.2
100	-	9.4	10.1	28.7

7.5.3 Preparation of Mono- and Bilayer Ni-YSZ Anode Films

In the same manner, monolayer and bilayer anodic films can also be prepared [62]. The Ni-YSZ cermet is a widely utilized anode material for SOFCs because of the high ionic conductivity and stability of YSZ. Figure 7.23 shows cross-sectional views of the 50 wt% Ni-YSZ (50Ni-YSZ) and 70 wt% Ni-YSZ (70Ni-YSZ) monolayer films after sintering at 1300°C and reduction under hydrogen at 600°C. Both anode films are about 30 μm in thickness, and they are porous and homogeneous.

Figure 7.23 SEM images of the cross sections of the Ni-YSZ anode films formed by EPD [62].

Figure 7.24 shows SEM-EDX images of a double-layer anode in a cross-sectional view observed after hydrogen reduction. From the Ni and Zr elemental mappings, the formation of the double layer with a 6.9/8.7 μm thickness was confirmed.

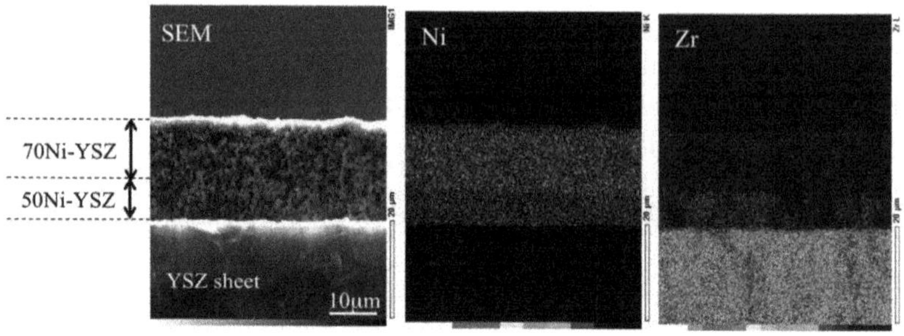

Figure 7.24 The cross-sectional SEM image and elemental mappings of a bilayered anode film [62].

Furthermore, it seems that the 70Ni-YSZ layer is more porous than the 50Ni-YSZ layer. This is because hydrogen reduction eliminates oxygen from NiO and the oxygen deficiency propagates into the pores. In this sense, the 50Ni-YSZ/70Ni-YSZ double layer is considered a porosity-graded structure as well as a concentration-graded structure. The porosity-graded structure is favorable for promoting gas diffusion [63].

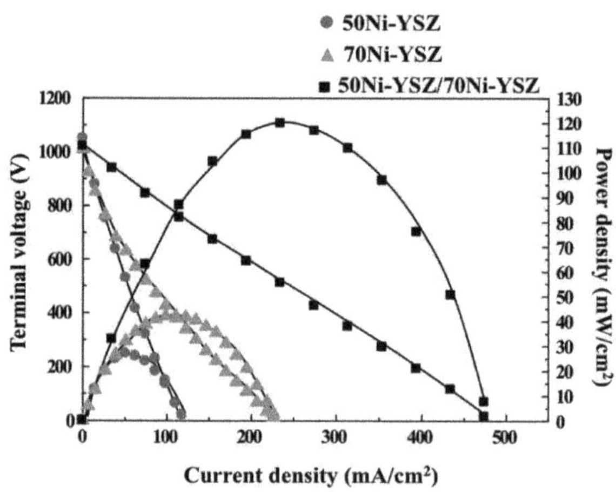

Figure 7.25 The *i-v* and *i-p* characteristics of cells with the mono- and bilayer anodes.

Figure 7.25 shows the *i-v* characteristics of the cells with 50Ni-YSZ and 70Ni-YSZ anodes. The film thickness was 13 and 16 μm, respectively. Apparently, the 70Ni-YSZ anode exhibited a higher performance than the 50Ni-YSZ anode. Figure 7.25 also shows the *i-v* characteristic of the cell with a 50Ni-YSZ/70Ni-YSZ anode (3 μm/14 μm). A remarkable enhancement in the *i-v* characteristic was observed for the double-layer anode. This result suggests that the 50Ni-YSZ and the 70Ni-YSZ layers act as active and current collecting layers, respectively, and the 50Ni-YSZ layer provides a longer TPB length than the 70Ni-YSZ layer.

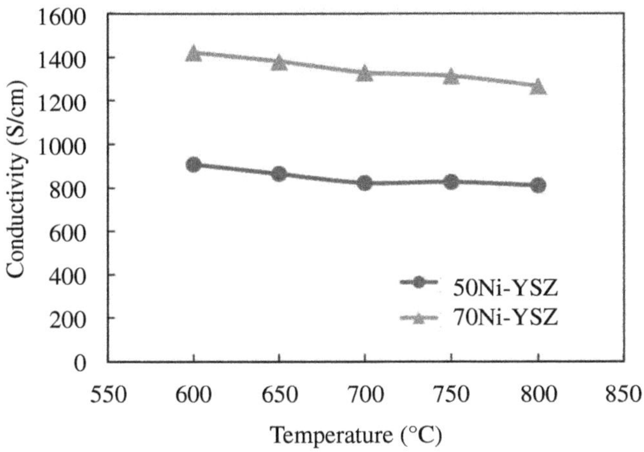

Figure 7.26 DC conductivity of the 50 and 70Ni-YSZ sinters measured by the four-terminal setup.

Figure 7.26 shows the results of the four-terminal dc conductivity measurements for the 50Ni-YSZ and 70Ni-YSZ compacts prepared by uniaxial pressure of 20 MPa. Both cermets exhibited sufficiently high conductivities, and 70Ni-YSZ exhibited a higher conductivity than 50Ni-YSZ due to a higher Ni concentration. The observed negative temperature dependence means that metallic conduction of the Ni is predominant. It is widely accepted that the conductivity of a Ni-YSZ cermet is controlled by the percolation effect and the conductivity threshold occurs around 33 vol% [64]. Both 50Ni-YSZ (40 vol%) and 70Ni-YSZ (60 vol%) are beyond the threshold and, therefore, their conductivities should be high enough to have a sufficient current collecting effect. Therefore, the difference in conductivities is not likely to be a reason for the difference in the anodic properties. Since the H_2 oxidation reaction mainly takes place at the TPB of H_2 (or H), Ni, and YSZ, the 70Ni-YSZ anode, with the higher concentration of Ni, seems to have a longer TPB length. However, this is not necessarily true, because the high concentration of Ni simultaneously induces Ni particle aggregation, which results in a reduced TPB length. The anodic performance of the double layers with different layer thicknesses was examined.

Figure 7.27 shows the anodic overpotential of the monolayer and bilayer anodes. On using the bilayer structure, the anodic overpotential dramatically decreased compared to the monolayer anodes. Figure 7.27 shows the anodic overpotential at different current densities. It is clear that the anodic property depends more on the thickness of the first layer. On increasing the thickness of the first layer from 3 μm to 6 μm, a significant increase in the overpotential was observed. This result suggests that an approximately 3 μm 50Ni-YSZ layer practically acts as the active layer and the residual part of the layer is only considered a low-conductivity layer. Therefore, it should be effective to replace the resistive layer by the high-conductivity 70Ni-YSZ layer. Antunes et al. [60] investigated the thickness effect of the bilayer cathodes, that is, LSM-YSZ/LSM. In their study, the critical thickness of the LSM-YSZ layer was 8 μm. The critical thickness of the active layer cannot be unambiguously determined, because it is also a function of the microstructure (grain size) and porosity [65]. It seems that the anodic property does not significantly depend on the thickness of the 70Ni-YSZ layer in the range of 9–20 μm. As mentioned in Section 7.5.3, the current collecting effect is governed by the in-plane resistance of the electrode film and the effect increases with the electrode thickness [57]. The $R_{\text{in-plane}}$ value of 70Ni-YSZ at 700°C was evaluated as follows. The $R_{\text{in-plane}}$ value at 700°C was 7.5×10^{-4} Ω·cm. The r value was used as an aperture of the Ni mesh (0.075 mm). When the th-value is 9 μm, the $R_{\text{in-plane}}$ value of 70Ni-YSZ at 700°C was evaluated to be 1.2×10^{-5} Ω·cm^2.

Assuming that the current collecting effect is saturated above the thickness of 9 μm, the $R_{\text{in-plane}}$ value of 1.2×10^{-5} Ω·cm^2 is low enough to give sufficient current collecting effect. The $R_{\text{in-plane}}$ values of 50Ni-YSZ with a 13 μm thickness was 1.3×10^{-5} Ω·cm^2. This value is very close to the evaluated value of 70Ni-YSZ with a 9 μm thickness, which suggests that 50Ni-YSZ with a 13 μm thickness also possesses a sufficiently large current collection effect. This result supports the fact that the difference between the anodic properties of 50Ni-YSZ and 70Ni-YSZ shown in Fig. 7.25 is not due to the difference in the conductivities. Thus, in this study, the EPD method provided relatively dense Ni-YSZ anode films. This tendency is one of the characteristics of the ceramic film produced by the EPD method and might be a positive factor for preparing a thin, active layer a few micrometers in thickness. However, the dense morphology may cause a high gas diffusion barrier, which can be a negative factor for a current collecting layer that is more than 10 μm in thickness. Further study will be required for optimizing the EPD conditions to achieve a moderately porous morphology of anodic films.

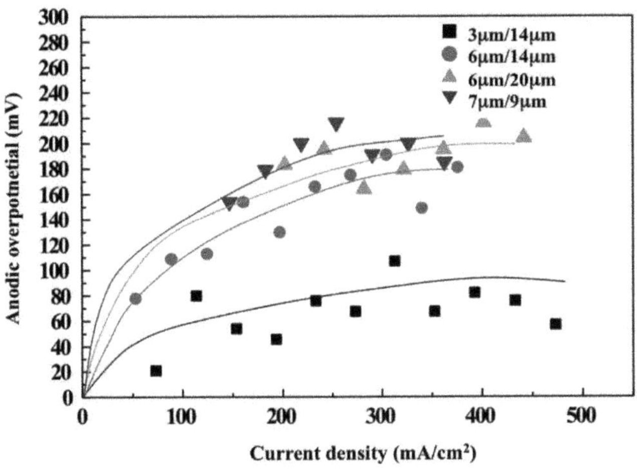

Figure 7.27 The anodic overpotential curves of the series of bilayer anodes.

7.6 EPD Coating on a Nonflat Surface

In a single-cell test, metal meshes are commonly used as a current collector and are attached to the surface of an electrode. For this reason, the effective contact area between mesh and electrode frequently affects the electrode overpotential [66–68]. Sasaki et al. [66] reported that the electrode overpotential was remarkably reduced with a decrease in the distance between the neighboring wires of the Pt mesh current collectors. They explained that the decrease in the distance, and hence an increase in the contact points, enhanced the homogeneous current density distribution and this increased the effective contact area between electrolyte and electrode. Jiang et al. [67] demonstrated that an increase in the number of contact points between a collector and an electrode surface could make the current distribution inside the cell uniform, resulting in an increase in the length of the TPB. In fact, the woven-shaped mesh does not come fully in contact with the electrode. Therefore, the coating of the mesh with the electrode material may increase the number of contact points. One of the most advantageous points of the EPD method is that film deposition can be achieved onto a nonflat surface without any methodological modifications.

As shown in Fig. 7.28, a metal mesh can also be a substrate for EPD: A LSM powder is deposited on the Pt mesh in the figure [69]. The metal wires are completely covered with the LSM powder. Figure 7.29 shows the metal wires before and after the EPD coating of a LSM powder. From the cross-sectional view, it can be seen that the electrode powder that gets deposited forms a layer that is about 20 μm thick. Furthermore, the coated mesh has a good contact with the electrode surface via the deposition layer. In the present case, it is suggested from SEM observations that the EPD treatment increased the effective contact area between the wires and the electrode.

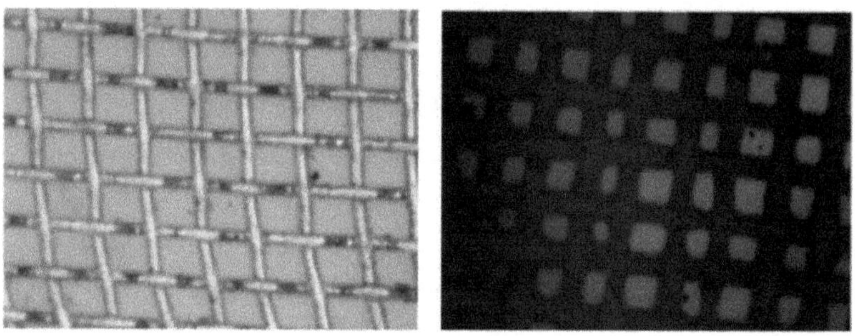

Figure 7.28 A Pt mesh (left) and that coated with a LSM powder by the EPD method (right). Reproduced from Ref. [69] with permission from the Electrochemical Society.

Of course, the metal mesh can be coated with an anode powder, for example, Ni-YSZ, with a similar procedure. Figure 7.30 shows the electrode overpotential curves with and without mesh coatings.

Assuming a linear relationship between the current density dependence and the voltage losses, the polarization (η) and ohmic (R) resistances were evaluated, as listed in Table 7.4. Apparently the EPD coatings for the wire meshes remarkably reduced both polarization and ohmic losses at each of the electrodes compared to those for noncoated wire meshes: reductions of 54.5% and 33.3% were achieved for polarization and ohmic resistances, respectively. EPD treatment for wire meshes increases the effective contact area between the mesh and the electrodes. This might also contribute to the reduction of polarization losses

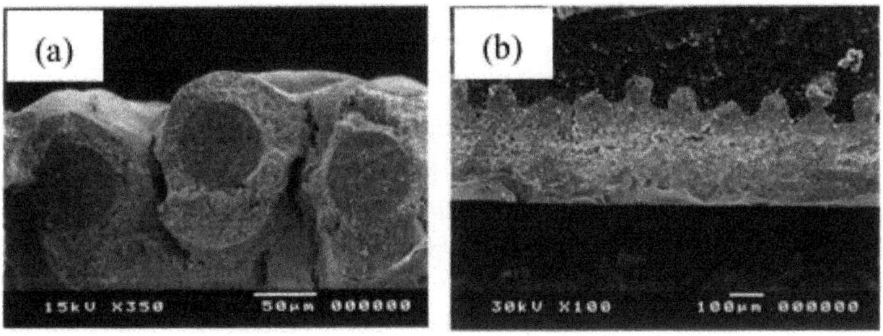

Figure 7.29 Cross-sectional SEM pictures of the woven Pt mesh with the EPD coating (a) and that embedded in the electrode layers (b). Reproduced from Ref. [69] with permission from the Electrochemical Society.

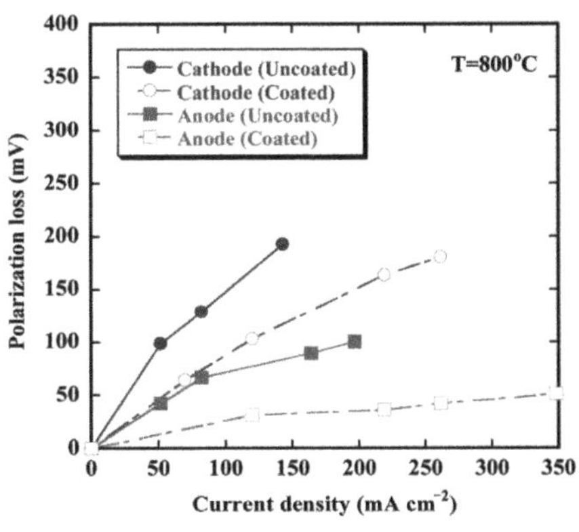

Figure 7.30 The polarization losses of the electrodes attached by the metal meshes with and without EPD coatings. Reproduced from Ref. [69] with permission from the Electrochemical Society.

Table 7.4 Polarization and ohmic resistances of the electrodes with the EPD-coated wire meshes and the uncoated wire meshes measured at 800°C

	Cathodic resistance (Ω·cm^2) η_c R_c	**Anodic resistance (Ω·cm^2) η_a R_a**	**Total resistances (Ω·cm^2) η_{c+a} R_{c+a} $\eta_{c+a} + R_{c+a}$**
With EPD coatings	0.68 0.66	0.14 0.90	0.82 1.56 2.38
Without coatings	1.32 0.91	0.48 1.43	1.80 2.34 4.14

Ohmic resistances, R_a and R_c, include a contribution from the solid electrolyte.

Figure 7.31 shows the *i-v* and *i-p* characteristics for the single cells attached with the EPD-coated wire meshes and the uncoated wire meshes measured at 800°C. The open-circuit voltage (OCV) was identical (0.95 V) for both cells. Apparently, a higher cell performance was obtained for the cell with the coated wire meshes compared to that with the uncoated ones. The maximum power densities were 9.3×10^{-2} (at 249 mA·cm^{-2}) and 5.0×10^{-2} Wcm^{-2} (at 155 mA·cm^{-2}), respectively.

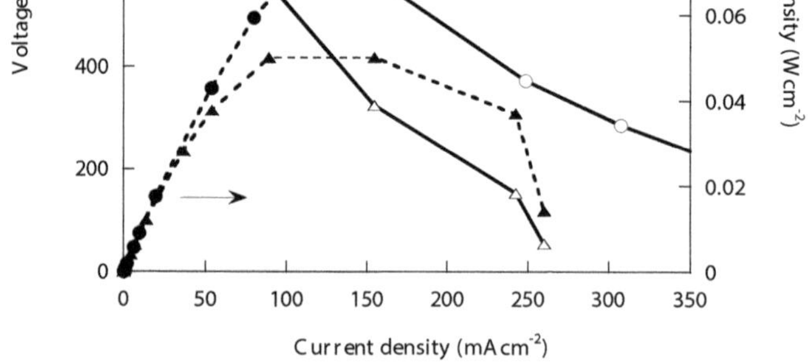

Figure 7.31 *i-v* (solid lines) and *i-p* (dashed lines) characteristics of the single cells with the EPD-coated metal wire meshes and uncoated metal wire meshes measured at 800°C. Reproduced from Ref. [69] with permission from the Electrochemical Society.

In SOFC stacks, metallic interconnectors are frequently used instead of a metal mesh. An interconnector is usually a wave-shaped metal plate to form the separated flow-channels of air and fuel. It is also expected that metallic interconnectors coated with electrode powders have good contact with the electrode surface. Therefore, the EPD technique for the nonflat surface is applicable to the real systems as well.

7.7 EPD Effect on Carbon Deposition in the Direct Methane SOFC

As mentioned in Section 7.1, a SOFC has high fuel flexibility: CO and hydrocarbons can also be used as direct fuels with the internal fuel reforming process. Such direct-fuel SOFC can eliminate the fuel reformer, which minimizes the scale of the SOFC unit. Methane, which is the main component of town gases, is accepted as the primary fuel in the current system. The steam reforming reactions are as follows [70]:

$$CH_4 + H_2O \rightarrow CO + 3H_2 \tag{7.9}$$

$$CO + H_2O \rightarrow CO_2 + H_2 \tag{7.10}$$

These processes should be conducted in a high steam/fuel ratio (>3). Otherwise, an undesired carbon deposition reaction occurs and leads to catalyst deterioration.

$$CH_4 \rightarrow C + 2H_2 \tag{7.11}$$

$$CO \rightarrow C + CO_2 \tag{7.12}$$

For the internal reforming SOFC, such excessive humidity may increase the oxygen partial pressure at the anode and decrease the cell performance. Therefore, it is desirable to operate in a low humidity. The authors found that the anode films prepared by the EPD method significantly suppress the carbon deposition [13].

Figure 7.32 shows the pictures of the anodic sides (Ni-SDC) of the cells operated with dry methane for 48 h. The anodes were formed by EPD and slurry coating. After the cell operation, the slurry-coated anode was totally covered with the deposited carbon. On the other hand, only a small amount of carbon was deposited on the anode coated by EPD. This carbon durability of the EPD anode could be due to the uniformly densified morphology of the EPD anode, which has better oxide ionic conductivity through the SDC phase. This good oxygen diffusibility may effectively burn out the carbon deposited at the anode. The carbon durability of the anode is also dependent on the film thickness.

Figure 7.32 Pictures of the anode sides of the cell operated for 48 h with dry methane fuel. The anodes were formed by the EPD method (left) and the slurry coating (right). The anodes are Ni-SDC.

Table 7.5 shows the amount of carbon deposition of the Ni-SDC films after 6 h cell operation with dry methane, of which the thickness is controlled by the EPD method.

Table 7.5 Amounts of O^{2-} supplied to the anode and carbon deposited on Ni-SDC anodes with different thicknesses during 6 h operation with dry methane at 973 K

Anode thickness/μm	O^{2-} supplied to the anode/10^{-5} mol	Amount of carbon deposition/mg
9	0 (OCV)	5.7
	1.07	~0
20	0 (OCV)	13.3
	1.06	0.8
	7.35	~0
36	0 (OCV)	19.0
	1.01	12.3

Carbon depositions were significant at OCV conductions and increased with an increase in thickness, while the carbon depositions were reduced by the current flow where the deposited carbon was burned out by the pumped oxygen. This carbon elimination effect is more remarkable at a smaller thickness. This result is probably due to the fact that the O^{2-} ions can reach the surface in the thin anode film and react with the deposited carbon.

7.8 Summary

This chapter introduced our recent activities related to the development of SOFC materials and the structural controls of the electrode films using the EPD technique. Several chemical routes of electrode powder preparation have enabled precise control of the particle size distribution. In an anode, size distribution of Ni particles is a decisive factor for anodic performance. A highly Ni-dispersed anode containing a relatively low concentration of Ni (20 wt%) was possible in Ni-SDC, where SDC particles form electronic conduction paths in a reducing atmosphere. The mean size of the Ni particles loaded on the SDC support was variable with the surface area of the SDC support.

The morphology of the electrode film is directly related to the length of the TPB and the current collecting ability. EPD, which is a colloidal wet process for oxide film formation, is applicable not only for electrolyte films but also for electrode films. The thickness and morphology are controlled by the EPD conditions, that is, the deposition time and the applied voltage. EPD electrode films tend to possess a moderately dense morphology, which enables a good electrode performance. In some cases, an EPD coating enabled a better electrode performance than the conventional slurry coating. The EPD technique is applicable in the formation of bilayered thin-film structures of precise thicknesses. Furthermore, the EPD technique is also applicable to a nonflat substrate. As a demonstration, electrode powder deposition onto a metal mesh as a current collector was introduced in this chapter. The authors believe that the EPD technique can be a candidate in SOFC fabrication processes. For the practical application of EPD in the fabrication of a SOFC single cell, long-term stability of the suspension should be ensured.

Acknowledgments

Several parts of our works were conducted in the COE formation project and the Research Unit for Power Generation and Storage Materials (PGeS) at Ehime University. The works were also subsidy supported by JSPS. We also appreciate the contribution from Prof. Y. Sadaoka and Dr. T. Yamaji, as well as from Dr. M. Asamoto at the university.

References

1. Badwal, S. P. S. (1984). Electrical conductivity of single crystal and polycrystalline yttria-stabilized zirconia, *J. Mater. Sci.*, **19**, pp. 1767–1776.
2. Xia, C. (2001). Electrolytes, in *Semiconductors and Semimetals*, Vol. 73, Fergus, J. W., Hui, R., Li, X., Wilkinson, D. P. and Zhang, J. (eds.), CRC Press, Boca Raton, pp. 1–72.
3. Gazzarri, J. I. and Kesler, O. (2007). Non-destructive delamination detection in solid oxide fuel cells, *J. Power Sources*, **167**, pp. 430–441.
4. Minh, N. Q. (1993). Ceramic fuel cells, *J. Am. Ceram. Soc.*, **76**, pp. 563–588.
5. Laurencin, J., Delette, G., Morel, B., Lefebvre-Joud, F. and Dupeux, M. (2009). Solid oxide fuel cells damage mechanisms due to Ni-YSZ re-oxidation: case of the anode supported cell, *J. Power Sources*, **192**, pp. 344–352.
6. Matsuzaki, Y. and Yasuda, I. (2000). The poisoning effect of sulfur-containing impurity gas on a SOFC Anode: Part I. Dependence on temperature, time, and impurity concentration, *Solid State Ionics*, **132**, pp. 261–269.
7. Stanislowski, M., Froitzheim, J., Niewolak, L., Quadakkers, W. J., Hilpert, K., Markus, T. and Singheiser, L. (2006). Reduction of chromium vaporization from SOFC interconnectors by highly effective coatings, *J. Power Sources*, **164**, pp. 578–589.
8. http://www.nedo.go.jp/content/100642944.pdf (2016/8/31 access).
9. Lee, J.-H., Moon, H., Lee, H.-W., Kim, J., Kim, J.-D. and Yoon, K.-H. (2002). Quantitative analysis of microstructure and its related electrical property of SOFC anode, Ni-YSZ cermet, *Solid State Ionics*, **148**, pp. 15–16.

10. Kharton, V. V., Figueireddo, F. M., Navarro, L., Naumovich, E. N., Kovalevsky, A. V., Yaremchenko, A. A., Viskup, A. P., Carneiro, A., Marques, F. M. B. and Frade, J. R. (2001). Ceria-based materials for solid oxide fuel cells, *J. Mater. Sci.*, **36**, pp. 1105–1117.

11. Yahiro, H., Eguchi, K. and Arai, H. (1989). Electrical properties and reducibilities of ceria-rare earth oxide systems and their application to solid oxide fuel cell, *Solid State Ionics*, **36**, pp. 71–75.

12. Wang, J. B., Jang, J.-C. and Huang, T.-J. (2003). Study of Ni-samaria-doped ceria anode for direct oxidation of methane in solid oxixde fuel cells, *J. Power Sources*, **122**, pp. 122–131.

13. Asamoto, M., Miyake, S., Itagaki, Y., Sadaoka, Y. and Yahiro, H. (2007) Electrocatalytic performances of Ni/SDC anodes fabricated with EPD techniques for direct oxidation of CH_4 in solid oxide fuel cell, *Catal. Today*, **139**, pp. 77–81.

14. Suzuki, S., Uchida, H. and Watanabe, M. (2006). Interaction of samaria-doped ceria anode with highly dispersed Ni catalysts in a medium-temperature solid oxide fuel cell during long-term operastion, *Solid State Ionics*, **177**, pp. 359–365.

15. Xie, Z, Zhu, W., Zhu, B. and Xia, C. (2006). $Fe_xCo_{0.5-x}Ni_{0.5}$-SDC anodes for low temperature solid oxide fuel cells, *Electrochem. Acta*, **51**, pp. 3052–3057.

16. Asamoto, M., Miyake, S., Sugihara, K., Yamaguchi, S. and Yahiro, H. (2009). Improvement of Ni/SDC anode by alkaline earth metal oxide addition for direct methane-SOFC, *ECS Trans.*, **25**, pp. 2155–2160.

17. Sugihara, K., Asamoto, M., Itagaki, Y., Takemasa, T., Yamaguchi, S., Sadaoka Y. and Yahiro, H. (2014). A quantitative analysis of influence of Ni particle size of SDC-supported anode on SOFC performance: effect of particle size of SDC support, *Solid State Ionics*, **262**, pp. 433–437.

18. Ishihara, T., Yan, J., Shinagawa, M. and Matsumoto, H. (2006). Ni-Fe bimetallic anode as an active anode for intermediate temperature SOFC using $LaGaO_3$ based electrolyte film, *Electrochem. Acta*, **52**, pp. 1645–1650.

19. Suzuki, T., Jasinski, P., Andersen, H. U. and Dogan, F. (2004). Role of composite cathodes in single chamber SOFC, *J. Electrochem. Soc.*, **151**, pp. A1678–A1682.

20. Lu, X. C. and Zhu, J. H. (2007). Ni-Fe + SDC composite as anode material for intermediate temperature solid oxide fuel cell, *J. Power Sources*, **165**, pp. 678–684.

21. An, W., Gatewood, D., Dunlap B. and Yurner, C. H. (2011). Catalytic activity of bimetallic nickel alloys for solid-oxide fuel cell anode reactions from density-functional theory, *J. Power Sources*, **196**, pp. 4724–4728.

22. Resini, C., Herrera Deigado, M. C., Prest, S., Alemany, L. J., Riani, P., Marazza, R., Ramis G. and Busca, G. (2008). Yttria-stabilized zirconia (YSZ) supported Ni-CO alloys (precursor of SOFC anodes) as catalysts for the steam reforming of ethanol, *Int. J. Hydrogen Energy*, **33**, pp. 3728–3735.

23. Grgicak, C. M., Pakulska, M. M., O'Brien, J. S. and Giorgi, J. B. (2008). Synergistic effects of $Ni_{1-x}Co_x$-YSZ alloyed cermet SOFC anodes for oxidation of hydrogen and methane fuel containing H_2S, *J. Power Sources*, **183**, pp. 26–33.

24. Itagaki, Y., Takemasa, T., Yamaguchi, S. and Yahiro, H. (2015). Impedance study of anodic properties of Ni-Fe impregnated SDC, *ECS Trans.*, **68**, pp. 1427–1434.

25. Reuter, K. B., Williams, D. B. and Goldstein, J. I. (1989). Determination of the Fe–Ni phase diagram below 400°C, *Metall. Trans. A*, **20**, pp. 719–725.

26. Mizusaki, J., Mori, N., Takai, H., Yonemura, Y., Minamiue, H., Tagawa, H., Dokiya, M., Inaba, H., Naraya, K., Sasamoto, T. and Hashimoto, T. (2000). Oxygen nonstoichiometry and defect equilibrium in the perovskite-type oxides $La_{1-x}Sr_xMnO_{3+d}$, *Solid State Ionics*, **129**, pp. 163–177.

27. Jiang, S. P. (2002). A comparison of O_2 reduction reactions on porus (La, Sr)MnO_3 and (La, Sr)(Co, Fe)O_3 electrodes, *Solid State Ionics*, **146**, pp. 1–22.

28. Tu, H. Y., Takada, Y., Imanishi, N. and Yamamoto, O. (1999). $Ln_{0.4}Sr_{0.6}Co_{0.8}Fe_{0.2}O_{3-d}$ (Ln=La, Pr, Nd, Sm, Gd) for the electrode in solid oxide fuel cells, *Solid State Ionics*, **117**, pp. 277–281.

29. Kindermann, L., Das, D., Nickel, H. and Hilpert, K. (1996). Chemical compatibility of the $LaFeO_3$ based perovskites $(La_{0.6}Sr_{0.4})_zFe_{0.8}M_{0.2}O_{3-?}$ (z=1, 0.9; M=Cr, Mn, Co, Ni) with yttria stabilized zirconia, *Solid State Ionics*, **89**, pp. 215–220.

30. Shiono, M., Kobayashi, K., Nguyen, T. L., Hosoda, K., Kato, T., Ota, K. and Dokiya, M. (2004). Effect of CeO_2 interlayer on ZrO_2 electrolyte/La(Sr) CoO_3 cathode for low-temperature SOFCs, *Solid State Ionics*, **170**, pp. 1–7.

31. Mai, A., Haanappel, V. A. C., Tietz, F. and Stöver, D. (2006). Ferrite-based perovskites as cathode materials for anode-supported solid oxide fuel

cells, Part II. Influenece of the CGO interlayer, *Solid State Ionics*, **177**, pp. 2103–2107.

32. Choi, J. H., Jang, J. H., Ryu, J. H. and Oh, S. M. (2000). Microstructure and cathodic performance of $La_{0.9}Sr_{0.1}MnO_3$ electrodes according to particle size of starting powder, *J. Power Sources*, **87**, pp. 92–100.
33. Pechini, M. P. (1967). Method of preparing lead and alkaline earth titanates and niobates and coating method using the same to form a capacitor, US Patent 3,330,697.
34. Tai, L. W. and Lessing, P. A. (1992). Modified-intermediate processing of perovskite powders: part I. Optimization of polymeric precursors, *J. Mater. Res.*, **7**, pp. 502–510.
35. Alber, K. S. and Cox, J. A. (1997). Electrochemistry in solids prepared by sol-gel processes, *Mikrochim. Acta.*, **127**, pp. 131–147.
36. Liu, J., Co, A. C., Paulson, S. and Birss V. I. (2005). Oxygen reduction at sol-gel derived $La_{0.8}Sr_{0.2}Co_{0.8}Fe_{0.2}O_3$ cathodes, *Solid State Ionics*, **177**, pp. 377–387.
37. Pelosato, R., Cristiani, C., Dotelli, G., Mariani, M., Donazzi, A. and Sora, I. N. (2011). Co-precipitation sysntheses of SOFC electrode materials, *Int. J. Hydrogen Energy*, **38**, pp. 480–491.
38. Yamaji, T., Itagaki, Y., Arakawa, K. and Sadaoka, Y. (2010). Formation of $La_{0.8}Sr_{0.2}MnO_3$ films as SOFC cathodes by electrophoretic deposition, *J. Ceram. Soc. Jpn.*, **118**, pp. 1202–1206.
39. Hart, N. T., Brandon, N. P., Day, M. J. and Shemilt, J. E. (2001). Functionally graded cathodes for solid oxide fuel cells, *J. Mater. Sci.*, **36**, pp. 1077–1085.
40. Sarkar, P. and Nicholson, P. S. (1996). Electrophoretic deposition (EPD): mechanism, kinetics, and application to ceramics, *J. Am. Ceram. Soc.*, **79**, pp. 1987–2002.
41. Ishihara, T., Sato, T. and Takita, Y. (1996). Electrophoretic deposition of Y_2O_3-stabilized ZrO_2 electrolyte films in solid oxide fuel cells, *J. Am. Ceram. Soc.*, **79**, pp. 913–919.
42. Chen, F. L. and Liu, M. L. (2001). Preparation of yttria-stabilized zirconia (YSZ) films on $La_{0.85}Sr_{0.15}MnO_3$ (LSM) and LSM–YSZ substrates using an electrophoretic deposition (EPD) process, *J. Eur. Ceram. Soc.*, **21**, pp. 127–134.
43. Besra, L., Compson, C. and Liu, M. (2007). Electrophoretic deposition on non-conducting substrates: the case of YSZ film on NiO–YSZ composite substrates for solid oxide fuel cell application, *J. Power Sources*, **173**, pp. 130–136.

44. Hosomi, T., Matsuda, M. and Miyake, M. (2007). Electrophoretic deposition for fabrication of YSZ electrolyte film on non-conducting porous NiO–YSZ composite substrate for intermediate temperature SOFC, *J. Eur. Ceram. Soc.*, **27**, pp. 173–178.

45. Jia, L., Lu, Z., Huang, X., Liu, Z., Chen, K., Sha, X., Li, G. and Su, W. (2005). Preparation of YSZ film EPD and its application in SOFCs, *J. Alloys Compd.*, **4224**, pp. 299–303.

46. Yamaji, K., Kishimoto, H., Xiong, Y., Horita, T., Sakai, N. and Yokokawa, H. (2004). Performance of anode-supported SOFCs fabricated with EPD techniques, *Solid State Ionics*, **175**, pp. 165–169.

47. Uchikoshi, T. and Suzuki, T. (2010). Ceramics, **45**, pp. 88–92 (in Japanese).

48. Santillán, M. J., Caneiro, A., Quaranta, N. and Boccaccini, A. R. (2009). Electrophoretic deposition of $La_{0.6}Sr_{0.4}Co_{0.8}Fe_{0.2}O_{3-\delta}$ cathodes on $Ce_{0.9}Gd_{0.1}O_{1.95}$ substrates for intermediate temperature solid oxide fuel cell (IT-SOFC), *J. Eur. Ceram. Soc.*, **29**, pp. 1125–1132.

49. Suzuki, H. T., Uchikoshi, T., Kobayashi, K., Suzuki, T. S., Sugiyama, T., Furuya, K., Matsuda, M., Sakka Y. and Munakata, F. (2009). Fabrication of GDC/LSGM/GDC tri-layers on polypyrrole-coated NiO-YSZ by electrophoretic deposition for anode-supported SOFC, *J. Ceram. Soc. Jpn.*, **117**, pp. 1246–1248.

50. Itagaki, Y., Watanabe, S., Yamaji, T., Asamoto, M., Yahiro, H. and Sadaoka, Y. (2012). Electrophoretic depostion of bi-layered LSM/LSM-YSZ cathodes for solid oxide fuel cell, *J. Power Sources*, **214**, pp. 153–158.

51. Kenjo, T. and Nishiya, M. (1992). $LaMnO_3$ air cathodes containing ZrO_2 electrolyte for high temperature solid oxide fuel cells, *Solid State Ionics*, **57**, pp. 295–302.

52. Murray, E. P., Tsai, T. and Barnett, S. A. (1998). Oxygen transfer processes in $(La,Sr)MnO_3/Y_2O_3$-stabilized ZrO_2 cathodes: an impedance spectroscopy study, *Solid State Ionics*, **110**, pp. 235–243.

53. Jørgensen, M. J., Primdahl, S. and Mogensen, M. (1999). Characterisation of composite SOFC cathodes using electrochemical impedance spectroscopy, *Electrochem. Acta*, **44**, pp. 4195–4201.

54. Kleitz, M. and Petitbon, F. (1996). Optimized SOFC microstructure, *Solid State Ionics*, **92**, pp. 65–74.

55. Kim, J. D., Kim, G. D., Moon, J. W., Lee, H. W., Lee, K. T. and Kim, C. E. (2000). The effect of percolation on electrochemical performance, *Solid State Ionics*, **133**, pp. 67–77.

56. Kim, D., Kim, G. D., Moon, J. W., Park, Y., Lee, W. H., Kobayashi, K., Nagai, M. and Kim, C. E. (2001). Characterization of LSM–YSZ composite electrode by ac impedance spectroscopy, *Solid State Ionics*, **143**, pp. 379–389.

57. Barbucci, A., Carpanese, M., Reverberi, A. P., Cerisola, G., Blanes, M., Cabot, P. L., Viviani, M., Bertei, A. and Nicolella, C. (2008). Influence of the electrode thickness on the performance of composite electrodes for SOFC, *J. Appl. Electrochem.*, **38**, pp. 939–945.

58. Holtappels, P. and Bagger, C. (2002). Fabrication and performance of advanced multilayer SOFC cathodes, *J. Eur. Ceram. Soc.*, **22**, pp. 41–48.

59. Liu, Y., Compson, C. and Liu, M. (2004). Nonostructure and functionally graded cathodes for intermediate temperature solid oxide fuel cells, *J. Power Source*, **138**, pp. 194–198.

60. Antunes, R., Golec, T., Miller, M., Kluczowski, R., Krauz, M. and Krzastek, K. (2010). Geometrical and microstructure optimization of double-layer LSM/LSM-YSZ cathodes by electrochemical impedance spectroscopy, *J. Fuel Cell Sci. Technol.*, **7**, pp. 1–6.

61. Perry Murray, E., Sever, M.J. and Barnett, S. A. (2002). Electrochemical perforemance of (La,Sr)(Co,Fe)O_3-(Ce,Gd)O_3 composite cathodes, *Solid State Ionics*, **148**, pp. 27–34.

62. Itagaki, Y., Shinohara, K., Yamaguchi, S. and Yahiro, H. (2015). Anodic performance of bilayer Ni-YSZ anodes formed by electrophoretic depostion, *J. Ceram. Soc. Jpn.*, **123**, pp. 235–238.

63. Holtappels, P., Sorof, C., Verbraeken, M. C., Rambert, S. and Vogt, U. (2006). Preparation of porosity-graded SOFC anode substrate, *Fuel Cells*, **2**, pp. 113–116.

64. Lee, J.-H., Moon, H., Lee, H.-W., Kim, J., Kim J.-D. and Yoon, K.-H. (2002). Quantitative analysis of microstructure and its related electrical property of SOFC anode, Ni-YSZ cermet, *Solid State Ionics*, **148**, pp. 15–26.

65. Virkar, A. V., Chen, J., Tanner, C. W. and Kim, J.-W. (2000). The role of electrode microsturcutre on activation and concentration polarization in solid oxide fuel cells, *Solid State Ionics*, **131**, pp. 189–198.

66. Sasaki, K., Wurth, J.-P., Gschwend, R., Gödickemeier, G. and Gauckler, L. J. (1996). Microstructure-property relations of solid oxide fuel cell cathodes and current collectors: cathodic polarization and ohmic resistance, *J. Electrochem. Soc.*, **143**, pp. 530–543.

67. Jiang, S. P., Love, J. G. and Apateanu, L. (2003). Effect of contact between electrode and current collector on the performance of solid oxide fuel cells, *Solid State Ionics*, **160**, pp. 15–26.

68. Wanzenberg, E., Tietz, F., Kek, D., Panjan, P. and Stöver, D. (2003). Influenece of electrode contacts on conductivity measurements on thin YSZ electrolyte films and the impact on solid oxide fuel cells, *Solid State Ionics*, **164**, pp. 121–129.

69. Itagaki, Y., Matsubara, F., Asamoto, M., Yamaura, H., Yahiro, H. and Sadaoka, Y. (2007). Electrophoretically coated wire mehes as current collectors for solid oxide fuel cell, *ECS Trans.*, **7**, pp. 1319–1325.

70. Armor, J. (1999). The multiple roles for catalysis in the production of H_2, *Appl. Catal., A*, **176**, pp. 159–176.

Chapter 8

Charging Up the Future by Organic Solar Cells

Tetsuo Okujima
Department of Chemistry and Biology, Graduate School of Science and Engineering, Ehime University, and Research Unit for Power Generation and Storage Materials, Ehime University, 2-5 Bunkyo-cho, Matsuyama, Ehime 790-8577, Japan
okujima.tetsuo.mu@ehime-u.ac.jp

8.1 Introduction

Photovoltaics are expected to be among next-generation energy technologies, together with wind power, biomass, geothermal power, and so on [1]. Solar cells are classified into two groups, organic and inorganic solar cells, as summarized in Fig. 8.1. Among the inorganic solar cells, Si-based solar cells exhibit a power conversion efficiency (PCE) of 15–25%, while GaAs-, InP-, CdS-, and CdTe-based solar cells exhibit a PCE of over 25%. They exhibit a high PCE, and most practical solar cells (>90%) belong to the group of inorganic solar cells. However, they have a demerit, that is, a high power cost.

On the other hand, organic solar cells are expected to have the advantages of flexibility, low weight, easy design, large area, and low

Functional Materials: Advances and Applications in Energy Storage and Conversion
Edited by Toshio Naito

ISBN 978-981-4800-09-9 (Hardcover), 978-0-429-46813-1 (eBook)
www.panstanford.com

cost. However, they exhibit a low PCE and low stability compared to Si-based solar cells. Organic solar cells are classified, as shown in Fig. 8.1, depending on type, molecular weight, and fabrication method: dye-sensitized solar cells (DSSCs) are composed of titania, organic dye, and electrolyte (solution-based solar cell), while organic photovoltaics (OPVs) are composed of active layers of organic semiconducting materials, buffer layers, and electrodes (solid-based solar cell). This chapter provides an overview of the recent development of organic solar cells in terms of synthesis chemistry and versatile synthesis methods of organic materials for solar cells and organic field-effect transistors (OFETs).

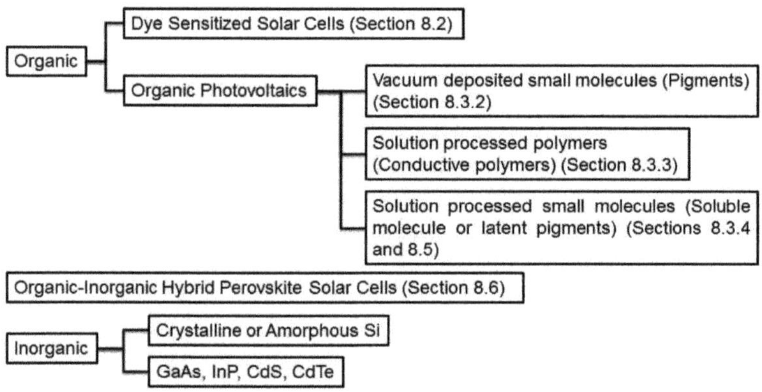

Figure 8.1 Solar cells.

8.2 Dye-Sensitized Solar Cells

8.2.1 Introduction

Since the breakthrough for DSSCs has been reported, in 1991, by Grätzel and O'Regan in the Swiss Federal Institute of Technology in Lausanne (EPFL), Switzerland, which was composed of a Ru complex (dye **1**) on titania, more than 2000 papers of the investigation on DSSCs have been published (Fig. 8.2) [2, 3]. Recent progress on various dyes resulted in cells with a PCE of 12.3% and composed of zinc porphyrin **YD2-*o*-C8/Y123**-dye as sensitizers and a Co(II)/Co(III) redox-based electrolyte [4].

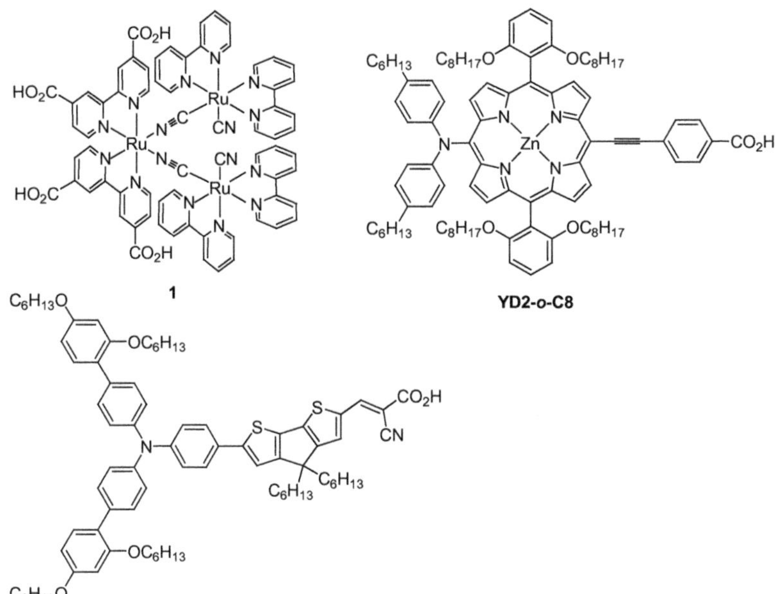

1

YD2-o-C8

Y123

Figure 8.2 Structures of sensitizers.

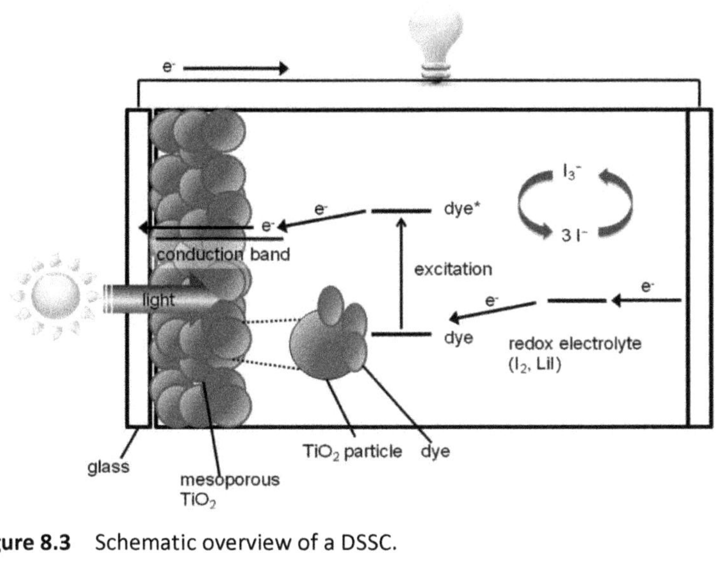

Figure 8.3 Schematic overview of a DSSC.

A schematic overview of a DSSC is shown in Fig. 8.3. Photoexcitation of dyes on the mesoporous oxide layer, composed of a network of TiO_2 nanoparticles (film thickness: 10 μm; nanoparticle diameter: 10–30 nm), results in the injection of an electron into the conduction band of the oxide. The dye is relaxed to the ground state by electron transfer from the I^- ion in the electrolyte to the I^-/I_3^- redox system in the organic solvent. The I^- ion is oxidized through the electrolyte to the cathode to afford the I_3^- ion. The advantages of a DSSC are as follows: (i) wide variation and easy design, (ii) light-harvesting region of 400–800 nm, (iii) low cost, and (iv) a PCE of ca. 10% (lower than that of an inorganic solar cell but still higher than that of an OPV). However, practical DSSCs are required for a long DSSC lifetime (of over 10 years) in outdoor conditions, high performance (a PCE of over 15%), low cost, and a large area for outdoor use.

8.2.2 Ru(II) Complexes for DSSCs

Ru(II) complexes show good photovoltaic properties: a wide light-harvesting region, suitable energy levels of excited and ground states, and relatively good stability. Several DSSCs show PCEs of more than 10% under standard measurement conditions.

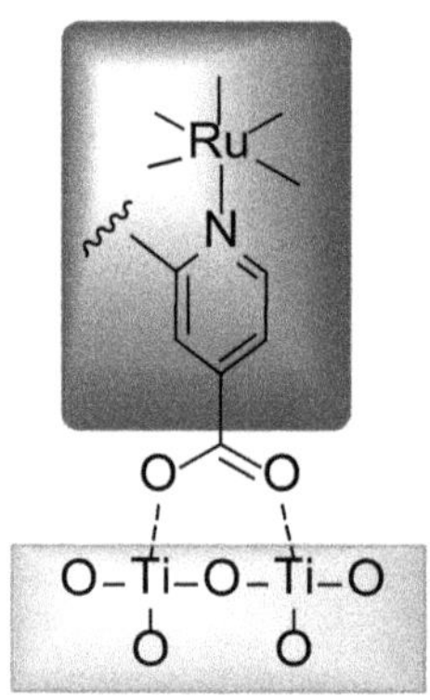

Grätzel and O'Regan reported the breakthrough DSSC with a PCE of 7.1–7.9% in 1991. The dye **1** was used in this DSSC [2]. In 1993, Grätzel and coworkers achieved 10% DSSCs using the **N3**-dye (Fig. 8.4.) [5]. The **N3**-dye shows good properties, such as a broad visible absorption spectrum, an incident photon-to-current conversion efficiency (IPCE) spectrum extending to 800 nm, a long excited-state lifetime, and strong absorption on the titania surface. The dyes are chemically bound on the TiO_2 surface as a monolayer by the formation of an ester group of hydroxyl groups on the TiO_2 surface with carboxyl groups of bipyridyl ligands. The ester groups provide efficient electron transfer from the dye to TiO_2. Much attention was focused on changing the ligands of Ru complexes in order to improve the efficiency of DSSCs. Grätzel and coworkers designed the **N749**-dye (black dye), in which the Ru center has three isothiocyanate (–N=C=S) ligands and one terpyridine ligand [6], and the **N719**-dye (red dye), in which the Ru center has two NCS ligands and two bipyridine ligands, the bis(tetrabutylammonium) salt of **N3** [7], as shown in Fig. 8.4. The **N719**-dye absorbs light in the visible region (~800 nm) whereas the absorbing onset is red-shifted for the **N749**-dye to 900 nm [5]. The DSSCs using **N719**-dyes showed high PCEs, η = 11–18%, with the short-circuit current I_{SC} = 17.73 mA/cm^2, the open-circuit voltage V_{OC} = 0.85 V, and the fill factor (FF) = 0.745 in a cell area of 0.158 cm^2; and the DSSCs using **N749**-dyes showed η = 10.4%, with I_{SC} = 20.53 mA/cm^2, V_{OC} = 0.72 V, and FF = 0.704 in a cell area of 0.186 cm^2 [8, 9].

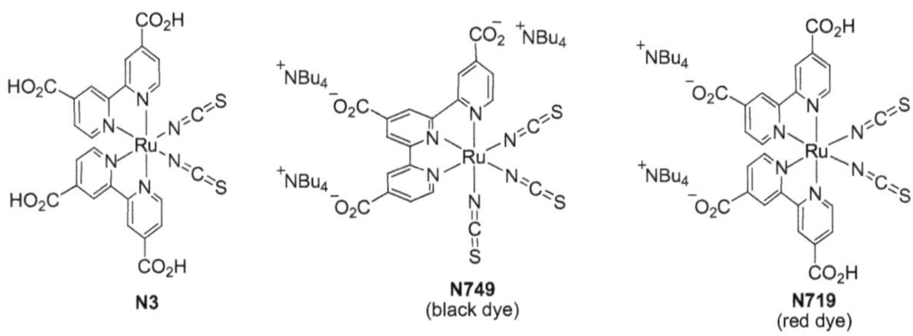

Figure 8.4 Structures of Ru complex photosensitizers.

Yanagida, Sugihara, and coworkers in the National Institute of Advanced Industrial Science and Technology (AIST), Japan, optimized tandem-structured DSSCs [10]. The DSSCs consist of a **N719**-based top cell absorbing the UV-visible light and a **N749**-based bottom cell absorbing the near-infrared light with transparent and high electromotive forced TiO_2-electrodes. The η value of the tandem DSSC is 10.6%. The I_{SC}, V_{OC}, and FF values are 20.0 mA/cm^2, 0.725 V, and 0.733, respectively. The efficiency of the tandem DSSC is stable for 500 h.

8.2.3 Organic Dyes for DSSCs

Organic dyes are attractive as novel sensitizers, with the following advantages compared to the expensive Ru complexes: (i) structures with an easy design that are easy to modify, (ii) low cost and environmental burden, and (iii) a large molar extinction coefficient. A donor-π-acceptor (D-π-A) structure is the common characteristic of these organic dyes, shown in Fig. 8.5. From among the many D-π-A sensitizers, selected examples are described in this section.

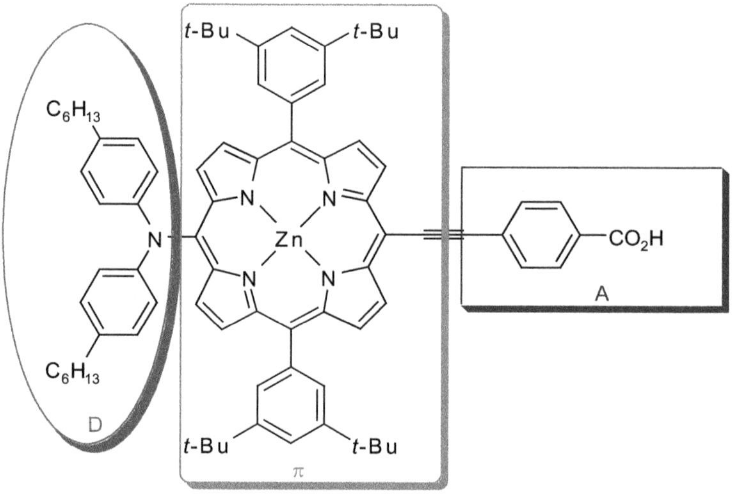

Figure 8.5 D-π-A organic dye **YD2**.

8.2.3.1 Porphyrin dyes

Porphyrin and related compounds are suitable photosensitizers for DSSCs due to the Soret band in the 400–450 nm region and the Q band in the 500–700 nm region. Porphyrins show efficient photoinduced electron injection into the conduction band of the TiO_2 layer. Initially, the performance of porphyrinoid-based DSSCs was relatively low compared to that of Ru complexes.

Recent progress in the performance of DSSCs is summarized in Table 8.1 and Figs. 8.6 and 8.7. Nazeeruddin and coworkers reported DSSCs using zinc phthalocyanine (Pc) **2** (Fig. 8.7), which gave I_{SC} of 6.5 mA/cm^2, V_{OC} of 0.635 V, and FF of 0.743, corresponding to an η value of 3.05% under standard global AM 1.5 solar conditions (corresponding to sunlight with a zenith angle of ca. 48° in a temperate zone) [11]. The η values were obtained under similar conditions unless otherwise noted. Torres and coworkers created a DSSC based on regioisomerically pure Pc **3** (Fig. 8.7) and cosensitizer **D35** [12]. The dye **D35** shows an η of 5.07% with the I^-/I_3^- redox system, which indicates a higher η of the DSSC with the combination of **3** and **D35** compared to that using each dye separately. The η value of 6.3% is obtained with I_{SC} of 15.6 mA/cm^2, V_{OC} of 0.584 V, and FF of 0.69 for the DSSC based on **3/D35**.

Table 8.1 Performance of DSSCs based on porphyrins

Dye	I_{SC} /mA/cm^2	V_{OC}/V	FF	η/%	Refs.
2	6.5	0.635	0.743	3.05	[11]
3/D35	15.6	0.584	0.69	6.3	[12]
4	12.95	0.60	0.66	5.14	[13]
5	14.0	0.680	0.74	7.1	[14]
YD2	18.6	0.77	0.764	11	[15]
XW11/WS5	20.33	0.760	0.744	11.5	[16]
6/N719	9.56	0.732	0.69	4.83	[17]
YD2-*o*-C8/Y123	17.66	0.935	0.74	12.3	[4]

Tan and coworkers reported DSSCs using thiophene-linked porphyrin **4** (Fig. 8.7), which gave I_{SC} of 12.95 mA/cm^2, V_{OC} of 0.60 V, and FF of 0.66, corresponding to an η value of 5.14% [13]. Grätzel, Officer, and coworkers reported DSSCs using zinc tetraarylporphyrin **5** (Fig. 8.7), which gave I_{SC} of 14.0 mA/cm^2, V_{OC} of 0.680 V, and FF of 0.74, corresponding to an η value of 7.1% [14]. Introduction of donor and acceptor substituents in porphyrin improves the light-harvesting ability of the porphyrin dye. The Grätzel group in EPFL achieved a high η value of 11% using the porphyrin D-π-A sensitizer **YD2** [15]. The I_{SC}, V_{OC}, and FF values are 18.6 mA/cm^2, 0.77 V, and 0.764, respectively.

Xie, Zhu, and coworkers designed a porphyrin sensitizer **XW11** (Fig. 8.7), which has a phenothiazine as an electron donor [16]. The DSSC based on **XW11** showed a good η value of 7.8%. The onset wavelength of the photocurrent response is red-shifted from 730 nm to 830 nm due to the introduction of ethynylene and benzothiadiazole units. The corresponding I_{SC} is improved to be 18.83 mA/cm^2. However, the V_{OC} is decreased due to the dye aggregation compared to the corresponding porphyrin sensitizer without a benzothiadiazole unit. To improve the V_{OC}, a coadsorption/cosensitization approach is utilized by using **WS5** (Fig. 8.7). As a result, the efficiency of the **XW11/WS5**-based DSSC was successfully improved to be 11.5%, with I_{SC} of 20.33 mA/cm^2, V_{OC} of 0.760 V, and FF of 0.744. More recently, Costa, Guldi, and coworkers have successfully synthesized tetraphenyltetrabenzoporhyrin (TPBP) **6** (Fig. 8.7) as a sensitizer for a DSSC [17]. The performance of a **6**-based DSSC cosensitized with **N719** (Fig. 8.4) was measured. The I_{SC} and V_{OC} are 9.56 mA/cm^2 and 0.732 V, respectively, with a FF value of 0.69. The η value is 4.83%, under which conditions the η of **N719** is 3.46%, with I_{SC} of 7.20 mA/cm^2, V_{OC} of 0.645 V, and FF of 0.72. In 2011, Diau, Yeh, Zakeeruccin, Grätzel, and coworkers reported DSSCs with a Co(II)/Co(III) redox-based electrolyte [4]. The porphyrin sensitizer **YD2-*o*-C8** (Fig. 8.2) generates a large photocurrent by using visible light (400–700 nm). The cosensitization of **YD2-*o*-C8** with **Y123** enhances the DSSC performance to show the best η value of 12.3%, which ensures that DSSCs still have the best performance. The I_{SC}, V_{OC}, and FF values are 17.66 mA/cm^2, 0.935 V, and 0.74, respectively.

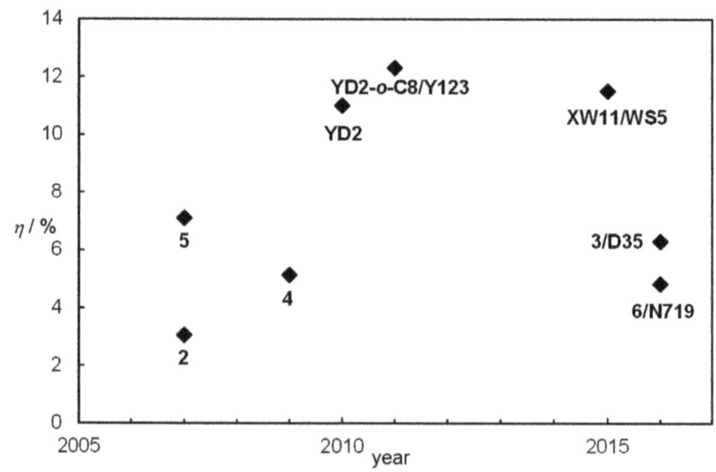

Figure 8.6 Recent progress of porphyrin-based DSSCs. **2**, η = 3.05% (Fig. 8.7); **3/D35**, η = 6.3% (Fig. 8.7); **4**, η = 5.14% (Fig. 8.7); **5**, η = 7.1% (Fig. 8.7); **YD2**, η = 11% (Fig. 8.5); **XW11/WS5**, η = 11.5% (Fig. 8.7); **6/N719**, η = 4.83% (Figs. 8.4 and 8.7); and **YD2-*o*-C8/Y123**, η = 12.3% (Fig. 8.2).

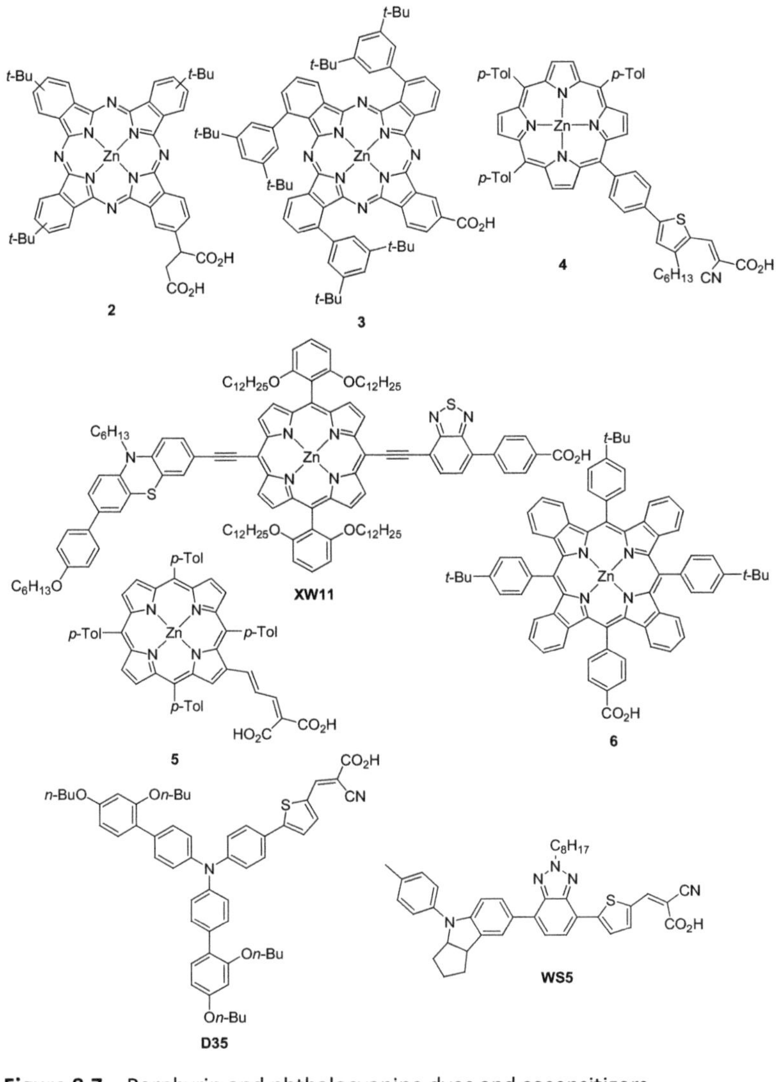

Figure 8.7 Porphyrin and phthalocyanine dyes and cosensitizers.

8.2.3.2 Carbazole dyes

Harima and coworkers designed and synthesized benzofuro[2,3-*c*]oxazolo[4,5-*a*]carbazole-type fluorescent dyes for DSSCs **7–10** (Fig. 8.8) and studied the effect of the position of the anchoring group on DSSC performance. The ICE values η of **7a-d**-based DSSCs are in the range of 0.72–0.89% [18]. Among them, the DSSC based on **2a** showed the best η (of 0.89%), with I_{SC} of 2.00 mA/cm^2, V_{OC} of 0.504 V, and FF of 0.54. The carboxyl group acts as an anchoring group similar to other D-π-A sensitizers. Harima and coworkers designed novel sensitizers with carboxyl groups at different positions of the chromophore **8** [19]. The distance between the acceptor moiety and the TiO_2 layer affects the efficiency of the electron injection. The performance of DSSCs is estimated to be η values of 1.06% (I_{SC} 2.10 mA/cm^2, V_{OC} 0.530 V, and FF 0.58) for **8a**, 0.34% (I_{SC} 0.84 mA/cm^2, V_{OC} 0.435 V, and FF 0.57) for **8b**, and 0.67% (I_{SC} 1.50 mA/cm^2, V_{OC} 0.470 V, and FF 0.58) for **8c**. The acceptor moiety (the cyano group) of **8a** would be close to the oxide layer by the anchoring carboxyl group compared to **8b** and **8c**. As a result, hydrogen and/or coordinate bonds of the cyano group with the TiO_2 surface lead to effective electron injection from the dye to the TiO_2 electrode, as shown in Fig. 8.9b. It was found that strong interaction between the acceptor moiety and the electrode was important for efficient electron injection. The carboxyl group does not solely act as the acceptor in a DSSC. Harima et al. reported novel D-π-A-sensitized dyes **9** and **10**, each having a pyridine ring as an electron-withdrawing-injecting anchoring group [20]. The pyridine ring forms a strong coordinate bond with Lewis acid sites (Ti^{n+}), as shown in Fig. 8.9c. The carboxyl anchoring dyes **11a** and **11b** provide the DSSC with η values of 0.91% (I_{SC} 2.96 mA/cm^2, V_{OC} 0.503 V, and FF 0.61) and 0.97% (I_{SC} 3.07 mA/cm^2, V_{OC} 0.520 V, and FF 0.61), respectively. Similar dyes with pyridinyl anchors **9a** and **9b** afford η values of 1.04% (I_{SC} 3.16 mA/cm^2, V_{OC} 0.524 V, and FF 0.63) and 1.15% (I_{SC} 3.35 mA/cm^2, V_{OC} 0.522 V, and FF 0.62) under the same amounts of dyes on the TiO_2 surface. The η and I_{SC} values of DSSCs based on **9** with a pyridinyl group are greater than those of **11** with a carboxyl group as an electron-withdrawing anchor. The coordinate bond between the pyridine ring of **9** and the Lewis acid sites of the TiO_2 surface leads to efficient electron injection due to a strong interaction between them compared to the ester linkage between the carboxyl group of **11** and the Brønsted acid sites of the TiO_2 surface. The best η values (of 2.35%) is obtained with I_{SC} of 7.04 mA/cm^2, V_{OC} of 0.568 V, and FF of 0.59 for the DSSC based on **10c**. Therefore, the pyridine ring of the dye sensitizer can be used not only as an electron-withdrawing anchoring group but also as an electron-injecting group. This result affords novel direction to design and synthesize D-π-A dye sensitizers.

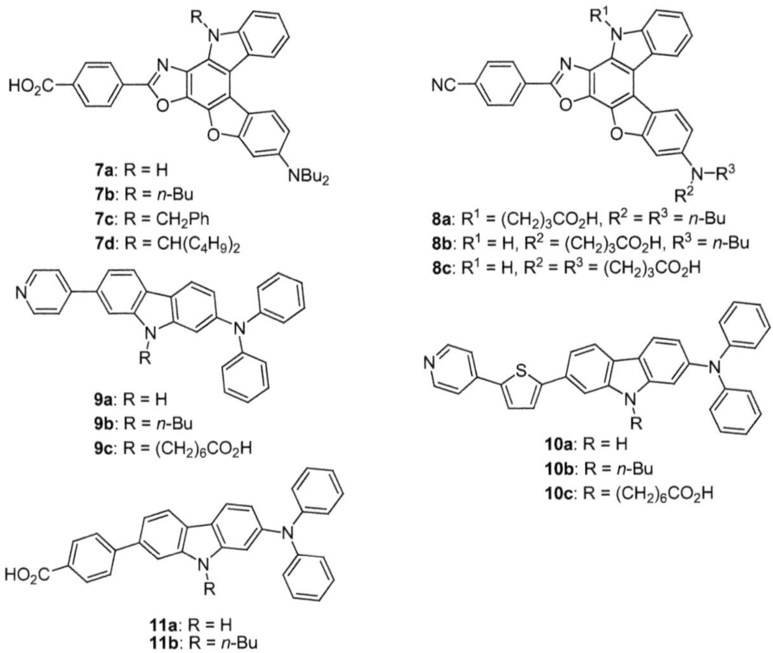

Figure 8.8 D-π-A fluorescent carbazole dyes **7–11**.

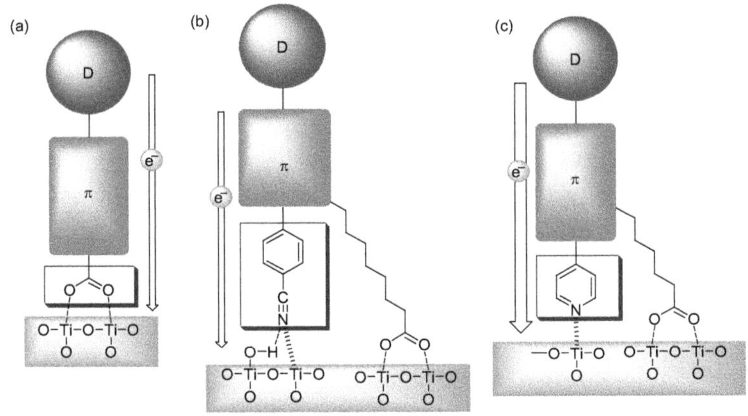

Figure 8.9 Possible configurations of carbazole dyes.

8.2.3.3 Other D-π-A dyes

Several D-π-A dye sensitizers with high η values are summarized in Fig. 8.10. Park, Kim, and coworkers prepared the dye **12**, which contained a π-conjugated oligophenylenevinylene unit with an electron donor of a triphenylamine moiety and an electron acceptor of a cinnamic acid moiety for intramolecular charge transfer and a carboxyl group as an anchor for the attachment of the dye on the TiO_2 surface [21]. The performance of the DSSC based on **12** was measured under standard global AM 1.5 solar conditions. The I_{SC} and V_{OC} of **12** are 18.1 mA/cm^2 and 0.743 V, respectively, with a FF value of 0.675. The η value for **12** is 9.1%, under which conditions the η of **N719** is 10.1%, with I_{SC} of 19.9 mA/cm^2, V_{OC} of 0.769 V, and FF of 0.657. A tandem cell of **12** with **N719** is expected to show a better performance because these cells showed a good performance over different wavelengths.

Wang and Hara's group at AIST designed and synthesized a thiophene-functionalized dye with an electron-donor moiety of a coumarin skeleton, **NKX-2700** [22]. Initially, the η value of the **NKX-2700**-based DSSC was 5.0%, with I_{SC} of 12.0 mA/cm^2, V_{OC} of 0.59 V, and FF of 0.71. Upon addition of deoxycholic acid (DCA) to the dye solution, incident photon-to-electron conversion efficiency (IPCE) was enhanced significantly. Coadsorption of DCA is effective in improving the solar cell performance, which reduces the dye coverage by ~50% but improves the I_{SC} by 30%. The breakup of π-stacking might improve electron injection yield and thus I_{SC}. The improvement of V_{OC} is attributed to the suppressed charge recombination. In the dye solution with DCA, the efficiency was increased to 8.2%, with I_{SC} of 15.9 mA/cm^2, V_{OC} of 0.69 V, and FF of 0.75, for the **NKX-2700**-sensitized DSSC. This efficiency is close to 9.0% for **N719**-sensitized DSSC under similar conditions. **NKX-2700** produces higher I_{SC} and FF but lower V_{OC} than **N719**. If higher V_{OC} is achieved for this kind of dye, much higher efficiency is expected.

In 2010, Wang and coworkers reported DSSCs based on **C219**, which contained ethylenedioxythiophene-dithienosilole conjugation as a π-spacer and triphenylamine as an electron donor (Fig. 8.10) [23]. The I_{SC}, V_{OC}, and FF of the DSSC made from **Z907** (Fig. 8.11) as the standard ruthenium sensitizer are 16.84 mA/cm^2, 0.773 V, and

0.715, respectively, yielding an η of 9.3%. In contrast, the η value of **C219**-based DSSC is 10.1%, with I_{SC} of 17.94 mA/cm^2, V_{OC} of 0.770 V, and FF of 0.730. Two batches of 10 cells show an efficiency range from 10.0 to 10.3%. On the other hand, a solvent-free ionic liquid cell based on **C219** achieved a high η (of 8.9%) under a low light intensity (of 14.39 mW/cm^2). This result indicates that **C219** should be favorable for the practical indoor application of flexible DSSCs.

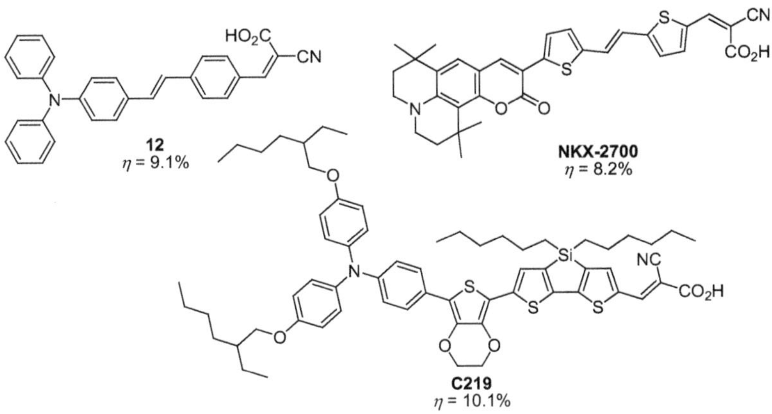

Figure 8.10 D-π-A sensitizer dyes.

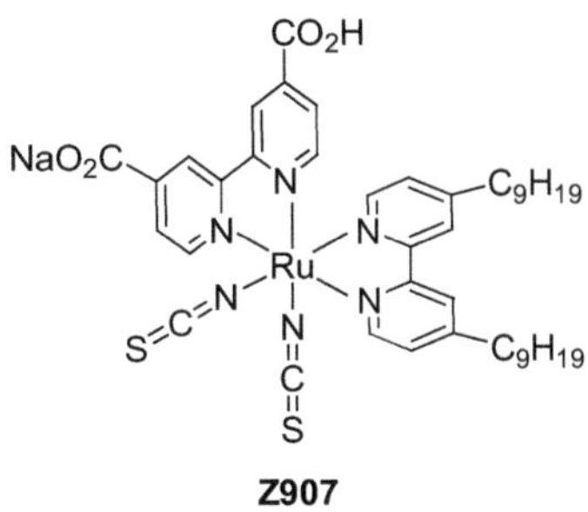

Figure 8.11 Structure of the Ru complex photosensitizer.

8.2.3.4 Oligothiophene dyes

Miyazaki, Takimiya, and coworkers designed and synthesized simple oligothiophene-based sensitizer dyes with a cyanoacrylic acid anchor **14**, as shown in Fig. 8.12 [24]. Although the molecular properties are quite similar to those of oligothiophene dyes **13**, the performance of DSSCs based on **14** is fairly improved in comparison with DSSCs based on **13** with carboxylic acid anchors (Table 8.2). A DSSC with an active area of 0.5 × 0.5 cm^2 square on a fluorine-doped tin oxide (FTO) substrate was fabricated using an oligothiophen-based sensitizer, TiO_2 nanoparticles, and an electrolyte solution consisting of iodine, lithium iodide, 1-propyl-2,3-dimethylimidazoliun iodide, and 4-*t*-butylpyridine in dry acetonitrile. The performance of the DSSC was measured under AM 1.5 illumination and is summarized in Table 8.1. The I_{SC}, V_{OC}, and FF of the DSSC containing **N719** as the standard ruthenium sensitizer are 14.68 mA/cm^2, 0.724 V, and 0.58, respectively, yielding an η of 6.15%.

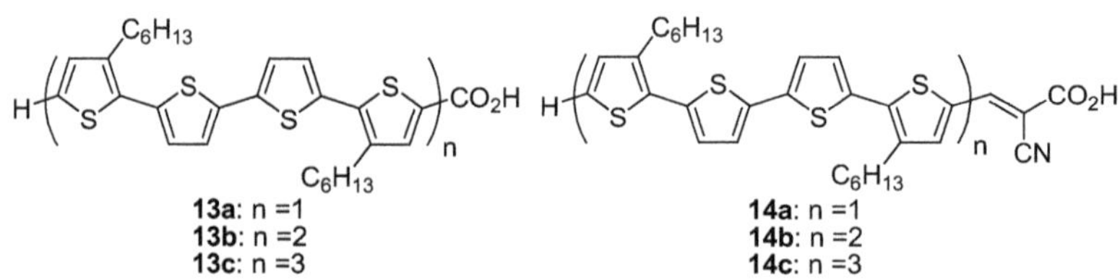

Figure 8.12 Oligothiophene-based photosensitizers.

Table 8.2 Performance of DSSCs based on **13**, **14**, and **N719**

Dye	I_{SC}/mA/cm^2	V_{OC}/V	FF	η/%
13a	3.38	0.496	0.55	0.92
13b	3.74	0.596	0.58	1.29
13c	2.98	0.408	0.45	0.55
14a	8.34	0.605	0.54	2.93
14b	11.9	0.710	0.58	4.86
14c	12.8	0.668	0.57	4.89
N719	14.68	0.724	0.58	6.15

The DSSCs based on **13** show an η range from 0.55 to 1.29%. Compared with **13**-based DSSCs, the performance of **14**-based DSSCs is apparently improved for all oligothiophene lengths, not only for the V_{OC} but also for the I_{SC}. This result indicates the importance of the anchoring group on the oligothiophene moiety. Thiophene octamer–based dye **14b** and thiophene dodecamer–based dye **14c** show η as high as 4.8%. The η value of a **2c**-based DSSC is 4.89%, with I_{SC} of 12.8 mA/cm^2, V_{OC} of 0.668 V, and FF of 0.57.

8.3 Organic Photovoltaics

8.3.1 Introduction

An OPV is a solid-type solar cell based on the semiconducting properties of organic materials, which is composed of active layers of organic semiconducting materials, buffer layers, and electrodes. The device structure is similar to that of a Si-based solar cell. In contrast, the DSSC is a liquid-type solar cell based on the photochemical redox system. The performance of an OPV depends on the balance between carrier mobility and generation. Breakthrough results have been reported by Tang at Eastman Kodak, who developed organic electroluminescent diodes [25, 26]. This OPV, composed of a copper complex of **Pc (CuPc)** and perylene, is fabricated by vacuum deposition as a p-n-heterojunction device, illustrated in Fig. 8.13.

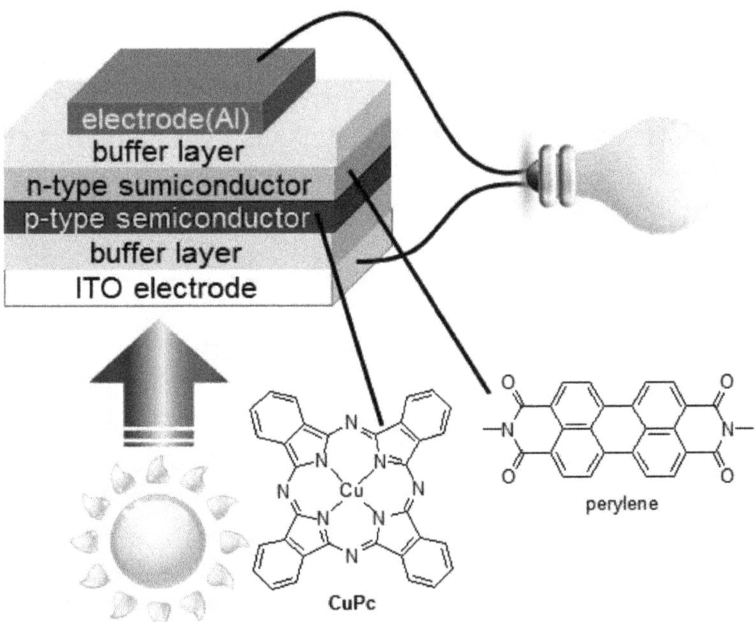

Figure 8.13 Structure of a p-n-heterojunction OPV.

A schematic overview is shown for a p-n-heterojunction OPV based on poly(3-hexylthiophene) (P3HT) as a p-type semiconductor and [6,6]-phenyl-C61 butyric acid methyl ester (PCBM) as an n-type semiconductor in Fig. 8.14. After light is absorbed on the p-layer followed by exciton dissociation, photogenerated holes and electrons are present in the highest occupied molecular orbital (HOMO) of the donor and the lowest unoccupied molecular orbital (LUMO) of the acceptor, respectively. The holes are carried and reach the anode, that is, the indium tin oxide (ITO) electrode, while the electrons reach the cathode (the Al electrode). To improve the performance of the OPV, a molecular design of semiconducting materials is required, as follows: a large energy difference between the HOMO of the p-layer material and the LUMO of the n-layer material for large V_{OC} and an energy difference of 0.2–0.3 eV between the LUMO of the p-layer material and the LUMO of the n-layer material for effective charge separation.

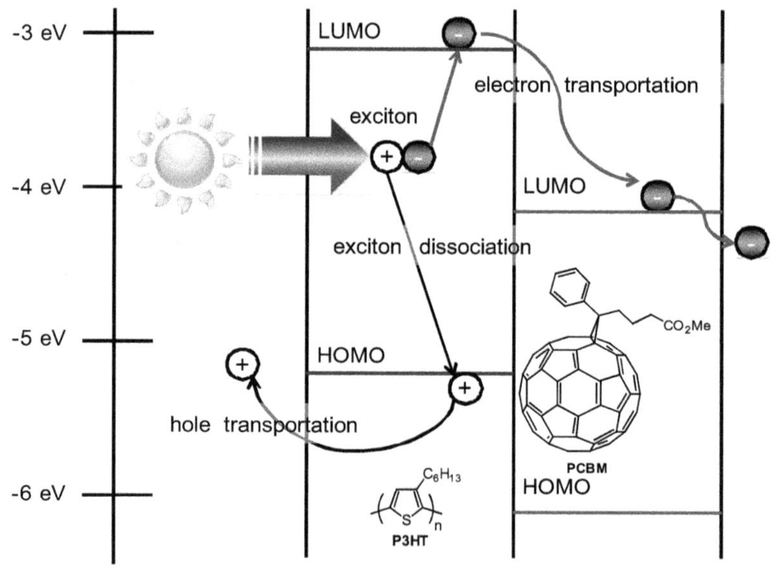

Figure 8.14 Schematic overview of an OPV.

In 1991, Hiramoto and coworkers reported a three-layered OPV, a p-i-n bulk heterojunction, with an interlayer (i-layer) of codeposited pigments of n-type perylene and p-type Pc [27, 28].

The PCE reached ca. 1%. Sariciftci and coworkers used p-type poly(phenylenvinylene) and n-type fullerene C60 for the bulk heterojunction OPV [29]. Forrest and coworkers developed a planar-mixed molecular heterojunction OPV consisting of p-type CuPc and n-type C60 [30]. The cell structure is ITO/CuPc(p-layer)/CuPc:C60 (1:1 by weight, i-layer)/C60(n-layer)/bathocuproine (BCP)/Ag. A maximum PCE η of 5.0% was achieved, with I_{SC} of 15.0 mA/cm^2, V_{OC} of 0.54 V, and FF of 0.54. In 2007, Lee and coworkers reported tandem solar cells, in which two cells with different absorptions are linked to use a wider range of the solar spectrum [31]. Power conversion efficiencies of an η more than 6% were achieved. The charge separation layer for the front cell is a bulk heterojunction composed of poly[2,6-(4,4-bis(2-ethylhexyl)-4*H*-cyclopenta[2,1-*b*;3,4-*b*′]dithiophene)-*alt*-4,7-(2,1,3-benzothiadiazole)] (PCPDTBT) and PCBM, while the charge separation layer for the back cell is a bulk heterojunction composed of P3HT and [6,6]-phenyl-C71 butyric acid methyl ester (PC70BM) (Figs. 8.15 and 8.16). The back cell mainly absorbs the light of 400–600 nm, and the front cell absorbs the light of 600–900 nm. The two polymer-fullerene layers are separated by a transparent TiO_x layer and a conductive hole transport layer, poly(3,4-ethylenedioxylenethiophene) and polystyrene sulfonic acid (PEDOT:PSS). The η value of the tandem cell is 6.5%, with I_{SC} of 7.8 mA/cm^2, V_{OC} of 1.24 V, and FF of 0.67. In contrast, the η value of the PCPDTBT:PCBM single cell is 3.0%, with I_{SC} of 9.2 mA/cm^2, V_{OC} of 0.66 V, and FF of 0.50. The η value of the P3HT:PC70BM single cell is 4.7%, with I_{SC} of 10.8 mA/cm^2, V_{OC} of 0.63 V, and FF of 0.69.

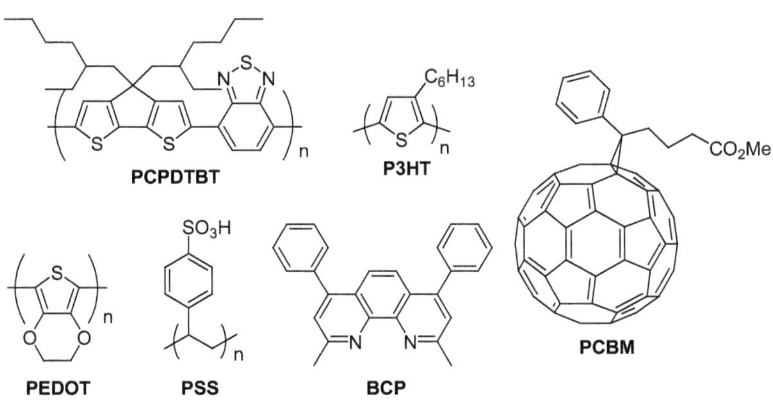

Figure 8.15 Structures of PCPDTBT, P3HT, PCBM, PEDOT, PSS, and BCP.

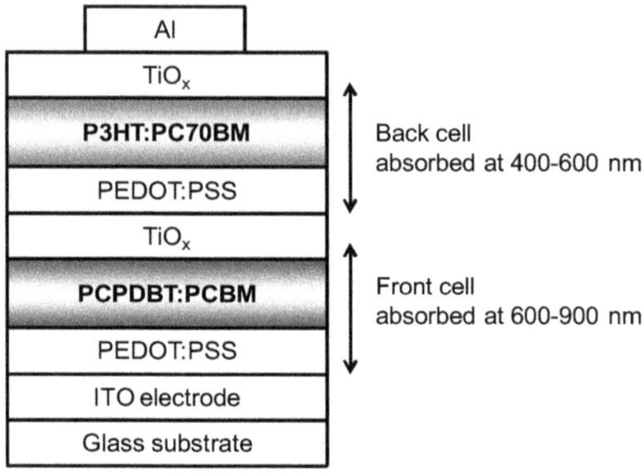

Figure 8.16 Device structure of the tandem OPV.

OPVs are classified into three groups depending on the fabrication methods and the types of semiconducting materials, (i) vacuum-deposited small molecules (pigments), (ii) solution-processed polymers (conductive polymers), and (iii) solution-processed small molecules (small molecules with solubilizing groups or small molecules with removable groups [latent pigments]), as shown in Fig. 8.1. In this section, these OPVs are described according to these classifications.

8.3.2 Vacuum-Deposited Small Molecules

Small molecules, such as Pcs, pentacenes, fullerenes, and perylenes, have the following characteristics as semiconducting materials: various purification methods can be used for high purity, and these molecules have well-defined structures compared to polymers; a crystalline film of such molecules can be fabricated by vacuum deposition; and they have a low solubility, not compatible with solution process.

For example, Taima and coworkers, at the AIST, have reported a p-i-n bulk heterojunction OPV fabricated by vacuum deposition [32]. In this cell, the i-layer was fabricated by the codeposition of ZnPc and C60. The AIST group optimized the OPV to change the proportion of the semiconducting layer thicknesses. When the thicknesses of p-, i-, and n-layers were 5 nm, 15 nm, and 30 nm, respectively, and the ratio of ZnPc and C60 in the i-layer was 1:1 by volume ratio, a PCE η of 3.6% was obtained, with I_{SC} of 8.35 mA/cm^2, V_{OC} of 0.56 V, and FF of 0.61 (Fig. 8.18a). Jabbour and coworkers have reported OPVs based on metallo-Pcs and 3,4,9,10-perylene tetracarboxylic bisbenzimidazole (PTCBI, Fig. 8.17) and C60 [33]. The device structure of MPc/PTCBI-based OPVs is ITO/MPc (20 nm)/PTCBI (20 nm)/BCP (15 nm)/Ag (100 nm), as shown in Fig. 8.18b. The device parameters are summarized in Table 8.3. Among these OPVs, PdPc/PTCBI-OPV showed the highest PCE η, of 1.3%, with a high I_{SC} of 4.0 mA/cm^2 and FF of 0.64, while CuPc, ZnPc, and PtPc showed I_{SC} of 2.9, 3.0, and 2.9 mA/cm^2, respectively. When fullerene C60 was used as an acceptor for PdPc- and CuPc-OPVs, I_{SC} was improved compared to PdPc/PTCBI-OPVs. The device structure of MPc/C60-based OPVs is ITO/MPc (20 nm)/C60 (30 nm)/BCP (10 nm)/Ag (100 nm) (Fig. 8.18c). The PdPc/C60-based OPV showed a larger I_{SC} (of 6.8 mA/cm^2), resulting in an η of 2.2%, with V_{OC} of 0.57 V and FF of 0.57, while a CuPc-based OPV shows an η of 1.6%, with I_{SC} of 4.5 mA/cm^2, V_{OC} of 0.56 V, and FF of 0.62. Shao and Yang reported a p-n-heterojunction OPV fabricated by the vacuum deposition of platinum octaethylporphyrin (PtOEP) and C60 as ITO/PEDOT/PtOEP (30 nm)/C60 (30 nm)/BCP (8 nm)/Al (100 nm) (Fig. 8.18c) [34]. This OPV shows an η of 2.1%, with I_{SC} of 5.6 mA/cm^2, V_{OC} of 0.66 V, and FF of 0.57.

Table 8.3 Performance of OPVs based on MPcs, PtOEP, and DTDCTB

Donor	Acceptor	Cell type	I_{SC}/mA/cm^2	V_{OC}/V	FF	η/%	Refs.
ZnPc	C60	p-i-n	8.35	0.56	0.61	3.6	[32]
CuPc	PTCBI	p-n	2.9	0.53	0.60	0.91	[33]
ZnPc	PTCBI	p-n	3.0	0.58	0.63	1.1	[33]
PdPc	PTCBI	p-n	4.0	0.52	0.64	1.3	[33]
PtPc	PTCBI	p-n	2.9	0.49	0.50	0.70	[33]
PdPc	C60	p-n	6.8	0.57	0.57	2.2	[33]
CuPc	C60	p-n	4.5	0.56	0.62	1.6	[33]
PtOEP	C60	p-n	5.6	0.66	0.57	2.1	[34]
DTDCTB	C60	p-n	11.40	0.80	0.48	4.41	[35]
DTDCTB	C70	p-n	14.68	0.79	0.50	5.81	[35]

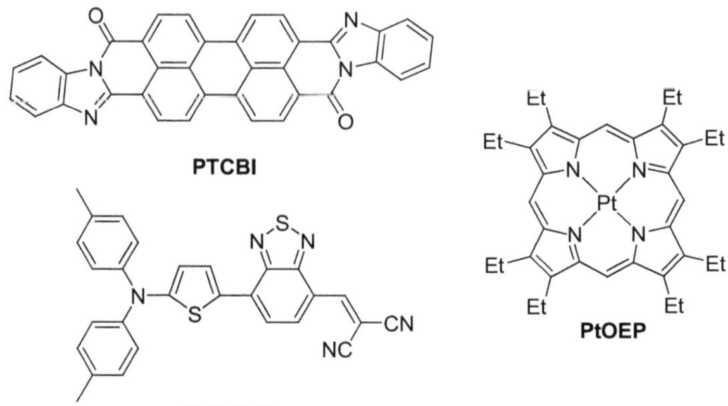

Figure 8.17 Structures of PTCBI, PtOEP, and DTDCTB.

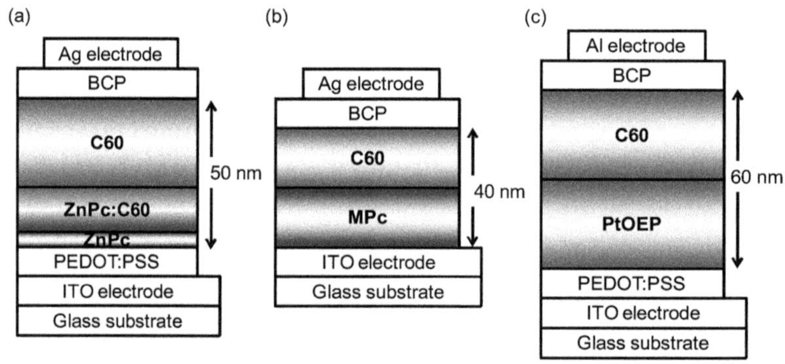

Figure 8.18 Device structures of OPVs based on (a) ZnPc/C60 (p-i-n), (b) MPc/PTCBI (p-n), and (c) PtOEP/C60 (p-n).

Lin, Wong, and coworkers designed a donor-acceptor-acceptor (D-A-A) donor molecule 2-{[7-(5-*N*,*N*-ditolylaminothiophen-2-yl)-2,1,3-benzothiadiazol-4-yl]methylene}malononitrile (DTDCTB) in which an electron-donating ditolylaminothienyl moiety and an electron-withdrawing dicyanovinylene moiety are bridged by another electron acceptor, 2,1,3-benzothiadiazole (Fig. 8.17) [35]. The p-n-heterojunction OPV based on DTDCTB/C60 afforded I_{SC} of 11.40 mA/cm^2, V_{OC} of 0.80 V, and FF of 0.48, yielding an η of 4.41%. Remarkably, the DTDCTB/C70-based OPV delivered a higher-performance η of 5.81%, with I_{SC} of 14.68 mA/cm^2, V_{OC} of 0.79 V, and FF of 0.50. The high V_{OC} values of these OPVs are due to the moderately low HOMO level of DTDCTB. The HOMO level of the DTDCTB thin film has a value of –5.30 eV, determined by UV photoelectron spectroscopy.

8.3.3 Solution-Processed Polymers

Conjugated polymers are candidates for use in low-cost and large-area OPVs since an amorphous thin film is readily obtained by solution process [36]. Polymeric materials have been extensively studied in spite of the statistically defined structure and difficulty in purification. Thiophene polymers and oligomers belong to the most promising organic semiconductors. For example, the AIST group has reported solution-processed OPVs based on P3HT as p-type semiconductors and PCBM and PC70BM as n-type semiconductors [37]. The device structure is ITO/PEDOT:PSS/photovoltaic layer/Al, as illustrated in Fig. 8.19. Solutions blending P3HT and PCBMs (1:0.7 by weight ratio) are prepared by dissolving the two in chlorobenzene and spin-coating onto the substrate. The film thickness is ca. 90 nm. The η and I_{SC} of the P3HT/PC70BM-OPVs are improved compared to those of P3HT/PCBM-OPVs due to the wide and intense light absorption of PC70BM. A maximum PCE η of 3.8% was achieved for the P3HT/PC70BM-OPV on thermal annealing, with I_{SC} of 9.31 mA/cm^2, V_{OC} of 0.65 V, and FF of 0.63. In contrast, the η value of the P3HT/PCBM-OPV is 3.6%, with I_{SC} of 8.46 mA/cm^2, V_{OC} of 0.64 V, and FF of 0.66 (Table 8.4).

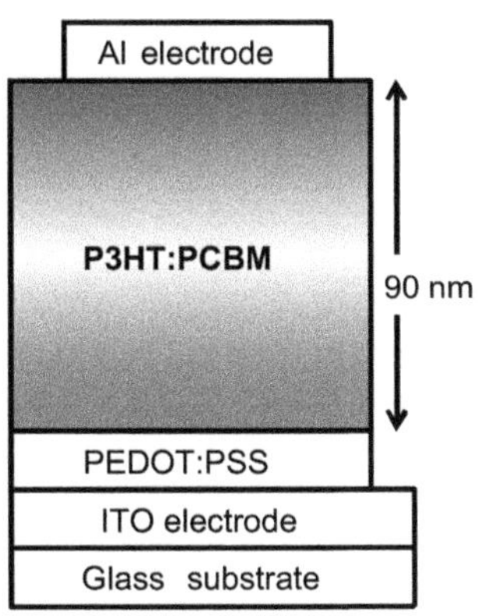

Figure 8.19 Device structures of OPVs based on P3HT/PCBM.

Table 8.4 Performance of bulk heterojunction OPVs based on polymers

Donor	Acceptor	I_{SC}/mA/cm^2	V_{OC}/V	FF	η/%	Refs.
P3HT	PCBM	8.46	0.64	0.66	3.6	[37]
P3HT	PC70BM	9.31	0.65	0.63	3.8	[37]
PDTSTTz-4	PC70BM	11.25	0.73	0.716	5.88	[38]
PTzBT-14HD	PCBM	10.6	0.84	0.64	5.7	[39]
PDBT-T1	IC-C6IDT-IC	15.05	0.89	0.65	8.71	[40]

Li and coworkers designed D-A copolymers based on a donor dithienosilole unit and an acceptor bithiazole unit [38]. The copolymer PDTSTTz-4, which is synthesized by the Stille coupling reaction, adopts a planar structure and shows a hole mobility of ca. 7.8×10^{-2} cm^2 V^{-1} s^{-1} (Fig. 8.20). The photovoltaic properties of PDTSTTz-4 were measured using PC70BM as an acceptor in bulk heterojunction OPVs with a device structure of ITO/PEDOT:PSS/PDTSTTz-4:PC70BM/Ca/Al. The active layer is fabricated by spin-coating at 100°C from an *o*-dichlorobenzene solution, followed by thermal annealing. The η value of the OPV based on PDTSTTz-4 as a donor and PC70BM as an acceptor (1:1.3 by weight ratio) reached 5.88%, with I_{SC} of 11.25 mA/cm^2, V_{OC} of 0.73 V, and FF of 0.716.

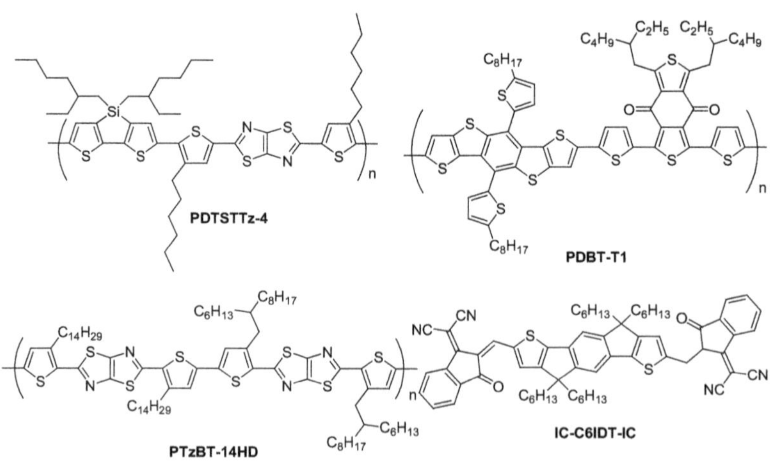

Figure 8.20 Structures of PDTSTTz-4, PTzBT-14HD, PDBT-T1, and IC-C6IDT-IC.

In 2012, Osaka and coworkers reported solution-processed OPVs based on a thiazolothiazole copolymer PTzBT-14HD (Fig. 8.20) [39]. The copolymer PTzBT-14HD shows a good hole mobility of ca. 0.1 $cm^2\ V^{-1}\ s^{-1}$. Photovoltaic properties of OPVs based on PTzBT-14HD synthesized by Stille coupling with a number-average molecular weight (M_N) of 13–73 kDa and a weight-average molecular weight (M_W) of 18–1450 kDa and PCBMs were examined with a device structure of ITO/PEDOT:PSS/PTzBT-14HD:PCBM/LiF/Al. The active layer thickness of the optimized devices is in the range of 150–170 nm. The V_{OC} slightly decreases as the molecular weight increases. The ISC increases gradually and is maximized at M_N = 33 kDa and decreases when the molecular weight further increases. As a result, the best η of 5.7% was achieved for the PTzBT-14HD/PCBM-based OPV with I_{SC} of 10.6 mA/cm^2, V_{OC} of 0.84 V, and FF of 0.64 when PTzBT-14HD at M_N = 33 kDa and M_W = 300 kDa was used with a p:n ratio of 1:2 by weight.

A planar fused-ring acceptor IC-C6IDT-IC was designed by Sun and coworkers in 2016 and used for the bulk heterojunction OPV based on the donor polymers PDBT-T1 and IC-C6IDT-IC (Fig. 8.20) [40]. The cells are fabricated with a device structure of ITO/ZnO/PDBT-T1:IC-C6IDT-IC/MoO_x/Ag. The best performance was obtained with the D:A ratio of 1:1 by weight. The η, I_{SC}, V_{OC}, and FF values are 8.71%, 15.05 mA/cm^2, 0.89 V, and 0.65, respectively. The average η value is 8.57% among over 20 devices.

8.3.4 Solution-Processed Small Molecules with Solubilizing Groups

Soluble polymers are attractive for large-area and low-cost OPVs due to easy fabrication by solution process, such as spin-coating or printing techniques, although they are difficult to purify and are well defined. On the other hand, small molecules are also attractive for OPVs based on highly purified materials of well-defined structures, leading to high mobilities compared to those of polymers. However, vacuum deposition techniques are not suitable for fabricating large-area and low-cost devices. Thus, there is a requirement to design and develop small molecules in order to provide large-area and low-cost OPVs fabricated by solution process. There are two classes of solution-processable small molecules: small molecules with solubilizing groups (Section 8.3.4) and small molecules with thermally or photochemically removable groups (latent pigments) (Section 8.5).

Solution-processable small molecules are attractive for application in photovoltaic cells [41]. In this section, recent progress in OPVs based on small molecules with solubilizing groups is reviewed. The performance of the OPVs is summarized in Table 8.5.

Table 8.5 Performance of bulk heterojunction OPVs based on soluble small molecules

Donor	Acceptor	I_{SC}/mA/cm^2	V_{OC}/V	FF	η/%	Refs.
DCAO3TSi	PCBM	11.51	0.80	0.64	5.84	[42]
DR3TBDT	PCBM	10.78	0.91	0.65	6.38	[43]
DR3TBDT	PC70BM	12.21	0.93	0.65	7.38	[43]
DERHD7T	PCBM	13.98	0.92	0.474	6.10	[44]
15	PCBM	7.74	0.83	0.466	2.98	[45]
BIT6F	PC70BM	12.44	0.89	0.76	9.09	[46]

Chen and coworkers demonstrated that the A-D-A molecule structure with a conjugated donor moiety and acceptor terminals had advantages for bulk heterojunction OPVs as follows: high charge carrier mobility, low bandgap resulting from the intramolecular charge transfer, and good film morphology [42]. A planar-conjugated donor with a dithienosilole core DCAO3TSi was designed and used for bulk heterojunction OPVs (Fig. 8.18). Photovoltaic properties of OPVs based on DCAO3TSi were measured with a sandwich device structure of ITO/PEDOT:PSS/DCAO3TSi:PCBM/LiF/Al using solution process from a $CHCl_3$ solution. The thickness of the DCAO3TSi:PCBM active layer is about 130 nm. The optimized OPV, with a ratio of 1:0.8 for the active layer, shows an η of 5.84%, with I_{SC} of 11.51 mA/cm^2, V_{OC} of 0.80 V, and FF of 0.64.

The same group reported a similar A-D-A molecule DR3TBDT for bulk heterojunction OPVs (Fig. 8.21) [43]. DR3TBDT shows a PCE as high as 7.38% under the illumination of AM 1.5G, 100 mW/cm^2. The active layers were fabricated by spin-coating from the solution of DR3TBDT and PCBMs in $CHCl_3$. η as high as 6.38% was obtained with the optimized weight ratio of DR3TBDT to PCBM at 1:0.8. The I_{SC},

V_{OC}, and FF values are 10.78 mA/cm^2, 0.91 V, and 0.65, respectively. In contrast, the active layer with DR3TBDT to PC70BMs yielded an increased η of 7.38% due to the higher absorption coefficient of PC70BM in the visible region and the addition of a small amount of dimethylpolysiloxane (PDMS) in the active layer. The I_{SC}, V_{OC}, and FF values are 12.21 mA/cm^2, 0.93 V, and 0.65, respectively. The average η is 7.18% for over a hundred devices under optimized conditions.

Rhodanine-based septithiophene DERHD7T was also reported by the same group [44]. The OPVs were fabricated using DERHD7T with a device structure of ITO/PEDOT:PSS/DERHD7T:PCBM/LiF/Al by solution process from a $CHCl_3$ solution. The best result, of η = 6.10%, was obtained from the $CHCl_3$ solution with a DERHD7T:PCBM ratio of 1:0.5. The I_{SC}, V_{OC}, and FF values are 13.98 mA/cm^2, 0.92 V, and 0.474, respectively. The average η is 5.88%.

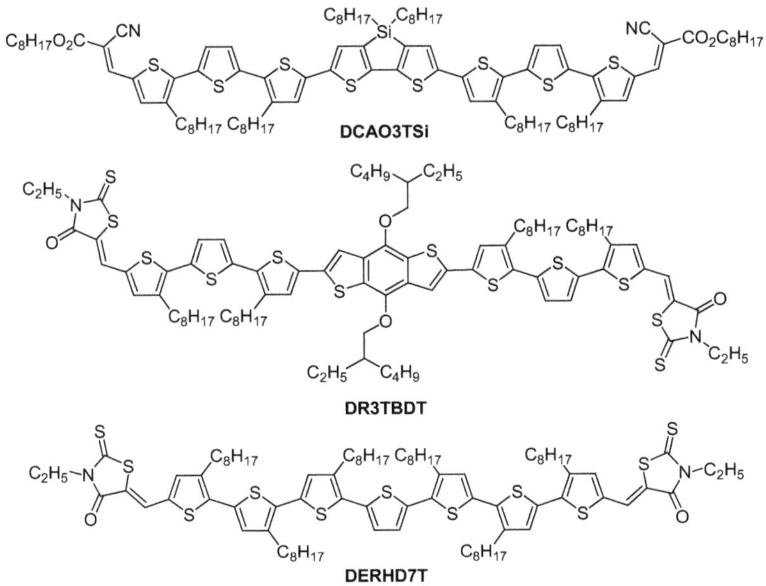

Figure 8.21 Structures of DCAO3TSi, DR3TBDT, and DERHD7T.

More recently, Lin, Yagai, and coworkers designed and synthesized benzodithiophene-functionalized oligothiphene with barbituric acid as a self-assembling unit by hydrogen bonding (Fig. 8.22) [45]. The bulk heterojunction OPV based on **15**/PCBM without thermal annealing shows an η value of 2.98%, with I_{SC} of 7.74 mA/

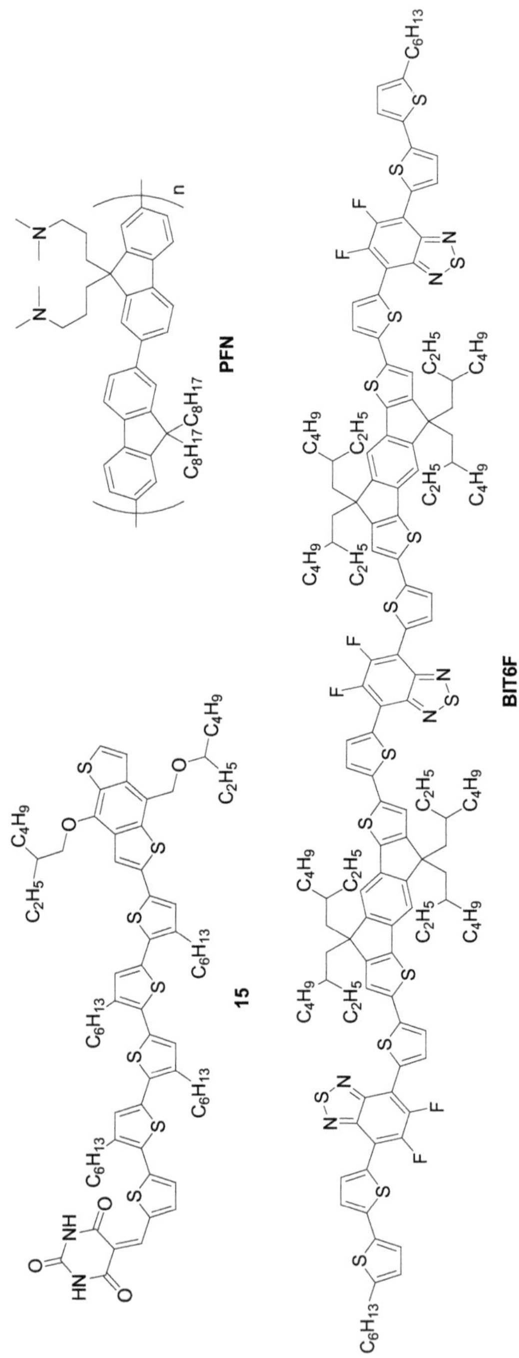

Figure 8.22 Structures of **15**, BIT6F, and PFN.

cm^2, V_{OC} of 0.83 V, and FF of 0.466. Wang and coworkers reported an indacenodithiophene-difluorobenzothiadiazole–based D-A oligomer BIT6F and other oligomers for OPVs with high efficiency [46]. Bulk heterojunction cells based on BIT6F as a donor were fabricated with a device structure of ITO/PEDOT:PSS/DIT6F:PC70BM/PFN/Al. The optimal weight ratio of BIT6F:PC70BM is 1:3. The OPV based on BIT6F/PC70BM without any treatment shows a high V_{OC}, of ca. 0.9 V, consistent with its deep-lying HOMO level, which shows an η value of 6.66%, with I_{SC} of 12.15 mA/cm^2 and FF of 0.61. With thermal annealing and solvent vapor annealing treatments, the best η value, of 9.09% (average η of 8.9%), was obtained with I_{SC} of 12.44 mA/cm^2, V_{OC} of 0.89 V, and FF of 0.76.

8.4 Organic Field-Effect Transistors Based on Small Molecules with Removable Groups

The introduction of a solubilizing group into small molecules is effective in achieving the fabrication of a thin film by solution process. However, the substituents could inhibit charge transfer between molecules depending on the sizes and positions of the substituents. For example, the Chikamatsu and Azumi group at AIST reported soluble fullerene derivatives bearing solubilizing alkyl chains at different positions and their semiconducting performances [47]. OFETs based on dodecyl-substituted fullerenes, as shown in Fig. 8.23, were fabricated by solution process. C60MC12 exhibits a better FET performance, with an electron mobility μ_e as high as 0.09 cm^2/Vs and a current on/off ratio I_{on}/I_{off} of 4×10^5 compared to μ_e of 8.1×10^{-3} cm^2/Vs for C60PC12, μ_e of 2.2×10^{-3} cm^2/Vs for **C60C12**, and μ_e of 1.5×10^{-3} cm^2/Vs for C60OC12. X-ray diffraction (XRD) measurements and atomic force microscopy (AFM) observation showed thin-film crystallinity of these dodecyl-substituted fullerenes as approximately C60MC12 > C60PC12 ≈ C60C12 > C60OC12, which agrees with the order of mobilities. The crystalline thin film of C60MC12 is based on a band transport of electrons, while the amorphous thin film of PCBM is based on a hopping transport. Thus, the properties of soluble small materials are different from those of the parent molecules.

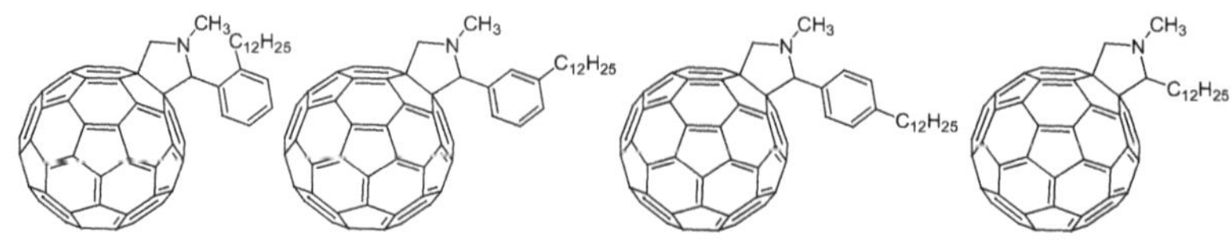

Figure 8.23 Structures of dodecyl-substituted fullerene derivatives.

Recently, a latent-pigment strategy has been developed using small molecules with solubilizing groups that are removed by thermal or photochemical conversion after spin-coating or printing. In 1997, a soluble precursor of diketopyrrolopyrrole (DPP) **16** was reported by Ciba Specialty Chemicals (Fig. 8.24) [48]. DPP is used as a stable red pigment not soluble for organic solvents due to the intermolecular hydrogen bonding network. A thin yellow film of **16** fabricated by solution process is readily converted into the red film of DPP by heating at 180°C to give a good yield. Thus, this strategy provides a method of fabricating a thin film of insoluble small molecules by solution process. Insoluble small molecules should be fabricated by vacuum deposition.

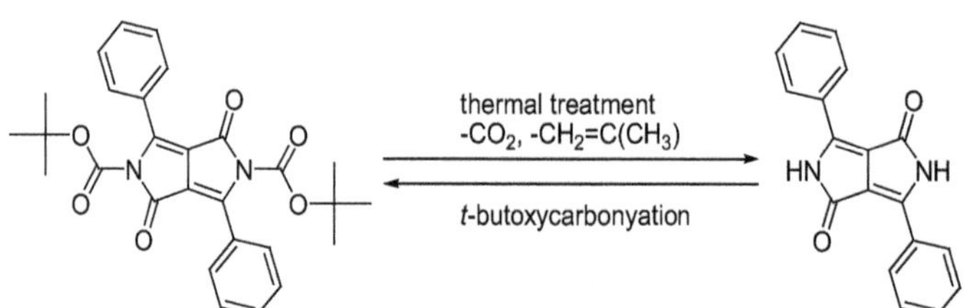

Figure 8.24 Soluble latent pigment **16** and insoluble parent pigment DPP.

For example, pentacene is one of the most widely investigated organic semiconductors. In many cases, a thin film of pentacene is fabricated by vacuum deposition. Solution-processed OFETs of parent pentacene have been reported based on thermal and photochemical conversion, as shown in Fig. 8.25. The OFETs were fabricated by spin-coating of **17** and **18**, followed by annealing. These devices exhibited good mobilities μ_h of 0.1–0.4 cm^2/Vs [49, 50] The photoconversion of precursor **19** into pentacene was reported by Ono, Uno, Yamada, and coworkers at Ehime University

[51, 52]. The device was fabricated as a top-contact OFET. The μ_h of the film is 0.34 cm^2/Vs and I_{on}/I_{off} is 2 × 10^6 [53]. Yamada and coworkers optimized the fabrication method by solution process and achieved a high performance, with μ_h of 0.86 cm^2/Vs and I_{on}/I_{off} is 4.3 × 10^6 [54]. This value is comparable to the vacuum-deposited OFETs of pentacene.

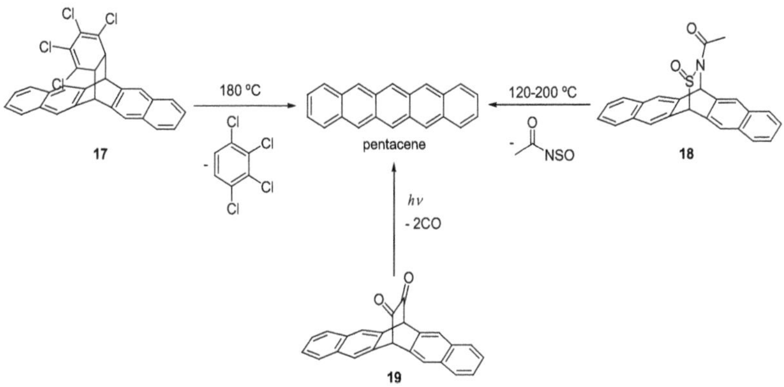

Figure 8.25 Thermal and photochemical conversions from **17–19** into pentacene.

Pcs and tetrabenzoporphyrins (TBPs) are promising semiconducting materials due to the facile modification of the molecular structures. A few examples of synthesis of Pcs and TBPs by precursor methods have been reported [55]. Two types of masked isoindoles, pyrroles fused with a nonaromatic ring, are used in the precursor methods for the synthesis of TBPs. Cyclohexene-fused porphyrins **20** are converted into the corresponding TBPs by oxidative aromatization (Fig. 8.26) [56, 57]. On the other hand, the thermal conversion of bicyclo[2.2.2]octadiene (BCOD)-fused porphyrins CPs affords TBPs by a retro Diels–Alder reaction, which was first reported by Ono and coworkers at Ehime University [58–62]. The thin film of TBP fabricated by spin-coating of CP and subsequent thermal conversion showed a good mobility, of ca. 1 cm^2/Vs^{-1} [63–67]. This value is better than those of the vacuum-deposited OFETs of TBPs. These TBP-based OFETs retain a high performance after the device is kept under air for over 500 h.

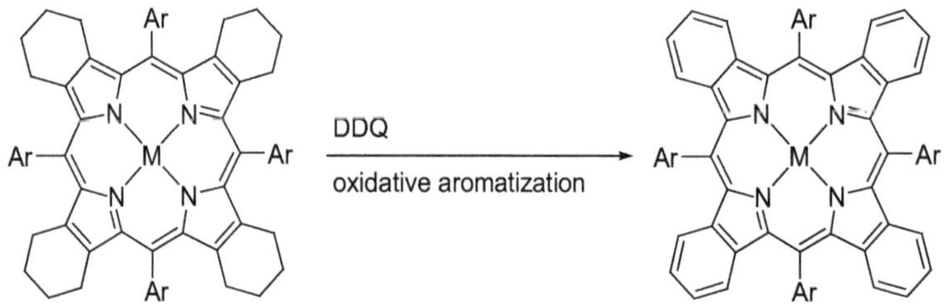

Figure 8.26 Synthesis of TBPs from the soluble precursors **20**.

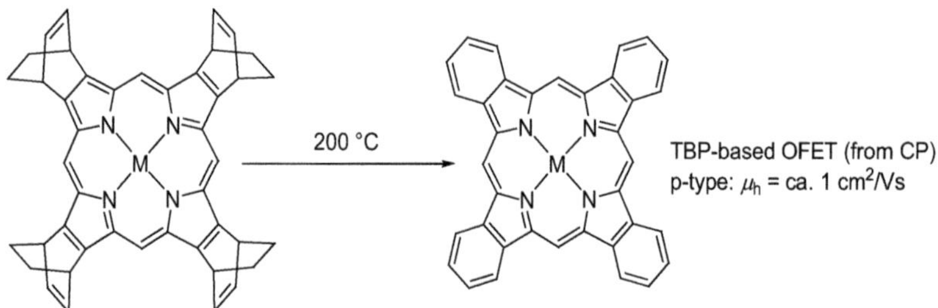

Figure 8.27 Thermal conversion of CP into TBP [63–67].

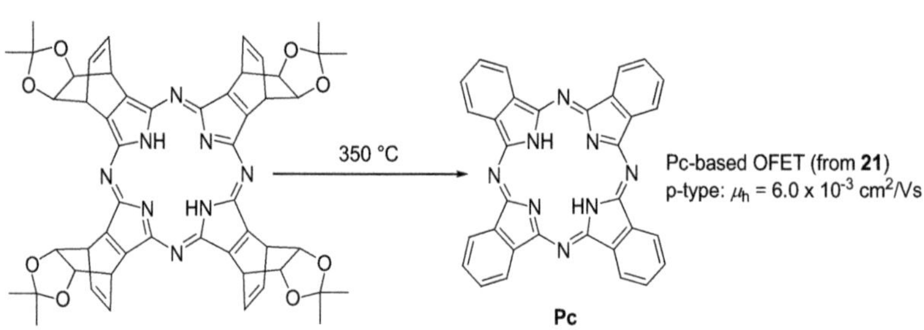

Figure 8.28 Thermal conversion of **21** into Pc.

In 2008, Okujima, Aramaki, Ono, and coworkers reported the first synthesis of a soluble Pc precursor and the first application in an OFET by solution process [68]. The Pc precursor **21** (Fig. 8.28) is soluble in many organic solvents and was used for the fabrication of Pc-based OFETs in a manner similar to that for TBP-based OFETs using CP. A solution of **21** in $CHCl_3$ was spin-coated onto the channel region, and the film of **21** was heated at 350°C to afford the Pc thin film. A good μ_h, of 6.0×10^{-3} cm^2/Vs, and a large I_{on}/I_{off} ratio, of 1.8×10^4, were accomplished, which is a higher performance compared to the vacuum-deposited OFETs of Pcs. Pcs are generally used as green and blue pigments and electrophotographic materials.

8.5 Organic Photovoltaics Based on Small Molecules with Removable Groups

The performance and stability of semiconducting materials are important for achieving a practical OPV with high photoconversion efficiency. TBPs and Pcs are promising materials for practical OPVs due to their thermal and chemical stability. A strategy based on soluble precursors is also expected to produce large-area and low-cost OPVs. The OPVs using soluble precursors are fabricated by solution process, for example, spin-coating and the subsequent thermal conversion, with the device structures of not only bulk heterojunction cells but also p-n- or p-i-n-heterojunction cells, while the OPVs using soluble small molecules or polymers are usually fabricated with device structures of bulk heterojunction cells.

The solution-processed fabrication followed for TBPs is applied to the OPVs with a high PCE by using the soluble precursors CPs. Matsuo, Nakamura, and coworkers have reported p-i-n heterojunction OPVs using the reported CP as a precursor of TBP and a fullerene derivative SIMEF, which was designed by them as a new acceptor [69, 70]. The fabrication process is illustrated in Fig. 8.29. After a solution of CP in $CHCl_3$/chlorobenzene is spin-coated on the glass/ITO/PEDOT:PSS substrate, heating it at 180°C affords the p-layer of TBP crystalline film on the substrate. In the next step, a mixture of CP and SIMEF1 in $CHCl_3$/chlorobenzene is spin-coated and the thermal treatment affords the i-layer of TBP/SIMIEF1 (Fig. 8.30). The subsequent spin-coating of SIMEF1 and heating to crystallize furnishes the p-i-n structure. Finally, a buffer material and an Al electrode are coated in vacuo to complete the cells. The device structure is ITO/PEDOT:PSS/TBP/TBP:SIMEF1/SIMEF1/BCP/Al, as shown in Fig. 8.29. OPV annealing at 65°C after n-layer formation showed an η of 4.1%, with I_{SC} of 9.1 mA/cm^2, V_{OC} of 0.76 V, and FF of 0.59. Annealing the n-layer at 180°C to form the crystalline film of SIMEF1 improves the I_{SC} and FF to show an η of 4.5%, with I_{SC} of 9.7 mA/cm^2, V_{OC} of 0.76 V, and FF of 0.62. Further optimization with the use of NBphen as a buffer layer affords a higher η, of 4.8%–5.2%. The best η value is 5.2%, with I_{SC} of 10.5 mA/cm^2, V_{OC} of 0.75 V, and FF of 0.65. On the other hand, when PCBM is used in place of SIMEF1, the η value decreases to 2.0%, with I_{SC} of 7.0 mA/cm^2, V_{OC} of 0.55 V, and FF of 0.51, due to the lower LUMO level of PCBM by ca. 0.1 V compared to that of SIMEF1. The same group has also reported a bulk heterojunction OPV based on a TBP/CABP mixture as a donor formed from a CP/CACP mixture by annealing and SIMEF2 as an acceptor (Fig. 8.30) [71].

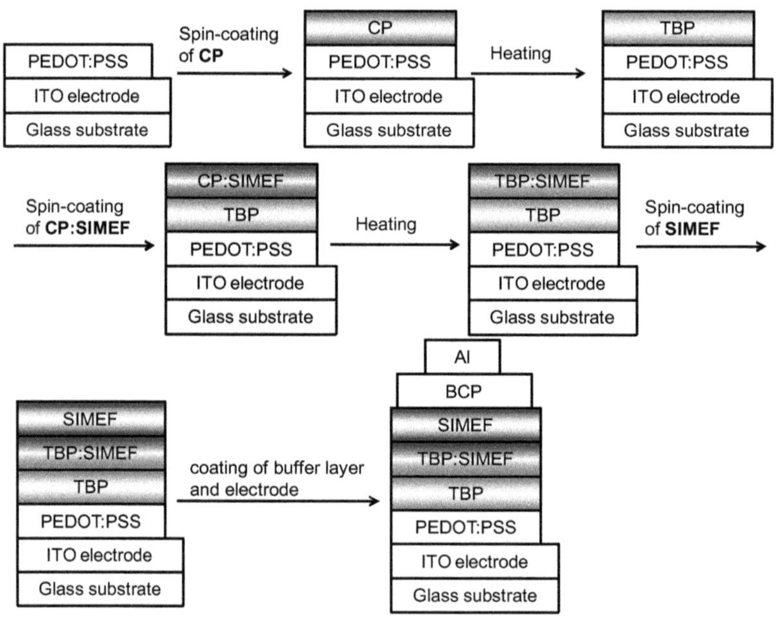

Figure 8.29 Fabrication process of a TBP/SIMEF-based OPV.

On the other hand, Chabinyc, Hawker, and coworkers and Nquyen and coworkers subsequently reported TBP/PCBNB-based p-i-n OPVs utilizing the precursor method for CP-to-TBP conversion [72,

73]. A printable OPV with a 10% conversion efficiency was reported by Mitsubishi Chemical [74, 75].

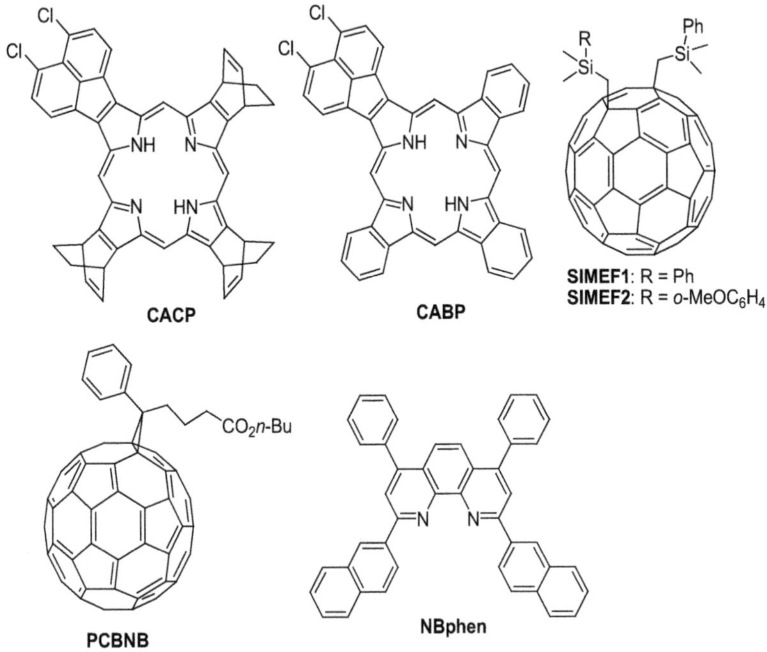

Figure 8.30 Structures of CACP, CABP, SIMEFs, PCBNB, and NBphen.

In 2014, Yamada and coworkers reported a new covalently-linked TBP-fullerene dyad for the i-layer [76]. The p-i-n OPV was fabricated by solution process with a device structure of ITO/PEDOT:PSS/TBP/BP-C60/PCBM/Al. The p-layer was prepared by spin-coating of a CP solution followed by heating at 200°C. The i-layer of BP-C60 was also prepared by spin-coating of CP-C60 and heating at 160°C (Fig. 8.31). The n-layer of PCBM was prepared by spin-coating and annealing at 195°C. The η value of 1.98% was obtained, with I_{SC} of 5.18 mA/cm^2, V_{OC} of 0.62 V, and FF of 0.61. As a reference device, an OPV with an i-layer of a mixture of BP and PCBM was prepared to exhibit the η value of 1.63%. The I_{SC}, V_{OC}, and FF values are 5.92 mA/cm^2, 0.59 V, and 0.46, respectively.

Yamada, Kobayashi, and coworkers reported the facile synthesis of tetrabenzoporphycene BPc, a constitutional isomer of TBP, based on the retro Diels–Alder strategy from the BCOD-fused precursors **22** (Fig. 8.32) [77]. In 2014, Yamada, Ishida, and coworkers successfully prepared BPc-based p-n-type OPVs by solution process [78]. The common structure of cells is as follows: ITO (200 nm)/PEDOT:PSS (30 nm)/BPc (50 nm)/PCBM (30 nm)/LiF (0.5 nm)/Al (60 nm). The η value of 1.49% was obtained from **22b**, with I_{SC} of 5.35 mA/cm^2, V_{OC} of 0.47 V, and FF of 0.59. In contrast, the OPV fabricated by the vacuum deposition of BPc exhibited a lower η value, of 0.99%, with I_{SC} of 4.70 mA/cm^2, V_{OC} of 0.52 V, and FF of 0.40.

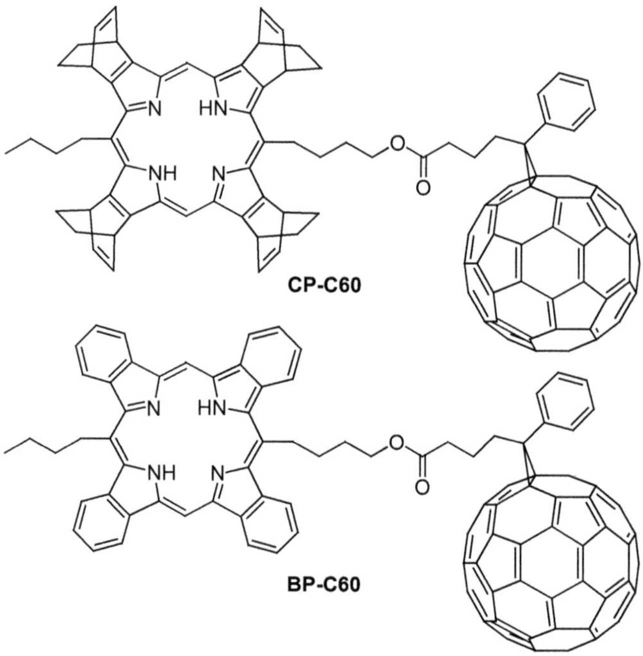

Figure 8.31 Structures of CP-C60 and BP-C60.

In 2011, Fukuda, Ishikawa, Kobayashi, and coworkers reported the new CuPc precursors **23** (Fig. 8.33) [79, 80]. Precursor **23a** was obtained by the reaction of phthalonitrile with lithium methoxide and was readily converted into CuPc by heating, to provide a quantitative yield. CuPc-based OPVs were fabricated by solution process from precursor **23a** [81]. The device structure is ITO/PEDOT:PSS/CuPc/PCBM/Ca/Al. A CuPc layer is fabricated on the PEDOT:PSS layer by spin-coating of a **23a** solution. The device is then heated at 165°C for 10 min. to form a p-layer of CuPc. The n-layer is fabricated by spin-coating of a PCBM solution, followed by annealing at 65°C for 10 min. The highest η, of 1.35%, was obtained for the p-n OPV based on CuPc/mix-PCBM (mixture of PCBM and PC70BM), with I_{SC} of 4.73 mA/cm^2, V_{OC} of 0.54 V, and FF of 0.53. The conversion efficiency of these devices showed a considerable solvent dependence. Good PCEs were attained using a 1:2 mixture of chlorobenzene and $CHCl_3$ by weight. The p-i-n devices were also fabricated to show an η of 1.00% by using PC70BM as an acceptor. The I_{SC}, V_{OC}, and FF values are 4.12 mA/cm^2, 0.51 V, and 0.48, respectively.

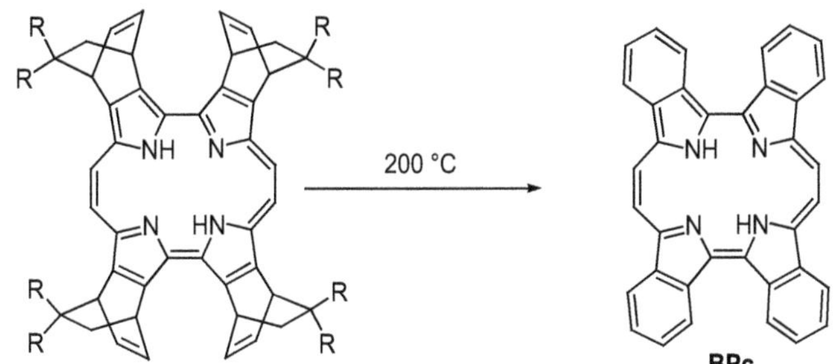

22a: R = H
22b: R = Me

Figure 8.32 Thermal conversion of **22** into BPc.

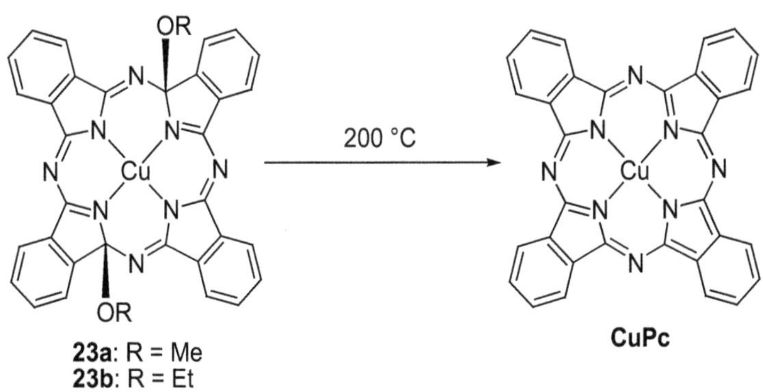

23a: R = Me
23b: R = Et
23c: R = Pr
23d: R = Bu

Figure 8.33 Thermal conversion of **23** into CuPc.

8.6 Organic-Inorganic Hybrid Perovskite Solar Cells

More recently, much attention has been focused on perovskite solar cells because of their high conversion efficiency, of over 15% [82]. The first paper was reported by Miyasaka and coworkers in 2009 [83]. The organic-inorganic lead halide perovskite compounds $CH_3NH_3PbX_3$ (X = Br or I) are used as visible light sensitizers in DSSCs. The crystal structure of the perovskite is shown in Fig. 8.34. Solar energy is converted, with an η of 3.81%, on a $CH_3NH_3PbI_3$-based cell with a high I_{SC} (of 11.0 mA/cm^2), a low V_{OC} (of 0.54 V), and FF of 0.53. In contrast, a high V_{OC}, of 0.96 V, is obtained with a $CH_3NH_3PbBr_3$-based cell, although a low I_{SC}, of 5.57 mA/cm^2, is obtained with an η of 3.13% and FF of 0.59. In these cells, the perovskite layer is not stable because a solution electrolyte is used as in DSSCs. They collaborated with the Snaith group to optimize the device structure [84]. A solid-state DSSC was fabricated by using n-type perovskite $CH_3NH_3PbI_2Cl$ as an electron transporting material and spiro-OMeTAD (Fig. 8.35) as a hole transporting material (HTM). The most efficient cell showed an η of 10.9%, with I_{SC} of 17.8 mA/cm^2, V_{OC} of 0.98 V, and FF of 0.63. Since then, several groups have improved the performance of perovskite-based solar cells as follows: η of 15.0%, with I_{SC} of 20.0 mA/cm^2, V_{OC} of 0.993 V, and FF of 0.73 [85]; η of 15.4%, with I_{SC} of 21.5 mA/cm^2, V_{OC} of 1.07 V, and FF of 0.67 [86]; η of 18.6%, with I_{SC} of 23.2 mA/cm^2, V_{OC} of 1.02 V, and FF of 0.79 (Trux-OMeTAD in Fig. 8.35) [87]; and η of 19.4%, with I_{SC} of 23.91 mA/cm^2, V_{OC} of 1.08 V, and FF of 0.750 [88].

Perovskite-based thin-film solar cells exhibit high conversion efficiency, of over 15%, while the η values of DSSCs and thin-film OPVs reach ca. 10% [1]. The $CH_3NH_3PbX_3$ layer absorbing all the visible light up to 800 nm and the high V_{OC} (over 1 V) resulted in such a high performance.

8.7 Summary and Outlook

The organic solar cells above exhibited a high performance of η values of 10%–15%. However, their efficiency and stability are still low compared to Si-based solar cells. The properties of these cells are summarized in Table 8.6.

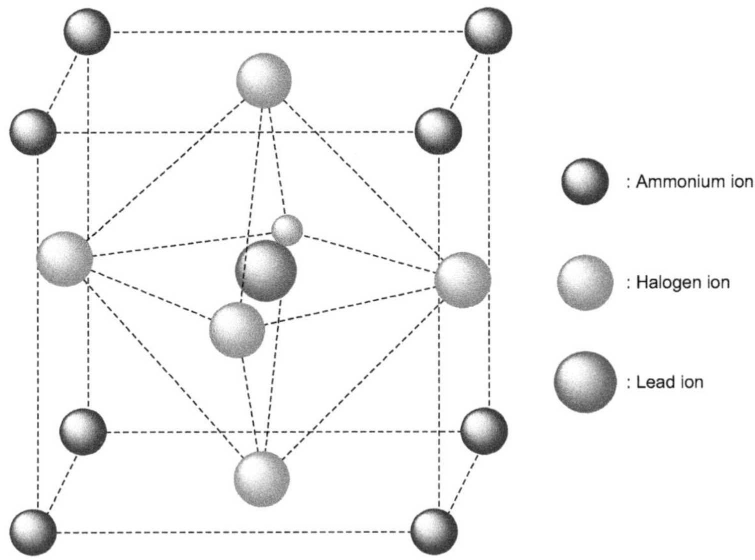

Figure 8.34 Crystal structure of perovskite compounds.

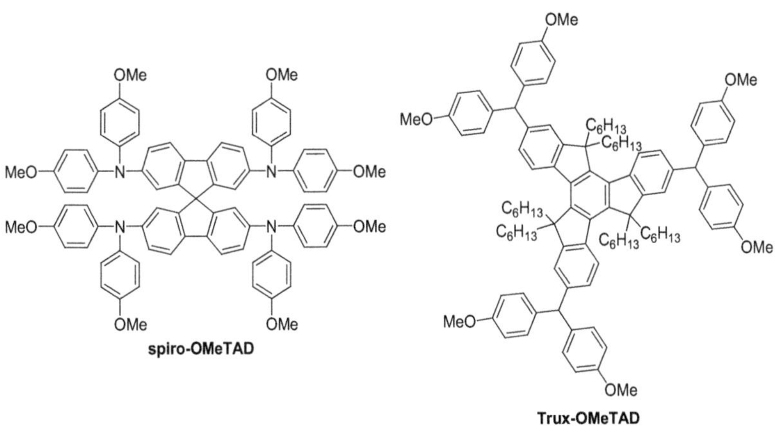

Figure 8.35 Structures of hole transporting materials for a perovskite solar cell.

The future outlook of these organic and organic-inorganic cells is shown in Fig. 8.36. Designing of new materials and optimization of the device structure are required in order to improve the performance. Improvement of the short-circuit current, the open-circuit voltage, and the FF could be achieved by the following factors:

- Absorption of light at the visible and near-infrared region effectively, a highly purifying technique, and a technique forming beneficial nanostructured film morphologies for the I_{SC} of over 30 mA/cm^2
- Control of the HOMO-LUMO levels for the V_{OC} of over 1 V
- Decrease of the internal resistance at the interface between metal electrode and organic materials for the FF of over 0.8

Table 8.6 Organic solar cells

Solar cell	Advantage	Challenge
DSSC	Low cost PCE: ~12%	Lifetime Expensive Ru Sealing of solution electrolyte
OPV	Low cost PCE: ~10% Thin film (100 – 300 nm)	Lifetime Low performance
Perovskite	Low cost High performance PCE: ~20%	Lifetime

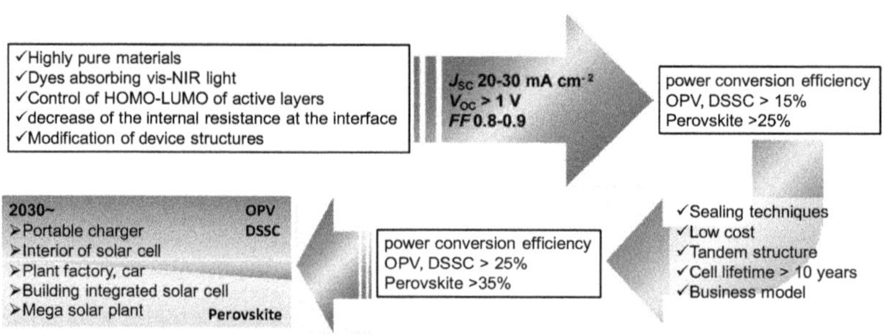

Figure 8.36 Future outlook of organic solar cells.

Organic materials are suitable for improving these problems since the desired compounds could be readily synthesized on the basis of the accumulated synthetic methods. The computational chemistry also helps to design the molecular structures. To prolong the cell lifetime to over 10 years, it is necessary to design and synthesize more stable organic materials, to produce a stable device structure, and to develop sealing techniques.

Organic solar cells have the advantages of being lightweight, flexible, and solution-processible (low cost and large area) compared to Si-based solar cells. Lightweight and flexible OPVs and DSSCs are expected to be suitable for indoor use. Large-area perovskite cells prepared at a low cost are expected to be suitable for outdoor use. For example, these solar cells could be used as portable chargers, in the interior of solar cells with high designability, as building-integrated solar cells, in plant factories as a combination of solar cells and organic light-emitting diodes, and in mega solar plants.

Acknowledgments

This work was partially supported by JSPS KAKENHI and by Grant-in-Aid for Research Promotion and the Research Unit for Power Generation and Storage Materials, Ehime University.

References

1. Green, M. A., Emery, K., Hishikawa, Y., Warta, W. and Dunlop, E. D. (2016). Solar cell efficiency tables (version 47), *Prog. Photovoltaics Res. Appl.*, **24**, pp. 3–11.
2. O'Regan, B. and Grätzel, M. (1991). A low-cost, high-efficiency solar cell based on dye-sensitized colloidal TiO_2 films, *Nature*, **353**, pp. 737–740.
3. Hagfeldt, A., Boschloo, G., Sun, L., Kloo, L. and Pettersson, H. (2010). Dye-sensitized solar cells, *Chem. Rev.*, **110**, pp. 6595–6663.
4. Yella, A., Lee, H.-W., Tsao, H. N., Yi, C., Chandiran, A. K., Nazeeruddin, Md. K., Diau, E. W.-G., Yeh, C.-Y., Zakeeruddin, S. M. and Grätzel, M. (2011). Porphyrin-sensitized solar cells with cobalt (II/III)-based redox electrolyte exceed 12 percent efficiency, *Science*, **334**, pp. 629–634.
5. Nazeeruddin, M. K., Kay, A., Rodicio, I., Humphry-Baker, R., Müller, E., Liska, P., Vlachopoulos, N. and Grätzel, M. (1993). Conversion of light to electricity by *cis*-X_2bis(2,2'-bipyridyl-4,4'-dicarboxylate) ruthenium(II) charge-transfer sensitizers (X = Cl^-, Br^-, I^-, CN^-, and SCN^-) on nanocrystalline titanium dioxide electrodes, *J. Am. Chem. Soc.*, **115**, pp. 6382–6390.

6. Nazeeruddin, Md. K., Péchy, P. and Grätzel, M. (1997). Efficient panchromatic sensitization of nanocrystalline TiO_2 films by a black dye based on a trithiocyanato-ruthenium complex, *Chem. Commun.*, pp. 1706–1706.

7. Nazeeruddin, Md. K., Zakeeruddin, S. M., Humphry-Baker, R., Jirousek, M., Liska, P., Vlachopoulos, N., Shklover, V., Fischer, C.-H. and Grätzel, M. (1999). Acid-base equilibria of (2,2'-bipyridyl-4,4'-dicarboxylic acid)ruthenium(II) complexes and the effect of protonation on charge-transfer sensitization of nanocrystalline titania, *Inorg. Chem.*, **38**, pp. 6298–6305.

8. Nazeeruddin, Md. K., Angelis, F. D., Fantacci, S., Selloni, A., Viscardi, G., Liska, P., Ito, S., Bessho, T. and Grätzel, M. (2005). Combined experimental and DFT-TDDFT computational study of photoelectrochemical cell ruthenium sensitizers, *J. Am. Chem. Soc.*, **127**, pp. 16835–16847.

9. Nazeeruddin, Md. K., Péchy, P., Renouard, T., Zakeeruddin, S. M., Humphry-Baker, R., Comte, P., Liska, P., Cevey, L., Costa, E., Shklover, V., Spiccia, L., Deacon, G. B., Bignozzi, C. A. and Grätzel, M. (2001). Engineering of efficient panchromatic sensitizers for nanocrystalline TiO_2-based solar cells, *J. Am. Chem. Soc.*, **123**, pp. 1613–1624.

10. Yanagida, M., Onozawa-Komatsuzaki, N., Kurashige, M., Sayama, K. and Sugihara, H. (2010). Optimization of tandem-structured dye-sensitized solar cell, *Sol. Energy Mater. Sol. Cells*, **94**, pp. 297–302.

11. Reddy, P. Y., Giribabu, L., Lyness, C., Snaith, H. J., Vijaykumar, C., Chandrasekharam, M., Lakshmikantam, M., Yum, J.-H., Kalyanasundaram, K., Grätzel, M. and Nazeeruddin, Md. K. (2007). Efficient sensitization of nanocrystalline TiO_2 films by a near-IR-absorbing unsymmetrical zinc phthalocyanine, *Angew. Chem. Int. Ed.*, **46**, pp. 373–376.

12. Tejerina, L., Martínez-Díaz, M. V., Nazeeruddin, Md. K. and Torres, T. (2016). The influence of substituent orientation of the phovoltaic performance of phthalocyanine-sensitized solar cells, *Chem. Eur. J.*, **22**, pp. 4369–4373.

13. Liu, Y., Xiang, N., Feng, X., Shen, P., Zhou, W., Weng, C., Zhao, B. and Tan, S. (2009). Thiophene-linked porphyrin derivatives for dye-sensitized solar cells, *Chem. Commun.*, pp. 2499–2501.

14. Campbell, W. M., Jolley, K. W., Wagner, P., Wagner, K., Walsh, P. J., Gordon, K. C., Shmidt-Mende, L., Nazeeruddin, Md. K., Wang, Q., Grätzel, M. and Officer, D. L. (2007). Highly efficient porphyrin sensitizers for dye-sensitized solar cells, *J. Phys. Chem. C*, **111**, pp. 11760–11762.

15. Bessho, T., Zakeeruddin, S. M., Yeh, C.-Y., Diau, E. W.-G. and Grätzel, M. (2010). Highly efficient mesoscopic dye-sensitized solar cells based on donor-acceptor-substituted porphyrins, *Angew. Chem. Int. Ed.*, **49**, pp. 6646–6649.

16. Xie, Y., Tang, Y., Wu, T., Wang, Y., Liu, J., Li, X., Tian, H. and Zhu, W.-H. (2015). Porphyrin cosensitization for a photovoltaic efficiency of 11.5%: a record for non-rutheniun solar cells based on iodine electolyte, *J. Am. Chem. Soc.*, **137**, pp. 14055–14058.

17. Lodermeyer, F., Costa, R. D., Malig, J., Jux, N. and Guldi, D. M. (2016). Benzoporphyrins: selective co-sensitization in dye-sensitized solar cells, *Chem. Eur. J.*, **137**, pp. 7851–7855.

18. Ooyama, Y., Ishii, A., Kagawa, Y., Imae, I. and Harima, Y. (2007). Dye-sensitized solar cells based on novel donor-acceptor π-conjugated benzofuro[2,3-*c*]oxazolo[4,5-*a*]carbazole-type fluorescent dyes exhibiting solid-state fluorescence, *New J. Chem.*, **31**, pp. 2076–2082.

19. Ooyama, Y., Shimada, Y., Kagawa, Y., Imae, I. and Harima, Y. (2007). Photovoltaic performance of dye-sensitized solar cells based on donor-acceptor π-conjugated benzofuro[2,3-*c*]oxazolo[4,5-*a*]carbazole-type fluorescent dyes with a carboxyl group at different positions of the chromophore skeleton, *Org. Biomol. Chem.*, **5**, pp. 2046–2054.

20. Ooyama, Y., Nagano, T., Inoue, S., Imae, I., Komaguchi, K., Ohshita, J. and Harima, Y. (2011). Dye-sensitized solar cells based on donor-π-acceptor fluorescent dyes with a pyridine ring as an electron-withdrawing-injecting anchoring group, *Chem. Eur. J.*, **17**, pp. 14837–14843.

21. Hwang, S., Lee, J. H., Park, C., Lee, H., Kim, C., Park, C., Lee, M.-H., Lee, W., Park, J., Kim, K., Park, N.-G. and Kim, C. (2007). A highly efficient organic sensitizer for dye-sensitized solar cells, *Chem. Commun.*, pp. 4887–4889.

22. Wang, Z.-S., Cui, Y., Dan-oh, Y., Kasuda, C., Shinpo, A. and Hara, K. (2007). Thiophene-fuctionalized coumarin dye for efficient dye-sensitized solar cells: electron lifetime improved by coadsorption of deoxycholic acid, *J. Phys. Chem. C*, **111**, pp. 7224–7230.

23. Zeng, W., Cao, Y., Bai, Y., Wang, Y., Shi, Y., Zhang, M., Wang, F., Pan, C. and Wang, P. (2010). Efficient dye-sensitized solar cells with an organic photosensitizer featuring orderly conjugated ethylenedioxythiophene and dithienosilole blocks, *Chem. Mater.*, **22**, pp. 1915–1925.

24. Miyazaki, E., Okanishi, T., Suzuki, Y., Ishine, N., Mori, H., Takimiya, K. and Harima, Y. (2011). Simple oligothiophene-based dyes for dye-

sensitized solar cells (DSSCs): anchoring group effects on molecular properties and solar cell performance, *Bull. Chem. Soc. Jpn.*, **84**, pp. 459–465.

25. Tang, C. W. (1986). Two-layer organic photovoltaic cell, *Appl. Phys. Lett.*, **48**, pp. 183–185.
26. Tang, C. W. and VanSlyke, S. A. (1987). Organic electroluminescent diodes, *Appl. Phys. Lett.*, **51**, pp. 913–915.
27. Hiramoto, M., Fujiwara, H. and Yokoyama, M. (1991). Three-layered organic solar cell with a photoactive interlayer of codeposited pigments, *Appl. Phys. Lett.*, **58**, pp. 1062–1064.
28. Hiramoto, M., Fujiwara, H. and Yokoyama, M. (1992). *p-i-n* like behavior in three-layered organic solar cells having a co-deposited interlayer of pigments, *J. Appl. Phys.*, **72**, pp. 3781–3787.
29. Sariciftci, N. S., Braun, D., Zhang, C., Srdanov, V. I., Heeger, A. J., Stucky, G. and Wudl, F. (1993). Semiconducting polymer-buckminsterfullerene heterojunctions: diodes, photodiodes, and photovoltaic, *Appl. Phys. Lett.*, **62**, pp. 585–587.
30. Xue, J., Rand, B. P., Uchida, S. and Forrest, S. R. (2005). A hybrid planar-mixed molecular heterojunction photovoltaic cell, *Adv. Mater.*, **17**, pp. 66–71.
31. Kim, J. Y., Lee, K., Coates, N. E., Moses, D., Nguyen, T.-Q., Dante, M. and Heeger, A. J. (2007). Efficient tandem polymer solar cells fabricated by all-solution processing, *Science*, **317**, pp. 222–225.
32. Taima, T., Toyoshima, S., Hara, K., Saito, K. and Yase, K. (2006). Control of measurement environments for high-efficiency organic photovoltaic cells, *Jpn. J. Appl. Phys.*, **45**, pp. L217–L219.
33. Kim, I., Haverinen, H. M., Wang, Z., Madakuni, S., Kim, Y., Li, J. and Jabbour, G. E. (2009). Efficient organic solar cells based on planar metallophthalocyanines, *Chem. Mater.*, **21**, pp. 4256–4260.
34. Shao, Y. and Yang, Y. (2005). Efficient organic heterojunction photovoltaic cells based on triplet materials, *Adv. Mater.*, **17**, pp. 2841–2844.
35. Lin, L.-Y., Chen, Y.-H., Huang, Z.-Y., Lin, H.-W., Chou, S.-H., Lin, F., Chen, C.-W., Liu, Y.-H. and Wong, K.-T. (2011). A low-energy-gap organic dye for high-performance small-molecule organic solar cells, *J. Am. Chem. Soc.*, **133**, pp. 15822–15825.
36. Mayer, A. C., Scully, S. R., Hardin, B. E., Rowell, M. W. and McGehee, M. D. (2007). Polymer-based solar cells, *Mater. Today*, **10**, pp. 28–33.

37. Tamanari, T., Taima, T., Sakai, J. and Saito, K. (2008). Highly efficient organic thin-film solar cells based on poly(3-hexylthiphene) and soluble C_{70} fullerene derivative, *Jpn. J. Appl. Phys.*, **47**, pp. 1230–1233.

38. Zhang, Z.-G., Min, J., Zhang, S., Zhang, J., Zhang, M. and Li, Y. (2011). Alkyl chain engineering on a dithieno[3,2-*b*:2',3'-*d*]silole-alt-dithienylthiazolo[5,4-*d*]thiazole copolymer toward high performance bulk heterojunction solar cells, *Chem. Commun.*, **37**, pp. 9474–9476.

39. Osaka, I., Saito, M., Mori, H., Koganezawa, T. and Takimiya, K. (2012). Drastic change of molecular orientation in a thiazolothiazole copolymer by molecular-weight control and blending with $PC_{61}BM$ leads to high efficiencies in solar cells, *Adv. Mater.*, **24**, pp. 425–430.

40. Lin, Y., He, Q., Zhao, F., Huo, L., Mai, J., Lu, X., Su, C.-J., Li, T., Wang, J., Zhu, J., Sun, Y., Wang, C. and Zhan, X. (2016). A facile planar fused-ring electron acceptor for As-cast polymer solar cells with 8.71% efficiency, *J. Am. Chem. Soc.*, **138**, pp. 2973–2976.

41. Lloyd, M. T., Anthony, J. E. and Malliaras, G. G. (2007). Photovoltaics from soluble small molecules, *Mater. Today*, **10**, pp. 34–41.

42. Zhou, J., Wan, X., Liu, Y., Long, G., Wang, F., Li, Z., Zuo, Y., Li, C. and Chen, Y. (2011). A planar small molecule with dithienosilole core for high efficiency solution-processed organic photovoltaic cells, *Chem. Mater.*, **23**, pp. 4666–4668.

43. Zhou, J., Wan, X., Liu, Y., Zuo, Y., Li, Z., He, G., Long, G., Ni, W., Li, C., Su, X. and Chen, Y. (2012). Small molecules based on benzo[1,2-b:4,5-b'] dithiphene unit for high-performance solution-processed organic solar cells, *J. Am. Chem. Soc.*, **134**, pp. 16345–16351.

44. Li, Z., He, G., Wan, X., Liu, Y., Zhou, J., Long, G., Zuo, Y., Zhang, M. and Chen, Y. (2012). Solution processable rhodanine-based small molecule organic photovoltaic cells with a power conversion efficiency of 6.1%, *Adv. Energy Mater.*, **2**, pp. 74–77.

45. Ouchi, H., Lin, X., Kizaki, T., Prabhu, D. D., Silly, F., Kajitani, T., Fukushima, T., Nakayama, K. and Yagai, S. (2016). Hydrogen-bonded oligothiophene rosettes with a benzodithiphene terminal unit: self-assembly and application to bulk heterojunction solar cells, *Chem. Commun.*, **52**, pp. 7874–7877.

46. Wang, J.-L., Liu, K.-K., Yan, J., Wu, Z., Xiao, F., Chang, Z.-F., Wu, H.-B., Cao, Y. and Russell, T. P. (2016). Series of multifluorine substituted oligomers for organic solar cells with efficiency over 9% and fill factor of 0.77 by combination thermal and solvent vapor annealing, *J. Am. Chem. Soc.*, **138**, pp. 7687–7697.

47. Chikamatsu, M., Itakura, A., Yoshida, Y., Azumi, R., Kikuchi, K. and Yase, K. (2006). Correlation of molecular structure, packing motif and thin-film transistor characteristics of solution-processed n-type organic semiconductors based on dodecyl-substituted C_{60} derivatives, *J. Photochem. Photobiol. A*, **182**, pp. 245–249.
48. Zambounis, J. S., Hao, Z. and Iqbal, A. (1997). Latent pigments activated by heat, *Nature*, **388**, pp. 131–132.
49. Herwig, P. T. and Müllen, K. (1999). A soluble pentacene precursor: synthesis, solid-state conversion into pentacene and application in a field-effect transistor, *Adv. Mater.*, **11**, pp. 480–483.
50. Afzali, A., Dimitrakopoulos, C. D. and Breen, T. L. (2002). High-performance, solution-processed organic thin film transistors from a novel pentacene precursor, *J. Am. Chem. Soc.*, **124**, pp. 8812–8813.
51. Uno, H., Yamashita, Y., Kikuchi, M., Watanabe, H., Yamada, H., Okujima, T., Ogawa, T. and Ono, N. (2005). Photo precursor for pentacene, *Tetrahedron Lett.*, **46**, pp. 1981–1983.
52. Yamada, H., Yamashita, Y., Kikuchi, M., Watanabe, H., Okujima, T., Uno, H., Ogawa, T., Ohara, K. and Ono, N. (2005). Photochemical synthesis of pentacene and its derivatives, *Chem. Eur. J.*, **11**, pp. 6212–6220.
53. Masumoto, A., Yamashita, Y., Go, S., Kikukchi, T., Yamada, H., Okujima, T., Ono, N. and Uno, H. (2009). Organic thin-film transistor from a pentacene photoprecursor, *Jpn. J. Appl. Phys.*, **48**, pp. 051505 (5 pages).
54. Nakayama, K., Ohashi, C., Oikawa, Y., Motoyama, T. and Yamada, H. (2013). Characterization and field-effect transistor performance of printed pentacene films prepared by phtoconversion of a soluble precursor, *J. Mater. Chem. C*, **1**, pp. 6244–6251.
55. Carvalho, C. M. B., Brocksom, T. J. and de Oliveira, K. T. (2013). Tetrabenzoporphyrins: synthetic developments and applications, *Chem. Soc. Rev.*, **42**, pp. 3302–3317.
56. Vicente, M. G. H., Tomé, A. C., Walter, A. and Cavaleiro, J. A. S. (1997). Synthesis and cycloaddition reactions of pyrrole-fused 3-sulfolenes: a new versatile route to tetrabenzoporphyrins, *Tetrahedron Lett.*, **38**, pp. 3639–3642.
57. Finikova, O. S., Cheprakov, A. V., Beletskaya, I. P., Carroll, P. J. and Vinogradov, S. A. (2004). Novel versatile synthesis of substituted tetrabenzoporphyrins, *J. Org. Chem.*, **69**, pp. 522–535.
58. Ito, S., Murashima, T., Uno, H. and Ono, N. (1998). A new synthesis of benzoporphyrins using 4,7-dihydor-4,7-ethano-2*H*-isoindole as a synthon of isoindole, *Chem. Commun.*, pp. 1661–1662.

59. Ito, S., Ochi, N., Murashima, T., Uno, H. and Ono, N. (2000). A new synthesis of benzoporphyrins using 4,7-dihydor-4,7-ethano-2*H*-isoindole as an isoindole equivalent, *Heterocycles*, **52**, pp. 399–411.

60. Uno, H., Ishikawa, T., Hoshi, T. and Ono, N. (2003). Soluble precursors convertible to tetrabensoporphyrins below room temperature, *Tetrahedron Lett.*, **44**, pp. 5163–5165.

61. Okujima, T., Hashimoto, Y., Jin, G., Yamada, H., Uno, H. and Ono, N. (2008). Synthesis of extremely soluble precursors of tetrabenzoporphyrins, *Tetrahedron*, **64**, pp. 2405–2411.

62. Jeong, S.-D., Min, B., Cho, S. Y., Lee, C., Park, B. K., An, K.-S. and Lim, J. (2012). Method for synthesis of tetrabenzoporphyrin precursor for use in organic electronic devices, *J. Org. Chem.*, **77**, pp. 8329–8331.

63. Aramaki, S., Sakai, Y. and Ono, N. (2004). Solution-processible organic semiconductor for transistor applications: tetrabenzoporphyrin, *Appl. Phys. Lett.*, **84**, pp. 2085–2087.

64. Shea, P. B., Kanicki, J., Pattison, L. R., Petroff, P., Kawano, M., Yamada, H. and Ono, N. (2006). Solution-processed nickel tetrabenzoporphyrin thin-film transistors, *J. Appl. Phys.*, **100**, pp. 034502 (7 pages).

65. Shea, P. B., Pattison, L. R., Kawano, M., Chen, C., Chen, J., Petroff, P., Martin, D. C., Yamada, H., Ono, N. and Kanicki, J. (2007). Solution-processed polycrystalline copper tetrabenzoporphyrin thin-film transistors, *Synth. Met.*, **157**, pp. 190–197.

66. Shea, P. B., Chen, C., Kanicki, J., Pattison, L. R., Petroff, P., Yamada, H. and Ono, N. (2007). Polycrystalline tetrabenzoporphyrin organic field-effect transistors with nanostructured channels, *Appl. Phys. Lett.*, **90**, pp. 233107 (3 pages).

67. Shea, P. B., Yamada, H., Ono, N. and Kanicki, J. (2012). Solution-processed zinc tetrabenzoporphyrin thin-film and transistors, *Thin Solid Films*, **520**, pp. 4031–4035.

68. Hirao, A., Akiyama, T., Okujima, T., Yamada, H., Uno, H., Sakai, Y., Aramaki, S. and Ono, N. (2008). Soluble precursors of 2,3-naphthalocyanine and phthalocyanine for use in thin film transistors, *Chem. Commun.*, pp. 4714–4716.

69. Matsuo, Y., Sato, Y., Niinomi, T., Soga, I., Tanaka, H. and Nakamura, E. (2009). Columnar structure in bulk heterojunction in solution-processable three-layered p-i-n organic photovoltaic devices using tetrabenzoporphyrin precursor and silylmethyl[60]fullerene, *J. Am. Chem. Soc.*, **131**, pp. 16048–16050.

70. Tanaka, H., Abe, Y., Matsuo, Y., Kawai, J., Soga, I., Sato, Y. and Nakamura, E. (2012). An amorphous mesophase generated by thermal annealing

for high-performance organic photovoltaic devices, *Adv. Mater.*, **24**, pp. 3521–3525.

71. Zhen, Y., Tanaka, H., Harano, K., Okada, S., Matsuo, Y. and Nakamura, E. (2015). Organic solid solution composed of two structurally similar porphyrins for organic solar cells, *J. Am. Chem. Soc.*, **137**, pp. 2247–2252.
72. Ku, S.-Y., Liman, C. D., Cochran, J. E., Toney, M. F., Chabinyc, M. L. and Hawker, C. J. (2011). Solution-processed nanostructured benzoporphyrin with polycarbonate binder for photovoltaics, *Adv. Mater.*, **23**, pp. 2289–2293.
73. Guide, M., Dang, X.-D. and Nguyen, T.-Q. (2011). Nanoscale characterization of tetrabenzoporphyrin and fullerene-based solar cells by photoconductive atomic force microscopy, *Adv. Mater.*, **23**, pp. 2313–2319.
74. Service, R. F. (2011). Outlook brightens for plastic solar cells, *Science*, **332**, p. 293.
75. Green, M. A., Emery, K., Hishikawa, Y., Warta, W. and Dunlop E. D. (2012). Solar cell efficiency tables (version 39), *Prog. Photovoltaics Res. Appl.*, **20**, pp. 12–20.
76. Tamura, Y., Saeki, H., Hashizume, J., Okazaki, Y., Kuzuhara, D., Suzuki, M., Aratani, N. and Yamada, H. (2014). Direct comparison of a covalently-linked dyad and a 1:1 mixture of tetrabenzoporphyrin and fullerene as organic photovoltaic materials, *Chem. Commun.*, **50**, pp. 10379–10381.
77. Kuzuhara, D., Mack, J., Yamada, H., Okujima, T., Ono, N. and Kobayashi, N. (2009). Synthesis, structures, and optical and electrochemical properties of benzoporphycenes, *Chem. Eur. J.*, **15**, pp. 10060–10069.
78. Saeki, H., Kurimoto, O., Nakaoka, H., Misaki, M., Kuzuhara, D., Yamada, H., Ishida, K. and Ueda, Y. (2014). Effect of crystallinity in small molecular weight organic heterojunction solar cells, *J. Mater. Chem. C*, **2**, pp. 5357–5364.
79. Kikukawa, Y., Fukuda, T., Fuyuhiro, A., Ishikawa, N. and Kobayashi, N. (2011). Facile one-pot preparation of thermally and photochemistry convertible soluble precursors of copper phthalocyanine and naphthalocyanine, *Chem. Commun.*, **47**, pp. 8518–8520.
80. Fukuda, T., Kikukawa, Y., Tsuruya, R., Fuyuhiro, A., Ishikawa, N. and Kobayashi, N. (2011). Effect of chain length on thermal conversion of alkoxy-substituted copper phthalocyanine precursors, *Inorg. Chem.*, **50**, pp. 11832–11837.
81. Wang, H., Fukuda, T., Ishikawa, N. and Matsuo, Y. (2014). Solvent-dependent morphology of thermally converted copper phthalocyanien

for solution-processed small molecule organic photovoltaic devices, *Org. Electron.*, **15**, pp. 139–143.

82. Miyasaka, T. (2015). Perovskite photovoltaics: rare functions of organo lead halide in solar cells and optoelectronic devices, *Chem. Lett.*, **44**, pp. 720–729.

83. Kojima, A., Teshima, K., Shirai, Y. and Miyasaka, T. (2009). Organometal halide perovskites as visible-light sensitizers for photovoltaic cells, *J. Am. Chem. Soc.*, **131**, pp. 6050–6051.

84. Lee, M. M., Teuscher, J., Miyasaka, T., Murakami, T. N. and Snaith, H. J. (2012). Efficient hybrid solar cells based on meso-superstructured organometal halide perovskites, *Science*, **338**, pp. 643–647.

85. Burschka, J., Pellet, N., Moon, S.-J., Humphry-Baker, R., Gao, P., Nazeeruddin, M. K. and Grätzel, M. (2013). Sequential deposition as a route to high-performance perovskite-sensitized solar cells, *Nature*, **499**, pp. 316–319.

86. Liu, M., Johnston, M. B. and Snaith, H. J. (2013). Efficient planar heterojunction perovskite solar cells by vapour deposition, *Nature*, **501**, pp. 395–398.

87. Huang, C., Fu, W., Li, C.-Z., Zhang, Z., Qiu, W., Shi, M., Heremans, P., Jen, A. K.-Y. and Chen, H. (2016). Dopant-free hole-transporting material with a C_{3h} symmetrical truxene core for highly efficient perovskite solar cells, *J. Am. Chem. Soc.*, **138**, pp. 2528–2531.

88. Kim H. D., Ohkita, H., Benten, H. and Ito, S. (2016). Photovoltaic performance of perovskite solar cells with different grain sizes, *Adv. Mater.*, **28**, pp. 917–922.

Index

For Product Safety Concerns and Information please contact our EU representative GPSR@taylorandfrancis.com
Taylor & Francis Verlag GmbH, Kaufingerstraße 24, 80331 München, Germany

www.ingramcontent.com/pod-product-compliance
Ingram Content Group UK Ltd.
Pitfield, Milton Keynes, MK11 3LW, UK
UKHW022200190926
475702UK00003B/121
* 9 7 8 9 8 1 4 8 0 0 0 9 9 *